Tribology in Germany: Latest Research and Development

Tribology in Germany: Latest Research and Development

Guest Editor

Dirk Bartel

Basel • Beijing • Wuhan • Barcelona • Belgrade • Novi Sad • Cluj • Manchester

Guest Editor
Dirk Bartel
Otto von Guericke University Magdeburg
Magdeburg
Germany

Editorial Office
MDPI AG
Grosspeteranlage 5
4052 Basel, Switzerland

This is a reprint of the Special Issue, published open access by the journal *Lubricants* (ISSN 2075-4442), freely accessible at: https://www.mdpi.com/journal/lubricants/special_issues/24C0U1OPFV.

For citation purposes, cite each article independently as indicated on the article page online and as indicated below:

Lastname, A.A.; Lastname, B.B. Article Title. *Journal Name* **Year**, *Volume Number*, Page Range.

ISBN 978-3-7258-2919-4 (Hbk)
ISBN 978-3-7258-2920-0 (PDF)
https://doi.org/10.3390/books978-3-7258-2920-0

© 2025 by the authors. Articles in this book are Open Access and distributed under the Creative Commons Attribution (CC BY) license. The book as a whole is distributed by MDPI under the terms and conditions of the Creative Commons Attribution-NonCommercial-NoDerivs (CC BY-NC-ND) license (https://creativecommons.org/licenses/by-nc-nd/4.0/).

Contents

About the Editor vii

Dirk Bartel
Tribology in Germany: Latest Research and Development
Reprinted from: *Lubricants* 2024, 12, 425, https://doi.org/10.3390/lubricants12120425 1

Ferdinand Schmid, Thomas Lohner and Karsten Stahl
Friction in Oil-Lubricated Rolling–Sliding Contacts with Technical and High-Performance Thermoplastics †
Reprinted from: *Lubricants* 2024, 12, 372, https://doi.org/10.3390/lubricants12110372 5

Dirk Bartel
Mixed Friction in Fully Lubricated Elastohydrodynamic Contacts—Theory or Reality
Reprinted from: *Lubricants* 2024, 12, 351, https://doi.org/10.3390/lubricants12100351 29

Paul Neubauer, Frank Kaulfuss and Volker Weihnacht
A Tribological Study of ta-C, ta-C:N, and ta-C:B Coatings on Plastic Substrates under Dry Sliding Conditions
Reprinted from: *Lubricants* 2024, 12, 331, https://doi.org/10.3390/lubricants12100331 42

Joachim Faller and Matthias Scherge
Running-In of DLC–Third Body or Transfer Film Formation
Reprinted from: *Lubricants* 2024, 12, 314, https://doi.org/10.3390/lubricants12090314 55

Maximilian Baur, Iyas Khader, Dominik Kürten, Thomas Schieß, Andreas Kailer and Martin Dienwiebel
Premature Damage in Bearing Steel in Relation with Residual Stresses and Hydrogen Trapping
Reprinted from: *Lubricants* 2024, 12, 311, https://doi.org/10.3390/lubricants12090311 66

Christian Orgeldinger, Manuel Reck, Armin Seynstahl, Tobias Rosnitschek, Marion Merklein and Stephan Tremmel
Process-Integrated Component Microtexturing for Tribologically Optimized Contacts Using the Example of the Cam Tappet—Numerical Design, Manufacturing, DLC-Coating and Experimental Analysis
Reprinted from: *Lubricants* 2024, 12, 291, https://doi.org/10.3390/lubricants12080291 80

Florian König, Florian Wirsing, Ankit Singh and Georg Jacobs
Machine-Learning-Based Wear Prediction in Journal Bearings under Start–Stop Conditions
Reprinted from: *Lubricants* 2024, 12, 290, https://doi.org/10.3390/lubricants12080290 102

Burhan Ibrar, Volker Wittstock, Joachim Regel and Martin Dix
Influence of Lubrication Cycle Parameters on Hydrodynamic Linear Guides through Simultaneous Monitoring of Oil Film Pressure and Floating Heights
Reprinted from: *Lubricants* 2024, 12, 287, https://doi.org/10.3390/lubricants12080287 123

Fabian Halmos, Sandro Wartzack and Marcel Bartz
Investigation of Failure Mechanisms in Oil-Lubricated Rolling Bearings under Small Oscillating Movements: Experimental Results, Analysis and Comparison with Theoretical Models
Reprinted from: *Lubricants* 2024, 12, 271, https://doi.org/10.3390/lubricants12080271 139

Nicolai Sprogies, Thomas Lohner and Karsten Stahl
Improved Operating Behavior of Self-Lubricating Rolling-Sliding Contacts under High Load with Oil-Impregnated Porous Sinter Material
Reprinted from: *Lubricants* 2024, 12, 259, https://doi.org/10.3390/lubricants12070259 151

Markus Golek, Jakob Gleichner, Ioannis Chatzisavvas, Lukas Kohlmann, Marcus Schmidt, Peter Reinke and Adrian Rienäcker
Numerical Simulations and Experimental Validation of Squeeze Film Dampers for Aircraft Jet Engines
Reprinted from: *Lubricants* 2024, 12, 253, https://doi.org/10.3390/lubricants12070253 188

Fabian Kaiser, Daniele Savio and Ravindrakumar Bactavatchalou
Modelling of Static and Dynamic Elastomer Friction in Dry Conditions
Reprinted from: *Lubricants* 2024, 12, 250, https://doi.org/10.3390/lubricants12070250 203

Géraldine Theiler, Natalia Cano Murillo and Andreas Hausberger
Effect of Hydrogen Pressure on the Fretting Behavior of Rubber Materials
Reprinted from: *Lubricants* 2024, 12, 233, https://doi.org/10.3390/lubricants12070233 220

Volker Schneider, Marius Krewer, Gerhard Poll and Max Marian
Effect of Harmful Bearing Currents on the Service Life of Rolling Bearings: From Experimental Investigations to a Predictive Model
Reprinted from: *Lubricants* 2024, 12, 230, https://doi.org/10.3390/lubricants12070230 237

Birthe Grzemba and Roman Pohrt
Ice-versus-Steel Friction: An Advanced Numerical Approach for Competitive Winter Sports Applications
Reprinted from: *Lubricants* 2024, 12, 203, https://doi.org/10.3390/lubricants12060203 252

Daniel Merk, Thomas Koenig, Janine Fritz and Joerg W. H. Franke
Tribological Behavior of Hydrocarbons in Rolling Contact
Reprinted from: *Lubricants* 2024, 12, 201, https://doi.org/10.3390/lubricants12060201 270

Andreas Keller, Knud-Ole Karlson, Markus Grebe, Fabian Schüler, Christian Goehringer and Alexander Epp
Practical Evaluation of Ionic Liquids for Application as Lubricants in Cleanrooms and under Vacuum Conditions
Reprinted from: *Lubricants* 2024, 12, 194, https://doi.org/10.3390/lubricants12060194 284

Shengqin Zhao, Rolf Merz, Stefan Emrich, Johannes L'huillier and Leyu Lin
Improved Tribological Performance of a Polybutylene Terephthalate Hybrid Composite by Adding a Siloxane-Based Internal Lubricant
Reprinted from: *Lubricants* 2024, 12, 189, https://doi.org/10.3390/lubricants12060189 303

Iakov A. Lyashenko, Thao H. Pham and Valentin L. Popov
Transition between Friction Modes in Adhesive Contacts of a Hard Indenter and a Soft Elastomer: An Experiment
Reprinted from: *Lubricants* 2024, 12, 110, https://doi.org/10.3390/lubricants12040110 315

Huanhuan Ding, Ümit Mermertas, Thomas Hagemann and Hubert Schwarze
Calculation and Validation of Planet Gear Sliding Bearings for a Three-Stage Wind Turbine Gearbox
Reprinted from: *Lubricants* 2024, 12, 95, https://doi.org/10.3390/lubricants12030095 332

Simon Graf and Oliver Koch
Changes in Surface Topography and Light Load Hardness in Thrust Bearings as a Reason of Tribo-Electric Loads
Reprinted from: *Lubricants* 2024, 12, 303, https://doi.org/10.3390/lubricants12090303 347

About the Editor

Dirk Bartel

Dirk Bartel is Head of the Chair of Machine Elements and Tribology at the Otto von Guericke University Magdeburg and Managing Director of Tribo Technologies GmbH. With more than 30 years of experience in tribology, his research focuses on the fundamentals of friction and wear, the simulation of machine elements (thermal elastohydrodynamics, computational fluid dynamics, finite element method), cross-domain system simulation (elastic mutibody simulation coupled with thermal elastohydrodynamics), as well as method development in simulation and testing. In 2018, he founded the internationally active Tribo Technologies GmbH, together with Otto von Guericke University Magdeburg and other shareholders, in order to quickly transfer the latest research findings into practice. He has published more than 70 peer-reviewed journal publications, one book and two book chapters, and is a member of several program committees. He is particularly involved in the German Society for Tribology (GfT), where he is currently Chairman of the Technical-Scientific Advisory Board and Head of the Education and Training Working Group.

Editorial
Tribology in Germany: Latest Research and Development

Dirk Bartel

Chair of Machine Elements and Tribology, Otto von Guericke University Magdeburg, 39106 Magdeburg, Germany; dirk.bartel@ovgu.de

Citation: Bartel, D. Tribology in Germany: Latest Research and Development. *Lubricants* **2024**, *12*, 425. https://doi.org/10.3390/lubricants12120425

Received: 24 November 2024
Accepted: 26 November 2024
Published: 2 December 2024

Copyright: © 2024 by the author. Licensee MDPI, Basel, Switzerland. This article is an open access article distributed under the terms and conditions of the Creative Commons Attribution (CC BY) license (https://creativecommons.org/licenses/by/4.0/).

Sixty-five years ago, in November 1959, the "Gesellschaft für Schmiertechnik" (GST, Society for Lubrication Technology), the predecessor organization of today's German Society for Tribology, was founded in the form of a non-profit technical scientific association. The society initially saw itself as a platform for the discussion and clarification of issues within lubrication technology in science and industry. The inclusion of the fundamental findings of friction and wear led to the renaming of the society in 1967 to "Gesellschaft für Schmierungstechnik und Tribologie" (Society for Lubrication Technology and Tribology), and then in 1974 to "Gesellschaft für Tribologie" (GfT, German Society for Tribology).

Since the beginning of the GST, regular annual conferences have been held on the topics of friction, lubrication, and wear; thus, The Tribology Conference celebrated its 65th anniversary in 2024.

To mark the occasion of our anniversary, this Special Issue intends to present the latest tribological results in research and development in Germany. However, the beginnings of systematic tribology research in Germany go back several centuries.

Let us start in the year 1557. In 1557, Georgius Agricola (Georg Bauer), who was a major of the city of Chemnitz in Saxonia, a medical doctor, a metallurgist, and a miner, published twelve printed books on "von Bergwergk" ("from the mine") with colored illustrations, in which the mining and metallurgy of the late Middle Ages were very well documented. The illustrations clearly illuminate and label an unlimited number of different types of tribosystems, but unfortunately no details are given.

After that, there is a gap until 1710. We know that this gap was filled by famous French and British scientists. In 1710, the philosopher and mathematician Gottfried Wilhelm Leibniz (1646 to 1716) published an essay "on the nature and means of overcoming resistance in machines caused by the interaction of bodies, on the occasion of a previous treatise on the same subject", in which he clearly described that in gears, the involutes are created by unrolling a straight line. "It is well known to those who combine the practical application of mechanics with the knowledge of its principles that–for a given driving force to be applied–the effect of machines can only be increased if unnecessary resistances are removed which require a superfluous application of forces; so that here any gain consists in saving effort". More than 300 years ago, Leibniz described the great economic benefits of reducing friction losses and energy consumption, which was very topical.

Regardless of the great individual achievements of Leibniz, Agricola, and a few others, scientists and engineers have only been working collectively and systematically on tribological issues in a broad sense for about 150–170 years. How do we define systematic research on tribology? Any research needs meaningful tools. If tribometry and viscometers are regarded as necessary tools, then their beginnings can be traced back to around this time.

In the second third of the 19th century, tribological problems increasingly arose in railroad locomotives and carriages. In 1862, Director Kirchweger of the General Directorate for Railway Carriages and Telegraphs in Hanover published a testing device for "Tests on journal friction on railroad carriage axles".

Theoretical reflections are as important as metrological devices. In 1881, Prof. Dr. Heinrich Hertz published his theory on contact mechanics, which is globally acknowledged as "Hertzian contact theory". He had not thought about tribology but was very concerned about how the beam path of optical lenses changes when they are pressed together.

In this decade, around 1885, Prof. Dr. Carl Oswald Viktor Engler of the University of Karlsruhe published an "Ausflußapparat" (outflow device) to measure and quantify the viscosity called, at this time, "Flüssigkeitsgrade" (the degree of fluidity). The so-called "Engler-Grade" (Engler degrees) were used in Europe until the 1960s and were superseded by the ISO-viscosity grades.

Prof. Dr. Adolf Martens, the eponym for "Martensite" or "Martens Universal Hardness", in 1885, published studies similar to later works of Stribeck. The progress made in Martens "Ölprobirmaschine" (Oil-tasting machine) was the ability to control speed, which was phenomenal at this time. Through publications in the semi-official journal of his royal Prussian institute, Martens had a much smaller reach than Stribeck, who published in the journal of the Association of German Engineers (VDI).

With Prof. Dr. Richard Stribeck's systematic studies on plain and roller bearings, which were published in 1901 and were acknowledged globally as the well-known Stribeck-type curves, these years were the beginning of systematic tribology research in Germany. It should be noted at this point that while Prof. Stribeck was honored for this work with the technical term "Stribeck-type curve", the assignment of his "curves" to lubrication regimes was carried out by Prof. L. Gümbel around 1913. On the other hand, Stribeck was the creator of the standard ball bearing steel 100Cr6 = AIS 52100 = SUJ2, which is still the work horse in the ball bearing industry today.

In today's world of ever-increasing energy demands, combined with an environmentally and cost-driven desire to conserve resources, friction and wear are becoming increasingly important. To avoid friction and wear, save energy, protect the environment, and improve the performance of machines and systems, lubrication is one of the most effective strategies. However, design measures, new materials and coatings, a better understanding of the fundamental tribological relationships, and the predictability of tribological systems also make a decisive contribution toward environmentally compatible and efficient machines and systems, with the goals of having lower energy consumption in the utilization phase, less wear, and the longer service life of mechanical systems.

The visibility of German tribology is internationally characterized by the automotive industry and mechanical engineering. Despite its basic orientation, tribological research is always focused on products and machine elements like rolling and journal bearings, gears, and elastomers. Currently, tribological work is concentrating on topics such as sustainability, raw material availability, the reduction of emissions in the use phase, recyclability, defossilization in mobility and industry, and digitalization.

With the 65th anniversary of the GfT, this Special Issue presents the latest results of tribological research and development in Germany with 21 contributions.

Five papers focus on aspects of rolling bearings. New results are presented on White Etching Cracks (contribution 5), the failure mechanisms in oil-lubricated rolling bearings with small oscillating movements (contribution 9), the tribological behavior of hydrocarbons in mixed friction areas with axial cylindrical roller bearings (contribution 16), the influence of damaging bearing currents on the lifetime of rolling bearings (contribution 14), and a review of the changes in the surface topography in axial bearings because of triboelectric loads (contribution 21).

Regarding plain bearings, there are two papers on the use of planetary gear plain bearings (instead of rolling bearings) in wind turbine gearboxes (contribution 20), and a study on a machine-learning-based approach for wear prediction in plain bearings (contribution 7).

Regarding gears, there is an interesting paper on gears made from high-performance thermoplastics (contribution 1).

With references to lubricants and lubrication, there are three papers on the improved operating behavior of self-lubricating rolling sliding contacts under high load with oil-impregnated porous sintered material (contribution 10), the practical evaluation of ionic liquids–as 'non-evaporating liquids'–for use as lubricants in clean rooms and under vacuum conditions (contribution 17), and an approach to "intrinsic lubrication" using high-molecular siloxane dispersion in a polybutylene terephthalate (contribution 18).

There are three papers on surface modifications of triboelements, namely a tribological study of ta-C, ta-C:N, and ta-C:B coatings on plastic substrates under dry sliding conditions (contribution 3), the run-in of amorphous carbon coatings and their transfer film formation (contribution 4), and the micro texturing of a cam follower contact (contribution 6).

Two papers deal with the modeling of static and dynamic elastomer friction in dry conditions (contribution 12) and the effect of hydrogen pressure on the fretting behavior of rubber materials (contribution 13).

Finally, four papers focus on fundamental tribological aspects, namely whether mixed friction can exist in fully lubricated elastohydrodynamic contacts (contribution 2), the influence of lubrication cycle parameters on hydrodynamic linear guides through the simultaneous monitoring of oil film pressure and float heights (contribution 8), the numerical simulations and experimental validation of squeeze film dampers for aircraft jet engines (contribution 11), and an experimental approach for friction modes in adhesive contacts of a hard, rigid steel indenter and a soft elastomer (contribution 19).

Following Germany's long tradition as an ice sports nation, one paper deals with an advanced numerical approach for competitive winter sports applications (contribution 15).

Acknowledgments: I would like to thank all authors and reviewers for their contributions. My special thanks go to my two esteemed colleagues, Rolf Luther, who is from Fuchs Lubricants Germany GmbH and is Chairman of the Board of the GfT, and Mathias Woydt, who is from Matrilub and is a member of the Board of the GfT, both of whom prepared the section on the beginnings of systematic tribology research in Germany and made it available for this Editorial.

Conflicts of Interest: The author declares no conflicts of interest.

List of Contributions:

1. Schmid, F.; Lohner, T.; Stahl, K. Friction in Oil-Lubricated Rolling–Sliding Contacts with Technical and High-Performance Thermoplastics. *Lubricants* **2024**, *12*, 372. https://doi.org/10.3390/lubricants12110372.
2. Bartel, D. Mixed Friction in Fully Lubricated Elastohydrodynamic Contacts—Theory or Reality. *Lubricants* **2024**, *12*, 351. https://doi.org/10.3390/lubricants12100351.
3. Neubauer, P.; Kaulfuss, F.; Weihnacht, V. A Tribological Study of ta-C, ta-C:N, and ta-C:B Coatings on Plastic Substrates under Dry Sliding Conditions. *Lubricants* **2024**, *12*, 331. https://doi.org/10.3390/lubricants12100331.
4. Faller, J.; Scherge, M. Running-In of DLC–Third Body or Transfer Film Formation. *Lubricants* **2024**, *12*, 314. https://doi.org/10.3390/lubricants12090314.
5. Baur, M.; Khader, I.; Kürten, D.; Schies, T.; Kailer, A.; Dienwiebel, M. Premature Damage in Bearing Steel in Relation with Residual Stresses and Hydrogen Trapping. *Lubricants* **2024**, *12*, 311. https://doi.org/10.3390/lubricants12090311.
6. Orgeldinger, C.; Reck, M.; Seynstahl, A.; Rosnitschek, T.; Merklein, M.; Tremmel, S. Process-Integrated Component Microtexturing for Tribologically Optimized Contacts Using the Example of the Cam Tappet—Numerical Design, Manufacturing, DLC-Coating and Experimental Analysis. *Lubricants* **2024**, *12*, 291. https://doi.org/10.3390/lubricants12080291.
7. König, F.; Wirsing, F.; Singh, A.; Jacobs, G. Machine-Learning-Based Wear Prediction in Journal Bearings under Start–Stop Conditions. *Lubricants* **2024**, *12*, 290. https://doi.org/10.3390/lubricants12080290.
8. Ibrar, B.; Wittstock, V.; Regel, J.; Dix, M. Influence of Lubrication Cycle Parameters on Hydrodynamic Linear Guides through Simultaneous Monitoring of Oil Film Pressure and Floating Heights. *Lubricants* **2024**, *12*, 287. https://doi.org/10.3390/lubricants12080287.
9. Halmos, F.; Wartzack, S.; Bartz, M. Investigation of Failure Mechanisms in Oil-Lubricated Rolling Bearings under Small Oscillating Movements: Experimental Results, Analysis and Comparison with Theoretical Models. *Lubricants* **2024**, *12*, 271. https://doi.org/10.3390/lubricants12080271.

10. Sprogies, N.; Lohner, T.; Stahl, K. Improved Operating Behavior of Self-Lubricating Rolling-Sliding Contacts under High Load with Oil-Impregnated Porous Sinter Material. *Lubricants* **2024**, *12*, 259. https://doi.org/10.3390/lubricants12070259.
11. Golek, M.; Gleichner, J.; Chatzisavvas, I.; Kohlmann, L.; Schmidt, M.; Reinke, P.; Rienäcker, A. Numerical Simulations and Experimental Validation of Squeeze Film Dampers for Aircraft Jet Engines. *Lubricants* **2024**, *12*, 253. https://doi.org/10.3390/lubricants12070253.
12. Kaiser, F.; Savio, D.; Bactavatchalou, R. Modelling of Static and Dynamic Elastomer Friction in Dry Conditions. *Lubricants* **2024**, *12*, 250. https://doi.org/10.3390/lubricants12070250.
13. Theiler, G.; Cano Murillo, N.; Hausberger, A. Effect of Hydrogen Pressure on the Fretting Behavior of Rubber Materials. *Lubricants* **2024**, *12*, 233. https://doi.org/10.3390/lubricants12070233.
14. Schneider, V.; Krewer, M.; Poll, G.; Marian, M. Effect of Harmful Bearing Currents on the Service Life of Rolling Bearings: From Experimental Investigations to a Predictive Model. *Lubricants* **2024**, *12*, 230. https://doi.org/10.3390/lubricants12070230.
15. Grzemba, B.; Pohrt, R. Ice-versus-Steel Friction: An Advanced Numerical Approach for Competitive Winter Sports Applications. *Lubricants* **2024**, *12*, 203. https://doi.org/10.3390/lubricants12060203.
16. Merk, D.; Koenig, T.; Fritz, J.; Franke, J.W.H. Tribological Behavior of Hydrocarbons in Rolling Contact. *Lubricants* **2024**, *12*, 201. https://doi.org/10.3390/lubricants12060201.
17. Keller, A.; Karlson, K.-O.; Grebe, M.; Schüler, F.; Goehringer, C.; Epp, A. Practical Evaluation of Ionic Liquids for Application as Lubricants in Cleanrooms and under Vacuum Conditions. *Lubricants* **2024**, *12*, 194. https://doi.org/10.3390/lubricants12060194.
18. Zhao, S.; Merz, R.; Emrich, S.; L'huillier, J.; Lin, L. Improved Tribological Performance of a Polybutylene Terephthalate Hybrid Composite by Adding a Siloxane-Based Internal Lubricant. *Lubricants* **2024**, *12*, 189. https://doi.org/10.3390/lubricants12060189.
19. Lyashenko, I.A.; Pham, T.H.; Popov, V.L. Transition between Friction Modes in Adhesive Contacts of a Hard Indenter and a Soft Elastomer: An Experiment. *Lubricants* **2024**, *12*, 110. https://doi.org/10.3390/lubricants12040110.
20. Ding, H.; Mermertas, U.; Hagemann, T.; Schwarze, H. Calculation and Validation of Planet Gear Sliding Bearings for a Three-Stage Wind Turbine Gearbox. *Lubricants* **2024**, *12*, 95. https://doi.org/10.3390/lubricants12030095.
21. Graf, S.; Koch, O. Changes in Surface Topography and Light Load Hardness in Thrust Bearings as a Reason for Tribo-Electric Loads. *Lubricants* **2024**, *12*, 303. https://doi.org/10.3390/lubricants12090303.

Disclaimer/Publisher's Note: The statements, opinions and data contained in all publications are solely those of the individual author(s) and contributor(s) and not of MDPI and/or the editor(s). MDPI and/or the editor(s) disclaim responsibility for any injury to people or property resulting from any ideas, methods, instructions or products referred to in the content.

Article

Friction in Oil-Lubricated Rolling–Sliding Contacts with Technical and High-Performance Thermoplastics [†]

Ferdinand Schmid *, Thomas Lohner and Karsten Stahl

Gear Research Center (FZG), Department of Mechanical Engineering, School of Engineering and Design, Technical University of Munich, Boltzmannstraße 15, D-85748 Garching, Germany; thomas.lohner@tum.de (T.L.); karsten.stahl@tum.de (K.S.)
* Correspondence: ferdinand.schmid@tum.de
[†] This work is an extended version of an abstract published in: Schmid, F.; Maier, E.; Lohner, T.; Stahl, K. Friction in Oil-lubricated Rolling-Sliding Contacts with Technical Thermoplastics. In Proceedings of the 64th German Tribology Conference 2023, Göttingen, Germany, 25–27 September 2023.

Abstract: Thermoplastics show great potential due to their lightweight design, low-noise operation, and cost-effective manufacturing. Oil lubrication allows for their usage in high-power-transmission applications, such as gears. The current design guidelines for thermoplastic gears lack reliable estimates for the coefficient of friction of oil-lubricated rolling–sliding contacts. This work characterizes the friction of elastohydrodynamic rolling–sliding contacts with technical and high-performance thermoplastics with oil lubrication. The influence of polyoxymethylene (POM), polyamide 46 (PA46), polyamide 12 (PA12), and polyetheretherketone (PEEK), as well as mineral oil (MIN), polyalphaolefin (PAO), and water-containing polyalkylene glycol (PAGW), was studied. Experiments were carried out on a ball-on-disk tribometer, considering different loads, speeds, temperatures, and surface roughness. The results show that, for fluid film lubrication, there is very low friction in the superlubricity regime, with a coefficient of friction lower than 0.01. Both sliding and rolling friction account for a significant portion of the total friction, depending on the contact configuration and operating conditions. In the mixed to boundary lubrication regime, the sliding friction depends on the thermoplastic and rises sharply, thus increasing the total friction.

Keywords: polymer; thermoplastic; elastohydrodynamic lubrication; friction; superlubricity; tribology; ball-on-disk tribometer; gears

Citation: Schmid, F.; Lohner, T.; Stahl, K. Friction in Oil-Lubricated Rolling–Sliding Contacts with Technical and High-Performance Thermoplastics. *Lubricants* **2024**, *12*, 372. https://doi.org/10.3390/lubricants12110372

Received: 15 August 2024
Revised: 15 October 2024
Accepted: 17 October 2024
Published: 28 October 2024

Copyright: © 2024 by the authors. Licensee MDPI, Basel, Switzerland. This article is an open access article distributed under the terms and conditions of the Creative Commons Attribution (CC BY) license (https://creativecommons.org/licenses/by/4.0/).

1. Introduction

The use of technical and high-performance polymers for machine element applications confers several advantages, including the ability to achieve low noise emissions through damping, the potential for lightweight design, and cost-effective manufacturing. Moreover, in oil-lubricated contacts, very low friction is possible [1–3]. The research into oil-lubricated rolling–sliding contacts in steel gears is extensive. However, for thermoplastic gears made of technical or high-performance polymers, significant gaps remain in the understanding of the influence of strongly dependent material properties and the relevant origins of friction. The standard VDI 2736 [4] for thermoplastic gears provides an estimated value for the coefficient of friction of oil lubrication of $\mu = 0.04$ for gear pairings, but no detailed calculation equations are included. As frictional power losses lead to bulk temperature increases and, consequently, a reduction in the load-carrying capacity, a detailed understanding of the friction in oil-lubricated thermoplastic gear contacts is essential for the development of advanced designs.

In general, oil lubrication is employed in the tribological contact of machine elements for oil film formation and cooling, facilitating low friction and wear. Elastohydrodynamically lubricated (EHL) contacts were classified by Johnson [5] using an elasticity and viscosity parameter. Consequently, any EHL contact can be classified into one of four distinct lubrication regimes, contingent upon the significance of pressure-induced viscosity increases and elastic deformation of the surfaces. Polymers are characterized by their high elasticity, which allows them to be classified in EHL contacts as either elastic isoviscous (IE) or elastic variable viscous (VE) [6,7]. The IE regime typically refers to soft or compliant EHL contacts [1,2,7–11]. Elastomers are typically classified within the IE regime, while thermoplastics, including technical and high-performance thermoplastics, can be classified within a transition regime (TR) between the IE and VE regime. The phenomenon was initially investigated in detail by Hooke [12], who developed an analytical calculation equation for the film thickness of highly loaded elastic contacts in the TR. Subsequently, Myers et al. [6] proposed that the TR should be regarded as an additional lubrication regime. The TR necessitates the consideration of pressure-dependent viscosity in the context of highly deformed contacts [13] in EHL contacts. Myers et al. [6] used a semi-analytical approach to derive an analytical equation for oil film thickness in the TR. Maier et al. [14] conducted numerical calculations of a thermoplastic EHL contact with polyoxymethylene (POM). They found good agreement with the analytical model of Myers et al. [6], with a maximum deviation of 5%. Ziegltrum et al. [15] performed thermal EHL simulations of oil-lubricated technical polymer contacts with polymethyl methacrylate (PMMA). It was observed that, when a polymer was paired with steel, local conformity in the contact occurred due to the significant difference in stiffness between the two bodies. Consequently, the thermoplastic body was subjected to a significantly greater degree of deformation than the steel material. This resulted in a curved contact area [14].

Besides strongly thermal-dependent and non-linear elasticity behaviors, polymers can exhibit a viscoelastic material behavior, which is characterized by a combination of viscous and elastic properties when deformed under a load. The first theoretical investigations of the rolling friction of dry soft contacts were performed by Hunter [16]. He demonstrated that rolling friction originates from an asymmetric pressure distribution, which is a consequence of viscoelastic material properties. Hooke and Huang [17] performed numerical calculations to investigate the impact of viscoelasticity on EHL contacts. They showed that viscoelasticity can influence the pressure distribution and oil film formation significantly. The pressure distribution shifted towards the contact inlet and showed a strong asymmetric distribution. The presence of viscoelastic behavior can be assessed using the Deborah number. This characteristic number indicates that the viscoelastic behavior is dependent not only on the material but also on the load frequency and is, therefore, dependent on the operating condition. Numerical and experimental investigations of the visco-EHL (VEHL) contact were carried out for PMMA by Putignano and Dini [18], Zhao et al. [19,20], and Krupka et al. [20]. They were able to measure the viscoelastic influence of the material on the oil film formation for the given experimental operating conditions.

An experimental analysis of the coefficient of friction of a soft EHL contact with an elastomer and non-technical liquids was performed by de Vicente et al. [1] for different lubrication regimes. Based on the measurements, they derived a calculation equation for the coefficient of friction of soft EHL contacts with ideal smooth surfaces. Furthermore, they performed numerical EHL simulations and evaluated the Couette as well as the Poiseuille friction portions. Subsequently, Vicente et al. [2] conducted experiments to measure the rolling and sliding friction of a lubricated elastomer contact using a ball-on-disk tribometer. Their findings indicated that, under fluid film lubrication, sliding friction originated from the shearing of the oil (Couette friction), while, in boundary lubrication, it mainly arose from the adhesion of the contacting asperities. Conversely, rolling friction could be attributed to the asymmetric contact pressure distribution (Poiseuille friction), and the hysteresis within the bulk material. Subsequently to this, Myant et al. [8] performed experimental investigations of a soft EHL contact with elastomers, employing a ball-on-disk tribometer. They investigated the influence of load and elasticity on rolling and sliding friction with non-technical liquids. In comparison to the analytical calculation equation of de Vicente et al. [1], they found, overall, a good accordance of the coefficient of friction for sliding and rolling. To investigate the viscoelastic friction of soft EHL contacts, Putignano et al. [21,22] carried out experimental investigations using a ball-on-disk tribometer under dry conditions. The measured rolling friction was evaluated as viscoelastic friction, while the sliding friction was considered to be Coulomb friction. This approach yielded good accordance with the developed model for viscoelastic friction proposed by Carbone and Putignano [23]. They later measured the temperature in the rolling contact, which they found to be in good accordance with their model. Numerical and experimental investigations on the rolling and sliding fluid friction in an EHL contact with thermoplastic POM was performed by Maier et al. [14]. They found that rolling and sliding fluid friction are of a comparable magnitude for the investigated conditions. The local conformity in the contact of steel/thermoplastic pairing results in a curved contact area, which can affect the friction evaluation for soft and hard contacts [16,24]. As a ball-on-disk tribometer considers contact with different body geometries, the influence of the contact configuration can be studied. For rubber elastomers, Sadowski and Stupkiewicz [25] found that the contact configuration does not affect the measured friction. Quinn et al. [26] also carried out experimental investigations on the influence of contact configuration for the low-stiffness polymer polydimethylsiloxane (PDMS). They found that the soft-disk/hard-ball configuration resulted in a higher friction than the hard-disk/soft-ball combination. This discrepancy was attributed to the differing hysteresis behaviors exhibited by the two configurations. Reitschuster et al. [3] performed experimental studies on the friction of oil-lubricated thermoplastics using a twin-disk tribometer. They examined the total friction for various rolling–sliding conditions when PEEK and polyamide 66 (PA66) were in contact with steel. They measured very low interfacial friction in the range of superlubricity for a high, relative film thickness. Friction by hysteresis losses was recognized through the temperature increase in the bulk material and was noticeably higher for polyamide 66 (PA66) than PEEK.

The above literature review shows that the tribological contact with polymers is a strong focus of research. EHL contacts with thermoplastics, including technical and high-performance polymers, are classified by the transition lubrication regime. When a polymer is paired with steel, local conformity in the contact occurs. Viscoelastic effects can strongly influence the pressure and film thickness distribution. Friction in polymer EHL contacts is shown to be influenced by interfacial friction due to rolling and sliding fractions and by hysteresis losses, mainly due to viscoelastic effects. However, many results are based on the analysis of soft contacts with elastomers. A systematic analysis of the friction in EHL contacts with technical and high-performance thermoplastics is lacking.

The objective of this study is to characterize the friction of EHL contacts with technical and high-performance thermoplastics under oil lubrication. The influence of the contact configuration, thermoplastic material, lubricant, load, oil temperature, and surface roughness are systematically investigated under rolling–sliding conditions on a ball-on-disk tribometer. Focus is put on the rolling and sliding friction portion. The results improve the understanding of friction in oil-lubricated rolling–sliding contacts, similar to those in cylindrical polymer gears. This work is an extended version of the conference abstract [27] that was originally presented at the 64th German Tribology Conference 2023.

2. Materials and Methods

In order to analyze the influences on the friction in thermoplastic EHL contacts under oil lubrication, different thermoplastics with different oils in various configuration conditions were investigated.

2.1. Experimental Configuration

The experiments were performed on an MTM2 ball-on-disk tribometer. Figure 1 shows the mechanical layout of the test chamber, wherein the ball and the disk are mounted on a rotating shaft each, which can be driven independently to adjust for any rolling–sliding conditions. Over a load arm, both specimens can be put into contact and enforced via a load spring.

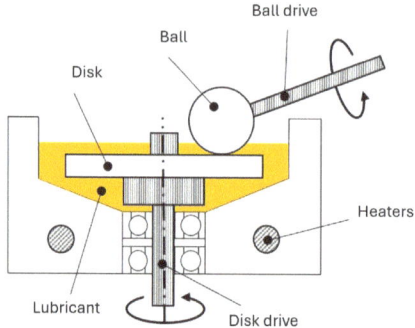

Figure 1. Schematic representation of the mechanical layout of the MTM2 test chamber from Hofmann et al. [28].

The surface speed of the ball is designated with v_1 and that of the disk with v_2, respectively. The mean velocity v_m, also referred to as rolling speed v_r, is defined as follows:

$$v_m = \frac{v_1+v_2}{2}. \tag{1}$$

The sliding speed v_s is defined as follows:

$$v_s = v_1 - v_2. \tag{2}$$

The slide-to-roll ratio SRR in the contact is then defined as follows:

$$SRR = \frac{v_s}{v_m} = 2\frac{v_1-v_2}{v_1+v_2}. \tag{3}$$

The friction force F_R and the normal force F_N are measured each with a load cell at the load arm so that a coefficient of friction μ can be evaluated, as given in Equation (4). The measured total friction force F_R is composed of different friction portions and can, therefore, be divided according to Vicente et al. [2] into the rolling friction $F_{R,r}$ and the sliding friction $F_{R,s}$. The total coefficient of friction μ is the sum of the rolling coefficient of friction μ_r and the sliding coefficient of friction μ_s.

$$\mu = \frac{F_R}{F_N} = \frac{F_{R,r}+F_{R,s}}{F_N} = \frac{F_{R,r}}{F_N} + \frac{F_{R,s}}{F_N} = \mu_r + \mu_s. \tag{4}$$

In order to differentiate between rolling and sliding friction, negative and positive SRRs were imposed on the MTM2. In both cases, the sliding friction differed in its direction, but the rolling friction direction remained the same. By measuring the friction force in both situations, the rolling and sliding friction could be evaluated afterward. For details of this procedure in MTM2, considering the elimination of the offset of the load cell, one can refer to Vicente et al. [2].

2.2. Thermoplastics

Jain and Patil [29] provide an overview of the range of polymers used in gear applications. In this study, four different technical and high-performance thermoplastics were considered, as shown in Table 1. Polyoxymethylene (POM) and polyamides PA12 and PA46 are typical materials for plastic gears [14,30–33]. The high-performance thermoplastic polyether ether ketone (PEEK) is becoming increasingly popular for use in gears that require the highest power density [31,34–36]. The investigated thermoplastics exhibit differences in stiffness and in the glass transition temperature ϑ_g, which primarily influences the impact of mechanical material properties, such as stiffness and viscoelasticity. All thermoplastics were non-reinforced, unfilled, and commercially available. The PEEK material used was Vestakeep 5000G (Evonik, Essen, Germany), the PA12 material was Vestamid L1940 (Evonik), the PA46 was Stanyl TW441 (Envalior, Dusseldorf, Germany), and the POM material used was Delrin 100 NC010 (DuPont, Wilmington, DE, USA). For POM, a homopolymer polyoxymethylene (POM-H) was used.

Table 1. Technical and high-performance thermoplastics considered according to [37].

	PEEK	POM	PA12	PA46	
E (40 °C)	3720	2615	1180	3260	N/mm^2
ν (40 °C)	0.4	0.42	0.37 [1]	0.37	-
ρ (40 °C)	1.32	1.39	1.01	1.17	g/cm^3
ϑ_g	145	−60	45	75	°C

[1] Estimated from PA46.

Figure 2 illustrates the dependence of Young's modulus E of the thermoplastics studied on the temperature for the dry state, as depicted in the VDI 2736 [37] standard, except for Poisson's ratio ν of PA12, which is assumed to be identical to that of PA46, since no value is given. It is a characteristic of thermoplastics that their stiffness decreases at elevated temperatures. The glass transition regimes are also indicated. In the glass transition regimes, there is a significant reduction in stiffness. As the glass transition temperature ϑ_g for POM is approximately at −60 °C, it can be seen that there is no visible step in the temperature decrease but rather a strong and nearly linear decrease over the indicated temperature range. Among the considered thermoplastics, PEEK has the highest thermal stability. It should be noted that polyamides change their stiffness and glass transition temperature behaviors with changing humidity. For details, see [37,38].

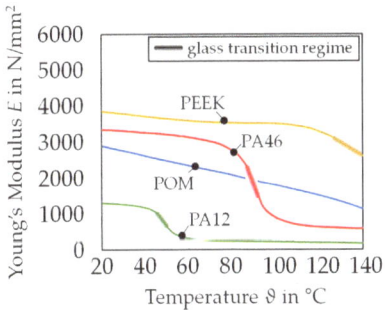

Figure 2. Temperature dependence of Young's modulus of the studied thermoplastics according to VDI 2736 [37] for a dry state.

2.3. Test Specimens

The ball of the MTM2 tribometer had a diameter of 19.05 mm. The disks had a diameter of 46 mm and a thickness of 6 mm at the running track surface. The thermoplastic disks were mounted on a carrier plate made of steel with a thickness of 2 mm. The tribological surfaces of the tested disks and balls were machined from semi-finished products, ground, and polished isotropically. The arithmetic mean roughness of the polished surfaces was $Ra < 0.02$ µm. In order to investigate the influence of the surface roughness of the thermoplastic material, quasi-isotopically ground PEEK disks with arithmetic mean roughness values of approx. $Ra = 0.25$ µm and $Ra = 0.50$ µm were considered. The roughness values were obtained from tactile measurement radially on the disk's running track with a measurement length of 4 mm and a cut-off wavelength of 0.08 mm (for $Ra < 0.02$ µm) and 0.8 mm (for $Ra = 0.25$ µm and $Ra = 0.50$ µm).

It is common practice to pair thermoplastic gears with steel gears, thereby creating a hybrid contact between the two materials. In order to investigate the influence of different contact configurations, polished PEEK and 100Cr6 specimens were paired in a combination of ball/disk configurations, including PEEK/PEEK, PEEK/100Cr6, 100Cr6/PEEK, and 100Cr6/100Cr6. The considered contact configurations are shown in Figure 3. In order to investigate the influence of thermoplastic materials, polished steel balls were paired with polished disks made from PEEK, POM, PA12, and PA46, as illustrated in Figure 4.

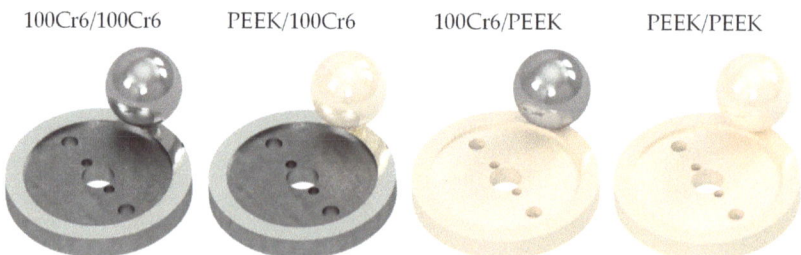

Figure 3. Considered contact configurations at the MTM2.

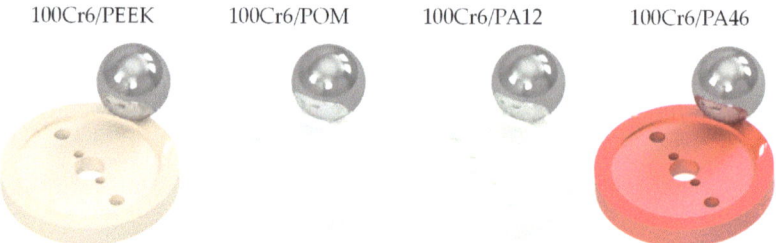

Figure 4. Considered thermoplastic disk materials for the contact configuration 100Cr6/thermoplastic at the MTM2.

2.4. Lubricants

The principal properties of the considered lubricants are presented in Table 2. The mineral oil MIN100, the polyalphaolefin PAO100, and the polyalkylene glycol PAGW100 with 20 wt.% water (see [28]) were of ISO viscosity grade (VG) 100. Additionally, a mineral oil MIN32 of ISO VG 32 was investigated. All oils were fully formulated with additives such as extreme pressure and anti-wear, foam-inhibiting, anti-freeze, and corrosion protection properties.

Table 2. Considered lubricants.

	MIN32	MIN100	PAO100	PAGW100	
ISO VG class	ISO VG 32	ISO VG 100	ISO VG 100	ISO VG 100	–
ν (40 °C)	32.6	94.5	104.6	111.5	mm/s^2
ν (100 °C)	5.5	9.8	15.5	18.9 [1]	mm/s^2
VI	104	77	157	191	–
ρ (15 °C)	0.887	0.885	0.852	1.126	g/cm^3

[1] Extrapolated.

2.5. Operating Conditions

Rolling–sliding conditions, such as in gears, were considered. On the one hand, friction curves with varying slide-to-roll ratios $SRRs = 0 \ldots 1$ were measured at a constant rolling speed of $v_m = 1.5$ m/s. This resulted in the regime of fluid film lubrication with separated surfaces for all polished surfaces. On the other hand, Stribeck-like curves with varying rolling speeds $v_m = 0.01 \ldots 2.5$ m/s were measured at a constant slide-to-roll ratio of $SRR = 0.5$. Depending on the surface roughness, the lubrication regime changed from fluid film to mixed to boundary lubrication with a decreasing v_m. Given the significant impact of temperature on the thermoplastic and lubricant behavior (see Figure 2), a range of temperatures around $\vartheta_\mathrm{oil} = 40\ldots80\ °\mathrm{C}$ was selected. The rolling–sliding contacts were loaded with normal forces of $F_\mathrm{N} = 10\ldots30$ N. Assuming the applicability of the Hertzian theory and using Young's modulus at $40\ °\mathrm{C}$ (see Table 1), this resulted in Hertzian pressures between 34.0 and 106.5 N/mm^2 depending on the contact configuration and thermoplastics (see Table 3).

Table 3. Hertzian pressures for the contact configurations 100Cr6/thermoplastic at $F_\mathrm{N} = \{10, 20, 30\}$ N at the MTM2 at $40\ °\mathrm{C}$.

	100Cr6/PEEK	100Cr6/POM	100Cr6/PA12	100Cr6/PA46	
$p_\mathrm{H}(F_\mathrm{N} = 10\ \mathrm{N})$	73.8	59.4	34.0	66.5	N/mm^2
$p_\mathrm{H}(F_\mathrm{N} = 20\ \mathrm{N})$	93.0	74.8	42.8	83.8	N/mm^2
$p_\mathrm{H}(F_\mathrm{N} = 30\ \mathrm{N})$	106.5	85.6	49.1	95.9	N/mm^2

In addition to the contact configurations 100Cr6/thermoplastic, further configurations were investigated. For the configuration 100Cr6/100Cr6, the calculated Hertzian pressure at $F_\mathrm{N} = 20$ N was $p_\mathrm{H} = 817.7$ N/mm^2. For PEEK/100Cr6 and 100Cr6/PEEK, the calculated Hertzian pressure at $F_\mathrm{N} = 20$ N and $40\ °\mathrm{C}$ was $p_\mathrm{H} = 93.0$ N/mm^2, while, for PEEK/PEEK, it was $p_\mathrm{H} = 59.4$ N/mm^2.

Prior to the start of testing, a ten-minute run-in procedure with $F_\mathrm{N} = 20$ N, $v_\mathrm{m} = 0.1$ m/s, $SRR = 0.5$, and an oil temperature of $\vartheta_\mathrm{oil} = 40\ °\mathrm{C}$ oil was performed for all configurations and tests. All measurements were repeated once and then averaged. In Appendix A, the relative change from the first to the second measurement is given, where the measurement value is averaged over the friction and Stribeck-like curves, respectively. An average deviation of 5.7% was found for all measurements.

3. Results

This Section presents the results of the measurements. In Sections 3.1 and 3.2, the results of the friction and Stribeck-like curves are presented in detail, while Sections 3.3–3.6 focus on important aspects of the investigated influences. To illustrate the results of the friction measurements, the contact configuration 100Cr6/PEEK with a polished specimen and MIN100 is used as a reference to show the effects of varying configuration and contact conditions. Based on this reference, Section 3.7 provides a summarized overview with the mean coefficients of friction for the influences shown in Sections 3.1–3.6.

3.1. Influence of Contact Configuration

Figure 5 shows the total μ, sliding μ_s, and rolling μ_r coefficients of friction over the slide-to-roll ratio SRR (Figure 5a–c) and rolling speed v_m (Figure 5a,d–f) for the contact configurations 100Cr6/100Cr6, PEEK/100Cr6, 100Cr6/PEEK, and PEEK/PEEK at a normal force $F_\mathrm{N} = 20$ N and oil temperature $\vartheta_\mathrm{oil} = 40\ °\mathrm{C}$ for MIN100.

The contact configuration 100Cr6/100Cr6 shows the highest total coefficient of friction μ in Figure 5a,d, whereas the contact configuration PEEK/PEEK shows the lowest friction. In between, there are contact configurations PEEK/100Cr6 and 100Cr6/PEEK with a very similar coefficient of friction μ. The friction levels are in alignment with the calculated contact pressures (see Table 3) and assume an increase in contact viscosity. All contact configurations with PEEK show a nearly linear increase in μ over SRR, whereas the contact configuration 100Cr6/100Cr6 shows linear, non-linear, and thermal regimes, as typically known from friction curves with steel EHL contacts [39]. The total friction at $SRR = 0$ is not zero, since rolling friction is also considered in this study. With decreasing rolling speeds v_m, the coefficient of friction μ for 100Cr6/100Cr6 is steadily increasing, while it has a minimum for the contact configurations with PEEK. Their behavior is similar to typical Stribeck-like curves.

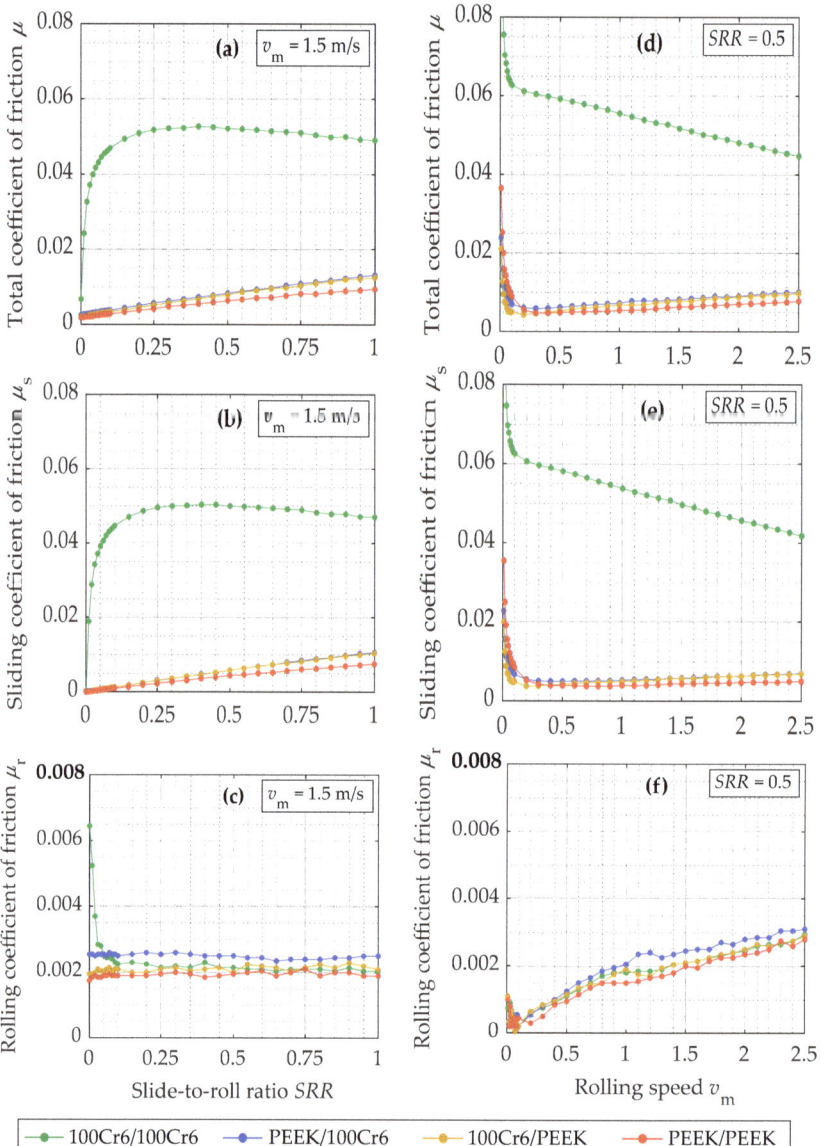

Figure 5. Total μ, sliding μ_s, and rolling μ_r coefficients of friction over the slide-to-roll ratio SRR (**a**–**c**) and rolling speed v_m (**d**–**f**) with polished disks for the contact configurations 100Cr6/100Cr6, PEEK/100Cr6, 100Cr6/PEEK, and PEEK/PEEK at $F_N = 20$ N and $\vartheta_{oil} = 40$ °C for MIN100.

Focusing on the coefficient of sliding friction μ_s in Figure 5b,e shows very similar trends to the total coefficient of friction μ. As the SRR increases, μ_s assumes a dominant role in μ. In contrast, as the rolling speed v_m increases, the coefficient of sliding friction μ_s declines for the contact configuration 100Cr6/100Cr6 and increases for the contact configurations with PEEK.

The rolling coefficient of friction μ_r in Figure 5c,f is approximately one magnitude smaller than μ_s and almost identical for all contact configurations for relevant SRR. This indicates that the relative share of μ_r in μ is higher for contact configurations with PEEK than for the contact configuration 100Cr6/100Cr6. Over SRR, a nearly constant rolling coefficient of friction μ_r is present. It should be noted that a pronounced increase in rolling friction is observed for the contact configuration 100Cr6/100Cr6 at a low-to-zero SRR, while, for increasing rolling speeds v_m, the rolling coefficient of friction μ_r increases.

3.2. Influence of Thermoplastic Material

Figure 6 shows the total μ, sliding μ_s, and rolling μ_r coefficients of friction over the slide-to-roll ratio SRR (a, b, c) and rolling speed v_m (d, e, f) for the contact configurations 100Cr6/PEEK, 100Cr6/POM, 100Cr6/PA12, and 100Cr6/PA46 at a normal force $F_N = 20$ N and oil temperature $\vartheta_{oil} = 40$ °C for MIN100.

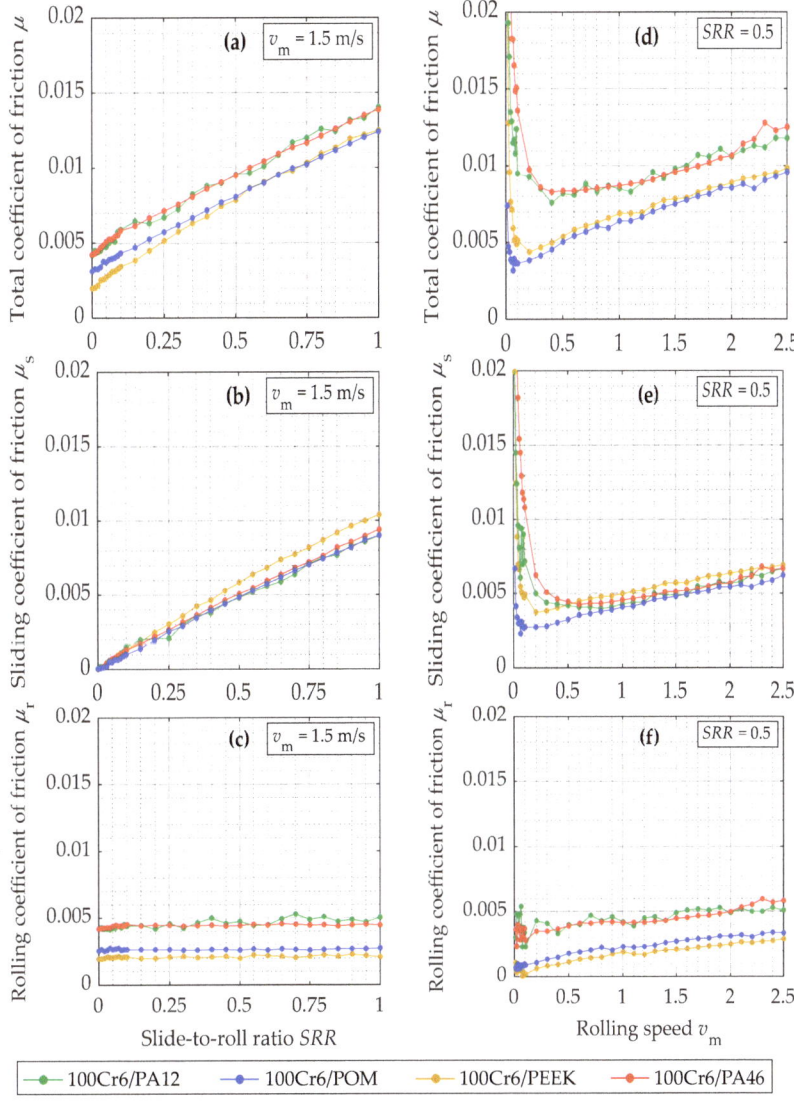

Figure 6. Total μ, sliding μ_s, and rolling μ_r coefficients of friction over the slide-to-roll ratio SRR (**a**–**c**) and rolling speed v_m (**d**–**f**) with polished disks for the contact configurations 100Cr6/PEEK, 100Cr6/POM, 100Cr6/PA12, and 100Cr6/PA46 at $F_N = 20$ N and $\vartheta_{oil} = 40$ °C for MIN100.

The contact configurations with PA12 and PA46 show the highest total coefficient of friction μ in Figure 6a,d at a similar level. The contact configurations with POM and PEEK have an overall lower μ but also on a nearly identical level.

The sliding coefficient of friction μ_s in Figure 6b shows a nearly linear increase over the SRR. Thereby, the contact configuration with PEEK shows the highest sliding coefficient of friction μ_s, which correlates with the highest contact pressure (see Table 3) and an assumed increase in contact viscosity. POM and PA46, which exhibit similar Young's moduli and contact pressures, show almost the same sliding coefficient of friction μ_s. The rolling coefficient of friction μ_r in Figure 6c is constant over the SRR for all thermoplastics. It can be observed that the contact configurations with POM and PEEK and those with PA12 and PA46 each have a similar μ_r. Thereby, the level of contact configurations with PA12 and PA46 is about two times higher. In the case of a low SRR, μ_r plays a dominant role. Conversely, for a high SRR, sliding friction is observed to exceed that of rolling friction. Overall, the rolling friction is a relevant portion of the total friction for the contact configurations with thermoplastics and leads to a higher total friction for contact configurations with polyamides.

Figure 6d shows the total coefficient of friction μ as a function of the rolling speed v_m. For very low rolling speeds v_m and, consequently, mixed to boundary lubrication, the sliding coefficient of friction μ_s in Figure 6e is very high. With increasing rolling speed v_m and fluid film lubrication, μ_s shows a nearly linear increase. Hence, a Stribeck-like behavior is observed. The rolling coefficient of friction μ_r in Figure 6f increases continuously with increasing rolling speeds v_m. Note that, for very low rolling speeds v_m, the rolling coefficient of friction μ_r of the contact configurations with POM and PEEK tends towards zero, while μ_r, of the contact configurations with PA12 and PA46, shows a minimal value larger than zero.

3.3. Influence of Lubricant

Figure 7 shows the influence of different oils on the total μ, sliding μ_s, and rolling μ_r coefficients of friction over the slide-to-roll ratio SRR (Figure 7a–c) and rolling speed v_m (Figure 7d–f) for the contact configuration 100Cr6/PEEK at a normal force $F_N = 20$ N and oil temperature $\vartheta_{oil} = 40\ °C$.

The total coefficient of friction μ over the SRR is depicted in Figure 7a, with MIN32 exhibiting a significantly lower level than MIN100, PAO100, and PAGW100. The sliding coefficient of friction μ_s is shown in Figure 7b. Although MIN32 also shows the lowest μ_s, a relevant portion of the lowest μ compared to the other lubricants is coming from the lowest rolling coefficient of friction μ_r in Figure 7c. Note that, out of the ISO VG100 oils MIN100, PAO100, and PAGW100, MIN100 shows the highest μ_s but the lowest μ_r. Contrary to this, PAGW100 shows the lowest sliding friction and the highest rolling friction.

Figure 7d–f show μ, μ_s, and μ_r over the rolling speed v_m. For the high rolling speed v_m, the comparison between the oils is overall similar to the relation between the oils in Figure 7a–c. However, it is noticeable that μ is strongly increasing at low rolling speeds v_m for both mineral oils MIN100 and MIN32. This mainly comes from the sliding coefficient of friction μ_s, as can be seen in Figure 7e. For higher rolling speeds, the difference regarding the total coefficient of friction μ between the mineral oil MIN32 and the three ISO VG100 oils MIN100, PAGW100, and PAO100 increases. Considering the friction portions, this comes from both μ_s and μ_r, which show the lowest increase for MIN32.

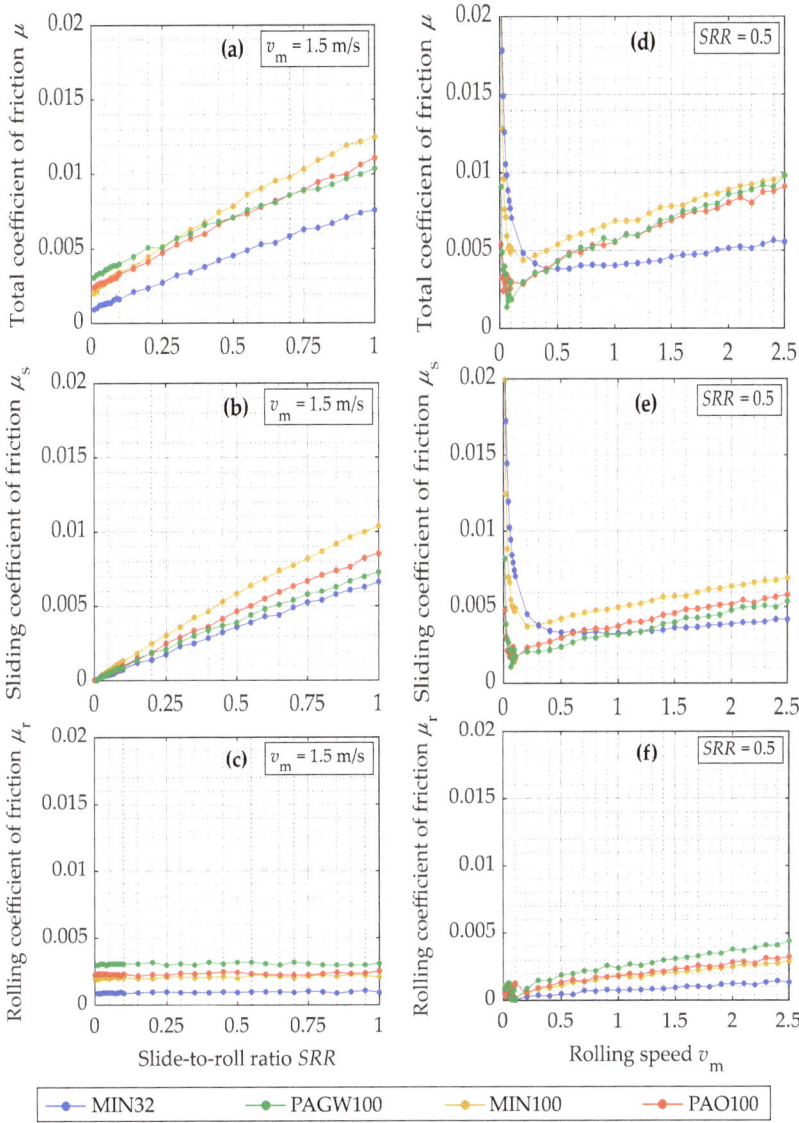

Figure 7. Total μ, sliding μ_s, and rolling μ_r coefficients of friction over the slide-to-roll ratio SRR (**a–c**) and rolling speed v_m (**d–f**) with polished disks for the configuration 100Cr6/PEEK lubricated with MIN100, PAGW100, PAO100, and MIN32 at $\vartheta_{oil} = 40\ °C$ and $F_N = 20\ N$.

3.4. Influence of Load

The influence of the normal force F_N on the total μ, sliding μ_s, and rolling μ_r coefficients of friction over the slide-to-roll ratio SRR (Figure 8a–c) and rolling speed v_m (Figure 8d–f) is illustrated in Figure 8 for the contact configuration 100Cr6/PEEK and an oil temperature $\vartheta_{oil} = 40\ °C$ for MIN100. Three different loads and, therefore, contact pressures (see Table 3) are compared.

The total coefficient of friction μ in Figure 8a,d decreases with increasing F_N. However, μ is very similar for $F_N = 10\ N$ and $F_N = 20\ N$. The differences in μ for the different loads increase with increasing rolling speeds v_m. For a very low v_m, no difference in the total coefficient of friction μ can be seen in Figure 8d.

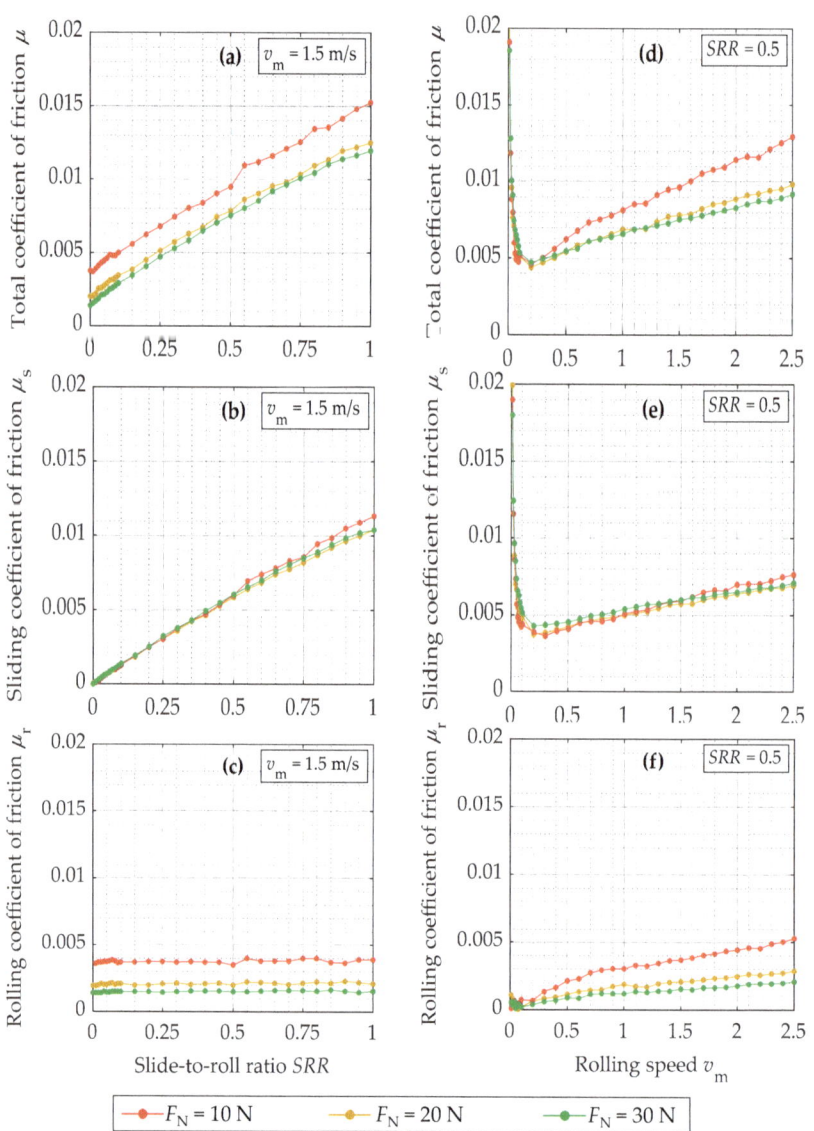

Figure 8. Total μ, sliding μ_s, and rolling μ_r coefficients of friction over the slide-to-roll ratio SRR (**a**–**c**) and rolling speed v_m (**d**–**f**) with polished disks for the configuration 100Cr6/PEEK at $F_N = \{10, 20, 30\}$ N and $\vartheta_{oil} = 40\ °C$ for MIN100.

The sliding coefficients of friction μ_s in Figure 8b,e are very similar for all the normal forces F_N considered. The rolling coefficient of friction μ_r in Figure 8c,f decreases with an increasing F_N and dominates the differences in the total coefficient of friction μ.

3.5. Influence of Oil Temperature

Figure 9 illustrates the influence of the oil temperature ϑ_{oil} on the total μ, sliding μ_s, and rolling μ_r coefficients of friction over the slide-to-roll ratio SRR (Figure 9a–c) and rolling speed v_m (Figure 9d–f) for the contact configuration 100Cr6/PEEK at a normal force $F_N = 20$ N for MIN100.

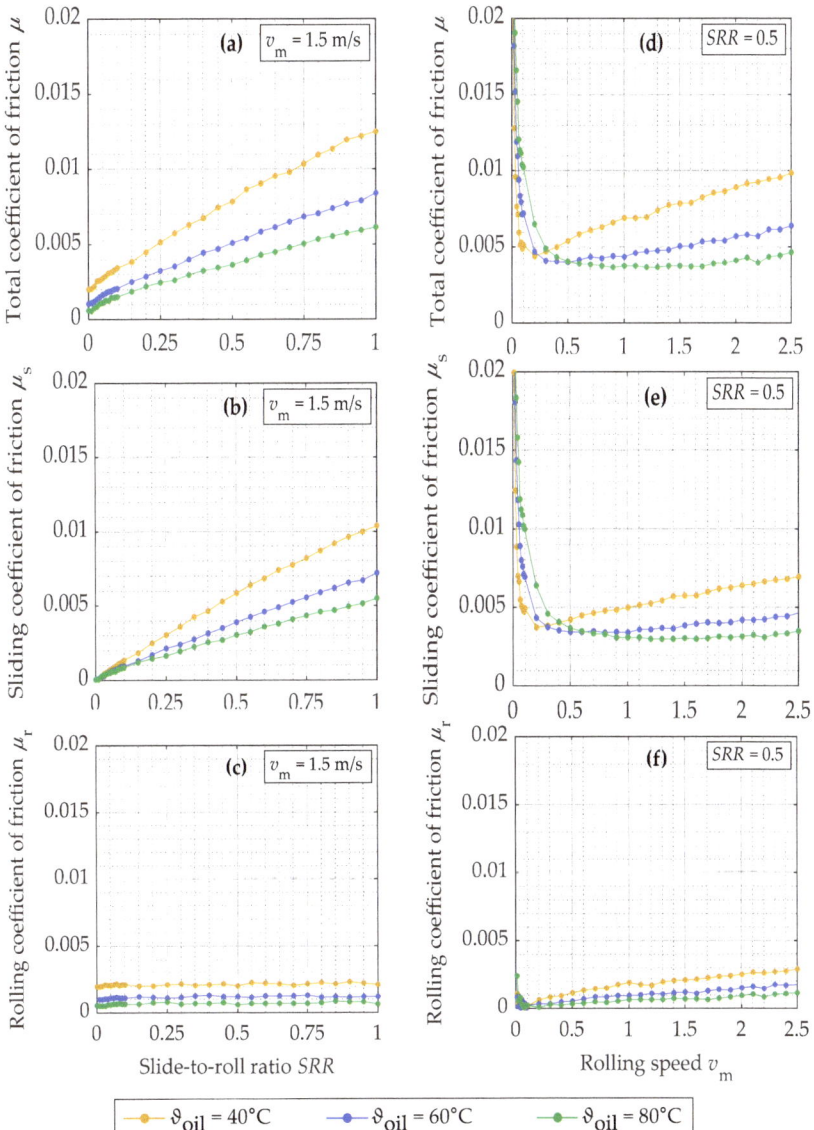

Figure 9. Total μ, sliding μ_s, and rolling μ_r coefficients of friction over the slide-to-roll ratio SRR (**a**–**c**) and rolling speed v_m (**d**–**f**) with polished disks for the configuration 100Cr6/PEEK at $F_N = 20$ N and $\vartheta_{oil} = \{40, 60, 80\}$ °C for MIN100.

Figure 9a shows that the total coefficient of friction μ over the SRR strongly decreases with increasing oil temperature ϑ_{oil}. This is a result of the decrease in both the sliding μ_s and rolling coefficient of friction μ_r in Figure 9b,c. Thereby, the sliding coefficient of friction μ_s shows a linear increase over the SRR for all three oil temperatures but can be distinguished by the gradients.

The coefficients of friction over v_m in Figure 9d–f confirm this trend for higher rolling speed v_m. For a low v_m with mixed-to-boundary lubrication, the influence of ϑ_{oil} is reversed.

Thereby, the sliding coefficient of friction μ_s in Figure 9e exhibits, for a low v_m, an increasing trend with increasing ϑ_{oil} and dominates the total coefficient of friction. The increase in the rolling coefficient of friction μ_r over v_m is smaller for higher oil temperatures (Figure 9f).

3.6. Influence of Surface Roughness

Figure 10 shows the influence of the surface roughness on the total sliding μ, sliding μ_s, and rolling μ_r coefficients of friction over the slide-to-roll ratio SRR (Figure 10a–c) and rolling speed v_m (Figure 10d–f) for the contact configuration 100Cr6/PEEK at a normal force $F_N = 20$ N and oil temperature $\vartheta_{oil} = 40$ °C for MIN100.

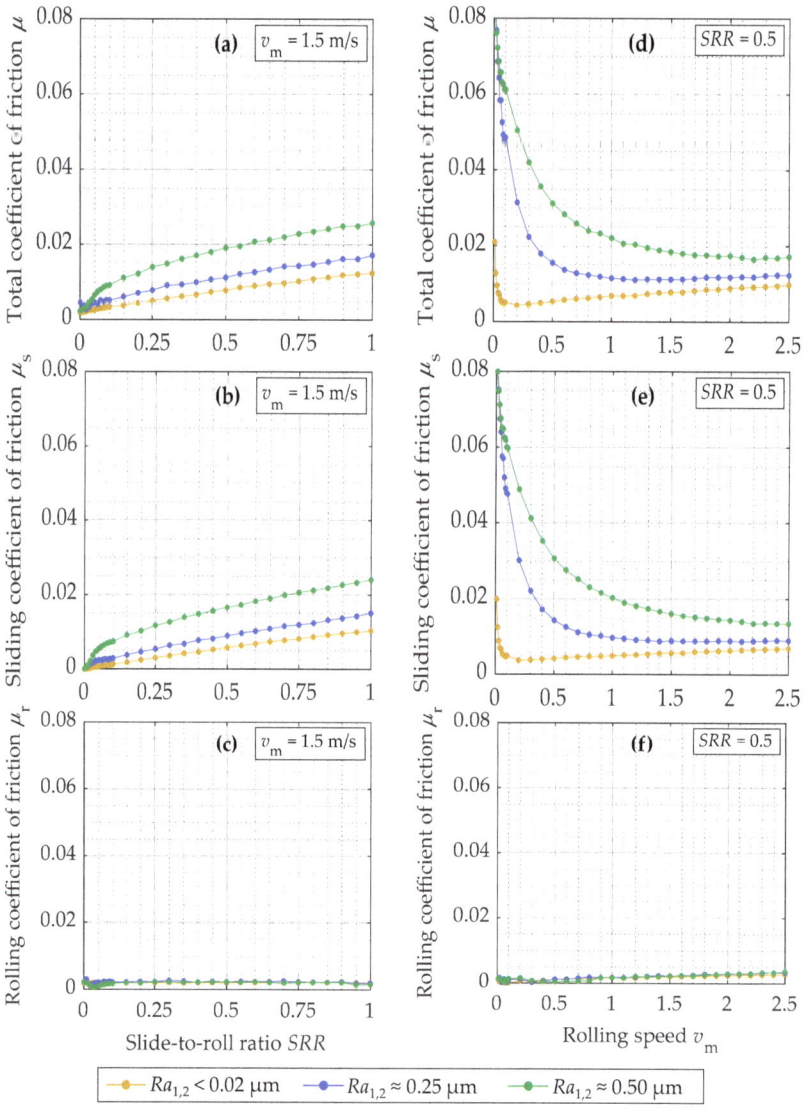

Figure 10. Total μ, sliding μ_s, and rolling μ_r coefficients of friction over the slide-to-roll ratio SRR (**a–c**) and rolling speed v_m (**d–f**) for the configuration 100Cr6/PEEK with surface roughness $Ra_{1,2} < 0.02$ µm, $Ra_{1,2} \approx 0.25$ µm, and $Ra_{1,2} \approx 0.50$ µm at $F_N = 20$ N and $\vartheta_{oil} = 40$ °C for MIN100.

A higher surface roughness results in a higher total coefficient of friction μ in Figure 10a,d. While the rolling coefficient of friction μ_r in Figure 10c,f is barely affected by the surface roughness, the sliding coefficient of friction μ_s in Figure 10b,e increases drastically with a higher surface roughness. The dependency of μ_s on the SRR in Figure 10b shows, for ground surfaces with $Ra_{1,2} \approx 0.50$ µm, approximately three times higher values than those observed with polished surfaces with $Ra_{1,2} < 0.02$ µm for most of the considered SRRs.

The dependency of the sliding coefficient of friction μ_s on v_m in Figure 10e shows, for decreasing rolling speed v_m, that μ_s increases strongly for a higher surface roughness with an increasing severity of mixed friction.

3.7. Summary of Results

Figure 11 gives an overview of the investigated influences on the coefficient of friction presented in Sections 3.1–3.6. The mean values $\overline{\mu_s}$ and $\overline{\mu_r}$ of the coefficient of sliding and rolling friction represent an average over the SRR at a constant rolling speed $v_m = 1.5$ m/s for each influence:

$$\overline{\mu_s} = \int_{SRR=0}^{SRR=1} \mu_s(SRR, v_m = 1.5 \text{ m/s}) \, dSRR, \tag{5}$$

$$\overline{\mu_r} = \int_{SRR=0}^{SRR=1} \mu_r(SRR, v_m = 1.5 \text{ m/s}) \, dSRR, \tag{6}$$

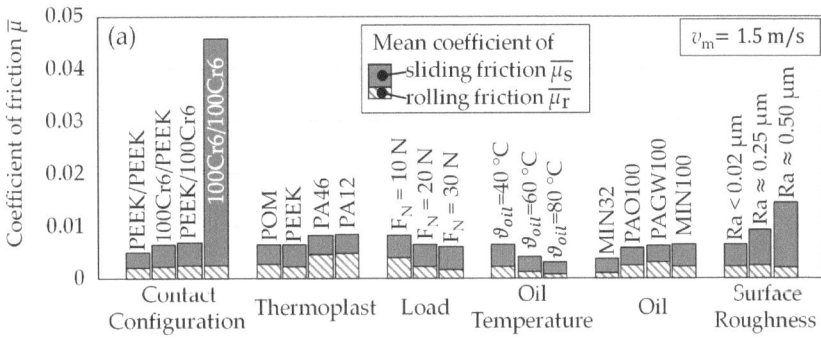

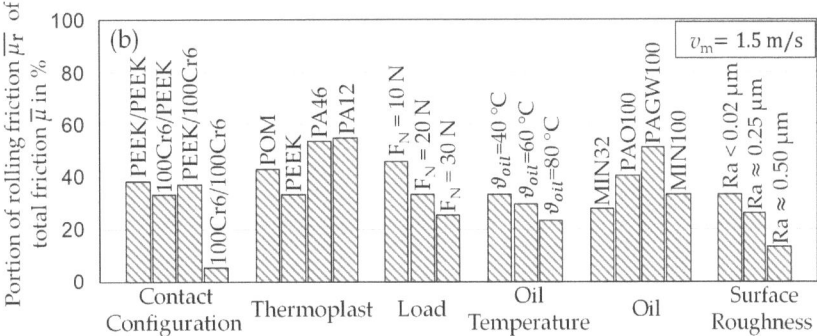

Figure 11. Mean coefficient of sliding friction $\overline{\mu_s}$ and rolling friction $\overline{\mu_r}$ (**a**) and relative portion of mean rolling friction $\overline{\mu_r}$ to the mean total coefficient of friction $\overline{\mu}$ in % (**b**) for the investigated influences.

The mean value $\bar{\mu}$ of the total coefficient of friction is determined by the following:

$$\bar{\mu} = \bar{\mu_s} + \bar{\mu_r}. \tag{7}$$

The mean total coefficient of friction $\bar{\mu}$ in Figure 11a is observed to decrease with increasing load F_N, increasing oil temperature ϑ_{oil}, and decreasing surface roughness Ra. The oil viscosity exerts a stronger influence than the type of oil. No obvious trend regarding Young's modulus of the thermoplastic materials is observed. However, the thermoplastic material exerts an influence on the total friction due to the presence of different rolling friction components. With regard to the contact configuration, PEEK/PEEK shows a lower friction compared to 100Cr6/100Cr6 or 100Cr6/PEEK. It can be observed that, for all contact configurations involving a thermoplastic body and polished surfaces, superlubricity is achieved, with a mean coefficient of friction $\bar{\mu}$ smaller than 0.01. However, for contact configurations involving a ground surface with $Ra_{1,2} \approx 0.50$ μm, $\bar{\mu}$ exceeds 0.01.

Figure 11b shows that a portion of over 50% of the total friction can be attributed to rolling friction. As the load F_N, oil temperature ϑ_{oil}, and surface roughness Ra increase, the portion of rolling friction decreases. With regard to the influence of the lubricant, the mineral oils MIN32 and MIN100 have, despite their different viscosity classes, a similar and lower portion of rolling friction than PAO100 and PAGW100. Considering the thermoplastic materials, the portion of rolling friction of the polyamides PA46 and PA12 is found to be similar and the highest, while that of PEEK is the lowest and that of POM is ranked as intermediate. Regarding the contact configuration, PEEK/PEEK and PEEK/100Cr6 seem to have a similar portion of rolling friction, while 100Cr6/PEEK has a slightly lower portion and 100Cr6/100Cr6 a distinctly lower one. A detailed analysis of the hybrid contacts, PEEK/100Cr6, and 100Cr6/PEEK reveals that the observed differences in the total coefficient of friction $\bar{\mu}$ can be attributed to the rolling friction. While the coefficient of sliding friction $\bar{\mu_s}$ is only 0.26% higher, the coefficient of rolling friction $\bar{\mu_r}$ is 15.5% higher for PEEK/100Cr6 compared to 100Cr6/PEEK. In total, $\bar{\mu}$ is about 8% higher for PEEK/100Cr6 contact compared to 100Cr6/PEEK contact.

4. Discussion

The results in Section 3 are discussed with respect to the influence of the contact configuration, thermoplastic material, lubricant, load, oil temperature, and surface roughness.

4.1. Influence of Contact Configuration

The contact configurations PEEK/100Cr6 and 100Cr6/PEEK showed no noticeable difference in the friction behavior, as illustrated in Figure 5. As demonstrated by Ziegltrum et al. [15], elastic deformation in a contact configuration with steel and a polymer occurs primarily at the polymer body. Accordingly, the contact geometry differs between the configurations PEEK/100Cr6 and 100Cr6/PEEK at the ball-on-disk tribometer. A schematic illustration of the deformed EHL contact geometry is provided in Figure 12, which is based on numerical calculations of Ziegltrum et al. [15].

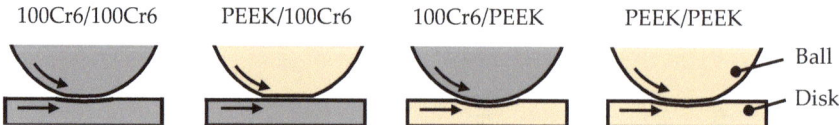

Figure 12. Schematic representation of the contact geometry (deformation) depending on the contact configurations 100Cr6/100Cr6, PEEK /100Cr6, 100Cr6/ PEEK, and PEEK/PEEK.

In the contact configuration PEEK/100Cr6 with the deformed PEEK ball, it is assumed that the contact area is nearly even. In the contact configuration 100Cr6/PEEK with the deformed PEEK disk, a curved contact area is observed. As the friction force is always measured at the ball specimen in the MTM2 tribometer (see Section 2.1), its impact on the friction measurement is not necessarily negligible. This is shown in Figure 11 and Section 3.7, where an 8% difference in $\overline{\mu}$ is observed, with the higher friction being measured for the contact configuration PEEK/100Cr6. An influence of configuration was also found by Quinn et al. [26]'s investigation of PDMS. Their results indicated that the contact configuration steel ball and PDMS disk exhibited higher friction than the contact configuration featuring a PDMS ball and a steel disk. The authors proposed that the observed difference in hysteresis behavior might have been a potential explanation for this. In this study with PEEK, it is proposed that the difference between the contact configurations originates mainly from the rolling friction portion. To gain a comprehensive understanding, numerical modeling and calculation could be conducted, as exemplified by Schmid et al. [24] for hard conformal contacts or Stupkiewicz et al. [9] for low-stiffness polymer contacts. For the configurations in this study containing PEEK, the EHL contact pressure distribution was expected to be almost symmetrical, as shown by Hofmann et al. [13]. They also did not find effects attributed to viscoelasticity or loading frequency for PEEK, which is supported by this study. Hence, it can be concluded that the main portion of rolling friction has its origin in the EHL oil film, related to Poiseuille friction, and the influence of the hysteresis behavior tends to be small. Moreover, it should be noted that the scale of the y-axis for rolling friction in Figure 5c,f in Section 3.1 is very low, which results in the overall values being very low. Therefore, the differences between the contact configurations are seen to be relatively small.

The increase in rolling friction for the contact configuration 100Cr6/100Cr6 at low SRR values (see Figure 5c) can be attributed to a micro-slip resulting from the difference in contact radii. This phenomenon was first observed by Heathcote [40] and discussed in detail by Schmid et al. [24]. In this study and in accordance with the evaluation of Vicente et al. [2], the additional friction portion due to micro-slip was evaluated as part of the rolling speed v_m at MTM2, as it is a consequence of rolling and dependent on the rolling direction.

The contact configuration PEEK/PEEK exhibited an even lower coefficient of friction μ than PEEK/100Cr6 and 100Cr6/PEEK. This was due to a larger elastically flattened area, which resulted in a higher EHL contact temperature. This phenomenon was also shown by Schmid et al. [24] for different contact radii, resulting in different elastically flattened area sizes. Further, this was amplified by a lower thermal effusivity $e = \sqrt{\lambda \cdot \rho \cdot c_\mathrm{p}}$ of PEEK ($e \approx 838\,\mathrm{J}/(\mathrm{m}^2 \cdot \mathrm{K} \cdot \sqrt{\mathrm{s}})$) compared to steel ($e \approx 12131\,\mathrm{J}/(\mathrm{m}^2 \cdot \mathrm{K} \cdot \sqrt{\mathrm{s}})$), resulting in a high thermal insulation effect for the PEEK/PEEK configuration.

4.2. Influence of Thermoplastic Material

The sliding friction of EHL contacts with technical and high-performance thermoplastics in a fluid film lubrication regime is dominated by Newtonian fluid behaviors, a consequence of the low contact pressures involved and in accordance with the results by Vicente et al. [1,2]. This can be seen in the linearly increasing friction curves illustrated in Figure 6b. In a comparison of the contact configurations in Section 3.2, the sliding coefficient of friction μ_s was largely consistent across the investigated thermoplastics, with the exception of 100Cr6/PEEK. When comparing the calculated Hertzian contact pressures in Table 3, all contact configurations in Section 3.2 exhibited similar values, with the exception of PA12. Hence, the frictional differences can be attributed to either a damping behavior with an increasing stiffness behavior for PA12 at the investigated conditions or to differences in Young's modulus regarding tensile or compressive forces, being, thus, a loading frequency-dependent behavior. Further, surface interactions between the oil and the thermoplastic material, such as sorption, can influence the thermoplastic material's mechanical properties, like its stiffness [38]. According to the results of Vicente et al. [1], the overall correlation between μ_s and E follows a bathtub curve according to their calculation regression. This means a decreasing coefficient of friction for very low stiffness values and an increasing coefficient of friction for a higher stiffness. It should be noted, however, that Vicente et al. [1] did not vary Young's modulus itself in their investigations. According to Myant et al. [8], the sliding friction decreases with an increasing stiffness. Assuming a higher stiffness for PEEK than POM (see Table 1), the results in this study show contrary results. However, it should be noted that Myant et al. [8] only focused on polymers with very low stiffness values, while this study focused on technical and high-performance thermoplastics with a higher stiffness, resulting in different areas of the bathtub curve. Also, the surface roughness of the polymers they studied was varied. The stiffness behavior of thermoplastics is complex and particularly dependent on temperature and load frequency.

For very low rolling speeds v_m, the highest sliding coefficient of friction μ_s was observed for the contact configurations with polyamides PA12 and PA46 (see Figure 6e). This phenomenon was attributed to the molecular structures of the interacting material surfaces in mixed and boundary lubrication.

Rolling friction depends on both the fluid flow behavior and the material behavior. Overall, the pressure distribution is decisive for the rolling friction. A high pressure (and therefore load) portion towards the contact inlet results in a higher rolling friction. This can be due to viscoelasticity or fluid flow characteristics. For the fluid flow characteristics, the Poiseuille flow indicates the amount of rolling friction. The pressure distribution is crucial and becomes more Hertzian as the stiffness increases and, therefore, more symmetrical, resulting in a lower rolling friction. The rolling friction is not only influenced by the lubricant but also by viscoelastic material behavior, which also influences the pressure distribution and shifts the pressure portions towards the contact inlet, as shown by Putignano and Dini [18], resulting in a higher rolling friction. The rolling coefficient of friction μ_r of the considered contact configurations with thermoplastics can be divided into two distinct categories. The contact configurations 100Cr6/POM and 100Cr6/PEEK show the identical rolling coefficient of friction μ_r, which is very similar to the rolling friction of 100Cr6/100Cr6. This suggests that, for these three contact configurations, the rolling friction in the EHL contacts originates mainly from the fluid and its pressure distribution, which is referred to as Poiseuille friction. This is supported by the observation of nearly zero rolling friction at very low rolling speeds v_m, given in Figure 6f. Given a minimal lubricant fluid film, the rolling friction of the fluid itself is also minimal for PEEK and POM. It should be noted that, for the considered operating conditions, PEEK operates clearly under and POM clearly above the glass transition temperature ϑ_g, depicted in Figure 2. Compared to 100Cr6/PEEK and 100Cr6/POM, the contact configurations 100Cr6/PA12 and 100Cr6/PA46 showed a higher rolling friction, even at very small rolling speeds v_m. This can be due to higher hysteresis friction within the polyamides. As the polyamides operate close to the glass transition temperature, where higher material damping occurs, an effect of ϑ_g on μ_r can be assumed.

Following the results of Vicente et al. [1], the rolling friction increases with increasing stiffness. Assuming a higher stiffness for PEEK than POM, the results in this study were contrasting. However, it should be noted that Vicente et al. did not vary Young's modulus in their investigations. Following the results of Myant et al. [8], the results of this study are in agreement regarding the influence of stiffness on rolling friction.

4.3. Influence of Lubricant

In the context of fluid film lubrication, friction in the EHL contact is typically dominated by the shearing of lubricant under a high pressure. As illustrated in Figure 7b, PAGW100 did not show the lowest sliding friction, which is typically observed for steel contacts with maximal contact pressures in the order of 1 GPa [41]. Thus, the high-pressure behavior of a lubricant showed little influence on EHL friction with technical and high-performance thermoplastics. Also, shear-thinning did not show a significant influence, as all sliding friction curves showed a nearly linear trend over the SRR (see Figure 7b). It could, therefore, be concluded that the influence of shear-thinning behaviors on friction in oil-lubricated thermoplastic rolling–sliding EHL contacts was small for the investigated lubricants. Rather than the lubricants' high-pressure and shear-thinning behaviors, it was the nominal viscosity to be of great consequence.

For mixed and boundary lubrication at a small rolling speed v_m, as shown in Figure 7d, the measured friction of the two mineral oils, MIN32 and MIN100, was predominantly higher in comparison to PAO100 and PAGW100. For MIN32, the transition from mixed to fluid film lubrication was observed at higher rolling speeds v_m (see Figure 7e). This was attributed to the lower viscosity of MIN32. Both mineral oils showed higher coefficients of friction μ in the mixed and boundary lubrication regimes. This could be related to high pressure and viscosity increases at micro-EHL contacts at surface asperities [42,43] or interactions and tribo-induced surface changes [38]. A difference in the rolling friction for the different oils can be observed (see Figure 7f). This shows that the rolling friction can strongly be driven by the lubricant itself and not only by the thermoplastic material.

4.4. Influence of Load and Oil Temperature

The sliding coefficient of friction μ_s did not change significantly when the load F_N varied, as seen in Figure 8b,e. This can be understood by considering the estimated contact pressures in Table 3. The calculated Hertzian pressures are low, and neither the pressure-related increase in contact viscosity nor the influence of contact temperature increases introduce significant differences or effects that cancel out for the considered contact configuration. Contrary, rolling friction decreased with higher loads. Hence, a higher contact pressure resulted in a lower rolling friction (see Figure 8c,f), which was in accordance with the findings of Vicente et al. [1], Myant et al. [8], and Lates et al. [44].

As discussed in Section 4.3, the viscosity at ambient conditions exerts a dominant influence on frictional behavior in EHL contacts with technical and high-performance thermoplastics. This is confirmed by the influence of oil temperature ϑ_{oil}, as friction decreases with increasing ϑ_{oil} (see Figure 9b), without a notable impact on the trend in the friction curves. Note that Hofmann et al. [13] have shown very small contact temperatures for EHL contacts with PEEK. The rolling friction was observed to decrease with a higher oil temperature ϑ_{oil} and increase with the rolling speed v_m (see Figure 9c,f). This finding is consistent with the results reported by Vicente et al. [1]. Given the test procedure and configuration of the MTM2 testing chamber (see Figure 1), it is reasonable to assume that the specimens have temperatures similar to those of oil. Therefore, an increase in oil temperature ϑ_{oil} results in an increase in the thermoplastic bulk temperature and, consequently, a decrease in stiffness and contact pressures (see Figure 2). Lower contact pressures mainly increase the rolling friction (see Figure 8b,c), a contrary effect to the overall decrease in rolling friction with higher temperatures which was observed. For the investigated oil temperature ϑ_{oil} range and the thermoplastic PEEK, with only a low stiffness reduction in this range, it could be concluded that the lubricant's behavior was the main influencing factor.

The viscosity of the oil is of great consequence with respect to the oil film thickness in EHL contacts. Consequently, at elevated oil temperatures ϑ_{oil}, the onset of the mixed lubrication regime shifted towards higher rolling speeds v_m (see Figure 9d).

4.5. Influence of Surface Roughness

In the case of mixed and boundary lubrication, it is essential to take into account the impact of roughness when assessing friction. The starting influence of this regime gives rise to the formation of local narrowing phenomena between roughness asperities at micro-EHL contacts [42,43]. In addition, the surface interaction and adhesive forces play a significant role [38,45]. The experimental results revealed that the coefficient of friction μ could increase significantly for the contact configurations with PEEK and 100Cr6 under boundary lubrication. A coefficient of friction exceeding $\mu = 0.08$ was observed, which was markedly higher than that recognized in the fluid film lubrication regime (see Figure 10d). It could be observed that the increase in friction was primarily attributable to sliding friction rather than rolling friction (see Figure 10e,f). Given that mixed and boundary lubrication are also accompanied by surface alteration and wear over time, the presented short-term measurements are quasi-stationary snapshots. Although a run-in procedure was conducted prior to the measurements, long-term testing is required to draw further conclusions.

5. Conclusions

In this study, different influence parameters on friction in oil-lubricated rolling–sliding EHL contacts with technical and high-performance thermoplastics were studied based on a phenomenological methodology using a ball-on-disk tribometer. The effects and relevance of the influence parameters were discussed. The general conclusions can be summarized as follows:

- In thermoplastic EHL contacts, both sliding and rolling friction contributed significantly to the total friction, contrasting plain steel contact configurations, in which sliding friction dominated.
- The sliding friction was low for all contact configurations with thermoplastics compared to the plain steel contact configuration at the same normal load, thus showing the potential frictional advantages of using thermoplastic materials.
- For the operating conditions studied, the rolling friction for the contact configurations with PA12 and PA46 was approximately twice that of the contact configurations with PEEK and POM.
- Despite the different contact geometries, the difference in sliding and rolling friction was small between the contact configurations of steel/PEEK and PEEK/steel.

For fluid film lubrication, the following conclusions can be drawn:

- The sliding friction is only a little influenced by the oil type, load, and thermoplastic material. A lower sliding friction is achieved by a lower Young modulus and a lower oil viscosity.
- The rolling friction is highly dependent on the thermoplastic material, oil type, oil viscosity, and load. A lower rolling friction is achieved by a higher normal load and a lower oil viscosity.
- Friction in oil-lubricated thermoplastic EHL contacts shows the potential for superlubricity with a coefficient of friction μ less than 0.01.

For mixed and boundary lubrication, the following conclusions can be drawn:

- The total friction increases drastically compared to fluid film lubrication.
- The rolling friction is only slightly influenced, but the sliding friction is strongly increased. This is attributed to the high solid friction caused by surface interactions.
- POM and PEEK have the lowest friction, and PA12 and PA46 have the highest friction in mixed and boundary lubrication.

In order to exploit the low-friction potential in machine elements with technical and high-performance thermoplastics, one should enhance the fluid film lubrication regime. This can be achieved with a low surface roughness or, in general, using oils that form a high oil film thickness. Nevertheless, mixed and boundary lubrication cannot be avoided for full operating maps. The interactions between surfaces and between surfaces and lubricants are complex and can be coupled by changing surface conditions with tribofilms and wear. This also affects the friction and can be the subject of further studies with regard to long-term behaviors.

Author Contributions: Conceptualization, F.S. and T.L.; methodology, F.S. and T.L.; experiments, F.S.; validation, F.S.; formal analysis, T.L.; writing—original draft preparation, F.S.; writing—review and editing, T.L. and K.S.; supervision, T.L. and K.S.; project administration, T.L. and K.S.; funding acquisition, K.S. All authors have read and agreed to the published version of the manuscript.

Funding: This research was funded by the research project CHEPHREN (03EN4005A), focusing on chemical and physical possibilities for friction reduction, and it is supported by the Federal Ministry for Economic Affairs and Climate Action (BMWK) and supervised by Project Management Jülich (PtJ).

Data Availability Statement: The raw data supporting the conclusions of this article will be made available by the authors upon request.

Acknowledgments: The authors are grateful for the sponsorship and support received from BMWK and PtJ.

Conflicts of Interest: The authors declare no conflicts of interest.

Nomenclature

E	Young's modulus in N/m²
F_N	Normal force in N
F_R	Friction force in N
SRR	Slide-to-roll ratio
v_m	Rolling speed in m/s
v_s	Sliding velocity in m/s
v	Surface velocity of solid body in m/s
Ra	Arithmetic average height of body in μm
E	Thermal effusivity in J/(K·m²·s^{1/2})
c_p	Specific thermal capacity in J/(kg·K)
VI	Viscosity index
p_H	Hertzian pressure in N/mm²

Greek symbols

λ	Thermal conductivity in W/(m·K)
μ	Coefficient of friction
ν	Poisson's ratio	kinematic viscosity in mm/s²
ϑ	Temperature in °C
ϑ_g	Glass transition regime in °C
ρ	Density in g/cm³
ϑ_{oil}	Oil temperature in °C
$\overline{\mu}$	Mean coefficient of friction
$\overline{\delta_\mu}$	Relative change in test run 2 with respect to test run 1

Indices

1	Test run 1
2	Test run 2 (repetition)
r	Rolling
s	Sliding
oil	Oil

Appendix A

In the following section, the repeatability of the measurements is evaluated. For this, the relative change $\overline{\delta_\mu}$ of the second run μ_2 compared to the first run μ_1 is evaluated for every measurement and then averaged for each curve after Equation (A1).

$$\overline{\delta_\mu} = \int_{SRR=0}^{SRR=1} \frac{\mu_2 - \mu_1}{\mu_1} \, dSRR, \quad \overline{\delta_\mu} = \frac{1}{(2.5 - 0.1) \, \text{m/s}} \int_{v_m=0.1 \, \text{m/s}}^{v_m=2.5 \, \text{m/s}} \frac{\mu_2 - \mu_1}{\mu_1} \, dv_m \quad \text{(A1)}$$

Table A1. Relative change $\overline{\delta_\mu}$ in the second test run with respect to the first test run, averaged over each friction and Stribeck-like curve.

	Figure 5a	Figure 5b	Figure 5c	Figure 5d	Figure 5e	Figure 5f
100Cr6/100Cr6	3.8%	7.9%	−14.9%	5.3%	5.9%	−10.5%
PEEK/100Cr6	−1.9%	8.7%	−9.8%	0.0%	4.1%	−12.9%
100Cr6/PEEK	−4.4%	−0.5%	−10.9%	0.2%	3.5%	−10.1%
PEEK/PEEK	0.7%	1.9%	−4.9%	14.5%	21.5%	−10.5%
	Figure 6a	Figure 6b	Figure 6c	Figure 6d	Figure 6e	Figure 6f
100Cr6/PEEK	−4.4%	−0.5%	−10.9%	0.2%	3.5%	−10.1%
100Cr6/PA12	2.8%	8.6%	3.9%	3.7%	−0.6%	9.4%
100Cr6/PA46	−9.5%	−3.8%	−13.5%	−10.3%	−5.9%	−14.6%
100Cr6/POM	1.8%	−2.4%	2.9%	5.1%	6.6%	1.9%

Table A1. Cont.

	Figure 7a	Figure 7b	Figure 7c	Figure 7d	Figure 7e	Figure 7f
MIN100	−4.4%	−0.5%	−10.9%	0.2%	3.5%	−10.1%
MIN32	0.6%	8.1%	−9.5%	22.9%	28.5%	−8.5%
PAO100	−6.5%	−2.4%	−10.8%	−5.2%	−3.7%	−8.0%
PAGW100	−5.6%	−2.2%	−6.9%	−3.9%	−3.7%	17.4%
	Figure 8a	Figure 8b	Figure 8c	Figure 8d	Figure 8e	Figure 8f
$F_N = 10$ N	−6.5%	2.2%	−12.2%	−1.8%	1.8%	−9.4%
$F_N = 20$ N	−4.4%	−0.5%	−10.9%	0.2%%	3.5%	−10.1%
$F_N = 30$ N	−2.1%	3.6%	−9.9%	5.5%%	7.6%	1.8%
	Figure 9a	Figure 9b	Figure 9c	Figure 9d	Figure 9e	Figure 9f
$\vartheta_{oil} = 40$ °C	−4.4%	−0.5%	−10.9%	1.5%	3.5%	−10.1%
$\vartheta_{oil} = 60$ °C	−4.8%	3.9%	−15.7%	7.8%	13.1%	−10.4%
$\vartheta_{oil} = 80$ °C	7.9%	19.7%	−7.2%	26.4%	30.4%	1.5%
	Figure 10a	Figure 10b	Figure 10c	Figure 10d	Figure 10e	Figure 10f
$Ra < 0.02$ μm	−4.4%	−0.5%	−10.9%	0.2%	3.5%	−10.1%
$Ra \approx 0.25$ μm	−15.1%	−24.6%	0.6%	−3.7%	−3.5%	−0.6%
$Ra \approx 0.50$ μm	−24.8%	−24.8%	38.9%	−4.1%	−4.7%	35.8%

References

1. de Vicente, J.; Stokes, J.R.; Spikes, H.A. The Frictional Properties of Newtonian Fluids in Rolling–Sliding soft-EHL Contact. *Tribol. Lett.* **2005**, *20*, 273–286. [CrossRef]
2. de Vicente, J.; Stokes, J.R.; Spikes, H.A. Rolling and sliding friction in compliant, lubricated contact. *Proc. Inst. Mech. Eng. Part J J. Eng. Tribol.* **2006**, *220*, 55–63. [CrossRef]
3. Reitschuster, S.; Maier, E.; Lohner, T.; Stahl, K. Friction and Temperature Behavior of Lubricated Thermoplastic Polymer Contacts. *Lubricants* **2020**, *8*, 67. [CrossRef]
4. VDI 2736 Blatt 2:2014-06; Thermoplastische Zahnräder. Stirnradgetriebe. Tragfähigkeitsberechnung. Verein Deutscher Ingenieure: Düsseldorf, Germany, 2014.
5. Johnson, K.L. Regimes of Elastohydrodynamic Lubrication. *J. Mech. Eng. Sci.* **1970**, *12*, 9–16. [CrossRef]
6. Myers, T.G.; Hall, R.W.; Savage, M.D.; Gaskell, P.H. The transition region of elastohydrodynamic lubrication. *Proc. R. Soc. Lond. A* **1991**, *432*, 467–479. [CrossRef]
7. Marx, N.; Guegan, J.; Spikes, H.A. Elastohydrodynamic film thickness of soft EHL contacts using optical interferometry. *Tribol. Int.* **2016**, *99*, 267–277. [CrossRef]
8. Myant, C.; Spikes, H.A.; Stokes, J.R. Influence of load and elastic properties on the rolling and sliding friction of lubricated compliant contacts. *Tribol. Int.* **2010**, *43*, 55–63. [CrossRef]
9. Stupkiewicz, S.; Lengiewicz, J.; Sadowski, P.; Kucharski, S. Finite deformation effects in soft elastohydrodynamic lubrication problems. *Tribol. Int.* **2016**, *93*, 511–522. [CrossRef]
10. Hooke, C.J.; O'Donoghue, J.P. Elastohydrodynamic Lubrication of Soft, Highly Deformed Contacts. *J. Mech. Eng. Sci.* **1972**, *14*, 34–48. [CrossRef]
11. Esfahanian, M.; Hamrock, B.J. Fluid-Film Lubrication Regimes Revisited. *Tribol. Trans.* **1991**, *34*, 628–632. [CrossRef]
12. Hooke, C.J. The Elastohydrodynamic Lubrication of Heavily Loaded Contacts. *J. Mech. Eng. Sci.* **1977**, *19*, 149–156. [CrossRef]

13. Hofmann, S.; Maier, E.; Lohner, T. In Situ Contact Analysis of Polyetheretherketone under Elastohydrodynamic Lubrication. *Polymers* **2022**, *14*, 4398. [CrossRef] [PubMed]
14. Maier, E.; Ziegltrum, A.; Lohner, T.; Stahl, K. Characterization of TEHL contacts of thermoplastic gears. *Forsch. Ingenieurwes.* **2017**, *81*, 317–324. [CrossRef]
15. Ziegltrum, A.; Maier, E.; Lohner, T.; Stahl, K. A Numerical Study on Thermal Elastohydrodynamic Lubrication of Coated Polymers. *Tribol. Lett.* **2020**, *68*, 71. [CrossRef]
16. Hunter, S.C. The Rolling Contact of a Rigid Cylinder with a Viscoelastic Half Space. *J. Appl. Mech.* **1961**, *28*, 611–617. [CrossRef]
17. Hooke, C.J.; Huang, P. Elastohydrodynamic lubrication of soft viscoelastic materials in line contact. *Proc. Inst. Mech. Eng. Part J J. Eng. Tribol.* **1997**, *211*, 185–194. [CrossRef]
18. Putignano, C.; Dini, D. Soft Matter Lubrication: Does Solid Viscoelasticity Matter? *ACS Appl. Mater. Interfaces* **2017**, *9*, 42287–42295. [CrossRef] [PubMed]
19. Zhao, Y.; Liu, H.C.; Morales-Espejel, G.E.; Venner, C.H. Effects of solid viscoelasticity on elastohydrodynamic lubrication of point contacts. *Tribol. Int.* **2022**, *171*, 107562. [CrossRef]
20. Krupka, J.; Dockal, K.; Krupka, I.; Hartl, M. Elastohydrodynamic Lubrication of Compliant Circular Contacts near Glass-Transition Temperature. *Lubricants* **2022**, *10*, 155. [CrossRef]
21. Putignano, C.; Reddyhoff, T.; Carbone, G.; Dini, D. Experimental Investigation of Viscoelastic Rolling Contacts: A Comparison with Theory. *Tribol. Lett.* **2013**, *51*, 105–113. [CrossRef]
22. Putignano, C.; Reddyhoff, T.; Dini, D. The influence of temperature on viscoelastic friction properties. *Tribol. Int.* **2016**, *100*, 338–343. [CrossRef]
23. Carbone, G.; Putignano, C. A novel methodology to predict sliding and rolling friction of viscoelastic materials: Theory and experiments. *J. Mech. Phys. Solids* **2013**, *61*, 1822–1834. [CrossRef]
24. Schmid, F.; Paschold, C.; Lohner, T.; Stahl, K. Characteristics in hard conformal EHL line contacts. *Issues Lang. Teach. (ILT)* **2023**, *75*, 730–740. [CrossRef]
25. Sadowski, P.; Stupkiewicz, S. Friction in lubricated soft-on-hard, hard-on-soft and soft-on-soft sliding contacts. *Tribol. Int.* **2019**, *129*, 246–256. [CrossRef]
26. Quinn, C.; Nečas, D.; Šperka, P.; Marian, M.; Vrbka, M.; Křupka, I.; Hartl, M. Experimental investigation of friction in compliant contact: The effect of configuration, viscoelasticity and operating conditions. *Tribol. Int.* **2022**, *165*, 107340. [CrossRef]
27. Schmid, F.; Maier, E.; Lohner, T.; Stahl, K. Friction in Oil-lubricated Rolling-Sliding Contacts with Technical Thermoplastics: 64. In Proceedings of the Tribologie-Fachtagung 2023, Reibung, Schmierung und Verschleiß, Göttingen, Germany, 25–27 September 2023.
28. Hofmann, S.; Lohner, T.; Stahl, K. Influence of water content on elastohydrodynamic friction and film thickness of water-containing polyalkylene glycols. *Front. Mech. Eng.* **2023**, *9*, 1128447. [CrossRef]
29. Jain, M.; Patil, S. A review on materials and performance characteristics of polymer gears. *Proc. Inst. Mech. Eng. Part C J. Mech. Eng. Sci.* **2023**, *237*, 2762–2790. [CrossRef]
30. Zhong, B.; Song, H.; Liu, H.; Wei, P.; Lu, Z. Loading capacity of POM gear under oil lubrication. *J. Adv. Mech. Des. Syst. Manuf.* **2022**, *16*, JAMDSM0006. [CrossRef]
31. Hriberšek, M.; Kulovec, S. Thermal and durability characterization of polyacetal and polyamide gear pairs. *J. Mech. Sci. Technol.* **2021**, *35*, 3389–3394. [CrossRef]
32. Md Ghazali, W.; Daing Idris, D.M.N.; Sofian, A.H.; Siregar, J.P.; Abdul Aziz, I.A. A review on failure characteristics of polymer gear. *MATEC Web Conf.* **2017**, *90*, 1029. [CrossRef]
33. Kalin, M.; Kupec, A. The dominant effect of temperature on the fatigue behaviour of polymer gears. *Wear* **2017**, *376–377 Pt B*, 1339–1346. [CrossRef]
34. Lagier, F.; Freund, N.; Bause, K.; Ott, S.; Albers, A. Simulation-based evaluation of high-speed PEEK gears in automotive powertrains and design of a validation environment for high-speed gears. In *Dritev, Proceedings of the 22nd International VDI Congress, Baden, Germany, 6–7 July 2022*; VDI Verlag: Düsseldorf, Germany, 2022; pp. 223–240, ISBN 9783181024010.
35. Zorko, D.; Kulovec, S.; Duhovnik, J.; Tavčar, J. Durability and design parameters of a Steel/PEEK gear pair. *Mech. Mach. Theory* **2019**, *140*, 825–846. [CrossRef]
36. Hoskins, T.J.; Dearn, K.D.; Chen, Y.K.; Kukureka, S.N. The wear of PEEK in rolling–sliding contact—Simulation of polymer gear applications. *Wear* **2014**, *309*, 35–42. [CrossRef]
37. *VDI 2736 Blatt 1:2016-07*; VDI-Richtlinie 2736—Blatt 1: Thermoplastische Zahnräder. Werkstoffe, Werkstoffauswahl, Herstellverfahren, Herstellgenauigkeit, Gestalten. Verein Deutscher Ingenieure: Düsseldorf, Germany, 2016.
38. Koplin, C.; Oehler, H.; Praß, O.; Schlüter, B.; Alig, I.; Jaeger, R. Wear and the Transition from Static to Mixed Lubricated Friction of Sorption or Spreading Dominated Metal-Thermoplastic Contacts. *Lubricants* **2022**, *10*, 93. [CrossRef]
39. Brandão, J.A.; Meheux, M.; Seabra, J.H.O.; Ville, F.; Castro, M.J.D. Traction curves and rheological parameters of fully formulated gear oils. *Proc. Inst. Mech. Eng. Part J J. Eng. Tribol.* **2011**, *225*, 577–593. [CrossRef]
40. Heathcote, H.L. (Ed.) The Ball Bearing: In the Making, Under Test and on Service. *Proc. Inst. Automob. Eng.* **1920**, *15*, 569–702.
41. Yilmaz, M.; Mirza, M.; Lohner, T.; Stahl, K. Superlubricity in EHL Contacts with Water-Containing Gear Fluids. *Lubricants* **2019**, *7*, 46. [CrossRef]
42. Chang, L. A deterministic model for line-contact partial elastohydrodynamic lubrication. *Tribol. Int.* **1995**, *28*, 75–84. [CrossRef]

43. Hultqvist, T. *Transient Elastohydrodynamic Lubrication: Effects of Geometry, Surface Roughness, Temperature, and Plastic Deformation*; Luleå University of Technology: Luleå, Sweden, 2020; ISBN 978-91-7790-604-9.
44. Lates, M.T.; Velicu, R.; Gavrila, C.C. Temperature, Pressure, and Velocity Influence on the Tribological Properties of PA66 and PA46 Polyamides. *Materials* **2019**, *12*, 3452. [CrossRef]
45. Jaeger, R.; Koplin, C.; Schluter, B. Lubricated polymer-steel-systems: Influence of the surface and interfacial energies of frictional partners on their tribological performance. In *International Conference on Gears 2022*; VDI Verlag: Düsseldorf, Germany, 2022; pp. 1223–1236, ISBN 9783181023891.

Disclaimer/Publisher's Note: The statements, opinions and data contained in all publications are solely those of the individual author(s) and contributor(s) and not of MDPI and/or the editor(s). MDPI and/or the editor(s) disclaim responsibility for any injury to people or property resulting from any ideas, methods, instructions or products referred to in the content.

Article

Mixed Friction in Fully Lubricated Elastohydrodynamic Contacts—Theory or Reality

Dirk Bartel

Chair of Machine Elements and Tribology, Otto von Guericke University Magdeburg, 39106 Magdeburg, Germany; dirk.bartel@ovgu.de

Abstract: Mixed friction in liquid-lubricated tribosystems is characterized by the simultaneous presence of liquid and solid friction. Liquid friction results from the shearing of the lubricant, and solid friction from deformation and adhesion. Elastic hysteresis and plastic deformation of the solids cause energy losses during deformation and the separation of molecular bonds between the solids causes energy losses during adhesion. The classic conception of mixed friction presupposes direct contact between rough solids for solid friction to exist. However, if hysteresis losses are fully accepted as a cause for solid friction, every fully lubricated elastohydrodynamic contact would ultimately be a mixed friction contact since the elastic deformations of the solids also cause a loss of energy induced by hysteresis. Thus, the classic conception of mixed friction should be expanded since mixed friction can occur even when solids do not have any direct contact.

Keywords: elastohydrodynamics; hysteresis friction; mixed friction; friction states; lubrication states

Citation: Bartel, D. Mixed Friction in Fully Lubricated Elastohydrodynamic Contacts—Theory or Reality. *Lubricants* **2024**, *12*, 351. https://doi.org/10.3390/lubricants12100351

Received: 22 August 2024
Revised: 3 October 2024
Accepted: 7 October 2024
Published: 14 October 2024

Copyright: © 2024 by the author. Licensee MDPI, Basel, Switzerland. This article is an open access article distributed under the terms and conditions of the Creative Commons Attribution (CC BY) license (https:// creativecommons.org/licenses/by/ 4.0/).

1. Introduction

To increase the lifetime of tribological systems, it is important to optimize friction and wear. This requires knowledge of the effective frictional forces. The frictional forces can be altered by changing the operating conditions, the material and lubricant, or the surface geometry. Tribological systems are often designed according to the trial-and-error method, based on many years of experience and many tests. This approach reaches its practical limits due to the high costs involved and the ever-shorter product development cycles. Virtual product development and thus the simulation of tribological systems offers a way out. However, calculating the friction behavior of tribological systems requires detailed consideration and a description of all mechanisms involved in friction.

1.1. Friction and Lubrication

Friction is due to interactions between contacting material areas of bodies, which counteract the relative movement of the bodies. The term "body" is representative of everything that has a mass and occupies a space. Bodies consist of substances that can be solid, liquid, or gaseous. Depending on the state of motion of the bodies, a distinction can be made between friction without relative motion (static friction) and friction with relative motion (dynamic friction). Depending on the affiliation of the material areas involved in the friction process, external or internal friction can be present. In the case of external friction, the contacting material areas belong to different bodies, whereas in the case of internal friction, they belong to one and the same body. Internal and external friction can occur simultaneously.

The friction in a tribological system can be specifically influenced by lubrication. Lubricants, which can be solid, liquid, consistent, or gaseous, are used for this purpose. The use of lubricants is often linked to the aim of achieving partial or complete separation of the friction bodies. The hydro-, elastohydro-, or aerodynamic as well as the hydro- or aerostatic effect can be utilized for this purpose. Whereas with hydro-, elastohydro-, and aerodynamics, the load capacity applied by the lubricant results from an internal pressure generation in the lubrication gap, with hydro- and aerostatics, this results from an external pressure generation in the lubrication gap by an external pump.

In German-speaking countries, a distinction tends to be made between friction states [1], whereas, outside German-speaking countries, lubrication states tend to be subdivided [2]. From a scientific point of view, it is necessary to distinguish between friction and lubrication states and to clearly separate them from one another. If the friction in a contact is considered, the friction state is decisive. If the type of lubrication of a contact is in the foreground, the lubrication state must be considered.

If friction is classified according to the aggregate state of the substances involved in the friction, the following friction states can be defined:

- *Solid friction* is friction between solid bodies in direct contact. If friction takes place between solid boundary layers, it is *boundary layer friction*. If the solids are covered by a very thin liquid film, this is *boundary friction*.
- *Liquid friction* is internal friction in the material area with liquid properties and is present in a hydrodynamically, elastohydrodynamically, or hydrostatically generated lubricating film.
- *Gas friction* is internal friction in the material area with gas properties and is present in an aerodynamically or aerostatically generated lubricating film.
- *Mixed friction* is any mixed form of the aforementioned friction states and can also result from more than two superimposed friction states.

Depending on the lubrication, the following lubrication states can be distinguished:

- *Boundary lubrication* is lubrication with solid boundary layers or very thin liquid films. A special case of lubrication with solid boundary layers is *superlubricity*. In the presence of very thin liquid films, hydrodynamic, elastohydrodynamic, or hydrostatic load capacity effects are negligible. Boundary lubrication is the subject of molecular dynamics.
- *Liquid lubrication* is the lubrication with a liquid in which a complete separation of the friction body surfaces by hydrodynamics, elastohydrodynamics, or hydrostatics is aimed for (full lubrication).
- *Gas lubrication* is lubrication with a gas in which complete separation of the friction surfaces is achieved by aerodynamics or aerostatics (full lubrication).
- *Mixed lubrication* or *partial lubrication* is any mixture of the aforementioned lubrication states.

The friction and lubrication states can be displayed in the Stribeck curve, as shown in Figure 1. Here, a representation with a linear or logarithmic axis for the rotational speed or sliding velocity is possible.

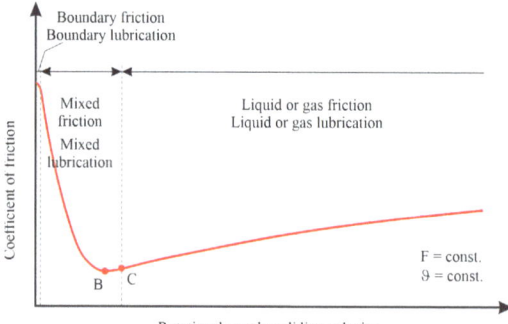

Figure 1. *Cont.*

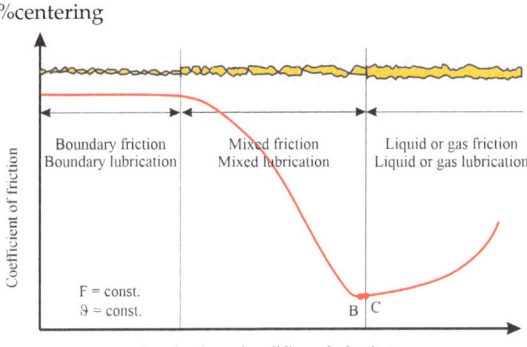

Figure 1. Stribeck curve in linear and logarithmic representation (B—friction minimum; C—lift-off point).

1.1.1. Friction Mechanism in Fluid Friction

Fluids are viscous media such as liquids and gases. The friction of liquids results from the contact and mutual displacement of molecules. Gas friction results from the collision of gas particles (atoms or small molecules). These friction mechanisms are effective when liquids and gases are sheared in the lubrication gap. One measure of shear resistance is the viscosity. The resulting friction or viscosity depends on the chemical structure of the fluid, the temperature, the pressure, and the shear rate. Fluid and gas friction are generally very low, with gas friction usually achieving much lower values than fluid friction. The calculation of fluid-lubricated systems is carried out using the basic equations of fluid mechanics, into which temperature-, pressure-, and shear gradient-dependent fluid properties are incorporated [3].

1.1.2. Friction Mechanism in Solid Friction

Solid friction is due to mechanical and atomic/molecular interactions. This was first formulated by Kragelskij [4] and Bowden et al. [5] and subsequently taken up and confirmed by many authors. Kragelskij spoke of the double nature of solid friction. Both parts can occur simultaneously but can be pronounced to different degrees.

If solids are subjected to stress, elastic and plastic deformation of the solids can occur. When the rough solids come into direct contact, the real contact surface is formed. The deformation of the solid bodies is associated with energy losses, which, in a tribosystem with kinetic friction, manifest themselves as continual solid friction losses and are assigned to the mechanical interactions. The plastic deformation work is irreversibly lost due to the lasting deformation and is therefore 100% frictionally effective. The elastic deformation work is only partially frictionally effective, as this is largely recovered by the elastic recovery of the solid bodies. The partial loss results from an incomplete elasticity of the materials and is called elastic hysteresis or mechanical damping. Hysteresis effects are reflected, for example, in the natural decrease in mechanical vibrations. A measure of the loss that occurs is the hysteresis factor H as the ratio of loss energy and applied elastic energy. The size of the hysteresis factor depends on the material, the deformation speed, the load level, the type of load (uniaxial or multiaxial), the load duration/frequency, and the temperature. The hysteresis tends to increase with an increase in the load parameters and temperature. Rubber, plastics, gray cast iron, or some ferromagnetic alloys exhibit a high hysteresis. Information on hysteresis factors of materials in standard tests can be found in [6] or [7]. Information on hysteresis factors of materials in the friction process, especially for the friction-relevant area near the surface, is not yet available to the desired extent.

When the rough solids come into direct contact, adhesive bonds are also formed in the real contact area A_r, which correspond to atomic/molecular interactions. These adhesive bonds are of a chemical or physical nature and can be more or less pronounced. The type and strength of the bond depend on the boundary layers/boundary films that form on the solids. For example, the adhesive bonding forces between two oxide-covered iron bodies (van der Waals bond) are lower than those resulting from direct iron/iron contact (metal bond). To maintain the relative movement between two adhesively interacting solids, energy must be permanently expended to separate the atomic/molecular bonds. Solid friction losses are the result. Depending on the bonding state, adhesion or cohesion bonds can be separated. The decisive factor here is which bonds fail first. When cohesive bonds, which ensure the cohesion of the solids, are separated, material is detached locally from the cohesively weaker bonded body, and material is transferred to the counter body. This is referred to as adhesion wear or more generally as scuffing. No material transfer occurs when adhesion bonds are separated.

To calculate solid friction, it makes sense to take an energetic approach [8–11]. While a force is defined by magnitude and direction (vectorial quantity), work or energy is described solely by a magnitude without direction (scalar quantity). Only the traceability of work to a force along a path leads to a directional consideration. This can be utilized in an energetic friction calculation. In G = general, the solid friction work W_{fs} results from

$$W_{fs} = \mu_s \cdot F_s \cdot s_f = F_{fs} \cdot s_f \tag{1}$$

The solid friction work can also be written as the sum of the energy components from deformation and adhesion in accordance with the solid friction mechanisms explained above:

$$W_{fs} = W_{fs,def} + W_{fs,ad} \tag{2}$$

The following applies to the deformation-related friction work, considering the hysteresis losses resulting from the elastic deformation and any plastic deformation losses, which are 100% frictionally effective:

$$W_{fs,def} = W^{el}_{fs,def} + W^{pl}_{fs,def} \tag{3}$$

with

$$W^{el}_{fs,def} = H_{red} \cdot W^{el}_{s,def} \tag{4}$$

and

$$W^{pl}_{fs,def} = W^{pl}_{s,def} \tag{5}$$

The reduced hysteresis factor H_{red} is introduced in Equation (4) considering different elastic hysteresis properties of the paired materials and can be calculated according to [11] using Equation (6).

$$H_{red} = \frac{H_1}{\frac{E_1(1-v_2^2)}{E_2(1-v_1^2)} + 1} + \frac{H_2}{\frac{E_2(1-v_1^2)}{E_1(1-v_2^2)} + 1} \tag{6}$$

The calculation of the adhesion-related solid friction work $W_{fs,ad}$ is more difficult, but can be determined, for example, from the specific work of adhesion γ_{ad} or the shear strength $\tau_{s,ad}$ of the atomic/molecular compound to be separated [11]:

$$W_{fs,ad} = \gamma_{ad} \cdot A_r = \tau_{s,ad} \cdot A_r \cdot s_f \tag{7}$$

If the solid body friction work W_{fs}, the friction distance s_f and the solid body load-bearing force F_s acting in the normal direction are known, the following can by inserting and converting to the solid body friction force:

$$F_{fs} = \frac{W_{fs}}{s_f} = \frac{W_{fs,def} + W_{fs,ad}}{s_f} = F_{fs,def} + F_{fs,ad} \tag{8}$$

or the solid friction coefficient μ_s can be concluded:

$$\mu_s = \frac{1}{F_s} \cdot \frac{W_{fs}}{s_f} = \frac{W_{fs,def} + W_{fs,ad}}{F_s \cdot s_f} \qquad (9)$$

1.2. Elastohydrodynamics

The term elastohydrodynamics (EHD) was introduced in the middle of the 20th century [12]. Initially, the term was only used for liquid-lubricated concentrated contacts (e.g., in rolling bearings or gears), but today, it is also used for liquid-lubricated non-concentrated contacts (e.g., in plain bearings or piston/cylinder pairings) if the gap flow calculation is coupled with an elastic gap deformation calculation. Elastohydrodynamics considers addition to hydrodynamics (HD), in which the pressure and shear gradient dependence of the viscosity must be considered, also the elastic deformation of the solids to be separated by a lubricating film. For the extension of the elastohydrodynamics by the temperature calculation, the term thermal elastohydrodynamics (TEHD) was coined.

Figure 2 shows a schematic illustration of the pressure distribution and the lubrication gap height profile of a concentrated EHD contact in the direction of movement. For comparison, the pressure distribution according to Hertz for the unlubricated contact is also shown. Two characteristic lubrication gap heights can be defined, namely the central film thickness h_c in the center of the contact and the minimum film thickness h_{min} in the gap constriction. Furthermore, a distinction can be made between a pressure area, in which the supporting effect of the lubricant is achieved, and a cavitation area, in which the gap that opens is not completely filled with lubricant.

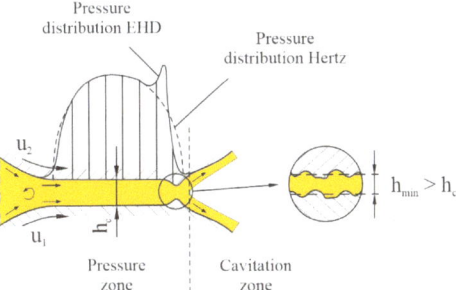

Figure 2. Pressure distribution and lubrication gap height profile in a concentrated EHD contact.

In the classic conception, the EHD contact shown in Figure 2 would be assigned to liquid friction or liquid lubrication if the minimum film thickness h_{min} is so large that the rough surfaces do not touch ($h_{min} > h_{cr}$). In fact, however, it is an EHD contact with mixed friction and liquid lubrication, provided that the definition of mixed friction in liquid-lubricated EHD contacts as a superposition of liquid and solid friction and the fact that hysteresis losses in the elastically deformed solids are to be assigned to solid friction as internal friction are accepted without restriction.

When calculating the friction of fully lubricated EHD contacts, hysteresis losses are often not considered. The friction is calculated solely from the shear of the lubricant. How large the "hysteresis friction" can be is intended to show the following explanations. The aim of this work is not to consider complex EHD contact and material models to be able to precisely determine the hysteresis friction in the solid bodies. Instead, a simplified approach is used to illustrate the possible orders of magnitude and the resulting findings.

2. Materials and Methods

According to Figure 2, the starting point should be a fully lubricated EHD contact in which two bodies curved on both sides roll against each other. The semi-axes of the elliptical contact area that forms and the pressure distribution acting there (Figure 3) should be calculated as a good approximation using Hertz's equations for the unlubricated normal contact [13].

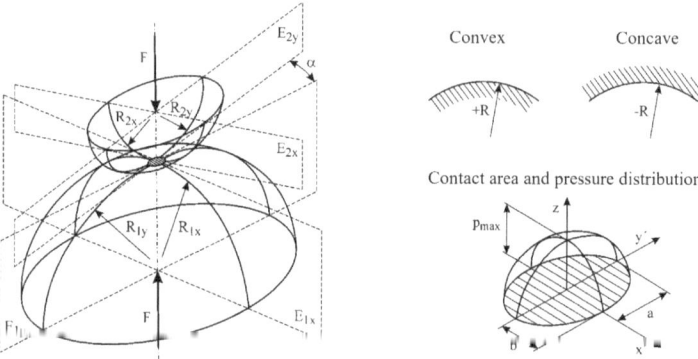

Figure 3. Hertzian contact and definition of geometric quantities.

According to Hertz, a reduced radius R_{red} is introduced, which can be interpreted as the radius of a fictional elastic sphere that is pressed against a rigid plane.

$$\frac{1}{R_{red}} = \frac{1}{R_{1x}} + \frac{1}{R_{1y}} + \frac{1}{R_{2x}} + \frac{1}{R_{2y}} \quad (10)$$

Here, the radius of a convex curvature must have a positive sign and that of a concave curvature a negative sign, as shown in Figure 3. Furthermore, it must be ensured that the radius of the convex body is smaller than that of the concave body in the case of a convex/concave pairing. If one of the solids is a plane, $1/R_{x,y} = 0$ is applied.

The elastic properties of the fictional sphere are determined from the elastic properties of the two elliptical bodies with the reduced Young's modulus E_{red}.

$$\frac{1}{E_{red}} = \frac{(1-\nu_1^2)}{E_1} + \frac{(1-\nu_2^2)}{E_2} \quad (11)$$

Furthermore, an auxiliary angle τ can be introduced, for which applies

$$\cos(\tau) = R_{red}\sqrt{\left(\frac{1}{R_{1x}} - \frac{1}{R_{1y}}\right)^2 + \left(\frac{1}{R_{2x}} - \frac{1}{R_{2y}}\right)^2 + 2\left(\frac{1}{R_{1x}} - \frac{1}{R_{1y}}\right) \cdot \left(\frac{1}{R_{2x}} - \frac{1}{R_{2y}}\right) \cdot \cos(2\alpha)} \quad (12)$$

or for a twist angle of $\alpha = 0$:

$$\cos(\tau) = R_{red} \cdot \left|\frac{1}{R_{1x}} - \frac{1}{R_{1y}} + \frac{1}{R_{2x}} - \frac{1}{R_{2y}}\right| \quad (13)$$

The auxiliary angle τ is also related to the half-axis ratio.

$$\kappa = \frac{b}{a} (a \geq b,\ 0 \leq \kappa \leq 1) \quad (14)$$

$$\cos(\tau) = \frac{E_{ell}(\kappa) \cdot (1+\kappa^2) - 2 \cdot K_{ell}(\kappa) \cdot \kappa^2}{E_{ell}(\kappa) \cdot (1-\kappa^2)} \quad (15)$$

If the $cos(\tau)$ value is known from Equation (12) or Equation (13), the half-axis ratio can be determined iteratively using Equation (15). The complete elliptic integrals $E_{ell}(\kappa)$ and $K_{ell}(\kappa)$ in Equation (15) can be calculated as follows:

$$E_{ell}(\kappa) = \int_0^{\pi/2} \sqrt{1 - (1 - \kappa^2) \cdot \sin^2(\phi)} \, d\phi \tag{16}$$

$$K_{ell}(\kappa) = \int_0^{\pi/2} \frac{1}{\sqrt{1 - (1 - \kappa^2) \cdot \sin^2(\phi)}} \, d\phi \tag{17}$$

With knowledge of κ, three coefficients can be determined which are used as correction factors in the equations for the fictional point contact to calculate the major and minor semi-axes of the real elliptical contact surface and the maximum deflection of the elliptical contact bodies. The correction factor for the major half-axis is

$$c_a = \sqrt[3]{\frac{2 \cdot E_{ell}(\kappa)}{\pi \cdot \kappa^2}} \tag{18}$$

The correction value for the minor half-axis is

$$c_b = \kappa \cdot c_a \tag{19}$$

The correction value for the maximum deflection of the contacting bodies is given by

$$c_w = K_{ell}(\kappa) \cdot \sqrt[3]{\frac{4 \cdot \kappa^2}{\pi^2 \cdot E_{ell}(\kappa)}} \tag{20}$$

Alternatively, the correction factors c_a and c_b can be calculated using the auxiliary value

$$f_\tau = \ln[1 - \cos(\tau)] \tag{21}$$

according to [14] with the following approximate solutions:

$$\begin{aligned} c_a &= exp\left(\frac{f_\tau}{-1.53 + 0.333 \cdot f_\tau + 0.0467 \cdot f_\tau^2}\right) \text{ for } 0 \leq \cos(\tau) < 0.949 \\ c_a &= exp\left(\sqrt{-0.4567 - 0.4446 \cdot f_\tau + 0.1238 \cdot f_\tau^2}\right) \text{ for } 0.949 \leq \cos(\tau) < 0.999998 \end{aligned} \tag{22}$$

$$\begin{aligned} c_b &= exp\left(\frac{f_\tau}{1.525 - 0.86 \cdot f_\tau - 0.0993 \cdot f_\tau^2}\right) \text{ for } 0 \leq \cos(\tau) < 0.949 \\ c_b &= exp\left(-0.333 + 0.2037 \cdot f_\tau + 0.0012 \cdot f_\tau^2\right) \text{ for } 0.949 \leq \cos(\tau) < 0.999998 \end{aligned} \tag{23}$$

In [14] only approximate equations for c_a and c_b are given. To be able to calculate c_w with little effort, an equation for c_w is required. This is achieved by rearranging Equation (15) to $K_{ell}(\kappa)$, Equation (18) to $E_{ell}(\kappa)$, and Equation (19) to κ, and by inserting the three resulting equations into Equation (20). The result is an equation with which c_w can be calculated as a function of c_a and c_b:

$$c_w = c_b^2 + \frac{1}{2} c_a^2 \cdot \left[1 - \left(\frac{c_b}{c_a}\right)^2\right] \cdot [1 - \cos(\tau)] \tag{24}$$

The deviations of the three approximate equations listed are not greater than 0.7% over the entire value range.

The major half-axis a and the minor half-axis b of the elliptical contact as well as the maximum deflection of the bodies follows from

$$a = c_a \cdot \sqrt[3]{\frac{3 \cdot F \cdot R_{red}}{4 \cdot E_{red}}} \tag{25}$$

$$b = c_b \cdot \sqrt[3]{\frac{3 \cdot F \cdot R_{red}}{4 \cdot E_{red}}} \tag{26}$$

$$w_{max} = c_w \cdot \sqrt[3]{\frac{9 \cdot F^2}{32 \cdot R_{red} \cdot E_{red}^2}} \tag{27}$$

The pressure distribution in the contact is calculated from

$$p(x,y) = p_{max} \cdot \sqrt{1 - \left(\frac{x}{b}\right)^2 - \left(\frac{y}{a}\right)^2} \tag{28}$$

where for the maximum pressure in the center of the contact area, the following is applied:

$$p_{max} = \frac{3 \cdot F}{2\pi \cdot a \cdot b} \tag{29}$$

The calculation of the elastic deformation work to be applied during rolling between the two bodies is achieved using an approach by Tabor [15–19], which was presented for the rolling of a rigid ball on an elastic plane or in an elastic groove and can be generalized. Within the contact area, the total deformation w at any point (x,y) can be described by the following function:

$$w(x,y) = -w_{max}(0,0) + \frac{x^2}{2R_{1x}} + \frac{x^2}{2R_{2x}} + \frac{y^2}{2R_{1y}} + \frac{y^2}{2R_{2y}} \tag{30}$$

The change in deformation when rolling forward by a small distance Δ in the x-direction results from the derivative of Equation (30) with respect to x as

$$\Delta \frac{\partial w}{\partial x} = \Delta \left(\frac{x}{R_{1x}} + \frac{x}{R_{2x}}\right) = \frac{\Delta \cdot x}{R_{red,x}} \quad \text{with} \quad \frac{1}{R_{red,x}} = \left(\frac{1}{R_{1x}} + \frac{1}{R_{2x}}\right) \tag{31}$$

The deformation work performed in all surface elements $dxdy$ (Figure 4) in the front half of the contact area when rolling forward by the distance Δ is given by

$$W_{s,def}^{el} = \frac{\Delta}{R_{red,x}} \int_{-a}^{a} \int_{0}^{\xi} p(x,y) x \, dx \, dy \quad \text{with} \quad \xi = b\sqrt{1 - \left(\frac{y}{a}\right)^2} \tag{32}$$

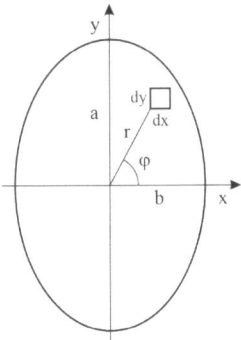

Figure 4. Hertzian contact area with surface element $dxdy$.

The solution of the integral is succeeded by substitution and subsequent transformation of the integral into polar coordinates. The result is the elastic deformation work to be applied when rolling in the direction of the minor half-axis (x-direction)

$$W^{el}_{s,def,x} = \frac{\Delta}{R_{red,x}} \cdot \frac{3 \cdot F \cdot b}{16} \tag{33}$$

and in an analogous derivation for rolling in the direction of the major half-axis (y-direction):

$$W^{el}_{s,def,y} = \frac{\Delta}{R_{red,y}} \cdot \frac{3 \cdot F \cdot a}{16} \tag{34}$$

When the two bodies roll, the friction bodies are loaded in the front contact area and unloaded again in the rear contact area. Hysteresis losses occur, which manifest themselves as solid friction losses according to Equation (4). The solid friction force resulting in the respective rolling direction can be calculated with a rolling or friction distance of $s_f = \Delta$ from

$$F_{fs,hys,x} = \frac{W^{el}_{fs,def,x}}{s_f} = \frac{H_{red} \cdot W^{el}_{s,def,x}}{\Delta} \tag{35}$$

$$F_{fs,hys,y} = \frac{W^{el}_{fs,def,y}}{s_f} = \frac{H_{red} \cdot W^{el}_{s,def,y}}{\Delta} \tag{36}$$

and the direction-dependent friction moment with

$$M_{fs,hys,x} = F_{fs,hys,x} \cdot R_{red,x} \tag{37}$$

$$M_{fs,hys,y} = F_{fs,hys,y} \cdot R_{red,y} \tag{38}$$

A solid coefficient of friction can be determined from the hysteresis-related solid friction force and the acting normal force. When there is a rolling in the direction of the minor half-axis, the result is

$$\mu_{s,hys,x} = \frac{F_{fs,hys,x}}{F} = H_{red} \cdot \frac{3}{16} \cdot \frac{b}{R_{red,x}} \tag{39}$$

and when rolling is in the direction of the major half-axis, the following is obtained:

$$\mu_{s,hys,y} = \frac{F_{fs,hys,y}}{F} = H_{red} \cdot \frac{3}{16} \cdot \frac{a}{R_{red,y}} \tag{40}$$

Analogous to the derivation for point contacts, it is also possible to derive a hysteresis-related coefficient of solid friction for line contacts

$$\mu_{s,hys,x} = H_{red} \cdot \frac{2}{3\pi} \cdot \frac{b}{R_{red,x}} \quad (41)$$

where the pressure distribution for line contacts must be used to derive Equation (41).

$$p(x) = p_{max} \cdot \sqrt{1 - \left(\frac{x}{b}\right)^2} \text{ with } p_{max} = \frac{2 \cdot F}{\pi \cdot b \cdot L} \quad (42)$$

3. Results and Discussion

Equations (39) and (40) show that all point contacts with the same axis-related rolling direction, the same ratio of half-axis and reduced radius, and the same reduced hysteresis factor will have identical hysteresis-related coefficients of solid friction. Figure 5 shows the possible magnitudes of these coefficients of friction. In Figure 5, the reduced hysteresis factor was varied in a range from 1% to 10%. The smaller values stand for steel/steel pairings and the larger values for plastic/steel or elastomer/steel pairings. The author is aware that the linear-elastic deformation approach, which is considered in Hertz's equations, does not adequately describe the deformation behavior of plastics or elastomers. In these cases, the linear-elastic approach would have to be supplemented by a viscoelastic component.

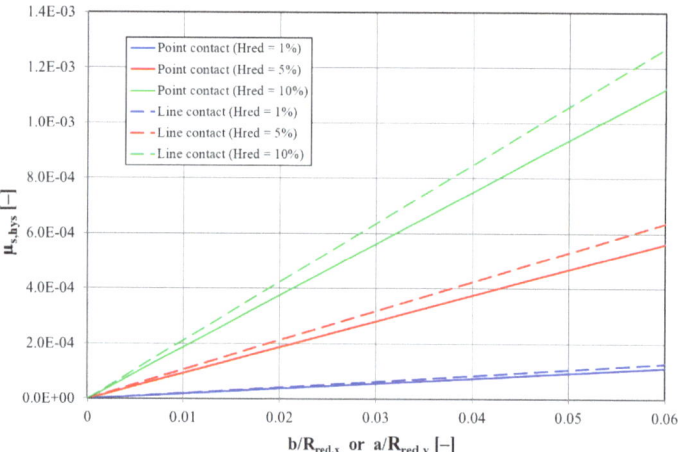

Figure 5. Hysteresis-related solid coefficients of friction of rolling contacts with point or line contact at different hysteresis factors.

The hysteresis loss when taking viscoelasticity into account will therefore be higher than that calculated using a purely linear-elastic approach. Independently of this, the range selected for the hysteresis factor in Figure 5 is probably not entirely unrealistic for rolling contacts as it can be assumed that the hysteresis factors will be higher than those determined in classical test procedures, such as dynamic mechanical thermal analysis (DMTA), on standard test specimens due to the high deformation speeds, loads, and temperatures in rolling contacts [7].

Furthermore, the range of values selected in Figure 5 for the ratio of half-axis and reduced radius corresponds to the values that can also occur in real applications. It is clear that the hysteresis-related coefficient of solid friction increases with increasing load, i.e., with increasing ratio of half-axis and reduced radius and increasing damping behavior of the material pairing.

If there is a completely separating lubricating film in an EHD contact between the two bodies, the adhesion component of the solid friction work according to Equation (2) is ineffective. If no plastic deformations occur according to Equation (3), friction losses are only effective due to hysteresis-related solid friction according to Equation (4), and additionally due to liquid friction. The frictional force F_f in the EHD contact then results from the liquid frictional force F_{fh} and a hysteresis-related solid friction force $F_{fs,hys}$.

$$F_f = F_{fh} + F_{fs} = F_{fh} + F_{fs,hys} \tag{43}$$

For the resulting friction coefficient μ, the following can be written:

$$\mu = \frac{F_f}{F} = \frac{F_{fh} + F_{fs,hys}}{F} \tag{44}$$

Figure 5 shows that the hysteresis-related coefficients of solid friction can be in the same order of magnitude as the coefficients of fluid friction of fully lubricated EHD contacts. Therefore, hysteresis friction can have a significant influence on the overall friction behavior of a fully lubricated EHD contact in the case of pure rolling or very small slip values and should not be neglected. With larger slip values or pure sliding, hysteresis friction should play a subordinate role. The above statements are equally valid for elastohydrodynamic point and line contacts.

The hysteresis-related solid friction can be calculated analytically with the equations previously listed for ideal Hertzian point and line contacts. The equations are no longer applicable for rolling elements with any geometries (rolling elements with logarithmic profiles, manufacturing deviations, wear profiles, etc.). Numerical calculations are an alternative here. In numerical EHD calculations, the hysteresis-related solid friction can be determined using the model for describing the lubrication gap deformations (elastic half-space or FEM) for stationary and transient operating conditions.

4. Conclusions

This paper showed that it is necessary to differentiate between friction and lubrication states and that these must be clearly separated from each other. If the friction in a contact is considered, the friction state is decisive. If the focus is on the type of lubrication of a contact, the lubrication state must be considered.

A calculation of the friction behavior of tribological systems requires a detailed consideration and description of all mechanisms involved in friction. Although hysteresis-related solid friction losses always occur in fully lubricated EHD contacts, these friction losses are often not considered. The friction losses acting in the EHD contact are attributed usually to fluid friction alone. Depending on the slip, the influence of hysteresis-related solid friction on the total friction of a fully lubricated EHD contact can be negligibly small, just as large, or even larger than the fluid friction and is more pronounced with plastics than with metals. Due to the always-existing hysteresis-related energy losses in the solids, which are attributable to solid friction and are superimposed on the liquid friction losses, it can be formulated that every fully lubricated EHD contact is a mixed friction contact. This would extend the classical concept of mixed friction, as mixed friction can also occur without direct contact between the solids. Furthermore, a fully lubricated EHD contact is traditionally assigned to the friction state of liquid friction and the lubrication state of liquid lubrication. However, in reality, this is an EHD contact with mixed friction and liquid lubrication.

Funding: This research received no external funding.

Data Availability Statement: All data can be found in this paper.

Conflicts of Interest: The author declares no conflicts of interest.

Nomenclature

A_r	Real contact area (m²)	w	Deformation (m)
a	Major half-axis (m)	x, y	Cartesian coordinates (m)
b	Minor half-axis (m)	α	Twist angle (°)
c_a, c_b, c_w	Correction factors (-)	γ	Specific work (Nm/m²)
E	Young's modulus (N/m²)	Δ	Distance (m)
E_{ell}	Elliptical integral 2nd kind	τ	Auxiliary angle (°)
F	Normal force (N)	φ	Angle (°)
F_s	Solid load capacity (N)	κ	Half-axis ratio (-)
F_f	Total friction force (N)	μ_h	Coefficient of liquid friction (-)
F_{fh}	Liquid friction force (N)	μ_s	Coefficient of solid friction (-)
F_{fs}	Solid friction force (N)	ν	Poisson's ratio (-)
f_τ	Auxiliary value (-)	τ_s	Shear strength (N/m²)
H	Hysteresis factor (-)		
h_c	Central film thickness (m)		
h_{cr}	Critical film thickness (m)	Frequently used indices	
h_{min}	Minimum film thickness (m)	1	Body 1
K_{ell}	Elliptical integral 1st kind	2	Body 2
L	Width of cylinder (m)	ad	Adhesion
p	Hertzian pressure (N/m²)	def	Deformation
R	Radius (m)	el	Elastic
s_f	Sliding distance (m)	hys	Hysteresis
u	Circumferential speed (m/s)	max	Maximum
W_s	Solid deformation work (Nm)	pl	Plastic
W_{fs}	Solid friction work (Nm)	red	Reduced

References

1. GfT Arbeitsblatt 7: Tribologie—Definitionen, Begriffe, Prüfung (GfT Worksheet 7: Tribology—Definitions, Terms, Testing). Gesellschaft für Tribologie e.V. 2002. Available online: www.gft-ev.de (accessed on 3 August 2024).
2. Hamrock, B.J.; Schmid, S.R.; Jacobson, B.O. *Fundamentals of Fluid Film Lubrication*, 2nd ed.; CRC Press: Boca Raton, FL, USA, 2004.
3. Bartel, D. *Simulation von Tribosystemen—Grundlagen und Anwendungen (Simulation of Tribosystems—Basics and Applications)*; Vieweg+Teubner: Wiesbaden, Germany, 2010.
4. Kragelskij, I.V. About the friction of unlubricated surfaces. All-Union Conference on Friction and Wear in Machines (О трении несмазанных поверхностей. Всесоюзная конференция по трению и износу в машинах); Academy of Sciences: Moscow, Russia, 1939; Volume 1, pp. 543–561.
5. Bowden, F.P.; Moore, A.J.W.; Tabor, D. The ploughing and adhesion of sliding metals. *J. Appl. Phys.* **1943**, *14*, 80–91. [CrossRef]
6. Lazan, B.J. *Damping of Materials and Members in Structural Mechanics*; Pergamon Press: Oxford, UK, 1968.
7. Menard, K.P. *Dynamic Mechanical Analysis: A Practical Introduction*, 2nd ed.; CRC Press: Boca Raton, FL, USA, 2008.
8. Fleischer, G. Energiebilanzierung der Festkörperreibung als Grundlage zur energetischen Verschleißberechnung. Teil I: (Energy balancing of solid friction as a basis for an energetical calculation of wear. Part I). *Schmierungstechnik* **1976**, *7*, 225–230.
9. Fleischer, G. Energiebilanzierung der Festkörperreibung als Grundlage zur energetischen Verschleißberechnung. Teil II: (Energy balancing of solid friction as a basis for an energetical calculation of wear. Part II). *Schmierungstechnik* **1976**, *7*, 271–279.
10. Fleischer, G. Energiebilanzierung der Festkörperreibung als Grundlage zur energetischen Verschleißberechnung. Teil III: (Energy balancing of solid friction as a basis for an energetical calculation of wear. Part III). *Schmierungstechnik* **1977**, *8*, 49–58.
11. Bartel, D. Berechnung von Festkörper- und Mischreibung bei Metallpaarungen (Calculation of Solid and Mixed Friction in Metal Pairs). Ph.D. Thesis, Otto von Guericke University, Magdeburg, Germany, 2000.
12. Dowson, D.; Higginson, G.R. A numerical solution to the elasto-hydrodynamic problem. *J. Mech. Eng. Sci.* **1959**, *1*, 6–15. [CrossRef]
13. Hertz, H.R. Über die Berührung fester elastischer Körper (About the contact of solid elastic bodies). *J. Die Reine Angew. Math.* **1881**, *92*, 156–171.
14. Grekoussis, R.; Michailidis, T. Stellung der Hertzschen Druckellipse auf der Oberfläche zweier einander in einem Punkt berührender Körper (Position of the Hertzian pressure ellipse on the surface of two bodies touching each other at one point). *Konstruktion* **1980**, *32*, 303–306.
15. Tabor, D. The mechanism of rolling friction. *Philos. Mag. Series* **1952**, *43*, 1055–1059. [CrossRef]
16. Eldredge, K.R.; Tabor, D. The mechanism of rolling friction I: The plastic range. *Proc. R. Soc. Lond. Ser. A* **1955**, *229*, 181–198.
17. Tabor, D. The mechanism of rolling friction II: The elastic range. *Proc. R. Soc. Lond. Ser. A* **1955**, *229*, 198–220.
18. Tabor, D. Elastic work involved in rolling a sphere on another surface. *Brit. J. Appl. Phys.* **1955**, *6*, 79–81. [CrossRef]

19. Greenwood, J.A.; Minshall, H.; Tabor, D. Hysteresis losses in rolling and sliding friction. *Proc. R. Soc. Lond. Ser. A* **1961**, *259*, 480–507.

Disclaimer/Publisher's Note: The statements, opinions and data contained in all publications are solely those of the individual author(s) and contributor(s) and not of MDPI and/or the editor(s). MDPI and/or the editor(s) disclaim responsibility for any injury to people or property resulting from any ideas, methods, instructions or products referred to in the content.

Article

A Tribological Study of ta-C, ta-C:N, and ta-C:B Coatings on Plastic Substrates under Dry Sliding Conditions

Paul Neubauer [1], Frank Kaulfuss [2,*] and Volker Weihnacht [2]

[1] Institut für Fertigungstechnik, Technische Universität Dresden, 01069 Dresden, Germany
[2] Fraunhofer Institute for Material and Beam Technology (IWS), Winterbergstr. 20, 01277 Dresden, Germany; volker.weihnacht@iws.fraunhofer.de
* Correspondence: frank.kaulfuss@iws.fraunhofer.de

Abstract: In this study, we analyze the extent to which thin hard coatings can serve as tribological protective layers for the selected plastic substrate materials PA12 (polyamide 12) und PEEK (polyetheretherketone), with and without fiber reinforcement. The approximately 1 μm thick coating variants ta-C, ta-C:N, and ta-C:B, which were applied using the laser arc process, are investigated. In oscillating sliding wear tests against a steel ball in an air atmosphere without lubricant, the wear of the coating and counter body is compared to analogous coating variants applied in parallel to AISI 52100 steel. The ta-C-based coatings show good adhesion strength and basic suitability as wear protection layers on the plastic substrates in the tribological tests. However, there are variations depending on the coating type and substrate material. The use of a Cr interlayer and its thickness also plays an important role. It is demonstrated that by coating under conditions where the uncoated plastic substrate would normally fail, a similarly good performance as with analogously coated steel substrates can be achieved by ta-C:N.

Keywords: ta-C; coating; plastic; wear; PA12; PEEK; friction; sliding; tribotest; doped ta-C

1. Introduction

Thin hard coatings, primarily composed of nitride hard materials (e.g., TiN and CrN) or diamond-like carbon (DLC, e.g., a-C:H and ta-C), have been proven to be effective for decades as tribological protective layers on sliding components made of steel. Due to their high hardness, they enhance the surfaces' resistance to micro-abrasive wear and, in some cases, significantly reduce friction. Such coatings are indispensable today, particularly for applications in internal combustion engines on piston-group (piston ring, piston pin), fuel-injection, and valve-train components (bucket tappets, finger follower) [1].

With the development of increasingly durable plastics, these materials are increasingly considered as alternatives for lightly to moderately loaded tribological components made of steel. Due to their unbeatable advantages such as their low cost, good machinability, flexibility in shaping, and low density, plastics are now increasingly considered for tribologically loaded components in bearings, gears, guide rails, etc. However, the abrasive wear of plastic surfaces under real operating conditions in the presence of dust, soot, and other wear particles emerges as a limiting factor more so than with steel.

Among the polymer types that are suitable for the mentioned applications, the PA12 (polyamide 12) und PEEK (polyetheretherketone) variants are of particular interest. Both of them possess the characteristics of high strength, good chemical resistance, high abrasion resistance, and low friction. PEEK is superior to PA12 in all respects, but is significantly more expensive to manufacture. PA12 is therefore considered a low-cost alternative to PEEK. In order to improve the mechanical properties of polymers, i.e., strength and rigidity, they are reinforced with fibers (usually glass fibers or carbon fibers). The type of fiber reinforcement depends on the specific requirements of the application, the costs, and the desired material properties. Even if the fiber reinforcement is advantageous for the mechanical behavior in many respects, the fiber content can have a detrimental effect in the tribological respect. When in tribological contact with soft mating bodies, fibers might be more abrasive than unreinforced polymers.

The use of thin, hard, wear-protective coatings on plastic substrates analogous to steel has hardly been considered to date in industrial applications. The difference in mechanical properties between the substrate and the coating is too large and it was assumed that the coating would fail quickly due to the so-called eggshell effect. There are a few tribological studies of DLC-coated polymers in the literature, e.g., on PEEK substrates under dry conditions [2–4] or under lubricated conditions [5]. Most research has been carried out on coated polymers in the context of medical technology applications, particularly for implants [6–9]. To the best of our knowledge, there is no published work in the literature on ta-C coatings on plastic substrates apart from our own work [10].

In this study, hard and superhard ta-C-based coatings, analogous to those used as tribological protective layers on steel, were deposited on various plastics and analyzed tribologically. For this purpose, the plastics PEEK (high-quality, relatively hard plastic), PA12 (relatively inexpensive, softer plastic), and both variants with carbon-fiber reinforcement were selected as substrate materials. A standard ta-C coating and two doped variants thereof (ta-C:N and ta-C:B) served as the coating. An important issue with coatings is the use of an inter- or adhesion layer. On the one hand, the interlayer is intended to achieve a better chemical bond or mediation between the metal and the coating material, and on the other hand, it is sometimes intended to accommodate strong differences between the mechanical properties or achieve a load-supporting effect. Cr or Ti is generally used as an interlayer for metallic substrates. In this study, the Cr interlayer was also used for the plastic substrates. In a further series, ta-C:B coatings were deposited without and with two different thicknesses of the Cr interlayer for comparison.

An initial setting of tribotest parameters is used to screen three different coating variants. In a second step, the most promising coating–substrate combination with modified test conditions is then selected. Finally, a comparative tribotest is carried out with the selected variant in comparison to uncoated plastic and a coated steel reference. Even though the potential use of the coated plastics will generally take place under lubricated conditions, this first study will test the wear behavior under extreme conditions, i.e., under sliding conditions without lubrication.

2. Materials and Methods

For the coating and subsequent investigations, five different substrates were used (see Table 1). Four of these were different plastics from Evonik (Evonik Operations GmbH, Kirschenallee, Darmstadt, Germany), which were compared with the steel substrate AISI 52100 as a reference. The polymers included two variants each of polyamide 12 (PA12) and polyetheretherketone (PEEK). In addition to the base version, glass-fiber-reinforced PA12 and carbon-fiber-reinforced PEEK substrates were used (PA12-GF30 and PEEK-CF30, respectively). The material content of glass and carbon fibers was 30 percent. The dimensions of the sample plates were 18 mm × 13.5 mm × 3 mm. The steel surface was polished, and the surfaces of the plastics were not mechanically post-processed after injection molding.

Table 1. Substrate materials used for coating and their mechanical properties.

Material	AISI 52100	PA12	PA12-GF30	PEEK	PEEK-CF30
Young's Modulus [GPa]	210 *	1.4 **	6.8 **	3.5 **	24 **
Yield Strength [MPa]	>835 *	43 **	120 **	95 **	251 **

* WIXSTEEL Industrial [11]. ** Evonik Industries AG product information [12].

The coatings were deposited in a commercial physical vapor deposition (PVD) chamber (VTD Vakuumtechnik Dresden GmbH, Dresden, Germany) with an attached LaserArc™

carbon evaporation source (Fraunhofer IWS, Dresden, Germany). An 8-axis planetary system was used as a sample holder and set in twofold rotation. Standard graphite targets from Plansee Composite Materials GmbH, Germany, were used as the cathode material for the generation of the ta-C and the nitrogen-gas-doped ta-C:N coatings. In the case of B-doped carbon, powder-pressed and sintered composite graphite targets with a nominal content of 5 at% of boron from the same company were used. Approximately the same content of 5 at% boron was found in the ta-C:B coating. The coating process was as follows: In the beginning, the deposition chamber was evacuated to a high vacuum at a pressure of about 10^{-4} Pa and an argon ion etching process was carried out prior to the coating deposition. Subsequently, in some cases, a chromium adhesion interlayer was deposited by magnetron sputtering, followed by deposition of the carbon coatings by pulsed arc discharge that generates the carbon plasma. The arc discharges were ignited using laser pulses from a Q-switched Nd-YAG laser. Detailed information on the LaserArc process has been published elsewhere [13,14]. In the case of the ta-C:N coatings, during the evaporation of carbon by LaserArc, nitrogen gas with a 40 sccm flow rate was introduced in the deposition chamber, resulting in a content of approximately 5 at% N in the ta-C:N coating. In all process steps (ion etching, interlayer, and carbon coating) no bias voltage was used due to the use of non-conductive polymer substrates.

The coating thickness was determined using the ball crater grinding method that is standardized in EN ISO 26423:2016 [15]. Three individual craters were ground on each specimen using the calotte grinding unit "KSG 117" (Inovap GmbH, Radeberg, Germany). Instrumented indentation was carried out with a Berkovich diamond indenter on the "ZHN-1" from Zwick/Roell GmbH, Germany, to determine hardness H and the Young's modulus E of the coatings according to EN ISO 14577-4 [16]. For this purpose, the QCSM technique and the sigmoid fitting method were used for data acquisition and evaluation [17]. The maximum normal loads were 40 mN. A Poisson's ratio of 0.19 was assumed for all coatings.

For tribological characterization, an SRV4 from Optimol Instruments with the standard oscillation setup was used, which corresponds to the translatory tester according to DIN 51834-1:2010-11 [18]. The coated flat substrates were fixed as lower samples against an uncoated, polished AISI 52100 steel ball with ⌀10 mm as the upper sample loaded with a normal load and oscillating movement. The counter-body wear was determined under the presumption of plain ball wear and was calculated via the diameter of the ball wear. The test specimen and holder were cleaned ultrasonically in high-purity benzine before the test. The tests were performed at room temperature (20 ± 5 °C) with no lubricant, in air with a controlled humidity of $50 \pm 5\%$.

In this study, three different parameter sets were used (see Table 2). Thereby, set I, with relatively harsh loading conditions but a short test duration, was used to obtain a quick general overview of the wear behavior on the different substrates. Set II used to simulate a load case that is closer to the application. The normal force is reduced to a tenth of the previous load and the test duration is extended by a factor of 10. Set III was only applied to selected substrates and is intended to show the limits in terms of load and load-bearing capacity.

Table 2. Parameter sets used for tribological characterization.

Parameter	Set I	Set II	Set III
Normal load [N]	10	1	10
Oscillation frequency [Hz]	10	10	10
Stroke [mm]	1	1	1
Testing time [min]	10	100	60

The white-light interferometry method was used to determine the surface roughness S_a using the Leica DCM 3D microscope according to EN ISO 25178 [19].

Particles and defects visible on the surface, i.e., carbon macroparticles incorporated in the coating, were quantified by the "surface defect fraction (SDF)" parameter that was determined using three individual light microscopy images from different positions at the surface at 500× magnification. More information about the method to determine the SDF parameter can be found elsewhere [20].

3. Results

First, a coating series of ta-C, ta-C:N, and ta-C:B with nominal 1 µm thicknesses were deposited on all substrates, using a Cr interlayer with about a 0.1 µm thickness. The compilation in Figure 1 shows the surface topography of the different substrates in the initial state and with a 1 µm ta-C coating. The PA12 and PEEK substrates are initially very smooth. The coating creates a crack pattern on the soft PA12. This has no negative effect on adhesion. With PEEK, on the other hand, the surface remains smooth. However, existing scratches are intensified by the coating, which can be seen in the interference contrast. The fiber-reinforced substrates are initially rougher, whereby the glass fibers in the soft PA12 matrix are not clearly visible. However, the fibers are significantly more pronounced due to the coating and the deformation of the surface. In the case of fiber-reinforced PEEK, the fibers are clearly visible on the surface. After coating the fibers are covered and their visibility is reduced.

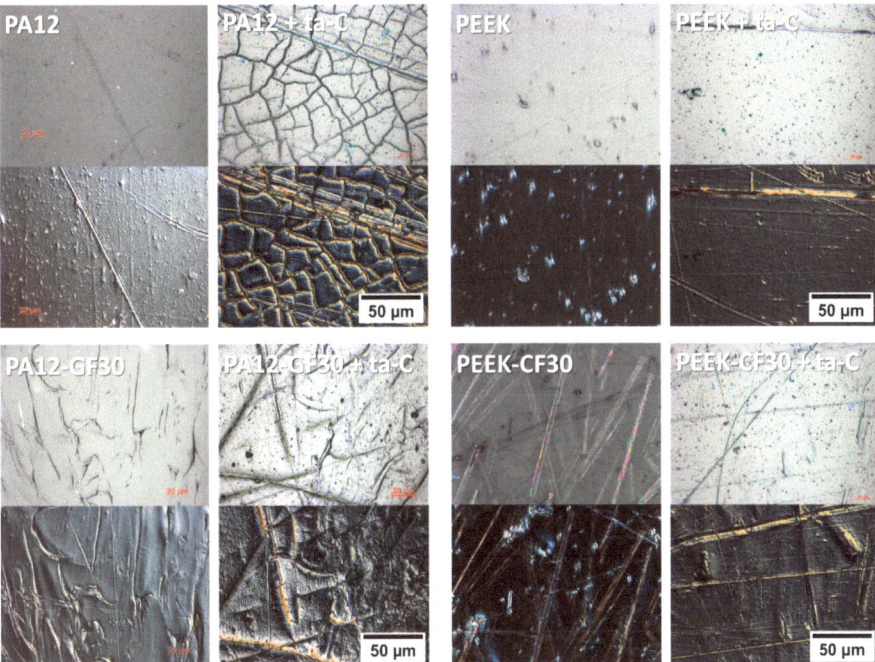

Figure 1. Microscopic images of the PA12, PA12-GF30, PEEK, and PEEK-CF30 uncoated (**left**), and 1 µm ta-C-coated (**right**) substrates in bright field (**top**) and with differential interference contrast (DIC) filter (**bottom**).

A decisive parameter is the surface defect fraction (SDF), which provides information on the extent to which the surface is covered with coating defects. These defects are unavoidable in arc coating processes and result in unfavorable roughness on the surface, which is associated with high wear in tribological contacts, especially on the counter body. The SDF parameter is more sensitive for detecting the effects of coating defects on the tribological behavior than surface roughness parameters. However, it is very instructive to compare the effect of the coating on both parameters.

As for the roughness parameter S_a, the steel surface showed very little change due to the coatings (Figure 2 (top)). PA12, on the other hand, was initially rather smooth, but showed a strong increase in S_a due to the coatings. The same applies to PA12-GF30, even if its initial roughness was significantly greater than that of PA12. In contrast, the two PEEK variants—similar to steel—did not show such strong changes in roughness as the PA12 variants. Here, the low roughness with the coating was retained, which was also the case for the fiber-reinforced PEEK.

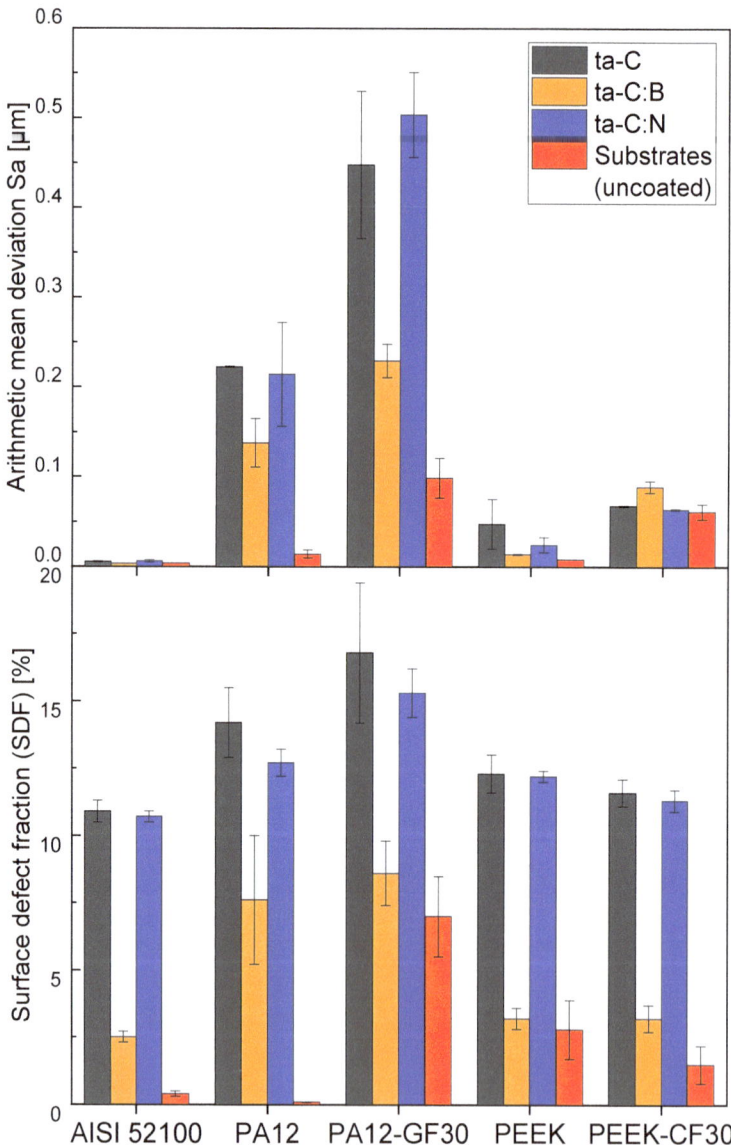

Figure 2. Arithmetic mean deviation Sa (**top**) and surface defect fraction (SDF) (**bottom**) for AISI 52100, PA12, PA12-GF30, PEEK, and PEEK-CF30 uncoated and coated with 1 µm ta-C, 1 µm ta-C:N, and 1 µm ta-C:B.

The SDF values (Figure 2 (bottom)) showed strong variations for the different coatings, but the nature of the changes depending on the coating type was the same for all substrate variants. This shows that the SDF parameter reacts very sensitively to coating-specific surface defects, with the ta-C:B film having the lowest SDF in all coated cases. The SDFs of ta-C and ta-C:N were very similar because the coating plasma is identical except for the N_2 gas inlet. In the case of ta-C:B, there were far fewer defects. This is already known from previous work [11], and therefore, this coating was favored over the other coating types.

The results of the basic characterization of these coatings are summarized in Table 3. Due to the large difference in properties between the coatings and the plastic substrates, it was not possible to determine any mechanical properties. Therefore, all values were determined on the coatings deposited on AISI 52100 steel substrates.

Table 3. Measured coating properties *.

Coating	Thickness [µm]	Hardness [GPa]	E-Modulus [GPa]	SDF [%]
ta-C	1.2 ± 0.1	37.2 ± 0.6	408 ± 12	10.9 ± 0.4
ta-C:N	1.4 ± 0.1	26.8 ± 1.3	303 ± 23	10.7 ± 0.2
ta:C:B	0.9 ± 0.2	48.1 ± 1.7	567 ± 14	2.5 ± 0.2

* Coatings deposited on polished AISI 52100 steel substrates.

It should be noted that the hardness of the ta-C coating was rather moderate compared to the common literature data for ta-C. This is due to the fact that no bias voltage was used in the processes to accelerate the ions in the coating plasma, as is typically done. In our case, the intrinsic kinetic ion energy of the carbon species in the plasma is crucial for the formation of sp^3 bonds, and hence, the hardness of ta-C. Consequently, the hardness decreases in the case of ta-C:N because the energy of some coating particles was slowed down by collisions with N_2 molecules due to the gas inlet. In the case of ta-C:B, on the other hand, the hardness increased compared to ta-C. This is presumably due to the fact that during the evaporation of B-doped graphite in the arc process, higher particle energies were present than in a pure carbon plasma. This is the subject of current investigations on plasma particle energy distribution.

3.1. Wear Behavior of Different Coating Types

Using parameter set I in tribotesting (see Table 2), the wear behavior of coatings on different substrates and the wear of uncoated steel-ball counter bodies were analyzed. In these tests, the three coating types, ta-C, ta-C:N, and ta-C:B including the 0.1 µm Cr interlayer (see Table 3), were investigated. The wear images are summarized in Table 4.

A volumetric determination of the wear abrasion from the coatings is not possible with the plastic samples due to the significant deformation. Therefore, only a qualitative assessment can be made here. However, it was possible to determine the counter-body wear on the steel balls and the result is summarized in Figure 3.

Table 4. Optical images of wear tracks on the coatings and counter body (steel ball) on different coatings after tribotests with set I.

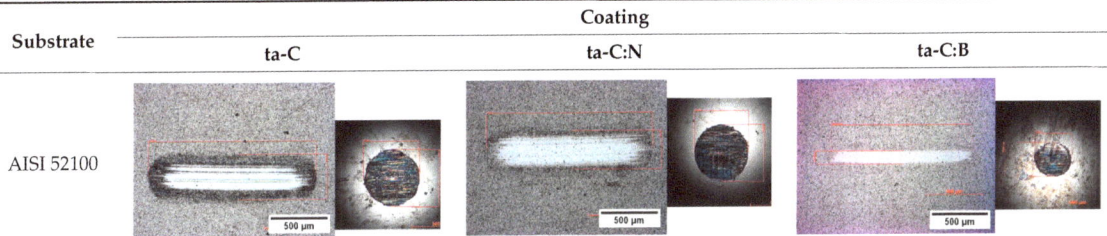

Table 4. *Cont.*

Substrate	Coating		
	ta-C	ta-C:N	ta-C:B
PA12	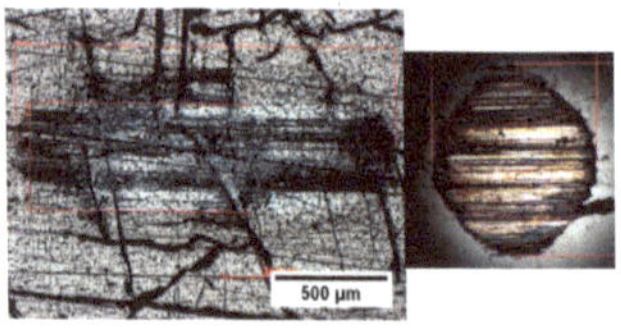	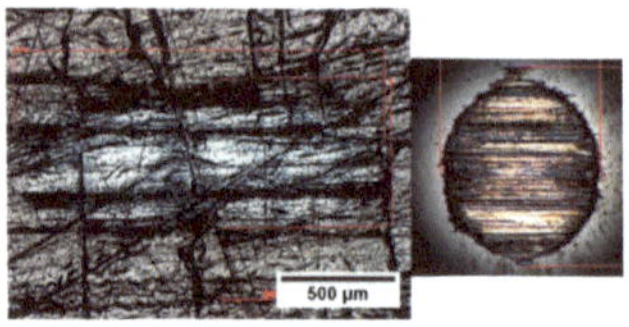	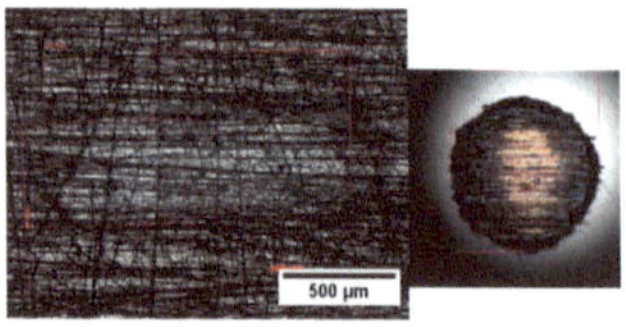
PA12-GF30	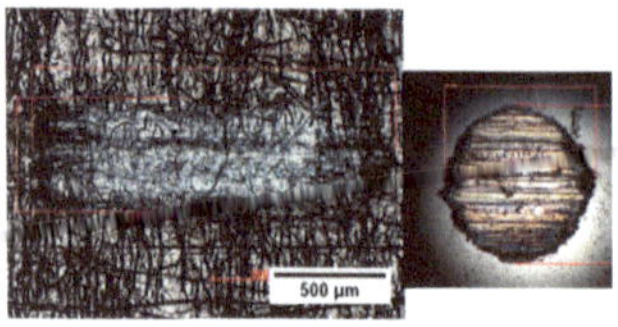	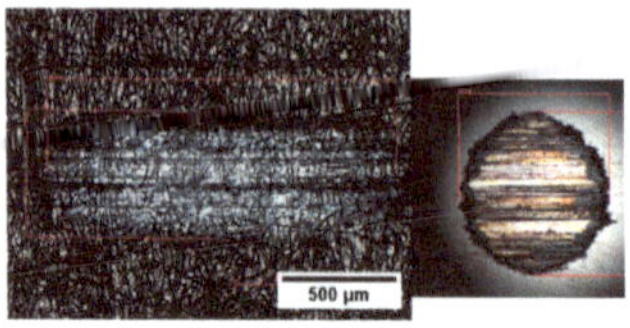	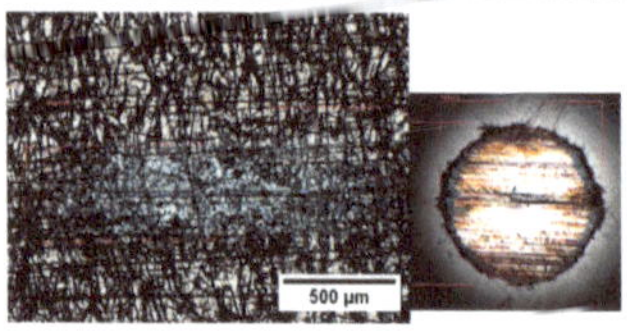
PEEK	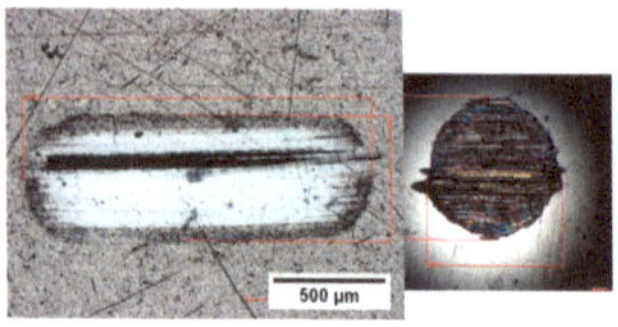	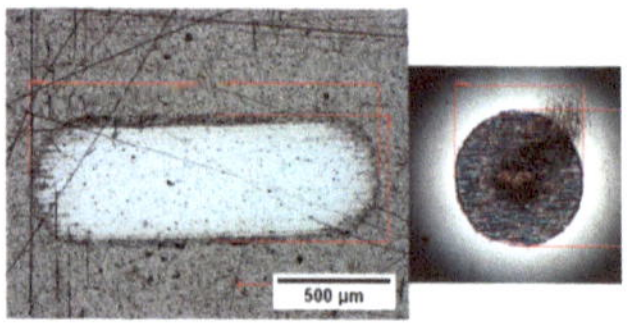	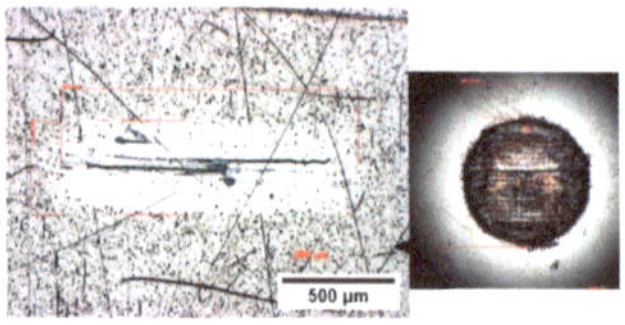
PEEK-CF30	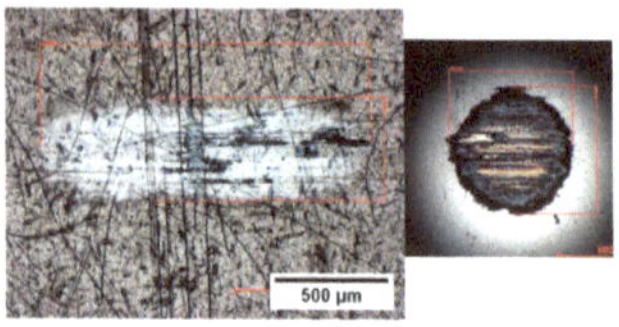	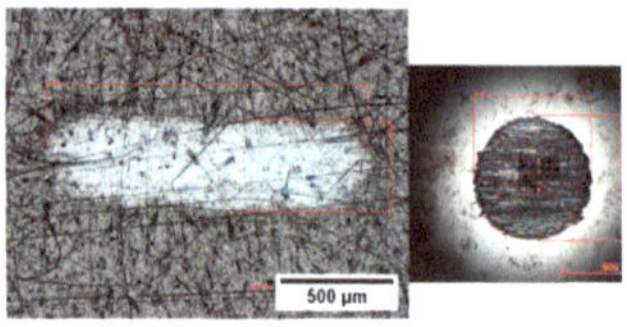	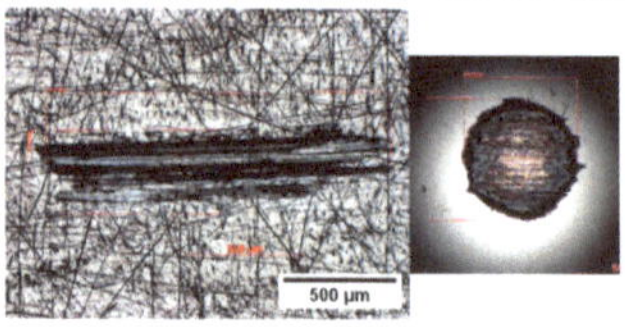

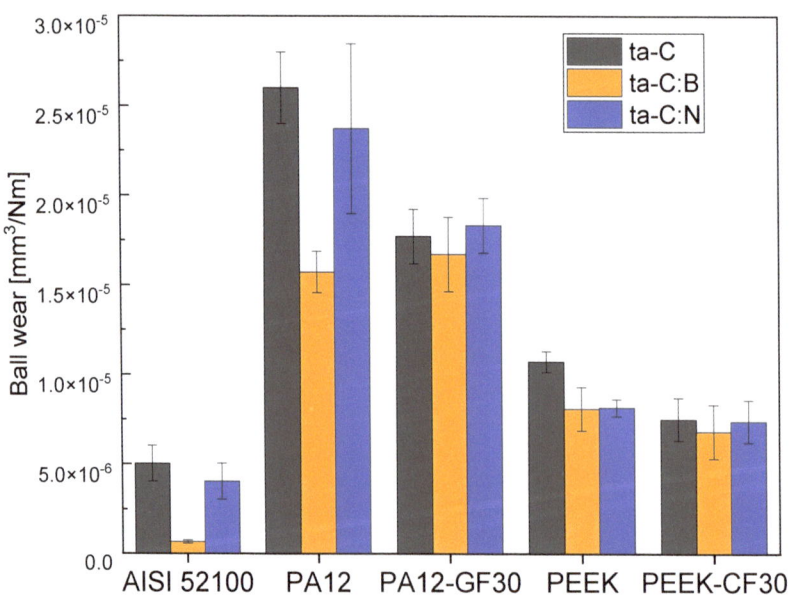

Figure 3. Steel-ball counter-body wear measured after tribotests with set I.

3.2. Influence of the Cr Interlayer on Wear Behavior

The investigations into the influence of the Cr interlayer were carried out with the ta-C:B (1 µm) coating variant. A variant without a Cr layer, a variant with 0.1 µm Cr, and a variant with 0.5 µm Cr were produced and tested in the tribometer. The tribotest parameters were the same as before (set I) and wear images are summarized in Table 5.

Table 5. Optical images of wear tracks on the coatings and counter body (steel ball) on ta-C:B-coatings with different interlayers after tribotests with set I.

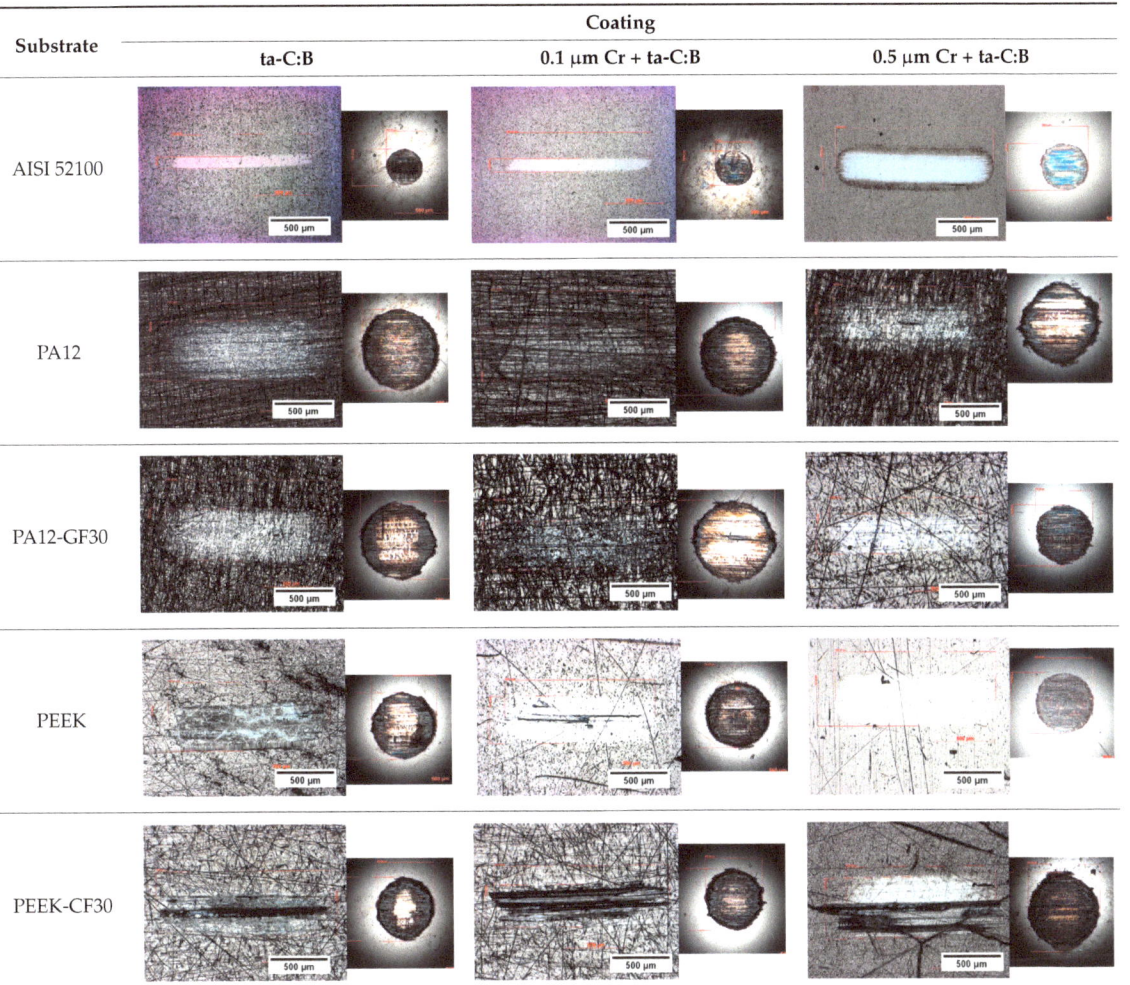

3.3. Wear Behavior under Long-Time Testing Conditions

In this series, using tribotest parameter set II, a reduced normal load but extended testing time of 100 min was used (see Table 2). The aim of this series of tests was to make a comparison between the coating variants under comparatively mild but application-related load conditions. Due to the fact that, in previous investigations, the mating body wear on the PA12 and PA12-GF30 substrates was rather high, these tests were limited to the PEEK and PEEK-CF30 substrates (Table 6).

Table 6. Optical images of wear tracks on coatings and counter body (steel ball) on selected coating–substrate systems after tribotesting with parameter set II.

Substrate	Coating			
	0.1 µm Cr + ta-C	0.1 µm Cr + ta-C:N	0.1 µm Cr + ta-C:B	0.5 µm Cr + ta-C:B
AISI 52100				
PEEK				
PEEK-CF30				

3.4. Comparison of the Load-Bearing Capacity

In this final tribotest, a comparison was made between a selected coated plastic variant with an analogously coated steel substrate and uncoated plastic. The high standard load of 10 N was again applied with the new parameter of set III, which, in contrast to set I, consisted of a test duration of 60 min. In these tests, the characteristics of the friction coefficients over the test duration were also examined.

The coating variant 0.1 µm Cr + 1 µm ta-C:N was selected as a suitable coating for this test. Although this did not have the lowest counter-body wear (like ta-C:B), it proved to be the most robust variant overall in the previous tests. The fact that no grooves or scratches were observed in the coating of ta-C:N after the wear tests was also considered favorable.

It can be seen that the ta-C:N-coated substrates were at a much lower friction level than the uncoated PEEK. This also shows damage that goes beyond abrasive wear removal. The surface appeared heavily deformed in the area of the wear track. Even if the coating wear was lowest on the steel substrate, the wear track of the ta-C:N on the PEEK substrate appeared basically similar in appearance. The coating generally withstood the load, but there were already coating breakthroughs in some areas. On one sample (see dashed arrow in Figure 4), the coating penetration also manifested itself in a sudden increase in the coefficient of friction.

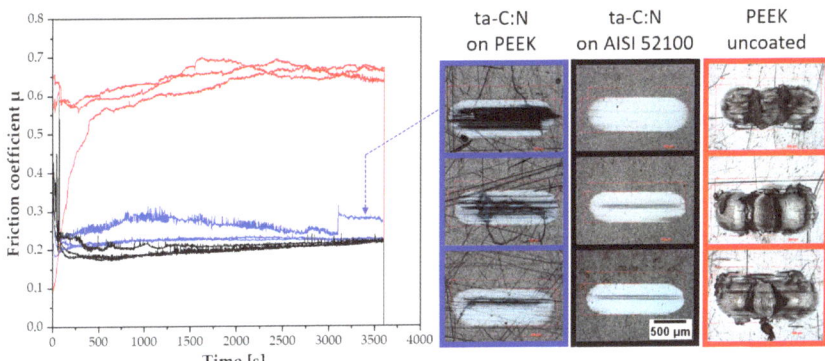

Figure 4. Comparison of friction coefficients (**left**) and coating/surface wear behavior between ta-C:N-coated steel (black) and PEEK (blue) vs. an uncoated PEEK (red) sample (**right**).

4. Discussion

In order to interpret the results of the tribological measurements, it is first necessary to consider the different mechanical contact situations of the various material pairings. The large differences in the modulus of elasticity between AISI 52100 steel and the plastics lead to very different Hertzian contact conditions. For example, the plastics have a much larger contact area, but smaller contact stress at the same load compared to steel. In order to analyze the stress distribution in terms of Hertz's contact theory, the stresses for all materials including a 1 µm thick ta-C topcoat were calculated using FilmDoctor software (SIO® Saxonian Institute of Surface Mechanics, Ummanz, Germany, https://siomec.com/software/filmdoctor-studio/, accessed on 12 September 2024) for the two normal loads of 10 N (parameter sets I and III) and 1 N (set II). The calculations were based on the assumption of purely elastic behavior of all materials. The parameters used for the calculations were the characteristic values of the materials from Tables 1 and 3. The results are shown as examples for steel, PEEK, and PEEK-CF30 in Figure 5.

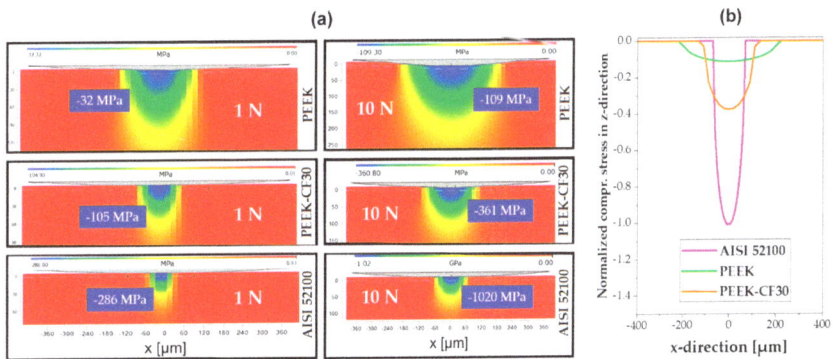

Figure 5. Visualization of the calculated stress distribution of the normal stress σ_z for three different substrate materials with 1 µm ta-C coating vs. steel ball with 1 N and 10 N normal load (**a**) and distribution of σ_z at $z = 0$ in x-directions at 10 N normal load (**b**).

As expected, the results show large differences in the contact surfaces, the level of compressive stress generated in the substrate by the ball indentation, and the different spatial distributions of the stress in the substrate. In this way, it becomes clear why the track widths of the coatings on the plastics in the wear tests are always wider than on the coated AISI 52100 steel (≈500 µm vs. ≈150 µm). It is now also clear why the counter-body wear values (ball abrasion) with the same coatings were significantly higher with the plastics than with steel (see Tables 4 and 5). With the softer substrates, the balls come into contact with a larger area of the abrasive coating surface. It is important to take a closer look at the special features of ta-C-based coatings deposited using arc technology: coatings deposited using a vacuum arc-evaporation process always exhibit unavoidable growth defects due to droplets or macroparticles, which are incorporated into the coating during deposition and increasingly cause roughness peaks on the ta-C-based coatings as the coating thickness increases. The SDF parameter (see Section 3) describes the phenomenon quantitatively. The higher the SDF value, the more abrasive the coating. Two things need to be mentioned in this context: (I) Due to the coating defects, the counter-body wear is initially very high. However, this is primarily a running-in effect. (II) The roughness peaks do not only wear the counter body. By breaking out some defects, a micro-abrasive material is generated, and the coating also suffers wear. The ta-C:B coating is distinguished from the other two coating types by a much lower SDF-value (see Table 3). This phenomenon, described in [14], makes this coating variant more advantageous compared to the other two variants, ta-C and ta-C:N, with high SDF values.

When looking at the results from Section 3.1, the comparison of the coating variants in an initial tribo-screening test, the previously discussed points are particularly important for the interpretation of the results. The counter-body wear is significantly less pronounced with ta-C:B than with the other two coatings. This is particularly evident with the steel variant, but also with PA12. Another noticeable aspect in Figure 4 is that the counter-body wear is much higher for the plastics and especially for the two PA12 variants than for steel. This is explained by the differences in the contact areas discussed above: the softer the substrate, the larger the contact area and the greater the amount of wear particles the steel ball is exposed to in the friction contact. Additionally, regarding the results in Section 3.1, it should be noted that despite the comparatively high load, complete coating failure occurred in only one case (ta-C on PA12). In most other cases of coated plastic substrates, the coatings exhibited grooves, scratches, and cracks in addition to abrasive wear, but remained largely intact. The ta-C:N coatings left the best impression, especially on PEEK and PEEK-CF30. After the tribotest, these coatings were still largely intact and showed a relatively smooth surface in the wear track. The behavior of the ta-C:N coating on the PEEK substrates was still worse than on steel, but it came closest to steel in the appearance of the wear behavior.

To test the influence of the adhesion layer on the tribological performance, the ta-C:B variant was selected due to its lowest coating roughness. The results presented in Section 3.2 show a differentiated picture. Basically, it can be initially said that with all adhesion layer variants, including without a Cr adhesion layer, there appears to be good adhesion strength of the coatings with one exception, the 0.5 µm thick Cr layer on the PEEK-CF30 substrate. In terms of wear behavior, both for the counter body and the coating, the thick adhesion layer (0.5 µm) shows a tendency to improve the tribological performance of the coated plastics. This is attributed to an improved load-supporting effect of the now approximately 1.5 µm thick coating system compared to the other coatings ($\approx$1 µm coating thickness). However, the advantages are not too significant and, as mentioned, in one case, there was also a failure with the 0.5 µm thick Cr adhesion layer.

The focus of the tribotests in the next step was to subject promising coating–substrate combinations to significantly longer test durations under more application-oriented conditions (lower load) with parameter set II (see Table 2). The results in Table 6 with the two PEEK and four coating variants compared to the analogously coated steel show a very good performance of the coatings. Overall, the wear of the counter body and the wear tracks in the coatings were now more similar to the coated steel samples than in the previous tests with the tribo-parameter set I. This is because, in these tests, the running-in process was obviously completed and a more stable tribological situation had established itself, which did not depend as much on the roughness and defects of the coating surface.

In the final tribotest, the overall most-promising plastic-coating system was compared to an uncoated plastic and an analogously coated steel substrate under demanding testing conditions. The PEEK substrate selected for this, with a 1 μm thick ta-C coating, showed similarly low friction values to the coated steel (see Section 3.4). However, it can be seen that the system has reached the limits of its load capacity after the 60 min test. Initial coating perforations indicated that the coating was about to fail, as can be seen in Figure 4. The uncoated PEEK suffered serious damage in the tribotest. Presumably, incipient plastic deformation coupled with abrasive wear could already be observed here.

Overall, it can be stated that the ta-C:N coating greatly improved the load-bearing capacity of the PEEK. The tribological behavior of the system is very similar to that of the ta-C:N-coated steel substrate, even if—as mentioned above—the service life is limited under these harsh conditions.

5. Conclusions

In this article, it is shown that, contrary to expectations, superhard ta-C-based coatings can be used as a tribological protective layer on soft-plastic substrates, even under unlubricated conditions. The adhesion of all the coating variants examined on the four plastic substrates was generally good, with no "eggshell effect" observed.

The ta-C, ta-C:N, and ta-C:B coatings showed similar tribological behavior, caused by the relatively high coating roughness. In all cases, a pronounced running-in behavior with high counter-body wear was observed. The ta-C:B coatings, with their naturally lower defect density, were advantageous in terms of counter-body wear. The studies also revealed that an adhesion-promoting interlayer of Cr is helpful in some cases (better load-supporting effect), although the coatings also work on plastic substrates without a Cr adhesion layer.

Overall, the study concludes that the performance of plastic substrates can be significantly improved by hard or superhard coatings of ta-C, ta-C:N, and ta-C:B, even under harsh tribological conditions, at least for relatively short testing times. Despite the extreme hardness difference and the roughness of the coatings, the systems show promising tribological behaviors in air, coming quite close to that of an analogously coated steel substrate.

In future, there are several possibilities to improve the coating performance due to reducing the defect density (through plasma filtering) and running-in layer concepts, i.e., graded soft top layers that are supposed to mitigate counter-body wear and thus improve the running-in behavior.

Author Contributions: Conceptualization, F.K. and V.W.; investigation, P.N.; writing—original draft preparation, V.W.; writing—review and editing, F.K.; supervision and project administration, F.K.; funding acquisition, V.W. All authors have read and agreed to the published version of the manuscript.

Funding: This research was funded by BMWK (Federal Ministry for Economic Affairs and Climate Action, Germany) within the project Chephren (03EN4005E), and by the Fraunhofer Internal Program SupraSlide under Grant No. PREPARE 840066.

Data Availability Statement: Data available on request due to restrictions.

Acknowledgments: The authors would like to thank Evonik Operations GmbH (Kirschenallee, Darmstadt, Germany) for providing the plastic substrates used in this research. We also acknowledge Stefan Makowski, Fabian Härtwig, and Paul Skorloff from the Fraunhofer IWS for their support during the experimental work.

Conflicts of Interest: The authors declare no conflicts of interest.

References

1. Kano, M. *Overview of DLC-Coated Engine Components*; Cha, S., Erdemir, A., Eds.; Coating Technology for Vehicle Applications; Springer International Publishing: Berlin/Heidelberg, Germany, 2015.
2. Kaczorowski, W.; Szymanski, W.; Batory, D.U.; Niedzielski, P. Tribological properties and characterization of diamond like carbon coatings deposited by MW/RF and RF plasma-enhanced CVD method on poly(ether-ether-ketone). *Plasma Process Polym.* **2014**, *11*, 878–887. [CrossRef]
3. Watanabe, Y.; Suzuki, H.; Nakamura, M. Improvement of the tribological property by DLC coating for environmentally sound high polymer materials. *Int. J. Mater. Prod. Technol.* **2001**, *2*, 787–792.

4. Tomaszewski, P.K.; Pei, Y.T.; Verkerke, G.J.; De Hosson, J. Improved tribological performance of PEEK polymers by application of diamond-like carbon coatings. *Eur. Cells Mater.* **2024**, *27*, 4.
5. Reitschuster, S.; Maier, E.; Lohner, T.; Stahl, K.; Bobzin, K.; Kalscheuer, K.; Thiex, M.; Sperka, P.; Hartl, M. DLC-Coated Thermoplastics: Tribological Analyses under Lubricated Rolling-Sliding Conditions. *Tribol. Lett.* **2022**, *70*, 121. [CrossRef]
6. Rothammer, B.; Marian, M.; Neusser, K.; Bartz, M.; Böhm, T.; Krauß, S.; Schroeder, S.; Uhler, M.; Thiele, S.; Merle, B.; et al. Amorphous carbon coatings for total knee replacements—Part II: Tribological behavior. *Polymers* **2021**, *13*, 1880. [CrossRef] [PubMed]
7. Puertolas, J.A.; Martinez-Nogues, V.; Martinez-Morlanes, M.J.; Mariscal, M.D.; Medel, F.J.; Lopez-Santos, C.U.; Yubero, F. Improved wear performance of ultra high molecular weight polyethylene coated with hydrogenated diamond like carbon. *Wear* **2010**, *269*, 458–465. [CrossRef]
8. Onate, J.I.; Comin, M.; Braceras, I.; Garcia, A.; Viviente, J.L.; Brizuela, M.; Garagorri, N.; Peris, J.L.U.; Alava, J.I. Wear reduction efect on ultra-high-molecular-weight polyethylene by application of hard coatings and ion implantation on cobalt chromium alloy, as measured in a knee wear simulation machine. *Surface Coat. Technol.* **2001**, *142–144*, 1056–1062. [CrossRef]
9. Wang, H.; Xu, M.; Zhang, W.; Kwok, D.T.K.; Jiang, J.; Wu, Z.; Chu, P.K. Mechanical and biological characteristics of diamond-like carbon coated poly aryl-ether-ether-ketone. *Biomaterials* **2010**, *31*, 8181–8187. [CrossRef] [PubMed]
10. Rothammer, B.; Schwendner, M.; Bartz, M.; Wartzack, S.; Böhm, T.; Krauß, S.; Merle, B.; Schroeder, S.; Uhler, M.; Kretzer, J.P.; et al. Wear Mechanism of Superhard Tetrahedral Amorphous Carbon (ta-C) Coatings for Biomedical Applications. *Adv. Mater. Interfaces* **2023**, *10*, 2202370. [CrossRef]
11. WIX WIXSTEEL Industrial, Product Information. Available online: https://www.wixsteel.com/products/alloy-steel-bar/ball-bearing-steel-bar/100cr6 (accessed on 12 September 2024).
12. EVO Evonik Industries AG, Product Information. Available online: https://www.plastics-database.com (accessed on 12 September 2024).
13. Kaulfuss, F.; Weihnacht, V.; Zawischa, M.; Lorenz, L.; Makowski, S.; Hofmann, F.; Leson, A. Effect of Energy and Temperature on Tetrahedral Amorphous Carbon Coatings Deposited by Filtered Laser-Arc. *Materials* **2021**, *14*, 2176. [CrossRef] [PubMed]
14. Krülle, T.; Kaulfuß, F.; Weihnacht, V.; Hofmann, F.; Kirsten, F. Amorphous Carbon Coatings with Different Metal and Nonmetal Dopants: Influence of Cathode Modification on Laser-Arc Evaporation and Film Deposition. *Coatings* **2022**, *12*, 188. [CrossRef]
15. DIN EN ISO 26423:2016-11; Fine Ceramics (Advanced Ceramics, Advanced Technical Ceramics)—Determination of Coating Thickness by Crater-Grinding Method. International Organization for Standardization: Geneva, Switzerland, 2016.
16. DIN EN ISO 14577-4:2017-04; Metallic Materials—Instrumented Indentation Test for Hardness and Materials Parameters—Part 4: Test Method for Metallic and Non-Metallic Coatings. International Organization for Standardization: Geneva, Switzerland, 2017.
17. Lorenz, L.; Chudoba, T.; Makowski, S.; Zawischa, M.; Schaller, F.; Weihnacht, V. Indentation modulus extrapolation and thickness estimation of ta-C coatings from nanoindentation. *J. Mater. Sci.* **2021**, *19*, 3. [CrossRef]
18. DIN 51834-1:2010-11; Testing of Lubricants—Tribological Test in the Translatory Oscillation Apparatus—Part 1: General Working Principles. German Institute for Standardization: Berlin, Germany, 2010.
19. DIN EN ISO 25178-2:2023-09; Geometrical Product Specifications (GPS)—Surface Texture: Areal—Part 2: Terms, Definitions and Surface Texture Parameters. International Organization for Standardization: Geneva, Switzerland, 2023.
20. Krülle, T.; Peritsch, P.; Kaulfuß, F.; Bui Thi, Y.; Zawischa, M.; Weihnacht, V. Investigation of surface defects on doped and undoped carbon coatings deposited by Laser Arc Technology using an optical surface quantification method. *Jahrb. Oberflächentechnik* **2021**, *77*, 233–247.

Disclaimer/Publisher's Note: The statements, opinions and data contained in all publications are solely those of the individual author(s) and contributor(s) and not of MDPI and/or the editor(s). MDPI and/or the editor(s) disclaim responsibility for any injury to people or property resulting from any ideas, methods, instructions or products referred to in the content.

Article

Running-In of DLC–Third Body or Transfer Film Formation

Joachim Faller and Matthias Scherge *

Fraunhofer IWM MikroTribologie Centrum, Rintheimer Querallee 2b, 76131 Karlsruhe, Germany; joachim.faller@iwm-extern.fraunhofer.de
* Correspondence: matthias.scherge@iwm.fraunhofer.de; Tel.: +49-721-2043-2712

Abstract: Amorphous carbon coatings are widely used due to their beneficial friction and wear characteristics. A detailed understanding of their behavior during running-in, apart from model tribosystems, has yet to be obtained. Multiple analytical methods were used to detect the physical and chemical changes in a ta-C coating and its thermally sprayed, metallic counterpart after a running-in procedure with pin-on-disk experiments. Both coatings exhibited changes in their surface and near-surface chemistry. The mechanisms in and on the metallic coating were identified to be a mixture of the third-body type, with the formation of gradients in the microstructure and chemistry and an additional carbon-rich tribofilm formation on top. The ta-C coating's changes in chemistry with sp^2 enrichment and lubricant element inclusions proved to be too complex to allocate them to tribofilm or third-body formation.

Keywords: running-in; lubricated sliding; ultra-low wear; DLC; third body

1. Introduction

Even though amorphous carbon coatings are already advanced (e.g., DuroGlide from Federal Mogul), the processes and mechanisms of the running-in are still largely not understood. Research has focused strongly on the mechanisms of model systems like diamond friction and wear behavior driven by amorphization and rehybridization to sp^2 bonding [1]. Another important role is played by the surface passivation by hydroxyls under water lubrication, which inhibits cold welding and thereby reduces friction [2–4]. Furthermore, recent discoveries suggest similar mechanisms of passivation for glycerol-lubricated DLC–DLC contacts [5] and a strong dependency between the lubricant chemistry and wear beahvior of super hard ta-C coatings [6,7]. The wear of hard ta-C coatings on the asperity scale shows a similar mechanism to that of the amorphization of diamonds [8,9]. Further studies with advanced analytical methods into the wear mechanism of DLC lubricated with oleic acid indicated the formation of graphene oxides on the surface [10]. Closest to the applications outlined in this publication is those of a study focused on the interplay of ZDDP additives with hard DLC coatings and the formation of tribofilms using XPS and TEM [11].

This is astonishing, as this knowledge represents a major lever for the application and optimization of tribosystems. This article therefore deals with the running-in behavior of amorphous carbon coatings in a lubricated tribological system with a metallic counter body. It focuses on the tribochemical changes happening during running-in and tries to discuss the prevalent mechanisms with regard to the well-known third-body model from Godet and Berthier [12–14] and to the model of tribofilm formation on surfaces [15,16].

Tribological tests are often carried out with high loads and pressures in order to generate wear in a short time. Since the transferability of the results to practice is questionable, samples generated under realistic boundary conditions were analyzed in this work that were energetically similar to the real system, see [17]. Furthermore, no model systems were used; instead, materials and lubricants with industrial applications were considered. This work is expressly not intended as research related to combustion engine technology but as a fundamental contribution to the friction- and wear-reducing use of amorphous carbon coatings and thus to a general increase in the efficiency of industry.

Citation: Faller, J.; Scherge, M. Running-In of DLC–Third Body or Transfer Film Formation. *Lubricants* **2024**, *12*, 314. https://doi.org/10.3390/lubricants12090314

Received: 30 July 2024
Revised: 19 August 2024
Accepted: 26 August 2024
Published: 4 September 2024

Copyright: © 2024 by the authors. Licensee MDPI, Basel, Switzerland. This article is an open access article distributed under the terms and conditions of the Creative Commons Attribution (CC BY) license (https://creativecommons.org/licenses/by/4.0/).

2. Tribological Measurements and Physico-Chemical Analysis

2.1. Tribometry

The tribological results performed in this study were generated from a highly modified pin-on-disk tribometer (SST) from Tetra GmbH (Ilmenau, Germany). As the name suggests, the tribological system consisted of a rotating disk and an eccentrically mounted cylindrical pin. The disk was driven by a motor via a toothed belt at an angular speed ω, and the pin was positioned radially and vertically by two stepper motors. A normal force F_N was applied by pretensioning a spring and, like the frictional force F_R, was measured via a multiaxis force sensor from ME Meßsysteme using strain gauges. Contact was continuously supplied with oil heated to temperature ϑ via a circulating lubrication system. The oil circuit was designed for use with radionuclide technology, which is described more in detail in [18].

2.2. Topography Measurement

The topography of the samples used was recorded with a confocal white light interferometer (WLI) (Bruker ContourGT-K, Bruker Corp., Billerica, MA, USA). Two magnifications (10× and 50×) were used to measure the surface. The former was used to determine shape deviations, waviness, and line-based roughness parameters on an area of 5×5 mm^2, and the latter was used to determine the surface-related roughness on an area of 100×100 µm^2.

2.3. Determination of the Chemical Composition near the Surface

X-ray photoelectron spectrometry (XPS) was carried out using a PHI 5000 Versaprobe II (ULVAC-PHI, Chigasaki, Japan)). In combination with an argon ion beam, it was possible to measure the element concentrations with depth resolution by ablating the surface step by step and then recording the spectrum. The ablation rate was calibrated to 2 nm/min on a Si-SiO$_2$ reference sample. For sputtering the DLC samples, the same procedure was used on a carbon coating on a silicon substrate reference sample provided by IWS. The removal of the 56 nm thick coating, which was measured in advance using X-ray reflectometry (XRR) at IWS, was carried out up to the SiC interface in 112 min. This corresponded to a removal rate of 0.5 nm/min. The lateral measuring range of the XPS was limited to a circle within a diameter of 200 µm. The photon energy of the Al-Kα X-ray source was 1486.7 eV, and the energy resolution was 0.2 eV.

2.4. Elastoplastic Material Characterization

A Hysitron Triboindenter TI 950 (Bruker Corp., Billerica, MA, USA) was used for elastoplastic material characterization. A cube corner was chosen as the tip, as this offered the smallest opening angle and tip radius with a self-similar tip shape [19]. The indenter and its tip were used as an AFM to image the surfaces of the sample and to position the tip. The loading function for the quasistatic indents went up to 1 mN with a constant strain rate and subsequent hold time to correct for creep. The area function of the tip is necessary for converting the measured forces and displacements into stresses and strains and then into modulus of elasticity and hardness. This requires the indentation area as a function of the indentation depth. The standard method for this is the Oliver–Pharr method on a test specimen with known and constant properties (usually quartz glass) [20]. A finer determination of the area function is possible by means of self-imaging of the tip, as shown by Saringer et al. [21]. For this purpose, a TGT1-AFM reference grating, which consisted of sharp needles and thus enabled in situ imaging of the tip, was scanned. The cross-sectional area was then calculated from the actual shape depending on the indentation depth [21].

2.5. Determination of Carbon Hybridization

Electron energy loss spectroscopy (EELS) makes use of the inelastic interaction of an electron beam with the electrons in a sufficiently thin sample [22]. Inelastic scattering can have several causes and accordingly occurs with different energy losses in the spectrum. In the low-loss range, for example, these are plasmons, while at higher loss energies, the electrons of the inner shells are excited and lead to a clearly detectable ionization edge [23]. EELS is one of the oldest methods of DLC analysis and is based on the evaluation of the carbon K-edge [24]. Here, the sp^2 content is determined via the intensity ratio of the π^* and σ^* states [25]. By considering the plasmon losses, relative changes in sample thickness are corrected [26,27]. Due to the TEM preparation, no exact sp^2 content of DLC coatings can be determined using EELS, as this is systematically overestimated [25,27]. Nevertheless, a semiquantitative statement and precise determination of sp^2 gradients in carbon-based materials is possible due to the high lateral resolution. An FEI Titan TEM (Thermo Fisher Scientific, Waltham, MA, USA) equipped with a Gatan Imaging Energy Filter Tridiem model 865 HR was used for the EELS measurements. The titanium microscope was operated at 300 kV in STEM mode to minimize radiation damage.

2.6. Nanostructural Investigation

Transmission electron microscopy (TEM) and high-resolution TEM (HRTEM) are powerful tools for investigating changes due to tribological loading [28]. In this work, both were used to observe the DLC near-surface volume, as a higher resolution and a much higher contrast than in an FIB cross-section in SEM are possible. The preparation of the lamellae was carried out with Ga^+-FIB, as the damage induced in the DLC caused by Ga^+ is well documented in the literature [29,30]. The fast Fourier transform (FFT) of HRTEM images enables an analysis of the near and far order, respectively, to separate them and thus make crystal orientations recognizable [25,31].

2.7. Chemical and Physical Analysis

Raman spectroscopy is a nondestructive method for the chemical and physical analyses of materials. It is widely used in the characterization of (amorphous) carbon coatings. This measuring method is based on the eponymous Raman effect, in which light is inelastically scattered by molecules. This energy transfer can be quantified using the wavelength shift of the scattered light. In the case of carbon, the D maximum at 1350 cm^{-1} and the G maximum at (1580–1600) cm^{-1} were used. These two modes are mainly influenced by the order of the sp^2-hybridized regions. A derivation of the sp^3 content of amorphous carbon is only possible indirectly [32]. For this purpose, their intensity ratio and the position of the G peak are used:

$$\Gamma = \frac{I(D)}{I(G)} \qquad (1)$$

In this study, the intensity ratio was calculated from the ratio of the heights of the individual maxima. For this purpose, the measured Raman spectrum in the relevant range was approximated with the sum of a Cauchy–Lorentz distribution for the D peak and a Breit–Wigner–Fano line for the G peak using the least squares method [32]. The background resulting from the fluorescence of the sample was removed via polynomial regression. An inVia confocal microscope (Renishaw, Wotton-under-Edge, UK) with a laser wavelength of 532 nm and 20 and 100× optical magnification was used. This resulted in a minimum lateral resolution of 1 µm, with the theoretical value based on a numerical aperture (NA) of 0.9 and the Rayleigh criterion being 361 nm. As Raman spectroscopy is not surface-sensitive, it was only used to characterize the carbon on the metallic counterpart. The surface was mapped using a grid and classified according to the occurrence of the characteristic carbon maxima.

3. Materials of the Tribosystem

3.1. Iron Spray Coating

Iron spray coating is a standard material for cylinder liners. This coating is applied to aluminum crankcases in a series production process using arc wire spraying (TWA) of two 13Mn6 wires [33].

The material for the tribometer tests was taken directly from an unhoned cylinder liner. The separated strips were ground plane-parallel, and the 5 mm pins were created from them using water jet cutting. Finishing was carried out in a cup grinding machine via a process similar to honing. This meant that both the coating thickness and the roughness were based on the honed cylinder bore. The Abbot parameters averaged over all samples were Svk 819 ± 160, Sk 656 ± 236, and Spk (232 ± 49) nm. The edges of all pins were rounded off manually due to possible burrs, which is why the actual diameter was 4.95 mm. Subsequently, all pins were measured in the WLI, and samples with burrs in general, as well as with waviness and shape deviations greater than 1 µm, were excluded in order to enable reproducible, full-surface contact in the tribometer.

3.2. DLC

Coatings with different sp^3 contents were deposited onto ground 100Cr6 disks using a PVD vacuum process at the Fraunhofer IWS. The so-called LaserArc process, in which a laser scans the graphite cathode and vaporizes the carbon, was used. A detailed description of the coating procedure, including substrate preparation and machine parameters, can be found in [34]. The plasma was filtered to reduce the particle implantation and thus the defect density [35]. Despite filtering, the coatings exhibited a large number of growth defects, especially within the investigated coating thickness range of (5.1–6.3) µm, measured in a calotte cut by IWS. These defects in turn led to the formation of a rough surface dominated by spikes (firmly anchored) and droplets (weakly bonded). These surfaces were not suitable for tribological use, which is why they had to be smoothed [36]. Two different methods were used for this purpose. One was short-stroke honing using a diamond belt, and the other was brush smoothing. The first method is also known as "superfinishing" and uses high-frequency oscillation perpendicular to the feed direction of the belt and while rotating. The resulting surface roughness is displayed in Table 1. This article focused on a ta-C coating with a Young's modulus of 523 GPa, a hardness of 50 GPa, and an sp^3 content of 66%.

Table 1. Abbot–Firestone roughness values for the tested DLC coatings. Index a denotes coatings in as-deposited state, g for ground state, and b brushed smoothed state.

Coating	Svk	Sk	Svk
ta-C_a	97	70	245
ta-C_{b1}	263	365	145
ta-C_{b2}	14	14	11
ta-C_g	68	150	110

3.3. Lubricant

All experiments in this work were carried out with the same lubricant at an identical temperature (80 °C). This was a fully additive engine oil of SAE viscosity class 0W20 from Fuchs Petrolub. On the one hand, this oil contained zinc dialkyl dithiophosphate (ZnDTP), which is a common additive for wear protection [16,37]. On the other hand, molybdenum dithiocarbamate (MoDTC) was used as a typical friction modifier [38]. The elemental composition of the oil was analyzed by ICP-OES.

4. Results and Discussion on the Differentiation between Third Body and Tribofilm

Based on the mechanisms of third body formation in metal-metal tribosystems, we discussed whether DLC is subject to a comparable change in the running-in or whether the tribochemical change is limited to a film structure. The behavior of the mating body was also considered in the discussion, as part of the film or third body also remains on the mating body when the contact is separated.

4.1. Tribometry

All coatings were tested in the same manner, which means that at least three experiments were conducted. The procedure included a preliminary test in which a parameter field dictated by the Stribeck curve was defined. Starting with low contact pressure and high velocity, followed by a decrease in velocity moving from hydrodynamic lubrication to boundary lubrication, the load was increased until the maximum load of the pin-on-disk-tribometer of 980 N was reached or the simultaneously measured wear rate went beyond 2 µm/h. From the initial experiment, so-called key stress levels were defined, which were in turn used for the parameterization in a new experiment. The procedure itself as well as the resulting coefficients of friction and wear rates are displayed and discussed in detail in [17].

The results of the tribometrical testing of the differently smoothed ta-C coatings are shown via Stribeck curves after running-in in Figure 1. To facilitate the comparison, all CoFs are shown independent of the pseudo λ calculated from the Hersey parameter and the resulting combined roughness Sq in the following manner:

$$\lambda^* = \frac{\eta v}{p S_q}. \qquad (2)$$

By doing so, the differences in the resulting roughness of the coatings and their counterbody were omitted from the comparison of friction coefficients under mixed lubrication. To give the reader an idea as to what happens during running-in, the results of the testing of the ground ta-C coating, which had the lowest CoFs in the Stribeck curve, are displayed in Figure 1 on the right. The test parameters were chosen in the same way as in the previously cited publication, with an initial contact pressure of 52 MPa and a relative velocity 1 m/s. Initial friction was high (CoF 0.9) and so were the linearized wear rates, which reached values of up to 1.5 µm/h. Passing the key stress levels, the CoF decreased below 0.01 after around twelve hours of stressing, and the pin wear rate stabilized below 10 nm/h.

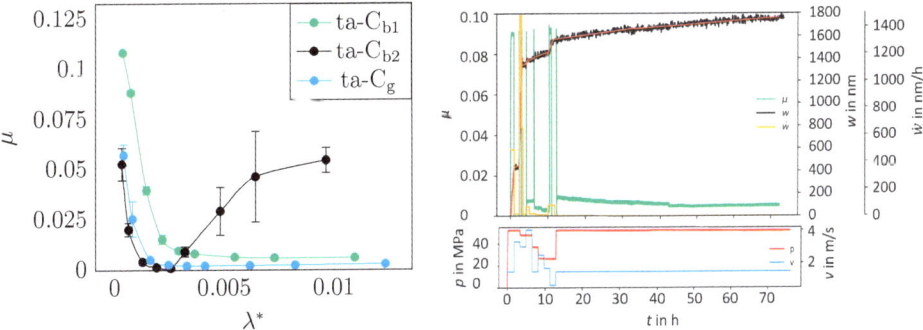

Figure 1. Stribeck curves acquired after running-in with coefficient of friction μ over pseudo λ (**left**) and the running-in of the ta-C_g coating with CoF in green, measured wear from the pin in black, and calculated wear rates in yellow (**right**).

All following investigations were carried out on specimens from the presented pin-on-disk experiment for a concise and focused discussion.

4.2. Gradient Formation

For the 13Mn6 spray coating, the formation of a gradient in the microstructure below the surface and its correlation with the wear rates were detected and are shown in Figure 2 on the left and are compared to those of the pristine material (right). The FIB cross-sections were placed in parallel to both the grooves and on top inside a ridge line.

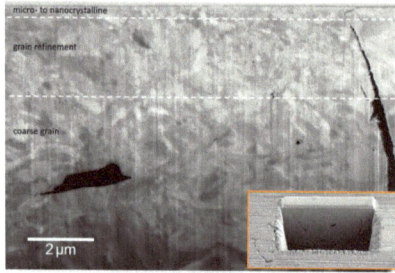
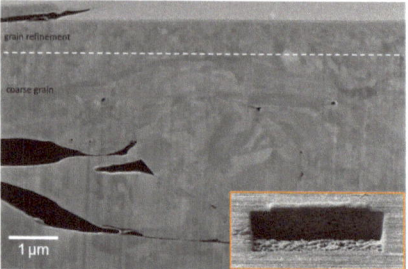

Figure 2. Near-edge microstructure of the spray coating after running-in against ta-C on the (**left**) and for comparison from an unworn, ground specimen on the (**right**).

The characteristics of near-surface microstructure are described in the literature [15,28]. A gradient formed in the grains from the microcrystalline structure above the bulk grains (between the lower and middle area, horizontal line) to the nanocrystalline near-surface region (above the upper line). This is in good agreement with the spray coatings that were rubbed against a metallic counter body [39], grey cast iron against chromium plated pins [40], and for steel–steel pairings under similar conditions [41]. This below-surface structure does not always develop in the same way but is influenced by the hardness of the DLC and its final processing with regard to grain size and the extent of the gradient.

The XP spectra and depth profiles (shown in Figure 3) were used to detect additive components from the wear protection (zinc and phosphorus) and the friction modifier (molybdenum and sulphur), which were present below the surface in decreasing concentrations. This form of gradient suggests the formation of a third body of natural and artificial material flow rather than an adsorbed film.

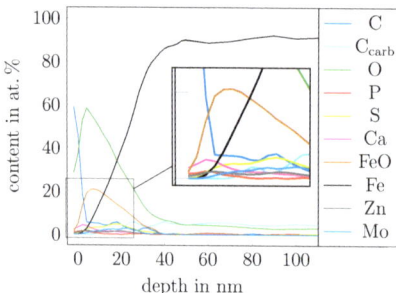
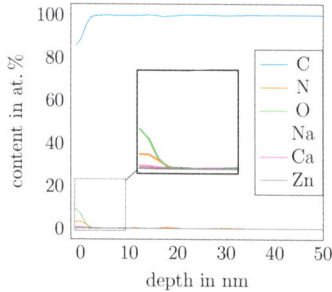

Figure 3. XPS depth profiles from a 13Mn6-coated pin (**left**) after running-in against a ta-C coating (**right**).

The search for gradients in amorphous carbon coatings was more difficult, as the tribochemically altered near-edge volume was a maximum of 100 nm in thickness. For the ta-C coating, the XPS indicated a tribochemical interaction only up to a depth of 20 nm. However, the gradients in XPS depth profiles are only of limited significance for two reasons. Firstly, they are superimposed by the sp^3-rich CH_x contamination of the surface. Secondly, the information depth of the C1s photoelectron is just under ten nanometers. However, this is not in the form of a uniform distribution but of a logarithmic decrease in the exit probability with increasing depth [42]. Ten nanometers represents a limit with the 99% information quantity, corresponding to 3σ of a normal distribution. As such, a thin film (20 nm) cannot be separated from a gradient at all with XPS and C1s evaluation.

The results of the nanoindentation, plotted as Young's modulus over height (ergo sample topography) in Figure 4a, must also be viewed critically, as the tip of the nanoindenter measures the elastic interaction with the material underneath and around the indentation surface, even at the smallest indentation depths. This means that the elastic modulus measured there, which is linearly related to the sp^3 content, is not suitable for determining the shape of a tribochemically modified volume, especially as the average value of (347 ± 17) GPa inside the wear track is not significantly different from the values outside $((340 \pm 14)$ GPa) of it. The only difference can be found in the slopes of the linear regression for both datasets, which indicate a lower Young's modulus on high points inside the wear track.

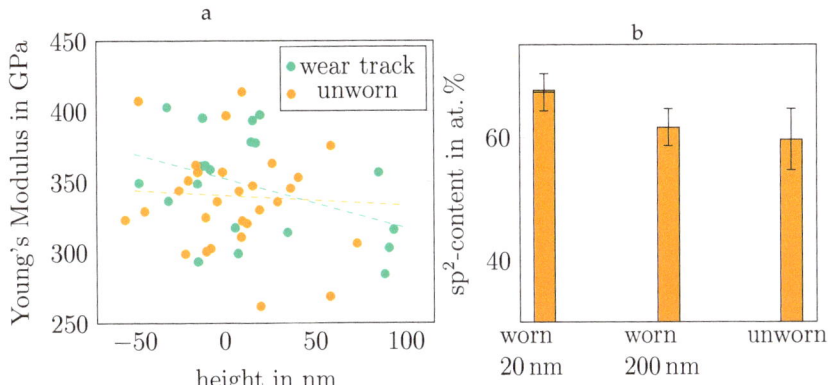

Figure 4. Comparison of the Young's moduli over measured topography height acquired on an unworn area on a ta-C disk in orange and data from inside the wear track in green (**a**) and EELS-measured sp^2 contents on two lamellae prepared from the wear track with windowing in two different depths and from an unworn area on the same DLC disk (**b**).

Although the results of the investigation using EELS in Figure 4b shows significant differences in the sp^2 content of the ta-C coating, in good agreement with the XPS, it is not possible to make a statement about the gradients due to the windowing that occurred when the lamella was scanned. Only the observations of the near-surface area using HRTEM and bright-field TEM in Figure 5 provide information about its composition. The original specimen surface is marked in both pictures with a white dashed line. Both lamellae show platinum contamination to some extent, even though the lamella from the worn ta-C sample was sputter-coated with gold beforehand. The lighter areas below the surface are sharp-edged (indicated with black dashed line) in the manner of a film. At first glance, this appears to be an argument in favor of the structure of a tribofilm.

In detail, however, there are inconsistencies with this model. With the surface showing traces of plastic flow in the SEM image and the presence of a metallic calcium phase (a Ca-L-edge was detected at 345 eV via EELS) below the surface, marked with the black dashed line in the HRTEM image in Figure 6, it could be assumed that the near-surface zone formed by mechanical mixing in frictional contact and thus was analogous to the formation of a third body. Salinas Ruiz et al. came to a similar conclusion. They used HRTEM and EDX to demonstrate the formation of a third body by incorporating the wear particles of the ta-C coating and lubricant components, in particular sulphur and zinc [11]. A tribofilm would form on the surface in the form of an adsorbent and would not allow the incorporation of material under the surface. The formation of a tribofilm in the form of graphene oxide, as demonstrated by Barros Bouchet et al. in [10], could not be demonstrated, as the ta-C coating was amorphous up to the surface.

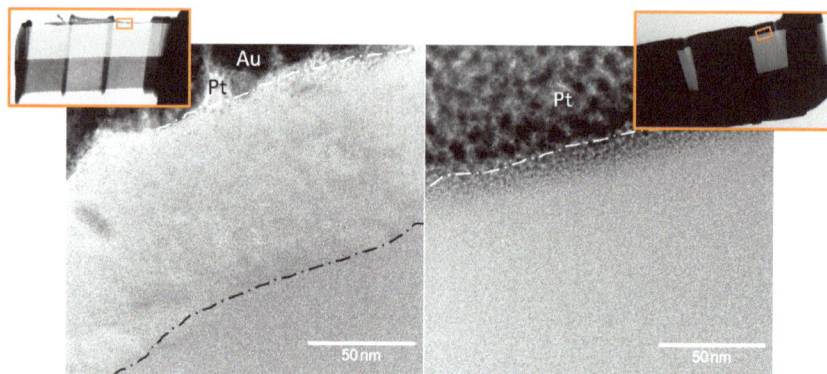

Figure 5. HRTEM image of the ta-C lamellae from under the surface of the wear track on the (**left**) and from an unworn area on the (**right**). The original surface is marked with a white dashed line.

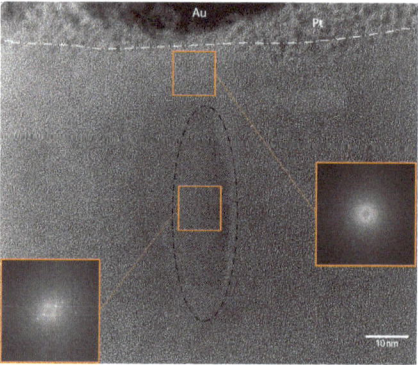

Figure 6. Magnified acquisition of Figure 5 left, directly beneath the surface, with the FFT of an amorphous area on the right and of an area with a long distance order (marked with the black dashed line) on the lower left.

4.3. Material Transfer

A third body comprises the material flows from the first two bodies. Since the third body remains on the two first bodies when the contact is released, material transferred from the other first body should be present on their surface. On the 13Mn6 spray coat, the detection of carbon transfer was not trivial, as this had to be separated from the lubricant residues in the form of CH_x contamination. The XPS shows a depth of approximately 5 nm in all depth profiles, which was dominated by sp^3-hybridized carbon. Below this, there was generally a gradient in the carbon content. The results of Raman spectroscopy in Figure 7 show a film-like distribution of carbon, with local variations in the D- and G-peak intensity ratio Γ and correlated G shift.

The average Γ value is 0.97 ± 0.05, with an average G shift of (1572 ± 1) cm^{-1}. The local variation in both values increased to an intensity ratio of 3.9 and a G shift of 1540 cm^{-1}. The average values indicated that the carbon film was between nanocrystalline graphite and a-C with an sp^2 content of 90%, according to [43]. The deviating spots should be treated with caution, since the deviations did not align with the visible geometry (patches, holes, and grooves) even though the marker size in the scatter plot was scaled according to the approximate laser spot diameter based on the airy disk calculated from the wavelength and NA. Additionally, the values measured at those spots with low G shifts and high Γ values do not match the data points from Ferrari and Robertson (neither for amorphous carbon nor for hydrocarbons) [32,43].

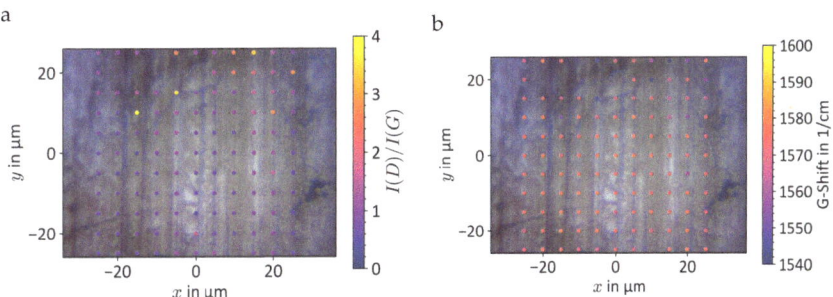

Figure 7. Raman mapping of a 13Mn6 pin after running-in against ta-C with intensity ratio Γ (**a**) and G shift (**b**). Values outside the limits were excluded.

On the DLC side, it was much easier to prove a transfer from the 13Mn6 spray coating by measurement. As this coating consisted largely of iron, it should be possible to detect iron on or in the DLC, be it in the form of oxides, carbides, or metallic iron. As all XPS measurements showed, no transfer of iron in any form could be detected. This is therefore a contradiction to the model concept of the third body, as this should consist of both internal material inflows Q_i as a natural third body.

5. Conclusions

In summary, it can be stated that tribochemical changes in near-surface volume cannot simply be attributed to the phenomenon of the formation of a third body or the formation of a tribofilm. For the metal side, there is evidence in the material that gradients develop in the microstructure and in the chemistry, as is typical for third bodies in metallic tribosystems. Furthermore, XPS, SEM, and Raman show that an additional carbonaceous tribofilm builds up. The amorphous carbon coatings appear to participate only to a limited extent in the formation of a third body. There is no transfer from the metallic counter body. However, there is the incorporation of foreign material, plastic flow, and rehybridization of the carbon. Thus, the tribological behavior of DLC–metal systems can neither be clearly assigned to the path of run-in via the formation of a third body nor to the formation of a transfer film.

A further investigation of the near-surface volume in the DLC requires in-depth measurements. The resolution of the possible sp^2–sp^3 gradient requires a smaller windowing in the EELS measurement, which will, however, be accompanied by greater measurement uncertainty. Furthermore, a qualification of the distribution of the tribochemical modification using EDX on the same TEM lamella would be desirable.

Gaining a deeper understanding of tribofilm formation on iron spray coating requires a similar approach. A three-dimensional mapping of the carbon hybridization state could be acquired using Auger electron spectroscopy coupled with a sputter etching technique that does not alter carbon bonding, like argon gas cluster ions.

Author Contributions: Conceptualization, M.S.; formal analysis, J.F.; funding acquisition, M.S.; investigation, J.F.; methodology, J.F.; project administration, J.F.; writing—original draft, J.F. and M.S. All authors have read and agreed to the published version of this manuscript.

Funding: This research was funded by Bundesministerium für Wirtschaft und Klimaschutz grant number 03ET1609.

Data Availability Statement: Dataset available on request from the authors.

Acknowledgments: The authors would like to thank Frank Kaulfuß and Stefan Makowski (Fraunhofer IWS, Dresden) for providing the DLC coatings. From Fraunhofer IWM, we thank Philipp Daum and Dominic Linsler for XPS and SEM. Erich Müller and Reinhard Schneider from the Laboratory for Electron Microscopy (LEM KIT, Karlsruhe) are thanked for the TEM preparation and EELS measurements.

Conflicts of Interest: The authors declare no conflicts of interest.

References

1. Pastewka, L.; Moser, S.; Gumbsch, P.; Moseler, M. Anisotropic mechanical amorphization drives wear in diamond. *Nat. Mater.* **2011**, *10*, 34–38. [CrossRef] [PubMed]
2. Tzeng, Y. Very low friction for diamond sliding on diamond in water. *Appl. Phys. Lett.* **1993**, *63*, 3586–3588. [CrossRef]
3. Zilibotti, G.; Corni, S.; Righi, M.C. Load-Induced Confinement Activates Diamond Lubrication by Water. *Phys. Rev. Lett.* **2013**, *111*, 146101. [CrossRef] [PubMed]
4. Kim, J.E.; Choi, J.I.J.; Kim, J.; Mun, B.S.; Kim, K.J.; Park, J.Y. In-Situ Nanotribological Properties of Ultrananocrystalline Diamond Films Investigated with Ambient Pressure Atomic Force Microscopy. *J. Phys. Chem. C* **2021**, *125*, 6909–6915. [CrossRef]
5. Kuwahara, T.; Romero, P.A.; Makowski, S.; Weihnacht, V.; Moras, G.; Moseler, M. Mechano-chemical decomposition of organic friction modifiers with multiple reactive centres induces superlubricity of ta-C. *Nat. Commun.* **2019**, *10*, 151. [CrossRef]
6. Makowski, S.; Weihnacht, V.; Schaller, F.; Leson, A. Ultra-low friction of biodiesel lubricated ta-C coatings. *Tribol. Int.* **2014**, *71*, 120–124. [CrossRef]
7. Makowski, S.; Schaller, F.; Weihnacht, V.; Englberger, G.; Becker, M. Tribochemical induced wear and ultra-low friction of superhard ta-C coatings. *Wear* **2017**, *392-393*, 139–151. [CrossRef]
8. Kunze, T.; Posselt, M.; Gemming, S.; Seifert, G.; Konicek, A.R.; Carpick, R.W.; Pastewka, L.; Moseler, M. Wear, Plasticity, and Rehybridization in Tetrahedral Amorphous Carbon. *Tribol. Lett.* **2014**, *53*, 119–126. [CrossRef]
9. von Lautz, J.; Pastewka, L.; Gumbsch, P.; Moseler, M. Molecular Dynamic Simulation of Collision-Induced Third-Body Formation in Hydrogen-Free Diamond-Like Carbon Asperities. *Tribol. Lett.* **2016**, *63*, 26. [CrossRef] [PubMed]
10. de Barros Bouchet, M.I.; Martin, J.M.; Avila, J.; Kano, M.; Yoshida, K.; Tsuruda, T.; Bai, S.; Higuchi, Y.; Ozawa, N.; Kubo, M.; et al. Diamond-like carbon coating under oleic acid lubrication: Evidence for graphene oxide formation in superlow friction. *Sci. Rep.* **2017**, *7*, 46394. [CrossRef]
11. Salinas Ruiz, V.R.; Kuwahara, T.; Galipaud, J.; Masenelli-Varlot, K.; Hassine, M.B.; Héau, C.; Stoll, M.; Mayrhofer, L.; Moras, G.; Martin, J.M.; et al. Interplay of mechanics and chemistry governs wear of diamond-like carbon coatings interacting with ZDDP-additivated lubricants. *Nat. Commun.* **2021**, *12*, 4550. [CrossRef] [PubMed]
12. Godet, M. The third-body approach: A mechanical view of wear. *Wear* **1984**, *100*, 437–452. [CrossRef]
13. Berthier, Y. Third-Body Reality—Consequences and Use of the Third-Body Concept to Solve Friction and Wear Problems. In *Wear—Materials, Mechanisms and Practice*; John Wiley & Sons Ltd.: Hoboken, NJ, USA, 2005; pp. 291–316. [CrossRef]
14. Fillot, N.; Iordanoff, I.; Berthier, Y. Wear modeling and the third body concept. *Wear* **2007**, *262*, 949–957. [CrossRef]
15. Scherge, M.; Brink, A.; Linsler, D. Tribofilms Forming in Oil-Lubricated Contacts. *Lubricants* **2016**, *4*, 27. [CrossRef]
16. Spikes, H. The History and Mechanisms of ZDDP. *Tribol. Lett.* **2004**, *17*, 469–489. [CrossRef]
17. Faller, J.; Scherge, M. The Identification of an Adequate Stressing Level to Find the Proper Running-In Conditions of a Lubricated DLC-Metal-System. *Lubricants* **2020**, *8*, 88. [CrossRef]
18. Scherge, M.; Pöhlmann, K.; Gervé, A. Wear measurement using radionuclide-technique (RNT). *Wear* **2003**, *254*, 801–817. [CrossRef]
19. Hintsala, E.D.; Hangen, U.; Stauffer, D.D. High-Throughput Nanoindentation for Statistical and Spatial Property Determination. *JOM* **2018**, *70*, 494–503. [CrossRef]
20. Oliver, W.C.; Pharr, G.M. Measurement of hardness and elastic modulus by instrumented indentation: Advances in understanding and refinements to methodology. *J. Mater. Res.* **2004**, *19*, 3–20. [CrossRef]
21. Saringer, C.; Tkadletz, M.; Kratzer, M.; Cordill, M.J. Direct determination of the area function for nanoindentation experiments. *J. Mater. Res.* **2021**, *36*, 2154–2165. [CrossRef]
22. Fink, J. Recent Developments in Energy-Loss Spectroscopy. *Adv. Electron. Electron Phys.* **1989**, *75*, 121–232. [CrossRef]
23. Egerton, R.F. Electron energy-loss spectroscopy in the TEM. *Rep. Prog. Phys.* **2008**, *72*, 016502. [CrossRef]
24. Berger, S.D.; McKenzie, D.R.; Martin, P.J. EELS analysis of vacuum arc-deposited diamond-like films. *Philos. Mag. Lett.* **1988**, *57*, 285–290. [CrossRef]
25. Zhang, X.; Schneider, R.; Müller, E.; Gerthsen, D. Practical aspects of the quantification of sp2-hybridized carbon atoms in diamond-like carbon by electron energy loss spectroscopy. *Carbon* **2016**, *102*, 198–207. [CrossRef]
26. Oh-ishi, K.; Ohsuna, T. Inelastic mean free path measurement by STEM-EELS technique using needle-shaped specimen. *Ultramicroscopy* **2020**, *212*, 112955. [CrossRef]
27. Mangolini, F.; Li, Z.; Marcus, M.A.; Schneider, R.; Dienwiebel, M. Quantification of the carbon bonding state in amorphous carbon materials: A comparison between EELS and NEXAFS measurements. *Carbon* **2021**, *173*, 557–564. [CrossRef]
28. Fischer, A.; Dudzinski, W.; Gleising, B.; Stemmer, P. Analyzing Mild- and Ultra-Mild Sliding Wear of Metallic Materials by Transmission Electron Microscopy. In *Microtechnology and MEMS*; Springer International Publishing: Berlin/Heidelberg, Germany, 2018; pp. 29–59. [CrossRef]
29. Mayer, J.; Giannuzzi, L.A.; Kamino, T.; Michael, J. TEM Sample Preparation and FIB-Induced Damage. *MRS Bull.* **2007**, *32*, 400–407. [CrossRef]
30. Tong, Z.; Jiang, X.; Luo, X.; Bai, Q.; Xu, Z.; Blunt, L.; Liang, Y. Review on FIB-Induced Damage in Diamond Materials. *Curr. Nanosci.* **2016**, *12*, 685–695. [CrossRef]
31. Chen, X.; Zhang, C.; Kato, T.; Yang, X.A.; Wu, S.; Wang, R.; Nosaka, M.; Luo, J. Evolution of tribo-induced interfacial nanostructures governing superlubricity in a-C:H and a-C:H:Si films. *Nat. Commun.* **2017**, *8*, 1675. [CrossRef]

32. Ferrari, A.C.; Robertson, J. Resonant Raman spectroscopy of disordered, amorphous, and diamondlike carbon. *Phys. Rev. B* **2001**, *64*, 075414. [CrossRef]
33. Biberger, J. Tribologisch Induzierte Oberflächenveränderung im Reib-Verschleiß-Kontakt Kolbenring Gegen Zylinderlaufbahn. Ph.D.Thesis, TU Berlin, Berlin, Germany, 2017. [CrossRef]
34. Kaulfuss, F.; Weihnacht, V.; Zawischa, M.; Lorenz, L.; Makowski, S.; Hofmann, F.; Leson, A. Effect of Energy and Temperature on Tetrahedral Amorphous Carbon Coatings Deposited by Filtered Laser-Arc. *Materials* **2021**, *14*, 2176. [CrossRef]
35. Schultrich, B. Structure and Characterization of Vacuum Arc Deposited Carbon Films—A Critical Overview. *Coatings* **2022**, *12*, 109. [CrossRef]
36. Schultrich, B. *Tetrahedrally Bonded Amorphous Carbon Films I*; Springer: Berlin/Heidelberg, Germany, 2018.
37. Mosey, N.J.; Muüser, M.H.; Woo, T.K. Molecular Mechanisms for the Functionality of Lubricant Additives. *Science* **2005**, *307*, 1612–1615. [CrossRef] [PubMed]
38. Vengudusamy, B.; Green, J.H.; Lamb, G.D.; Spikes, H.A. Behaviour of MoDTC in DLC/DLC and DLC/steel contacts. *Tribol. Int.* **2012**, *54*, 68–76. [CrossRef]
39. Linsler, D.; Kümmel, D.; Nold, E.; Dienwiebel, M. Analysis of the running-in of thermal spray coatings by time-dependent stribeck maps. *Wear* **2017**, *376–377*, 1467–1474. [CrossRef]
40. Shakhvorostov, D.; Gleising, B.; Büscher, R.; Dudzinski, W.; Fischer, A.; Scherge, M. Microstructure of tribologically induced nanolayers produced at ultra-low wear rates. *Wear* **2007**, *263*, 1259–1265. [CrossRef]
41. Brink, A.; Lichtenberg, K.; Scherge, M. The influence of the initial near-surface microstructure and imposed stress level on the running-in characteristics of lubricated steel contacts. *Wear* **2016**, *360–361*, 114–120. [CrossRef]
42. Jablonski, A.; Powell, C.J. Practical expressions for the mean escape depth, the information depth, and the effective attenuation length in Auger-electron spectroscopy and x-ray photoelectron spectroscopy. *J. Vac. Sci. Technol. Vacuum, Surfaces, Film.* **2009**, *27*, 253–261. [CrossRef]
43. Ferrari, A.C.; Robertson, J. Interpretation of Raman spectra of disordered and amorphous carbon. *Phys. Rev. B* **2000**, *61*, 14095–14107. [CrossRef]

Disclaimer/Publisher's Note: The statements, opinions and data contained in all publications are solely those of the individual author(s) and contributor(s) and not of MDPI and/or the editor(s). MDPI and/or the editor(s) disclaim responsibility for any injury to people or property resulting from any ideas, methods, instructions or products referred to in the content.

Article

Premature Damage in Bearing Steel in Relation with Residual Stresses and Hydrogen Trapping

Maximilian Baur [1,*], Iyas Khader [1,2], Dominik Kürten [1], Thomas Schieß [3], Andreas Kailer [1,*] and Martin Dienwiebel [4]

[1] Fraunhofer Institute for Mechanics of Materials IWM, Wöhlerstraße 11, 79108 Freiburg, Germany; iyas.khader@gju.edu.jo (I.K.); dominik.kuerten@iwm.fraunhofer.de (D.K.)
[2] Department of Industrial Engineering, German Jordanian University, P.O. Box 35247, Amman 11180, Jordan
[3] OSK-Kiefer GmbH, Göppertshausen 5-6, 85238 Petershausen, Germany
[4] Institute for Applied Materials—Reliablity and Microstructure, Karlsruher Institute of Technology, Straße Am Forum 7, 76131 Karlsruhe, Germany; martin.dienwiebel@iwm.fraunhofer.de
* Correspondence: maximilian.baur@iwm.fraunhofer.de (M.B.); andreas.kailer@iwm.fraunhofer.de (A.K.)

Abstract: In this study, premature damage in cylindrical roller bearings made of 100Cr6 (SAE 52100) was investigated. For this purpose, full bearing tests were carried out using two different lubricant formulations with similar viscosities. Published research has pointed out the occurrence of tribochemical reactions that cause lubricant degradation and the release of hydrogen in tribo-contact. Hydrogen content measurements were conducted on tested samples, and these measurements showed dependence on the lubricant formulations. Hydrogen diffusion and trapping were identified as significant factors influencing premature damage. The measurement of trapping energies was conducted by thermal desorption spectroscopy, whereas residual stresses, which influence hydrogen diffusion and accumulation, were measured using X-ray diffraction. The measured trapping energies indicated that rolling contact caused the creation and release of hydrogen traps. Over-rolling resulted in changes in residual stress profiles in the materials, demonstrated by changes in stress gradients. These can be directly linked to subsurface hydrogen accumulation. Hence, it was possible to determine that the location of the microstructural damage (WEC) was correlated with the residual stress profiles and the subsurface von Mises stress peaks.

Keywords: rolling bearings; hydrogen; residual stresses; trapping energies

Citation: Baur, M.; Khader, I.; Kürten, D.; Schieß, T.; Kailer, A.; Dienwiebel, M. Premature Damage in Bearing Steel in Relation with Residual Stresses and Hydrogen Trapping. *Lubricants* **2024**, *12*, 311. https://doi.org/10.3390/lubricants12090311

Received: 29 June 2024
Revised: 27 August 2024
Accepted: 29 August 2024
Published: 3 September 2024

Copyright: © 2024 by the authors. Licensee MDPI, Basel, Switzerland. This article is an open access article distributed under the terms and conditions of the Creative Commons Attribution (CC BY) license (https://creativecommons.org/licenses/by/4.0/).

1. Introduction

Rolling bearings are used in a wide range of applications ranging from their use in miniature bearings to wind turbines. Premature failures manifested by white etching cracks (WEC) have been identified in bearings of various sizes, applications, and service lives [1–3]. Premature failure is usually unpredictable, as it does not follow classical rolling-contact fatigue (RCF) calculation methods, leading to costly and time-consuming downtimes for maintenance.

WEC are characterized by wide networks of cracks associated with white crack flanks, known as white etching areas (WEA), in the subsurface region of the raceways [4–6]. Previous studies have argued that WEA are a form of nanocrystalline BCC iron microstructure [7–9]. Under cyclic load, WEC usually propagate and progress to brittle flaking of the surface of the components and a loss of functionality.

Various factors have been argued to be the root causes of this type of damage [10], one of which is the choice of bearing material. The conventional and affordable rolling bearing steel 100Cr6/SAE 52100 has shown poor resistance to WEC [11,12]. Literature findings indicate that hydrogen diffusion into bearing steel affects early failures [13], which is the focus of the present work. Hydrogen-assisted rolling-contact fatigue (HARCF) is associated with the occurrence of white etching cracks in bearing components. Besides hydrogen, the combination of stresses and microstructural aspects [14,15] are found in the literature to be correlated with the formation of WEC. Moreover, the research has also suggested that a negative slide–roll ratio significantly influences the formation of WEC [16]. It was demonstrated that a combination of mechanical stresses and accumulated hydrogen in the microstructure is decisive in triggering HARCF [17,18]. Possible sources of hydrogen in a bearing system may be water contamination of the lubricant [6,12], as well as tribochemical reactions resulting in lubricant degradation and the eventual release and increased accumulation of hydrogen [19], especially in the presence of certain lubricant additives such as zinc dithiophosphate (ZDDP) and overbased calcium sulfonate (OBCaSul) [20,21]. The passage of current through the bearing was also shown to trigger the formation of WEC [22]. It should be noted that the damage cannot only be attributed to hydrogen but also to the combination of load, overrolling, and plastic deformation [14]. The formation of pores and newly formed grain boundaries in tested samples were associated with the usage of specific lubricants [3,23].

The influencing factors affecting hydrogen accumulation in bearing steel are hence important to investigate. These mainly include concentration gradients, stress gradients, and residual stresses in the material [24–26]. Hydrogen trapping in steel may also considerably affect the accumulation of hydrogen. In this study, RCF tests were carried out using two lubricant mixtures. Consequently, residual stresses were measured using X-ray diffraction analysis (XRD), and the hydrogen content and trapping energies were measured using thermal desorption analysis (TDA) of the tested samples. The trapping energy was measured to investigate the correlation between microstructural alterations and the absorbed hydrogen. Trapping energies can give insight into the type of hydrogen traps being filled in the process of overrolling using two different lubricant formulations. A distinction can also be made as to whether reversible (<35 kJ/mol) or irreversible traps (>35 kJ/mol) are being occupied in the process. Free diffusing hydrogen only has around 10 kJ/mol. Grain boundaries, martensite lath boundaries, and dislocations are considered reversible traps, with trapping energies of up to 35 kJ/mol. Irreversible traps, such as vacancies, titanium carbides, and MnS interfaces, act as traps, with higher trapping energies of more than 50 kJ/mol [27–29]. Furthermore, the findings of this study will help in identifying the influence of residual stresses and hydrogen trapping affecting HARCF.

2. Materials and Methods

2.1. Full Bearing Tests

In this study, axial cylindrical roller thrust bearings (CRTBs) were tested under flooded oil lubrication [26]. For the bearing experiments, the test setup in Figure 1 was mounted on a biaxial universal test rig. The measurement of normal force and torque is performed with a Instron Dynacell load cell (±25 kN/100 Nm) and an Instron FastTrack 8800 controller, monitored by LabView 6, with a storage frequency of 1 Hz. The tests were carried out at 700 rpm and 90 °C. In this study, two different lubrication mixtures were tested: a low reference oil (low-ref), which is known to be associated with WEC damage, and a high reference (high-ref) oil; both oils possess similar viscosity. The lubricants are comparable to those in the Research Association for Drive Technology (FVA) reference oil catalogue in terms of the FVA reference oil classifications. The low ref oil contains the additives ZDDP and OBCaSul, and the high ref contains an FVA test additive, which prevents premature failure.

The test rig is designed for testing type 81104-TV CRTBs made of 100Cr6/SAE 52100 steel, undergoing a hardened martensitic heat treatment. The surface hardness was found to be 754 ± 9 (HV10) on the roller and 760 ± 9 (HV10) on the raceway. The roughness R_a was measured using a stylus profilometer and was determined to be 0.04 ± 0.01 μm on the roller and 0.09 ± 0.01 on the washer. The bearings have an inner diameter of 20 mm and an outer diameter of 35 mm. Each bearing contains 13 rollers, 4.5 mm in length and diameter. Due to the geometry of the bearing, a negative slip of ~0.16 occurs at the outer edge of the roller and positive slip of ~0.16 at the inner edge. Three rollers were removed from the bearing cage before testing, leaving ten rollers in each tested bearing. Consequently, the normal force is distributed over ten rollers only; the maximum Hertzian contact pressure is 1.7 GPa; and the C/P ratio, which is the ratio of dynamic load rating to dynamic equivalent load, used to estimate the service life, is approximately 1.8. The viscosity ratio κ is 0.4. These conditions lead to a predicted nominal lifetime of approximately 500 h [30]; hence, premature damage (associated with WEC formation) is expected to occur under reasonable test durations.

Figure 1. Temperature-controlled test setup for full bearing tests in oil bath: (a) rolling bearing, (b) cup, filled with 20 mL of the oil, and the thermocouple inserted for temperature measurement, (c) double-walled container for controlling the temperature of the oil at 90 °C, (d) load cell to measure the normal force as well as the friction torque, (e) shaft to apply the rotation and normal load.

Test runs with increasing run times were carried out to study the difference in hydrogen accumulation and trapping using the two lubricant formulations. The tests were carried out for 25 h (6.2×10^6 cycles on the rollers) and 50 h (12.4×10^6 cycles). Tests that did not show signs of bearing damage or failure were extended up to 300 h (74.4×10^6 cycles).

2.2. Metallographic Investigations

The microstructural analysis was carried out on the cylinder rollers of the tested bearings. For this purpose, three rollers per test duration were studied via sectioning analysis, in which a cross-sectional analysis in increments of 0.3 to 0.5 mm, starting from the outer edge with negative slip, were performed. The process is demonstrated in Figure 2. The mirror-smooth surface was then etched with a solution of picric acid to reveal the grain boundaries. Thus, not only the depth under the surface can be seen, but also the degree of damage in different locations in the axial direction can be achieved. For this investigation, three rollers per test duration were investigated, and both negative and positive slip zones were identified and examined.

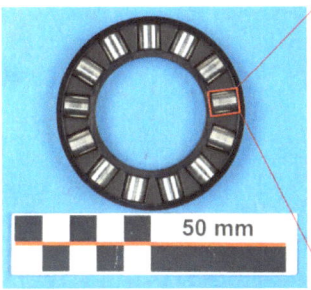
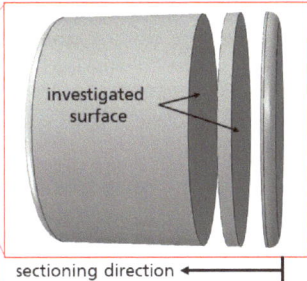

Figure 2. Illustration of the metallographic investigations on the cylindrical rolling elements, starting from the outer edge of the component.

2.3. X-ray Diffraction Analysis (XRD)

The residual stress profile is relevant for the accumulation of hydrogen because the diffusion of hydrogen is driven by both concentration gradients and stress gradients (i.e., stress-assisted diffusion) [25,26].

The measurements were conducted using Xstress 3000, Type G2, from Stresstech (Jyväskylä, Finland), using two linear detectors. The physical principle, on which the X-ray residual stress measurement is based, is the diffraction of monochromatic X-rays on crystal lattices. If X-rays of a certain wavelength λ hit a crystal lattice with a specific lattice plane spacing D at a certain angle θ, constructive interference results, according to Bragg's equation [31].

$$n \cdot \lambda = 2D \cdot \sin\theta \quad (1)$$

In the case of ferrite, this diffraction peak occurs at an angle of 156.4° for the (211) lattice planes in a stress-free state with Cr-Kα_1-radiation (λ = 2.2897 Å). A total of 14 measurements, at 13 different tilt-angles between +45° and −45° in the rolling direction (0°) and against the rolling direction (90°), were measured. Since the measured materials are usually not available as single crystals, this relatively simple measuring principle becomes considerably more complex when considering different crystal orientations measured simultaneously. The measured X-ray peaks were then evaluated using the $\sin^2 \psi$ method [32]. A Young's modulus of 211,000 MPa was taken as the reference.

The measurement was performed on the raceway. The residual stresses were determined to a depth of 200 µm under the contact surface and in finer increments directly under the surface. The material removal was achieved via electrochemical etching using a slightly modified KRISTALL 650 from QATM (Mammelzen, Germany), which ensures minimal influence on residual stresses.

2.4. Thermal Desorption Analysis (TDA)

For measuring the accumulated hydrogen content in the bearing elements, thermal desorption analysis was conducted using a Bruker G40 and a connected mass-spectrometer by InProcess Instruments (Bremen, Germany). After the test, the bearing rollers were instantly immersed in liquid nitrogen to decelerate hydrogen effusion. To make sure that there was no residual layer oil or its additives on the surface, the surface of the sample was slightly ground using 320 grit sanding paper directly after clearing it in ethanol and cleaned with acetone afterwards. Then, the prepared sample was placed in the quartz tube, and the measurement began. The sample was heated as fast as possible to a temperature of 800 °C, and the effused hydrogen was measured with the mass spectrometer. The amount of hydrogen was measured against a reference sample, and the calculated value was divided by the sample weight to give the hydrogen content in the sample.

TDA was chosen as the analysis method because it also provides the opportunity to obtain trapping energies [28,33]. To determine trapping energies, multiple measurements with various heating rates must be acquired. Different heating rates result in a peak shift of the hydrogen signal, which can be used to determine a binding energy by calculating the slope of the regression line of these data points, implemented in an Arrhenius plot. Equation (1) was used to determine the trapping energy, as follows:

$$\frac{d \ln\left(\varphi/T_p^2\right)}{d(1/T_p)} = -\frac{E_A}{R} \qquad (2)$$

where φ represents the heating rate in K/s, T_p is the temperature at which the peak forms, R is the ideal gas constant, and E_A is the activation or binding energy of the corresponding trapping site. Thus, multiplying the slope of the regression line with the ideal gas constant yields the binding energy of the particular site. In this study, the heating rates of 20, 40, and 80 K/min were used. Exemplary TDA measurement of one roller at a heating rate of 20 K/min is showed in Figure 3. When plotting the intensity over the temperature, it becomes clear that there is more than one peak to analyze. Through deconvolution of the curve, three different peaks are identified [28,34]. The smoothed curve is approximated with non-linear curve fitting according to Gaussian peaks, used because they achieved the best fit to the curve. As seen in Figure 3, Peak 1 appears at about 400 °C, Peak 2 forms at about 550 °C, and Peak 3 is the peak at the highest temperature of about 650 °C. Obviously, the peak depends on the applied heating rate. The intensity can be calculated to a desorption rate through multiplication with the calibration factor, which was consistent at the different measurements.

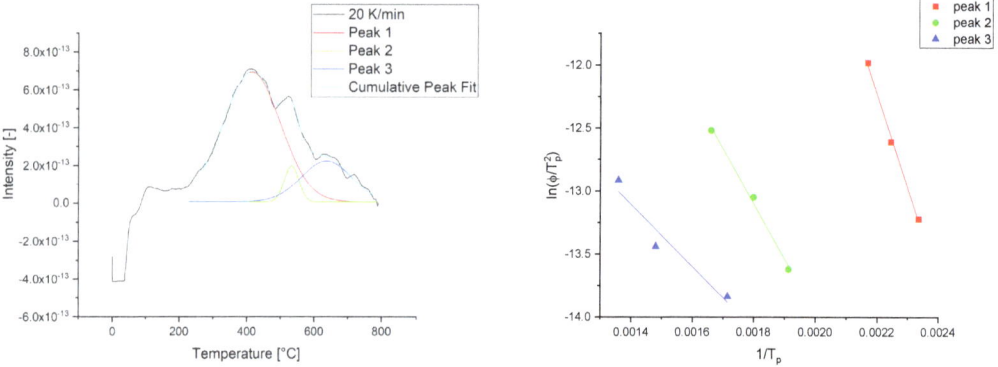

Figure 3. Deconvolution of the hydrogen intensity curve measured on the mass spectrometer (**left**) and the slope corresponding to the trapping energy of the different peaks (**right**).

3. Results

3.1. Experimental Results

The friction torque is shown in Figure 4. The friction torque was very consistent for the ten tests using each lubricant; however, there is a difference noted for the running-in behavior of the oils. The high-ref oil shows little to no running-in behavior, while a strong running-in behavior was observed with the low-ref oil. This shows, by way of example, that the mean value of the coefficient of friction converges to a very similar level. The difference between the lubricants lies in the fact that there is a strong running-in behavior with the low ref oil, which does not exist with the high ref oil.

Table 1 lists the test durations using both low ref and high ref oil. It should be noted that the maximum achievable duration with the low ref oil was about 70 h. The longer tests with high ref oil were initially stopped after 150 h; the tested samples did not reveal any signs of visible microstructural alternations and hence, the test duration was adjusted to 300 h.

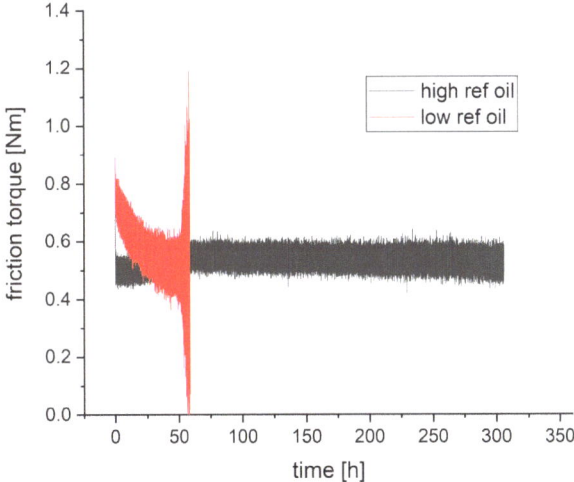

Figure 4. Exemplary COF vs time curve for the two different tests, showing the difference in running-in behavior between the lubricants.

Table 1. Tests completed, with the respective test duration and mean friction torque.

Number of Tests	Duration Low Ref Oil (h)	Friction Torque (Nm)	Duration High Ref Oil (h)	Friction Torque (Nm)
3	10 h	0.73 ± 0.06	25 h	0.57 ± 0.02
3	25 h	0.63 ± 0.03	50 h	0.59 ± 0.03
3	50 h	0.56 ± 0.05	150 h	0.60 ± 0.03
3	~70 h	0.58 ± 0.10	300 h	0.54 ± 0.03

3.2. Microsection of Rollers

In Figure 5, microsections of selected bearing rollers are shown after different testing durations. The distance to the outer edge of the roller is shown (in mm) in the images below. Independent of the runtime, at a distance of 0.5 mm from the outer edge, no damage can be identified. With the low-ref oil, a slight subsurface damage can be found after just 25 h of runtime. At a 1 mm axial distance, the first signs of damage appeared 10 to 50 μm below the surface after a test duration of 25 h. In contrast to these results, severe subsurface damages occurred after a 50 h runtime. After 50 h, severe propagation of the material cracks were observed in the subsurface region up to 200 μm under the surface. Also, WEC was found in these samples, although there was no damage visible on the contact surface. After approximately 70 h, the bearing rollers failed and showed severe flaking damage on the surface. There were large areas exhibiting WEAs, which were surrounded by long networks of cracks. At 1.5 mm to the outer edge of the roller, there were also some subsurface cracks after only a 25 h runtime, with severe damage occurring in this region after a longer runtime. After 70 h, the rollers showed flaking damage on the surface. Due to the chipped off surface and subsurface region, the number of observed WECs and WEAs was lower than that noted in the test at 50 h. However, subsurface damage was still found in the damaged samples.

In contrast, there was no damage observed using the high ref. oil, as shown in Figure 6. After 300 h, dark etching regions (DER) were observed in the cross-sectional cuts of the rollers at a distance of 1.5 mm from the outer side of the rollers in the subsurface region, which indicates classical rolling-contact fatigue.

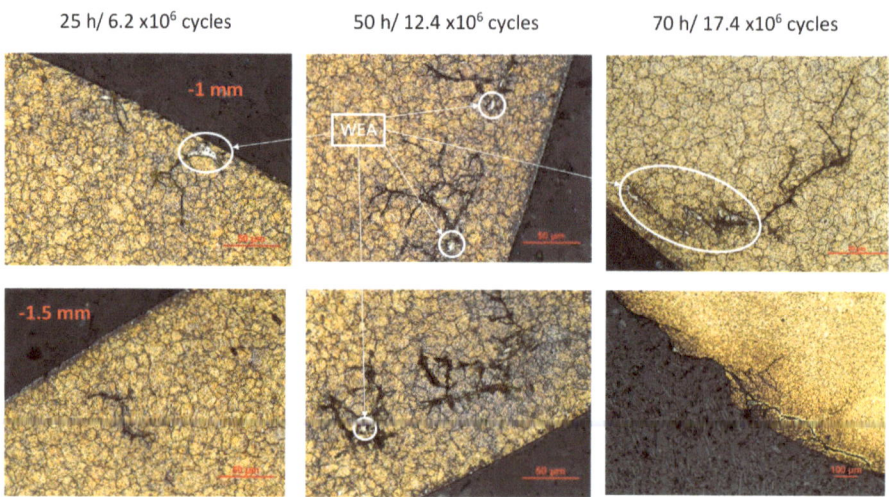

Figure 5. Cross-sectional cuts and metallographic examination of the rollers run with the low ref oil.

Figure 6. Cross-sectional cuts and metallographic examination of the rollers run with the high ref oil.

3.3. Residual Stress Measurements

The residual stresses were measured on the washers. The flat surface of the washer ensures better repeatability of the measurements, which would be otherwise affected by the crown of the roller (cylindrical rollers are actually manufactured with a slight camber that is subject to manufacturing tolerances). Nevertheless, to ensure transferability of residual stress results between the roller and washer, comparison measurements were carried out on virgin samples, as shown in Figure 7. The stress profiles obtained from the washer and the roller indicate very close results.

The residual stresses in the subsurface region were measured using gradually larger depth increments. The residual stress profiles were similar in samples tested with both oil formulations. Here, it is important to note that stress gradients (rather than stress magnitudes) are decisive for hydrogen diffusion [26].

In Figure 8, the residual stress profile of the washer was measured in the virgin state, as well as after 25 h and 50 h runtime, using the low-ref oil. In virgin samples, only compressive stresses were measured. With increasing depth, the compressive residual stresses decrease sharply; below 50 μm, the stress gradient decreases; however, stresses remain compressive throughout the whole measurement range. The test samples showed compressive stresses of approx. 600–700 MPa on the surface, compared to approx. 1100 MPa in the virgin samples. The compressive stresses decrease as the depth increases, changing to light compressive or even tensile stresses, with a peak at approx. 20 μm, below which the stress decreases and reaches a plateau at approximately 50 μm.

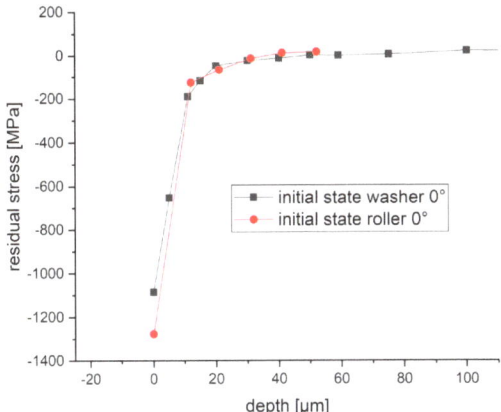

Figure 7. Comparison of the residual stresses of virgin bearing components.

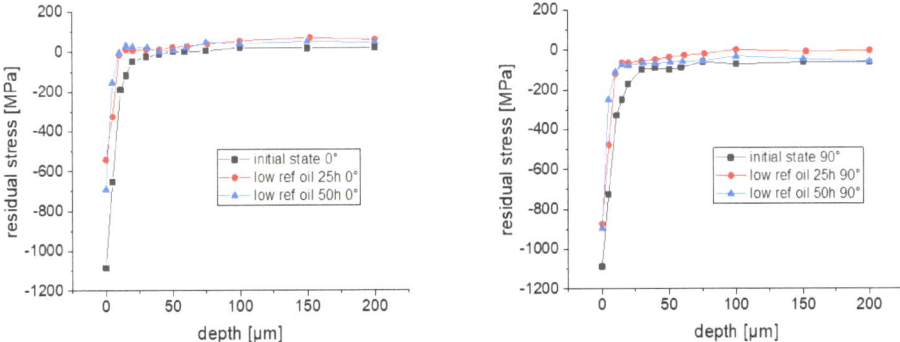

Figure 8. Residual stress profile of tests using low ref oil after 25 h and 50 h, measured in the rolling direction (**left**) and perpendicular to the rolling direction (**right**).

Figure 9 shows the residual stresses using the high ref oil. All samples tested for 25 h and 50 h showed similar stress profiles, with compressive stresses on the surface. The stress decreases sharply in the first micrometers of depth with a positive gradient, below which the stress reaches a peak, and the slope becomes negative. For the sample tested for 25 h using the high-ref oil, the peak in the residual stress profile is obvious.

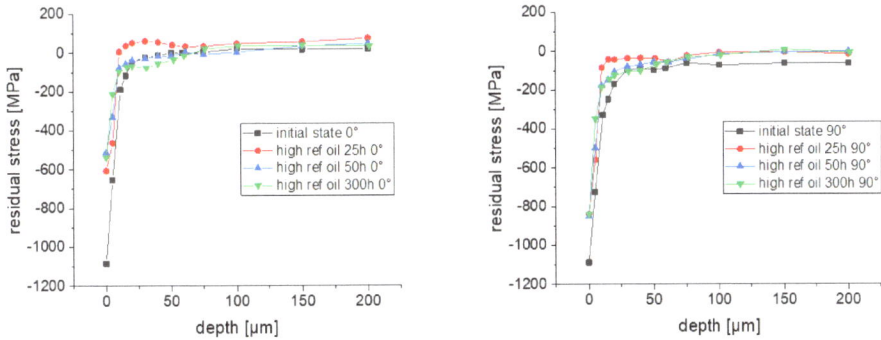

Figure 9. Residual stress profile of tests using high ref oil after 25 h, 50 h, and 300 h, measured in the rolling direction (**left**) and perpendicular to the rolling direction (**right**).

After a test duration of 300 h, the formation of a classical fatigue zone, with compressive residual stresses in the subsurface region, was observed. This is in line with the findings in the microstructural analysis, where a dark etching region (DER) is formed after 300 h of testing.

3.4. Thermal Desorption Analysis

3.4.1. Hydrogen Content Measurement

The measurement of hydrogen content using the TDA (Figure 10) shows the difference in the hydrogen accumulation in the bearing rollers using the different lubricants and test durations. The test runs using low ref oil showed a higher hydrogen content in the samples. The samples tested using high ref oil showed the maximum hydrogen content of 0.6 ppm, reached after approximately 100 h. Longer test durations did not result in higher hydrogen content in the samples. These runtimes were not even achieved with the low-ref oil due to the occurrence of damage. After a 25 h runtime, a hydrogen content over 0.6 ppm was measured, and after 50 h, an average of 0.8 ppm was measured in the roller samples. A measurement showing very high hydrogen content was obtained from a sample tested for 70 h with the low-ref oil; nevertheless, it was not shown on this graph. The cross-sectional analysis of the 70 h test run showed a network of cracks that propagated to the surface of the rollers. Thus, even after grinding the superficial area of the rollers, some lubricant might have been trapped in deeper cracks, which would have affect the measured hydrogen content.

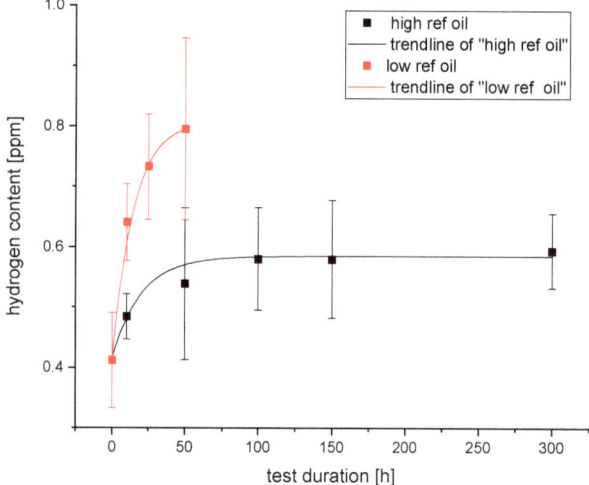

Figure 10. Hydrogen accumulation with different lubricants over the runtime.

3.4.2. Measurement of Trapping Energies

TDS measurements of trapping energies were carried out for three different rollers of one bearing. The measured trapping energies are listed in Tables 2 and 3.

Table 2. Measured trapping energies using high ref oil, with coefficient of determination.

	Peak 1		Peak 2		Peak 3	
	E_B (kJ/mol)	R^2	E_B (kJ/mol)	R^2	E_B (kJ/mol)	R^2
Initial state	10.28	0.93	28.82	0.89	50.57	0.98
High ref oil 150 h	7.08	0.92	10.03	0.89	13.05	0.82
High ref oil 290 h	38.83	0.90	38.23	0.98	41.67	0.92

Table 3. Measured trapping energies using low ref oil, with coefficient of determination.

	Peak 1		Peak 2		Peak 3	
	E_B (kJ/mol)	R^2	E_B (kJ/mol)	R^2	E_B (kJ/mol)	R^2
Initial state	10.28	0.93	28.82	0.89	50.57	0.98
Low ref oil 10 h	12.87	0.95	14.92	0.98	10.36	0.94
Low ref oil 10 h	17.59	0.97	47.65	0.98	43.99	0.95
Low ref oil 25 h	13.25	0.93	6.29	0.94	28.75	0.77
Low ref oil 25 h	6.28	0.92	25.54	0.99	34.58	0.98
Low ref oil 50 h	39.19	0.90	78.88	0.96	35.35	0.81
Low ref oil 50 h	14.39	0.78	13.81	0.77	24.09	0.85

In Table 2, it can be seen that after a 150 h runtime, the trapping energies of all three peaks are quite consistent, at around 10 kJ/mol. After 290 h, the trapping energies of the peaks are also consistent, at about 40 kJ/mol.

The trapping energies using the low ref oil vary widely, as shown in Table 3, and not just in regards to the runtime; it is also clear that the number of the peaks does not yield a consistent trend. Moreover, the smaller R^2 with a longer runtime shows the wider variation in the filled trapping sites. After a 50 h runtime, there was one particularly outstanding trapping site showing a high trapping energy, in which almost 80 kJ/mol were measured.

4. Discussion

In this work, rolling-contact fatigue behavior was studied for CRTBs lubricated using two different formulated lubricant mixtures (low ref and high ref). The metallographic examinations indicated the formation of WEC in samples tested with the low ref oil after 25 h (Figure 5); crack propagation resulted in substantial subsurface damaged after 50 h. Typical surface spalling was observed after around 70 h. In contrast, the samples tested with the high ref oil did not show any signs of premature damage, even after 300 h (Figure 6); signs of classical rolling contact fatigue, such as DER, were recorded.

Figure 7 shows almost identical residual stress profiles in the rollers and washers. To ensure better repeatability, the residual stresses in the tested samples were measured on the washers (refer to the explanation in Section 3.3). Moreover, the contact stresses, as well as material composition and properties, are almost identical in both contact counterparts, which ensures the transferability of the residual stress profiles between both components.

The measured residual stress profiles (refer to Figures 8 and 9) are in line with previous measurements obtained by Voskamp et al. [35,36]. It is worth noting that cyclic rolling contact changes the residual stress profile so that, just below the surface, a crest point is formed, indicating a positive-to-negative change in the stress gradient. This zone would act as an accumulation zone for hydrogen, as argued in Khader et al. [26,37], within which hydrogen concentrations 150% to 175% higher than those in the bulk material were computed [37]. WEC were observed at depths of 25 to 50 μm under the contact surface, which coincide with this region of accumulated hydrogen [37]. Hence, it is believed that the elevated hydrogen concentrations within the accumulation zone can be correlated with the formation of WEC [38]. Although the residual stresses were measured in the washers, the same profiles are expected to develop in the rollers. In 25 h and 50 h tests, the stress profile shows a peak just below the surface, resulting in a change in the stress gradient from positive to negative. The stress profile of the 300 h test also showed similar behavior, with higher compressive stresses below the surface. The difference in residual stresses using different oil formulations is attributed to the running-in behavior (Figure 4), which is mainly influenced by the oil additives.

Figure 11 shows the measured residual stress profile of the high ref oil after 25 h, measured in rolling direction, with a crest point in the tensile stress region. The graph also shows the von Mises stresses developed due to contact. The von Mises stress was extracted from a finite element model developed elsewhere [37]. The region marked by the ellipse in the figure indicates the zone of WEC formation in the first 25 h of the test time (with low ref oil). The subsurface hydrogen accumulation is believed to detrimentally affect the integrity of steel by means of hydrogen enhanced decohesion (HEDE) [39]. This processes [39,40] and the overlapped mechanical stresses (as indicated by the peak von Mises equivalent stress) is therefore believed to lead to the initiation of cracks in the subsurface region.

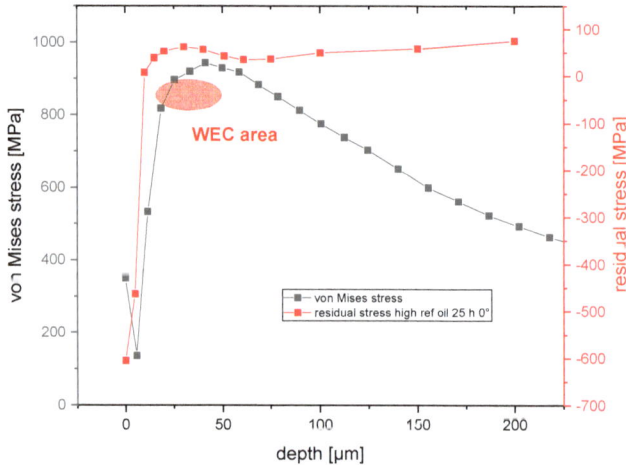

Figure 11. Overlap of gradient in residual stresses, with high von Mises stress relative to the observed WEC damage.

It is also worth mentioning that the damage predominantly occurred on the outer edge of the roller. It is assumed that slip in the rolling contact leads to the removal of the tribo-chemical layer, thus rendering the surface more susceptible to hydrogen diffusion. The nascent metal surface is more susceptible to the adsorption of hydrogen, which then gets absorbed by the bulk material [41,42]. It is known that higher slip can be associated with the occurrence of WEC damage [16].

The measured trapping energies (Tables 2 and 3) indicate the saturation of some sites and the creation of new ones. There are clear differences in the trapping energies, and the energies are more consistent in the tests using the high ref oil. With the high ref oil, it seems that hydrogen initially becomes more diffusible in the lattice due to over-rolling, as seen in Table 2. In the 150 h test, all three peaks were very close to each other at around 10 kJ/mol. After 290 h, the trapping energy is also consistent at approx. 40 kJ/mol, which may indicate saturation, revealing the formation of DER, which is associated with very high dislocation density [43]. The measured trapping energies did not show a clear trend for samples tested with the low ref oil (Table 3). This may be attributed to the unevenly distributed subsurface damage in steel, which is a reflection of a different localized load at the microscopic level. The literature findings for similar types of steels indicate that the lower energy traps (around 10 kJ/mol) are associated with interstitial sites or carbides, e.g., (Fe, Cr, Mo)$_2$C. Trapping sites of around 28–40 kJ/mol indicate dislocations, grain boundaries, or martensitic laths [29,44]. Higher energy traps would be considered irreversible traps [34,44]. Very high binding energy sites (similar to those in the test using low ref oil for 50 h) are assumed to be strained Fe$_3$C interfaces or an interface between α-phase and an MnS inclusion [29]. It is thus believed that more data are needed to be able to achieve a trend of trapping energies.

5. Conclusions

The following points summarize the work:

- The tests showed that a running-in behavior occurs using the low ref oil, and practically no running-in occurs using the high ref oil due to different additive mixtures.

- In the metallographic cuts, it was observed that the initiation of the damage begins at a depth of approximately 25 to 50 μm; cyclic rolling contacts result in damage accumulation and crack propagation.
- The hydrogen content measured in the rollers was higher for the low ref oil when compared to the high ref oil, which indicates that a higher hydrogen content and WEC damage are correlated.
- The trapping energies showed that the trapping sites with higher trapping energy become occupied after a longer runtime; it is assumed that at first the lower energy traps corresponding to dislocations, and subsequently, after 50 h, the high energy traps are occupied by hydrogen atoms.
- The residual stresses ranged from compressive stresses on the surface of the washers, to slight tensile stresses at depths of 20 to 50 μm, to lower tensile stresses at greater depths.

The correlation of the elevated hydrogen content, the measured residual stress, the observed failure due to subsurface cracks, and the subsequent spalling of the surface of the roller indicate that the emerging profile of residual stresses, in combination with the high von Mises stresses, play a significant role in the damage evolution of the WEC. Slip also has some effect, leading to damage occurring mainly on the outer edge of the rollers.

Author Contributions: Conceptualization, M.B., D.K. and I.K.; methodology, M.B. and D.K.; software, M.B.; validation, D.K. and I.K.; investigation, M.B. and T.S.; data curation, M.B. and T.S.; writing—original draft preparation, M.B., M.D., D.K., I.K. and A.K.; writing—review and editing, M.B., D.K, M.D. and I.K.; supervision, M.D., A.K. and I.K.; project administration, D.K. and A.K. All authors have read and agreed to the published version of the manuscript.

Funding: This work received funding from the German Research Foundation (DFG) (project: 450828000).

Data Availability Statement: The data presented in this study are available on request from the corresponding author due to privacy reasons.

Acknowledgments: The authors would like to thank Klüber Lubrication GmbH & Co. KG, Schaeffler AG, and Robert Bosch GmbH for their cooperation and for the provision of materials.

Conflicts of Interest: Author Thomas Schieß was employed by the company OSK-Kiefer GmbH and authors Maximilian Baur, Iyas Khader, Dominik Kürten and Andreas Kailer were employed by Fraunhofer Institute for Mechanics of Materials IWM. The remaining author declares that the research was conducted in the absence of any commercial or financial relationships that could be construed as a potential conflict of interest.

References

1. Greco, A.; Sheng, S.; Keller, J.; Erdemir, A. Material wear and fatigue in wind turbine Systems. *Wear* **2013**, *302*, 1583–1591. [CrossRef]
2. Gould, B.; Greco, A.; Stadler, K.; Xiao, X. An analysis of premature cracking associated with microstructural alterations in an AISI 52100 failed wind turbine bearing using X-ray tomography. *Mater. Des.* **2017**, *117*, 417–429. [CrossRef]
3. Spille, J.; Wranik, J.; Barteldes, S.; Mayer, J.; Schwedt, A.; Zürcher, M.; Lutz, T.; Wang, L.; Holweger, W. A study on the initiation processes of white etching cracks (WECs) in AISI 52100 bearing steel. *Wear* **2021**, *477*, 203864. [CrossRef]
4. Stadler, K.; Lai, J.; Vegter, R.H. A Review: The Dilemma with Premature White Etching Crack (WEC) Bearing Failures. In *Bearing Steel Technologies: 10th Volume, Advances in Steel Technologies for Rolling Bearings*; Beswick, J.M., Ed.; ASTM International: West Conshohocken, PA, USA, 2014; pp. 1–22.
5. Kürten, D.; Khader, I.; Raga, R.; Casajús, P.; Winzer, N.; Kailer, A.; Spallek, R.; Scherge, M. Hydrogen assisted rolling contact fatigue due to lubricant degradation and formation of white etching areas. *Eng. Fail. Anal.* **2019**, *99*, 330–342. [CrossRef]
6. Evans, M.-H.; Wang, L.; Wood, R. Formation mechanisms of white etching cracks and white etching area under rolling contact fatigue. *Proc. Inst. Mech. Eng. Part J J. Eng. Tribol.* **2014**, *228*, 1047–1062. [CrossRef]
7. Diederichs, A.M.; Barteldes, S.; Schwedt, A.; Mayer, J.; Holweger, W. Study of subsurface initiation mechanism for white etching crack formation. *Mater. Sci. Technol.* **2016**, *32*, 1170–1178. [CrossRef]
8. Holweger, W.; Schwedt, A.; Rumpf, V.; Mayer, J.; Bohnert, C.; Wranik, J.; Spille, J.; Wang, L. A Study on Early Stages of White Etching Crack Formation under Full Lubrication Conditions. *Lubricants* **2022**, *10*, 24. [CrossRef]
9. Šmelova, V.; Schwedt, A.; Wang, L.; Holweger, W.; Mayer, J. Microstructural changes in White Etching Cracks (WECs) and their relationship with those in Dark Etching Region (DER) and White Etching Bands (WEBs) due to Rolling Contact Fatigue (RCF). *Int. J. Fatigue* **2017**, *100*, 148–158. [CrossRef]

10. Gesellschaft für Tribologie e.V., White Etching Cracks: Positionspapier. Available online: https://www.gft-ev.de/wp-content/uploads/Position-paper-en-Web.pdf (accessed on 10 June 2024).
11. Bhadeshia, H. Steels for bearings. *Prog. Mater. Sci.* **2012**, *57*, 268–435. [CrossRef]
12. Gould, B.; Paladugu, M.; Demas, N.G.; Greco, A.C.; Hyde, R.S. Figure the impact of steel microstructure and heat treatment on the formation of white etching cracks. *Tribol. Int.* **2019**, *134*, 232–239. [CrossRef]
13. Evans, M.-H.; Richardson, A.; Wang, L.; Wood, R. Effect of hydrogen on butterfly and white etching crack (WEC) formation under rolling contact fatigue (RCF). *Wear* **2013**, *306*, 226–241. [CrossRef]
14. Gould, B.; Demas, N.; Erck, R.; Lorenzo-Martin, M.C.; Ajayi, O.; Greco, A. The effect of electrical current on premature failures and microstructural degradation in bearing steel. *Int. J. Fatigue* **2020**, *145*, 106078. [CrossRef]
15. López-Uruñuela, F.J.; Macareno, L.M.; Pinedo, B.; Aguirrebeitia, J. Confirming dark groove microstructural alterations as WEC initiation first stage in cylindrical roller thrust bearings lubricated with WEC critical oil. *Wear* **2024**, *550*, 205396. [CrossRef]
16. Gould, B.; Greco, A. The Influence of Sliding and Contact Severity on the Generation of White Etching Cracks. *Tribol. Lett.* **2015**, *60*, 29. [CrossRef]
17. Ruellan, A.; Ville, F.; Kleber, X.; Arnaudon, A.; Girodin, D. Understanding white etching cracks in rolling element bearings: The effect of hydrogen charging on the formation mechanisms. *Proc. Inst. Mech. Eng. Part J J. Eng. Tribol.* **2014**, *228*, 1252–1265. [CrossRef]
18. Szost, B.; Rivera-Díaz-Del-Castillo, P. Unveiling the nature of hydrogen embrittlement in bearing steels employing a new technique. *Scr. Mater.* **2012**, *68*, 467–470. [CrossRef]
19. Kohara, M.; Kawamura, T.; Egami, M. Study on Mechanism of Hydrogen Generation from Lubricants. *Tribol. Trans.* **2006**, *49*, 53–60. [CrossRef]
20. Steinweg, F.; Mikitisin, A.; Janitzky, T.L.; Richter, S.; Weirich, T.E.; Mayer, J.; Broeckmann, C. Influence of additive-derived reaction layers on white etching crack failure of SAE 52100 bearing steel under rolling contact loading. *Tribol. Int.* **2023**, *180*, 108239. [CrossRef]
21. Gould, B.; Demas, N.G.; Pollard, G.; Rydel, J.J.; Ingram, M.; Greco, A.C. The Effect of Lubricant Composition on White Etching Crack Failures. *Tribol. Lett.* **2018**, *67*, 7. [CrossRef]
22. Mikami, H.; Kawamura, T. Influence of Electrical Current on Bearing Flaking Life. In *Influence of Electrical Current on Bearing Flaking Life*; SAE Technical Paper; SAE: Warrendale, PA, USA, 2007.
23. Dogahe, K.J.; Guski, V.; Mlikota, M.; Schmauder, S.; Holweger, W.; Spille, J.; Mayer, J.; Schwedt, A.; Görlach, B.; Wranik, J. Simulation of the Fatigue Crack Initiation in SAE 52100 Martensitic Hardened Bearing Steel during Rolling Contact. *Lubricants* **2022**, *10*, 62. [CrossRef]
24. Frappart, S.; Feaugas, X.; Creus, J.; Thebault, F.; Delattre, L.; Marchebois, H. Study of the hydrogen diffusion and segregation into Fe–C–Mo martensitic HSLA steel using electrochemical permeation test. *J. Phys. Chem. Solids* **2010**, *71*, 1467–1479. [CrossRef]
25. Winzer, N.; Rott, O.; Thiessen, R.; Thomas, I.; Mraczek, K.; Höche, T.; Wright, L.; Mrovec, M. Hydrogen diffusion and trapping in Ti-modified advanced high strength steels. *Mater. Des.* **2016**, *92*, 450–461. [CrossRef]
26. Khader, I.; Kürten, D.; Raga, R.; Winzer, N.; Kailer, A. Modeling hydrogen diffusion in a tribological scenario: A failure analysis of a thrust bearing. *Wear* **2019**, *438-439*, 203054. [CrossRef]
27. Siegl, W. Hydrogen Trapping in Iron and Iron-Based Alloys. Ph.D. Thesis, Montanuniversitaet Leoben, Leoben, Austria, 2020.
28. Zafra, A.; Peral, L.; Belzunce, J. Hydrogen diffusion and trapping in A 42CrMo4 quenched and tempered steel: Influence of tempering temperature. *Int. J. Hydrog. Energy* **2020**, *45*, 31225–31242. [CrossRef]
29. Peral, L.; Amghouz, Z.; Colombo, C.; Fernández-Pariente, I. Evaluation of hydrogen trapping and diffusion in two cold worked CrMo(V) steel grades by means of the electrochemical hydrogen permeation technique. *Theor. Appl. Fract. Mech.* **2020**, *110*, 102771. [CrossRef]
30. Medias Bearing Calculation. Available online: https://medias.schaeffler.de/de/produkt/rotary/waelz--und-gleitlager/rollenlager/axial-zylinderrollenlager/81104-tv/p/395849#Calculation (accessed on 10 June 2024).
31. Bragg, W.H. The Reflection of X-rays by Crystals. *Nature* **1913**, *91*, 477. [CrossRef]
32. Spieß, L. *Moderne Röntgenbeugung: Röntgendiffraktometrie für Materialwissenschaftler, Physiker und Chemiker*, 2nd ed.; Vieweg + Teubner: Wiesbaden, Germany, 2009.
33. Simoni, L.; Falcade, T.; Ferreira, D.C.; Kwietniewski, C.E. An integrated experimental and modeling approach to determine hydrogen diffusion and trapping in a high-strength steel. *Int. J. Hydrogen Energy* **2021**, *46*, 25738–25751. [CrossRef]
34. Yamabe, J.; Yoshikawa, M.; Matsunaga, H.; Matsuoka, S. Hydrogen trapping and fatigue crack growth property of low-carbon steel in hydrogen-gas environment. *Int. J. Fatigue* **2017**, *102*, 202–213. [CrossRef]
35. Voskamp, A.P.; Österlund, R.; Becker, C.; Vingsbo, O. Gradual changes in residual stress and microstructure during contact fatigue in ball bearings. *Met. Technol.* **1980**, *7*, 14–21. [CrossRef]
36. Voskamp, A.P.; Mittemeijer, E.J. State of residual stress induced by cyclic rolling contact loading. *Mater. Sci. Technol.* **1997**, *13*, 430–438. [CrossRef]
37. Khader, I.; Kürten, D.; Kailer, A. *The Influence of Mechanical Stresses on the Diffusion and Accumulation of Hydrogen in a CRTB*; Bearing World: Frankfurt, Germany, 2020.
38. Richardson, A.D.; Evans, M.-H.; Wang, L.; Wood, R.J.K.; Ingram, M. Thermal Desorption Analysis of Hydrogen in Non-Hydrogen-Charged Rolling Contact Fatigue-Tested 100Cr6 Steel. *Tribol. Lett.* **2017**, *66*, 4. [CrossRef]

39. Bae, D.-S.; Sung, C.-E.; Bang, H.-J.; Lee, S.-P.; Lee, J.-K.; Son, I.-S.; Cho, Y.-R.; Baek, U.-B.; Nahm, S.-H. Effect of highly pressurized hydrogen gas charging on the hydrogen embrittlement of API X70 steel. *Met. Mater. Int.* **2014**, *20*, 653–658. [CrossRef]
40. Nguyen, T.T.; Park, J.; Nahm, S.H.; Tak, N.; Baek, U.B. Ductility and fatigue properties of low nickel content type 316L austenitic stainless steel after gaseous thermal pre-charging with hydrogen. *Int. J. Hydrogen Energy* **2019**, *44*, 28031–28043. [CrossRef]
41. López-Uruñuela, F.J.; Fernández-Díaz, B.; Pagano, F.; López-Ortega, A.; Pinedo, B.; Bayón, R.; Aguirrebeitia, J. Broad review of "White Etching Crack" failure in wind turbine gearbox bearings: Main factors and experimental investigations. *Int. J. Fatigue* **2020**, *145*, 106091. [CrossRef]
42. Guzmán, F.G.; Oezel, M.O.; Jacobs, G.; Burghardt, G.; Broeckmann, C.; Janitzky, T. Influence of Slip and Lubrication Regime on the Formation of White Etching Cracks on a Two-Disc Test Rig. *Lubricants* **2018**, *6*, 8. [CrossRef]
43. El Laithy, M.; Wang, L.; Harvey, T.J.; Schwedt, A.; Vierneusel, B.; Mayer, J. Mechanistic study of dark etching regions in bearing steels due to rolling contact fatigue. *Acta Mater.* **2023**, *246*. [CrossRef]
44. Chen, W.; Zhao, W.; Gao, P.; Li, F.; Kuang, S.; Zou, Y.; Zhao, Z. Interaction between dislocations, precipitates and hydrogen atoms in a 2000 MPa grade hot-stamped steel. *J. Mater. Res. Technol.* **2022**, *18*, 4353–4366. [CrossRef]

Disclaimer/Publisher's Note: The statements, opinions and data contained in all publications are solely those of the individual author(s) and contributor(s) and not of MDPI and/or the editor(s). MDPI and/or the editor(s) disclaim responsibility for any injury to people or property resulting from any ideas, methods, instructions or products referred to in the content.

Article

Process-Integrated Component Microtexturing for Tribologically Optimized Contacts Using the Example of the Cam Tappet—Numerical Design, Manufacturing, DLC-Coating and Experimental Analysis

Christian Orgeldinger [1,*], Manuel Reck [2], Armin Seynstahl [1], Tobias Rosnitschek [1], Marion Merklein [2] and Stephan Tremmel [1]

[1] Engineering Design and CAD, Universität Bayreuth, Universitätsstr. 30, 95447 Bayreuth, Germany; armin.seynstahl@uni-bayreuth.de (A.S.); tobias.rosnitschek@uni-bayreuth.de (T.R.); stephan.tremmel@uni-bayreuth.de (S.T.)

[2] Institute of Manufacturing Technology, Friedrich-Alexander-Universität Erlangen-Nürnberg, Egerlandstr. 13, 91058 Erlangen, Germany; manuel.reck@fau.de (M.R.); marion.merklein@fau.de (M.M.)

* Correspondence: christian.orgeldinger@uni-bayreuth.de

Citation: Orgeldinger, C.; Reck, M.; Seynstahl, A.; Rosnitschek, T.; Merklein, M.; Tremmel, S. Process-Integrated Component Microtexturing for Tribologically Optimized Contacts Using the Example of the Cam Tappet—Numerical Design, Manufacturing, DLC-Coating and Experimental Analysis. *Lubricants* **2024**, *12*, 291. https://doi.org/10.3390/lubricants12080291

Received: 19 June 2024
Revised: 6 August 2024
Accepted: 11 August 2024
Published: 16 August 2024

Copyright: © 2024 by the authors. Licensee MDPI, Basel, Switzerland. This article is an open access article distributed under the terms and conditions of the Creative Commons Attribution (CC BY) license (https://creativecommons.org/licenses/by/4.0/).

Abstract: To meet the demand for energy-efficient and, at the same time, durable, functional components, the improvement of tribological behavior is playing an increasingly important role. One approach to reducing friction in lubricated tribological systems is the microtexturing of the surfaces tailored to the application, but in most cases, this leads to increased manufacturing costs and thus often makes their use in industry more difficult. In this work, we, therefore, present an approach for an efficient design and fully integrated production process using a cam tappet as an example. For the used cam tappet contact, we first determined the optimal texture geometries using two differently complex EHL (elastohydrodynamic lubrication) simulation models. Based on these, textured tappets were manufactured in a combined manner using sheet-bulk metal-forming and deposition with a diamond-like-carbon (DLC) coating for additional wear protection without further post-processing of the coating. We show that the simulation approach used has a rather subordinate influence on the optimization result. The combined forming of components with textured surfaces is limited by the local material flow, the resulting texture distortion, and tool wear. However, a targeted process design can help to exploit the potential of single-stage forming. The applied DLC coating has good adhesion and can completely prevent wear in subsequent reciprocal pin-on-disc tests, while the friction in the run-in behavior is initially higher due to the soothing effects of the coating. The experiments also show a tendency for shallow textures to exhibit lower friction compared to deeper ones, which corresponds to the expectations from the simulation.

Keywords: EHL simulation; microtextured surfaces; optimization; micro-coining; sheet metal; DLC-coating

1. Introduction

To enable the transition to a sustainable industry, tribological improvements will play an increasingly important role in the future. A reduction in friction and wear is directly linked to greater efficiency and material savings and thus leads to lower energy losses, material waste and air pollution. According to Ciulli [1], this goal is currently being increasingly pursued by various tribological research areas. Holmberg and Edemir [2] quantified the share of energy consumption due to friction and wear at around 23% of global energy consumption, although there is still considerable long-term savings potential here. There are various technical approaches to improving tribological contacts, such as the development of better and more sustainable lubricants [3], adapting the materials used [4], or the surface treatment of components, for example, using amorphous carbon coatings (DLC-coatings) [5].

Another approach is the defined microtexturing of the surfaces. In addition to other positive effects, such as the inclusion of wear particles [6], this can achieve improved frictional behavior in hydrodynamic contacts (HD) [7] and elastohydrodynamic contacts (EHL). In the latter case, texturing can be used to technical advantage, for example, in cam tappet contacts [8], in gear contacts [9], or in the field of medical technology for knee [10] or hip implants [11]. However, since the exact shape and arrangement of the textures are decisive for the resulting friction behavior and, in the worst case, an unsuitable texture leads to a deterioration of the tribological behavior [12], a design tailored for the application is absolutely necessary. As the experimental determination of optimal textures is very time-consuming, simulation-based approaches have become increasingly popular in recent decades. For HD contacts, a good summary of the models is presented in [13], while Marian et al. [14] provided a comprehensive overview of EHL contacts. Some works already deal with the numerical design of the texture geometry itself [15].

In addition to the tailored design of the textures, it must also be ensured that they can be applied to the surfaces easily and economically. Common methods include laser texturing [16], micro-coining [17], or micromachining [18]. An overview of recent approaches is compiled in Costa and Hutchings [19], for example. To address sustainability aspirations, resource-efficient manufacturing processes are shifting to the focus of scientific research. Due to the high material utilization and good economic efficiency achieved through short cycle times, metal-forming processes are particularly suitable for the production of microtextured components [20]. Besides the extrusion of filigree functional elements, metal-forming processes are also suitable for imprinting microscopic surface textures [21]. In addition to the near-net-shape production of the component geometry, it is also possible to positively influence the mechanical properties of the workpiece during the forming process [22]. In this context, the combined cup-backward extrusion in conjunction with the micro-coining of lubrication pockets for the production of textured tappets was investigated [23]. The focus was on tribological influences on the accuracy of the microtextured components. It was found that the radial material flow in the area of the textured flat surface of the tappet leads to pronounced flank angles in the microtextures, which reduces the manufacturing accuracy.

Especially in relation to the counterbody, textured surfaces can help reduce wear in the system [24]. However, the textures themselves can also wear out quickly [25], which means that the positive effects are increasingly canceled out over time. A possible approach would be to additionally apply a tribological coating to the textured surfaces. For example, there has been work done with MoS_2 coatings on steel substrates [26], TiN and DLC on silicon [27], and DLC on copper and steel [28]. Texturing cemented carbides using a DLC layer also demonstrated improved adhesion of the coating to the substrate [29]. However, most of the aforementioned works examine the underlying mechanisms and do not make any application reference to real machine elements.

As part of our work, we have set the goal of considering the complete process for the selected example of a cam tappet contact, starting from the numerical design of the microtextures through integrated production to the subsequent DLC-coating and testing. On the numerical side, we have investigated for the first time how the simplification of such a simulation model affects the result of the optimization. Another new feature is the integrated production of the tappets and their subsequent coating without further post-processing. In doing so, we examined both the production aspects and the tribological behavior of the resulting textures in the model contact.

The cam tappet contact was selected because it represents an already well-understood and frequently investigated application [8] and has high production requirements. Based on the existing contact conditions, suitable textures are first determined using two numerical models of varying complexity. These textures, as well as textures that are rather unsuitable from a numerical point of view, are then produced in a combined cam tappet process.

The aim is to identify process limitations of the production of textured components used in different applications. For this purpose, a single-stage process sequence consisting of deep-drawing, ironing and micro-coining is designed. This sequence is used to identify and evaluate influencing variables on the component and process. Experimental and numerical test plans are used to investigate the influence of texture geometry and process control on the dimensional accuracy of the coined component surface. Geometric analyses using profile measurements of the textures provide the basis for researching the forming limits. Possible limitations are distortions of the individual texture indentations and shifts in the texture pattern. In addition, measures to extend the process limits are discussed. On this basis, generally, valid design guidelines for single-stage texturing processes are derived. The formed parts are furthermore coated with a DLC-coating using a combined physical vapor deposition (PVD) and plasma enhanced chemical vapor deposition (PECVD) process without subsequent finishing of the surface. DLC-coating systems are often used to reduce friction or wear in contrast to steel–steel contact. Model tests on a ball-on-disc tribometer are then used to analyze the potential of the investigated procedure, both in terms of the numerical design and the resulting tribological behavior of the system. In addition, the component and tool-side application behavior, which plays a decisive role in efficient manufacturing in large-scale production, is evaluated from a manufacturing point of view.

2. Materials and Methods

2.1. Application and Contact Conditions

The cam tappet contact was selected as an application example because it involves complex and widely varying contact conditions with a high proportion of sliding, and the contact has already been well investigated in preliminary work [30]. In addition, the topic might become more relevant in the coming years since even more efficient internal combustion engines [31] or engines with alternative fuels [32] could be required as a transitional solution.

The contact conditions occurring in the cam tappet contact are typically very dynamic and change significantly with changing engine speeds and over the cam angle. For this reason, the simplified 2D simulations were carried out at six selected load cases to determine the influence on the optimum textures. The dynamic contact conditions were assumed as in [33] and are summarized in more detail in [34]. The cam geometry is based on profilometric measurements of a BMW K48 camshaft. The selection of the material, geometry, and lubricant was based on the industrial standards. The constant material and fluid parameters used for the simulation are summarized in Table 1, while the varying contact conditions are shown in Table 2.

Table 1. Constant material and fluid parameters of the cam tappet contact used in the simulation.

Lubricant	FVA 3
Base density ρ_0	805 kg/m^3
Base viscosity η_0	0.03 Pa · s
Pressure viscosity coefficient α_η	$1.31 \cdot 10^{-8}$ Pa^{-1}
Critical shear stress G_c	6 MPa
Second plateau viscosity η_∞	$0.2 \cdot \eta_0$
Carreau parameter a_c	2.2
Carreau parameter n_c	0.8
Oil temperature t_{oil}	70 °C
Cam material	100Cr6/1.3505
Cam geometry	Based on BMW K48 camshaft
Young's modulus of the cam E_c	209 GPa
Poisson's ratio of the cam ν_c	0.3
Tappet material	16MnCr5/1.7131

Table 1. Cont.

Lubricant	FVA 3
Young's modulus of the tappet E_t	216 GPa
Poisson's ratio of the tappet ν_t	0.3
Coefficient of friction for boundary friction μ	0.1
Combined surface roughness σ_c	0.1 μm

Table 2. Contact conditions at load cases (a)–(f) as a function of cam angle φ and camshaft speed n_{cam} based on the reference cam BMW K48.

Parameter	Cam Tip Contact ($\varphi = 0°$)			Cam Flank Contact ($\varphi = -45°$)		
Camshaft Speed n_{cam}	500 rpm (a)	1000 rpm (b)	2000 rpm (c)	500 rpm (d)	1000 rpm (e)	2000 rpm (f)
Normal force F_N	365 N	347 N	279 N	214 N	221 N	250 N
Mean entrainment speed u_m	0.46 m/s	0.92 m/s	1.84 m/s	0.88 m/s	1.76 m/s	3.51 m/s
Contact line speed u_1	−0.13 m/s	−0.26 m/s	−0.52 m/s	1.32 m/s	2.64 m/s	5.27 m/s
Rolling speed u_2	1.05 m/s	2.10 m/s	4.20 m/s	0.44 m/s	0.88 m/s	1.76 m/s
Cam radius R	2.36 mm	2.36 mm	2.36 mm	25.1 mm	25.1 mm	25.1 mm

For the comparative full 3D simulation, only the cam tip contact at 500 rpm was considered. The load case was selected because the greatest influence of the cam edges is to be expected there with the minimum lubrication gap height and the most solid friction, which cannot be considered in the 2D simulation.

2.2. Numerical Modeling and Optimization Approach

The contact problem was modeled for both simulation approaches using the full system finite element approach, according to Habchi [35], with implementation in the commercial finite element software Comsol Multiphysics (Version 6.1). The calculation is dimensionless, whereby all variables are normalized to the variables of the dry herzian line contact [36] p_{Hertz} and b_{Hertz} as well as the fluid properties in the reference state ρ_0 and η_0. The elasticity problem is solved in a coupled system of equations together with the hydrodynamics. On the mechanical side, the linear elasticity equations already available in Comsol

$$\nabla \cdot \sigma = 0, \text{ with } \sigma = C \cdot \varepsilon \tag{1}$$

are solved on the contact area Ω_c, whereby the displacement boundary condition $\delta(x,y,z,t) = 0$ is defined on the side of the area facing away from the contact. The pressure distribution resulting from the force equilibrium with the hydrodynamics is applied as the load boundary condition with the total load

$$\int_{\Omega_c} P_{tot}(X,Y,T) d\Omega_c = \int_{\Omega_c} P(X,Y,T) + P_{solid}(X,Y,T) d\Omega_c \tag{2}$$

The hydrodynamics in the lubrication gap are described by the Reynolds equation [37] adapted by Tan [38] and are calculated in dimensionless form

$$\nabla \cdot \left(\frac{H^3}{\psi} \frac{\overline{\rho}}{\overline{\eta}} \nabla P \right) - \frac{\partial(\overline{\rho}H)}{\partial X} - \frac{\partial(\overline{\rho}H)}{\partial T} = 0 \text{ with } \psi = \frac{12\mu_0 u_m R^2}{b_{Hertz}^3 p_{Hertz}}, \tag{3}$$

whereby at the edges of the calculation area the boundary condition

$$p = 0, \frac{\partial p}{\partial x} = 0, \frac{\partial p}{\partial y} = 0 \tag{4}$$

applies. Here, the density is modeled pressure-dependently according to the model of Dowson and Higginson [39], and the viscosity is described according to the Roelands model [40], whereby the non-Newtonian fluid properties are considered according to the

Carreau model [41] modified by Bair [42] in accordance with the fluid parameters from Table 1. Cavitation in contact is modeled with mass conservation according to the model by Marian et al. [43]. Since the mixed lubrication in the investigated contact cannot be neglected, a stochastic approach (cf. [44]) is used to divide the contact pressure into a hydrodynamic pressure P and a solid contact pressure P_{solid} as shown in Equation (2). The flow factor method developed by Patir and Cheng [45] corrects the pressure distribution in the Reynolds equation depending on the direction of the surface topography.

Since the highly dynamic contact conditions do not allow a truly transient simulation of a cam rotation, the texture variants are simulated in a quasi-static simulation. Here, the load cases described in Table 2 are initially calculated stationary, and then the textures, which are applied to the flat surface of the tappet and numerically coupled with its speed through the contact point, run through the calculation area as a function of time. The discrete microtextures are modeled via the lubrication gap height equation as a function $S(x, y, t)$. In the 2D model, the gap equation in dimensionless form is given by

$$H(X, T) = H_0(T) + \frac{X^2}{2} + \bar{\delta}(X, T) + S(X, T) \tag{5}$$

and, in the case of the 3D model, expands in the Y direction to

$$H(X, Y, T) = H_0(T) + \frac{X^2}{2} + E_g(Y) + \bar{\delta}(X, Y, T) + S(X, Y, T). \tag{6}$$

Here, $\bar{\delta}$ describes the elastic deformation of the surface, and in the 3D model, the edge geometry in the scaled calculation area is defined by $E_g(Y)$, which is described in more detail based on the concept of Winkler et al. [46] for the Cam tappet contact in [34]. For numerical reasons, the textures themselves are described as Gaussian pulses, whereby the texture amplitude A, the texture width w_x and the texture distance d_x are defined as parameters. In the 3D model, the distance in the Y direction d_y and its length w_y are also added, whereby a more detailed description of the modeling can be found in [47].

The objective of the optimization is to minimize the average total friction at the individual load points; the mathematical optimization problem is thus given by

$$\min \int_0^{t_{max}} F_R(A, w_x, d_x, t) dt \cdot \frac{1}{t_{max}}. \tag{7}$$

The frictional force, which depends on the parameters A, w_x and d_x is divided into a solid component and a fluid friction component according to Equation (2) and is time-dependent, as described by

$$F_R(t) = F_{R,fluid}(t) + F_{R,solid}(t) = \int_{\Omega_c} \eta \cdot \dot{\gamma}(x, y, t) d\Omega_c + \int_{\Omega_c} \mu \cdot p_{solid}(x, y, t) d\Omega_c. \tag{8}$$

The parameters of the simulated textures were defined using a Latin hypercube sampling (LHS) experimental design in Python, whereby 40 test points were used for the cam tip contact and 50 for the flank contact. The parameters were determined based on work [30] on the same cam tappet contact. Due to the wider contact area, a larger parameter space was chosen for the cam flank contact. The parameter spaces for both cases are summarized in Table 3.

Table 3. Parameter space of the LHS experimental design used in the simulation.

Parameter	Cam Tip Contact ($\varphi = 0°$)	Cam Flank Contact ($\varphi = -45°$)
Texture amplitude A	0 µm–4 µm	0 µm–15 µm
Texture distance d_x	10 µm–120 µm	10 µm–200 µm
Texture width w_x	10 µm–80 µm	10 µm–200 µm
Texture distance d_y (3D)	1500 µm	-
Texture length w_y (3D)	2000 µm	-

The results of the LHS were then processed using regression according to the Gaussian method [48] AI-based with the Python library scikit-learn (version 1.3.0) to determine the numerical optimum in the parameter space. Three-quarters of the test points were used for training the Gaussian progress regression (GPR) model, and the remaining simulations were used to validate the model. The prediction quality was evaluated using the coefficient of determination (CoD). The computational effort for the identical test plan in both models initially examined was very different. The 2D model had a computing time of approx. 4 h, while the 3D full model was significantly more computationally intensive at almost 16 days. For this reason and due to the very similar tendency of the results, the simplified model was used for all further simulations.

2.3. Combined Manufacturing Process

In the studies presented in this paper, the chromium-manganese alloyed case-hardening steel 1.7131 in the initial sheet thickness of 2.4 mm was used as a reference material. The steel was characterized by a low carbon alloy content of a maximum of 0.19% and a dual-phase structure of martensite and ferrite. The production of the tappet with the required microscopic texture geometry was realized using the tool setup shown in Figure 1. This enables the three required process steps, deep drawing, ironing, and cam tappet, to be combined in a single press stroke. The punch was positioned in the upper tool holder and transferred the pressing force to the workpiece. A hole in the extractor allows the punch to be inserted into it. To ensure high positioning accuracy, the die plates were connected with column guides. The forces from the reinforced die were transferred to the press table via the lower tool holder.

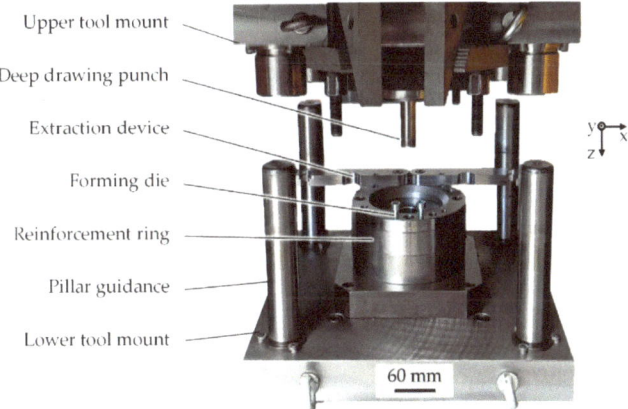

Figure 1. Tool setup of the combined forming process for the manufacturing of microtextured cam tappets.

The cam tappet was implemented using a structured coining cylinder, which was provided with the negative geometry of the subsequent indentations on the component surface. The structuring of the tool for the forming tests was carried out on an erosion machine (SX-200-HBM EDM system, Sarix SA, Sant'Antonino, Switzerland). This works according to the principle of spark erosion, in which a pulsed electrical high-frequency discharge is introduced into the workpiece by a tungsten electrode [49]. Small amounts of material particles are thereby released from the workpiece. During the discharge, a dielectric fluid flows around the engaged electrode. Its tasks are the control of the discharge pulsation, the absorption of heat, and the removal of dissolved particles [50].

Due to the rear turning of the deep drawing punch, the manufactured component did not adhere to the die during the subsequent return stroke. The textured cup was released from the punch at the extraction device. The corresponding process kinematics for the single-stage forming is shown in Figure 2.

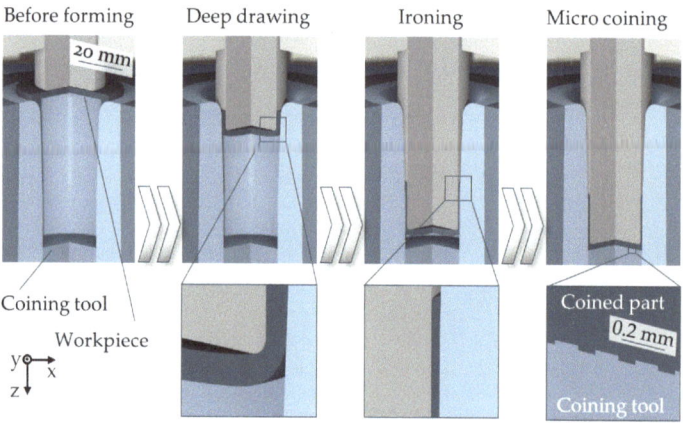

Figure 2. Process kinematics for the single-stage forming.

The single-stage process of kinematics comprises three steps: deep-drawing, ironing and micro-coining. Initially, the float-mounted punch contacts the oiled steel blank. Due to the drawing ratio of 1.5, no blank holder is required, which reduces tool complexity and enables the use of a single-acting press. As the process continues, the conical die leads to a uniformly reduced wall thickness and the desired cup height of the workpiece. Finally, the workpiece is compressed between the punch and the coining cylinder, whereby microstructures are embossed into the base surface. The ram retraction releases the component, which is then removed from the punch by the removal device, preventing renewed contact with the tool textures.

Texture patterns were selected for production that are interesting from a production engineering point of view on the one hand and, on the other hand, also fall within a range that can be considered rather favorable in terms of simulation (variants T_l1 and R_g). At the same time, unfavorable variants (T_g, T_l2 and R_l) were deliberately selected. The texture patterns labeled R were executed as concentric line textures arranged to the center of the tappet, while the textures labeled T represent discrete textures. On the one hand, global texture patterns were produced, which should be identical across the entire tappet (labeled g for global). In addition, different local textures were created (labeled l for local). The idea of local optimization is because optimal textures can be defined in the areas of the cam tip and cam flank contact.

2.4. Amorphous Carbon Coating

After the tappets were microtextured, the DLC-coating was applied without further post-treatment to test the feasibility of the coating on the microtextured surface and to identify the effect on the tribological behavior. Prior to deposition, the tappets were cleaned ultrasonically in acetone and isopropanol for 10 min each and afterward blown dry with nitrogen. An industrial-scale PVD deposition unit (TT 300 K4, H-O-T Härte- und Oberflächentechnik GmbH & Co. KG, Nuremberg, Germany) was utilized for coating fabrication. The DLC-coating was deposited under 2-fold substrate rotation on the tappets' microtextured surface using a combined PVD/PECVD process. Prior to the coating process itself, the deposition chamber was evacuated to high vacuum conditions with an initial pressure of 5.0×10^{-3} Pa and then heated to 250 °C for 40 min. Afterward, the tappet's surface was (Ar+) ion plasma etched for 40 min with an argon gas flow (Ar purity 99.999%) of 500 sccm and a bipolar pulsed bias voltage of -500 V (pulse frequency 40 kHz, reverse recovery time 5 µs). The same cleaning process was carried out for the Cr and WC target (both a purity of 99.9%) with a dimension of 267.5 × 170 mm and closed shutters for 3 min to remove impurities from the target. The coating architecture consisted of a thin Cr adhesion layer, a graded CrWC intermediate layer and a WC support layer, which were deposited by PVD using unbalanced magnetron sputtering (UBM) in an Ar atmosphere. The relevant deposition parameters are listed in Table 4. Subsequently, the functional layer of hydrogen-doped amorphous carbon (a-C:H) was deposited in a PECVD process utilizing a mixed acetylene-Ar (C_2H_2-Ar) plasma at 220 and 40 sccm, respectively, a bias voltage of -450 V, a chamber temperature of 100 °C and a substrate rotation speed of 3 rpm for 8580 s deposition time.

Table 4. Relevant deposition parameters for the DLC-coating without the functional layer.

Layer	Power	Pulse Frequency	Reverse Recovery Time	Duration	Chamber Temperature	Bias Voltage	Ar Flow
Cr	5.0 kW	70 kHz	4 µs	240 s	140 °C	−100 V	180 sccm
CrWC	5.0 ǀ 0.3 ↗ 1.2 kW	40 kHz	5 µs	1260 s	140 °C	−100 V	180 sccm
WC	1.2 kW	40 kHz	5 µs	1080 s	120 °C	−100 V	195 sccm

2.5. Characterization of Tools and Textured Tappets

The confocal microscope (VK-X 200, Keyence Deutschland GmbH, Neu-Isenburg, Germany) was used to characterize the microtextures on the component and tool sides. The measurement result was achieved by combining many focal planes with an incremental distance of 0.5–1.0 µm. To ensure a high resolution of the measured topography, the measurements were carried out at 500× magnification. In addition, height measurements of the texture surface were required to assess the accuracy of the micro-coining and to investigate the wear behavior. The analysis was carried out by defining the two measurement areas of texture and base. This procedure is necessary because the measurement method cannot give absolute height values, only relative values. The height of the texture results from the difference between the measurement area of the texture and the measurement area of the soil. With the aim of eliminating edge effects, the size of the texture measurement areas was defined as 70% of the nominal area of the top of the texture. A distance of 15 µm from the texture had to be considered for the placement of the measuring surface of the floor. An area corresponding to five times the target area of the top of the texture was used for the bottom area to achieve a sufficiently large measuring area for referencing the texture height. In addition, scanning electron microscopy (Merlin Gemini II, Carl Zeiss AG, Oberkochen, Germany) was used to assess the wear mechanisms of the tool-side textures and to evaluate selected textures on the tappets. To analyze the resulting texture patterns more closely regarding their geometry, representative laser microscopic images were taken (VK-X 3100, Keyence Deutschland GmbH, Neu-Isenburg, Germany) at 200× magnification.

2.6. Characterization of the Coating

To characterize the coating structure and coating hardness, round samples of tool steel (1.2379) were coated in the same process and then examined. The indentation hardness H_{IT} and indentation modulus E_{IT} were measured via nanoindentation (Picodentor HM500 and WinHCU, Helmut Fischer, Sindelfingen, Germany) in accordance with Oliver and Pharr [51,52]. Therefore, a Vickers indenter was utilized, and 17 indentations were produced with a force of 18 mN. The measurements were carried out using the standard application until the indenter attained the predetermined penetration force. In this case, the evaluation of the mechanical properties was performed at a penetration depth of <10% of the respective coating thickness to minimize the influence of the substrate [53]. A FEI Helios Nanolab 600i (ThermoFisher Scientific, Waltham, MA, USA) with a focused ion beam workstation (FIB) was used to generate cross-sections to characterize the coatings. An accelerating voltage of 5 kV and a probe current of 0.69 nA were selected to image the microstructure. To demonstrate the layered architecture with the elemental composition of each layer, energy-dispersive X-ray spectroscopy (EDS) mappings were also performed. The same electron microscope with an Xmax^{50} detector (Oxford Instruments, Abingdon, UK) was used for this purpose.

The surface roughness of the coating was measured directly on the non-textured tappets at 200× magnification using laser scanning microscopy (VK-X 3100, Keyence Deutschland GmbH, Germany) and compared with the roughness of the uncoated tappets. Four areas of 250 µm × 250 µm were measured on each of the three images, using an S-filter of 2 µm to match the measurement resolution. To examine the uniform coating thickness in the textures, the coated tappets were separated in the texture area, embedded, and then metallographically polished. As the epoxy resin used for embedding, the DLC-coating, and the substrate is abraded differently, images were taken and analyzed using focal stacking light microscopy (VK-X 3100, Keyence Deutschland GmbH, Germany) at 500× and 1000× magnification.

2.7. Tribological Characterization

To fundamentally characterize the resulting friction and wear behavior, tribometer tests were carried out in a linear oscillating pin-on-disc configuration, as shown in Figure 3. It should be mentioned at this point that, as a pure model test, these cannot be used for comparison with the simulations. 100Cr6 bearing balls with a diameter of d = 4 mm were used as counterbody, and a test was carried out over 20,000 cycles for all textured cam tappets with the different variants.

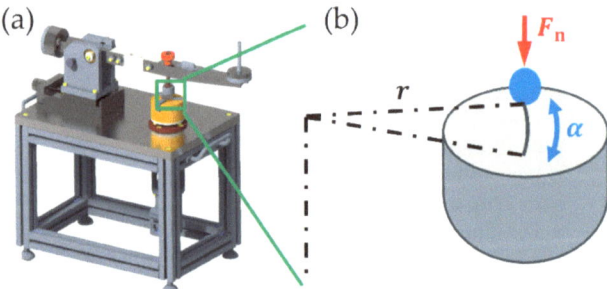

Figure 3. Used pin-on-disc tribometer (**a**) with linear reciprocal kinematics with excentric clamping (**b**).

In addition, two long-term tests with 200,000 cycles each were carried out on the DLC-coated tappets. The complete test conditions are summarized in Table 5.

In addition to the average frictional force present in the contact, the wear of the counterbodies was also examined to determine the influence of the coating. The wear on the textured tappets, on the other hand, was only recorded qualitatively, as a quantitative evaluation of the wear volume is not possible or is associated with excessive external uncertainties. The evaluation of the counterbody wear was carried out using light microscopy (Metallux I, Leica, Wetzlar, Germany) at 200× magnification. The wear on the tappets was qualitatively analyzed using laser scanning microscopy (VK-X 3100, Keyence Deutschland GmbH, Germany) at 200× magnification. The differences in friction and wear behavior were evaluated using a two-sided t-test for unpaired samples.

Table 5. Setup of the pin on disc tribometer tests.

Lubricant	10 µL PAO-40
Temperature t_{amb}	20.6–21.9 °C
Humidity	42–50%
Counterbody	100Cr6/1.3505 bearing balls (grade G10, DIN 5401, Ra $\leq$ 0.02 µm, hardness $\geq$ 61 HRC) with $d = 4$ mm
Tappet variants	Textured 16MnCr5/1.7131 (+DLC)
Normal load F_N	5 N
Max. Hertzian pressure p_{Hertz}	1472 MPa
Sliding speed u	0.092 m/s
Sliding radius r	45 mm
Cycles n	20,000/200,000
Oscillation angle α	30°

3. Results

3.1. Simulation Process and Numerical Optimization

To investigate the extent to which the presented simulation approaches in 2D and 3D differ in their results, both simulations were carried out for the cam tip contact at 500 rpm (load case (a)) as an example. Figure 4 shows the mean friction force per test point for both models, with the force shown relative to each modeling approach. For improved visualization, the results are shown as a function of two of the three parameters.

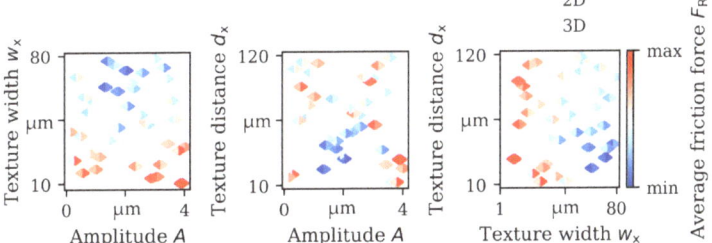

Figure 4. Resulting relative mean frictional force as a function of the selected design parameters for all test points ($n = 40$) and as a function of the simulation approach (2D and 3D) for the cam tip contact at 500 rpm.

It turned out that the simplified 2D simulation model led to similar results regarding the optimum textures. In particular, the globally clearly recognizable trend can be determined equally in both simulation approaches. For the load case (a) shown in Figure 4, textures with a medium texture amplitude, a rather large texture width, and a medium to rather small texture distance can be qualitatively evaluated as advantageous.

The optimum textures determined using the GPR models for all load cases investigated are summarized in Table 6.

Table 6. Optimization results for the different load cases at the cam tip (T) and flank (F) contacts including the quality of the optimization model.

Loadcase	A	w_x	d_x	CoD
(a) 500 rpm T	1.4 µm	77.4 µm	43.9 µm	93.3%
(b) 1000 rpm T	1.4 µm	22.7 µm	75.8 µm	19.4%
(c) 2000 rpm T	3.2 µm	11.3 µm	100.6 µm	64.5%
(d) 500 rpm F	1.4 µm	20.3 µm	193.5 µm	81.1%
(e) 1000 rpm F	14.6 µm	53.0 µm	102.1 µm	88.8%
(f) 2000 rpm F	14.7 µm	154.5 µm	14.0 µm	91.5%

The example of load case (a) shows that the numerically determined optimum falls within the optimum parameter range described above in purely qualitative terms, and the prediction quality is also very good in this case. It is also noticeable that the very different load cases lead to significantly different optimum textures. For example, while the optimum texture width at the cam tip contact decreases with the speed and the optimum texture distance increases, this is exactly the opposite for the flank contact. Apart from load case (b), the prediction quality of the developed GPR models can be rated as good. The model quality is particularly influenced by how well the test points describe the behavior in the parameter space. The prediction is particularly difficult in the case of ambiguous trends, as there may be several optimal parameter ranges for each load case instead of a specific optimum.

For this reason, the purely quantitative evaluation of the numerical optimum is not particularly meaningful without an additional consideration of the underlying model. However, the graphical evaluation of the results becomes more complex with just three parameters, as was the case in this work. To illustrate the trends in the parameter space, all six models developed are shown in Figure 5. The predicted mean friction forces were evaluated and displayed for 1000 randomly generated parameter combinations.

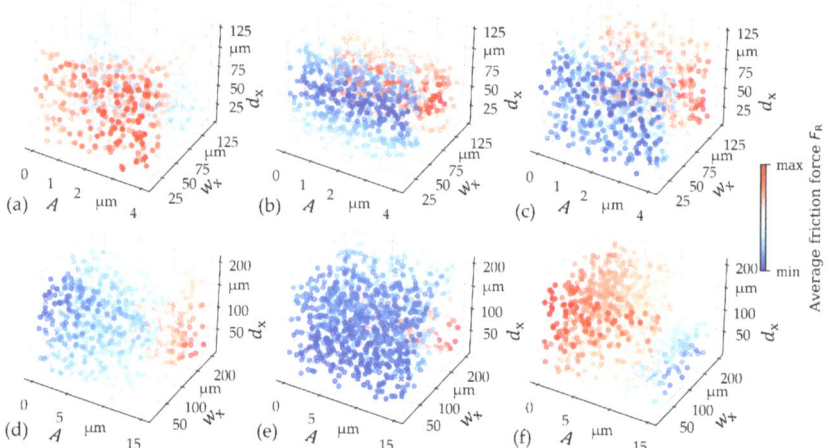

Figure 5. Average friction forces predicted by the GPR models for load cases (**a**–**f**).

The dependency on the parameters can be clearly recognized here. It is noticeable that for some load cases (e.g., (a) and (e)), there is a clearly recognizable optimization direction in the parameter range with a point optimum, whereas for others (e.g., (b) and (c)) there are larger, similarly optimum parameter ranges. Consequently, the prediction quality of these models is somewhat poorer.

3.2. Manufacturing Process Challenges and Solutions

Regarding the manufacturing process of the textured tappets, the challenges of the components, as well as the tools, are analyzed. The challenges were identified from internal forming tests carried out with the tool shown in Figure 1. The distortion of the textures depending on the positions is a significant challenge of single-stage forming in combination with micro-coining. The textures created in the outer areas are shown in Figure 6.

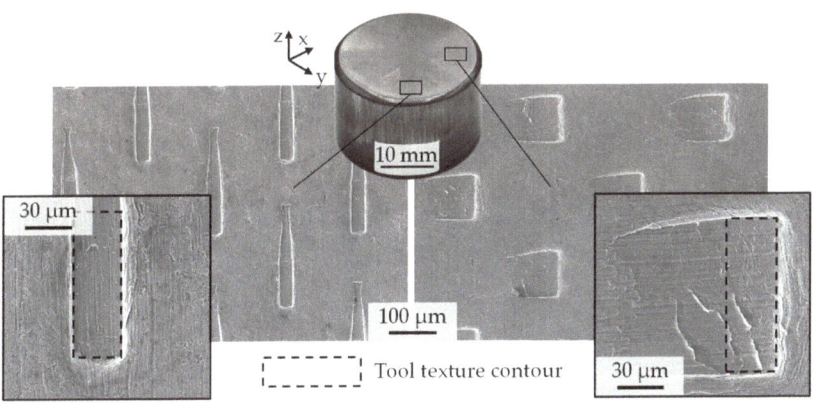

Figure 6. Distortion of the textures depending on the positions shown in the SEM. The rectangle texture corresponds to the texture used for the manufacturing tests.

The explanation for this component error is the radial material flow that occurs in the last process step, during the compression of the component base between the punch and the coining cylinder. The material flow is lowest in the bottom center, which explains the good reproduction accuracy of the textures in this area. Only a slight superposition of the local material flow in micro-coining and the global material flow due to the compression process of the tappet base is, therefore, advantageous for achieving a low texture distortion. In contrast, higher radial displacements are found in the direction of the bottom edge. This suggests that the radial material flow overlaps the material flow from the micro-coining to a large extent and, therefore, has a negative effect on the dimensional accuracy of the textures on the workpiece side. In addition to the texture distortion caused by the radial material flow, further challenges were identified, which are summarized in Table 7.

Table 7. Challenges and solutions in the single-stage manufacturing of textured tappets.

	Textured Component		Textured Tool	
Process limitation	Wall necking	Multiple texture pattern	Die wear	Texture wear
Characteristic				
Cause	Adhesive friction with die	Contact loss with coining tool	Contact pressure in ironing	Shear forces in micro coining
Counter measures	Friction reduction of die; Increase of wall thickness	Adaption of punch geometry; Use of active counterholder	Application of tool coatings; Adaption of tool geometries	Radial texture orientation; Coating of tool surface

After only a few strokes, a local necking of the material forms on the outer wall of the base body near the radius. As the number of strokes increases, the necking spreads until it finally surrounds the cup wall. A further increase in the number of strokes results in material fracture, whereby the base below the radius is detached from the cup wall. This stage follows the states described in the literature [54], consisting of crack initiation, crack propagation and crack arrest. As the boundary conditions are kept constant during component manufacture, a wear-related change in the tribological system can be assumed.

An explanation for this is the increase in wall friction due to wear on the inner wall of the die. This increases the static friction that must be overcome when the finished component is removed from the die. If the static friction in the lower wall area exceeds the yield point of the material, necking can form at these points, which can lead to complete bottom tears if friction increases further. To counteract this, the die could be finished with a friction-reducing PVD or CVD coating, or the wall thickness of the cup could be increased.

In addition to macroscopic component failure, macroscopic component defects are identified. A major challenge in single-stage texturing is the creation of more than one texture pattern. Additional undesired texture patterns occur slightly offset from the original location. This is due to the inhomogeneous contact formation during the cam tappet. Partial loss of contact occurs between the semi-finished product and the embossing cylinder. The simultaneous radial material flow causes a relative displacement between the component and the tool. This leads to a second embossing pattern, which occurs at a different location compared to the first pattern. Possible remedies are provided by adapting the punch geometry, which prevents contact detachment and, therefore, relative displacement. In addition, an actively controlled counterholder can be used, which clamps the sheet before deep drawing and stretching and thus prevents the convex geometry of the component.

On the tool side, severe wear is detected on the inner radius of the die. Numerical analyses identify a particularly high contact pressure in this area. This is caused by the ironing and the resulting material build-up due to the reduction in wall thickness. For this reason, cold welding of the semi-finished material occurs on the die surface. As a countermeasure, a PVD coating containing titanium is applied to the inside of the die. This measure enables the complete elimination of this process error. A small draft angle, for example, offers the potential to reduce the contact pressures by distributing the forming zone over a larger area.

In addition to the wear of the die, the change in the texture surface on the tool side represents a significant process challenge. As the negative geometry of the tool is transferred to the component during micro-coining, texture wear on the tool side impairs the coining accuracy of the process. The causes and remedial measures are therefore described separately in the following section.

3.3. Application Behavior on the Tool Side

An abstracted coining process is used to investigate the wear behavior of the tool textures. This is introduced in [55]. Since the deep drawing and ironing only have a minor influence on the work hardening in the cup bottom and thus on the tool load, the process for the wear investigation focuses on the cam tappet itself. The abstracted process, therefore, reproduces the indentation of the tool textures into the workpiece as well as the radial material flow with a sheet thickness reduction of 0.1 mm. The tool textures exhibit a height of 30 µm for the analysis. Although these are significantly higher than the structures identified as ideal in the simulation, wear mechanisms can already be identified at lower stroke rates. As fewer components must be produced, this measure contributes to more sustainable research.

The analysis is based solely on the texture height, as the main effects of wear on this dimension were shown in the previous investigations. Figure 7a shows the development of the texture height at the three positions on the punch surface with three measured punches and three scanned textures per position.

For measuring Positions 1 and 2, a considerable decrease in texture height can already be seen after 250 strokes. Further strokes no longer lead to a continuous decrease in height. This is an indication that resistance to low-cycle fatigue is reached at the residual height of 10.8 ± 2.0 µm at Position 1 and 12.0 ± 1.6 µm at Position 2. The running-in phase at Position 1 is, therefore, already completed after 250 strokes. The degressive progression of the amount of wear is an indication of the state of the incubation period, in which no or very little wear is usually detected [56]. However, a progressive increase in the amount of wear, which heralds the failure of the textures, cannot be detected even after 4000 strokes.

The incubation period, therefore, continues, although its end is only to be expected after a large number of further strokes. The measurable amount of wear is almost constant during this time, as processes such as crack formation and growth, as well as microstructural changes, occur gradually and slowly.

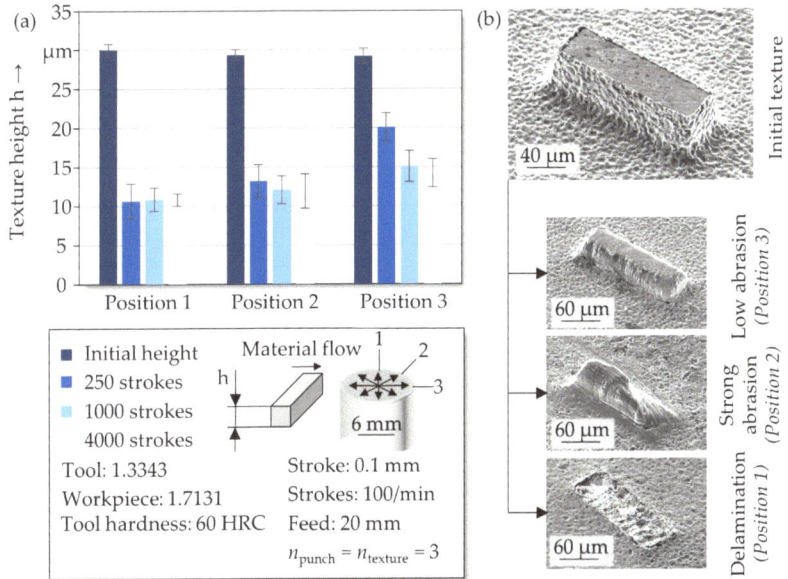

Figure 7. Texture height with respect to the position on the punch (**a**) and the corresponding signs of wear (**b**).

In Position 3, a gradual wear behavior of the texture surface can be observed. This is the degressive increase in the amount of wear in the running-in phase. The decrease in texture height appears to run into saturation and thus into the incubation period in the range of 4000 strokes. Due to the supporting effect of the longitudinal side of the texture, textures that are arranged in the direction of the material flow are consequently less affected by wear.

Initial textures before forming in Figure 7b have a crater-shaped surface on the side faces. This is caused by the incremental discharge pulses in the micro-erosion manufacturing process. As the electrode is not in contact with the top surface of the textures, this surface remains in its original state. Only the characteristic fusion build-up forms at the edge between the side surface and the top surface. As shown in the previous section, the wear behavior is highly inhomogeneous and depends on the orientation of the texture. Therefore, to identify the wear mechanisms, SEM images of the orientations at 0°, 45° and 90° are examined. Textures at 0° are only affected by slightly abrasive wear. Particle removal is less pronounced in the bonding zone and increases towards the top surface. This behavior can be explained by the localization of stress peaks in tools at their edges and tight radii [57]. Consequently, the abrasive wear at Position 3 is detected particularly on the side surfaces near the edge to the top surface. Due to the higher shear forces, highly abrasive wear occurs at Position 2 at an orientation of 45°. The material flow of the sheet metal strip causes particles to be detached from the surface of the base body by micro-chipping processes and repeated scoring [56]. It is to be expected that these detached tool particles cause furrow wear on the outer tool surface due to the high hardness. They are transported further by the radial material flow. This also accelerates the abrasive wear of the textures located there.

In summary, it can be stated that the use of non-rotationally symmetrical texture geometries on forming tools for embossing processes leads to highly anisotropic wear behavior due to the relative orientations to the material flow. Depending on this alignment, the running-in phase or the incubation period still prevails after 4000 strokes. As the textures are wear-resistant at a height of at least 10 µm, the suitability of the process for the application under investigation has been proven. Generally, the texture depth of structured surfaces is mainly lower than 10 µm [14]. The lower the required texture depth, the less critical the tool wear in combined forming processes. The manufacturing process presented is, therefore, suitable for most applications from a wear technology perspective.

3.4. Manufactured Textures

The resulting textures on the tappets are shown in Figure 8. The production-related deviations from the ideal geometry described above can also be recognized here to some extent. The deviations from the ideal shape are particularly pronounced in the deeper patterns (see Figure 8b,d). In contrast, the flat texture patterns R_g (Figure 8a), T_l1 (Figure 8f) and T_l2 (Figure 8g) deviate only slightly.

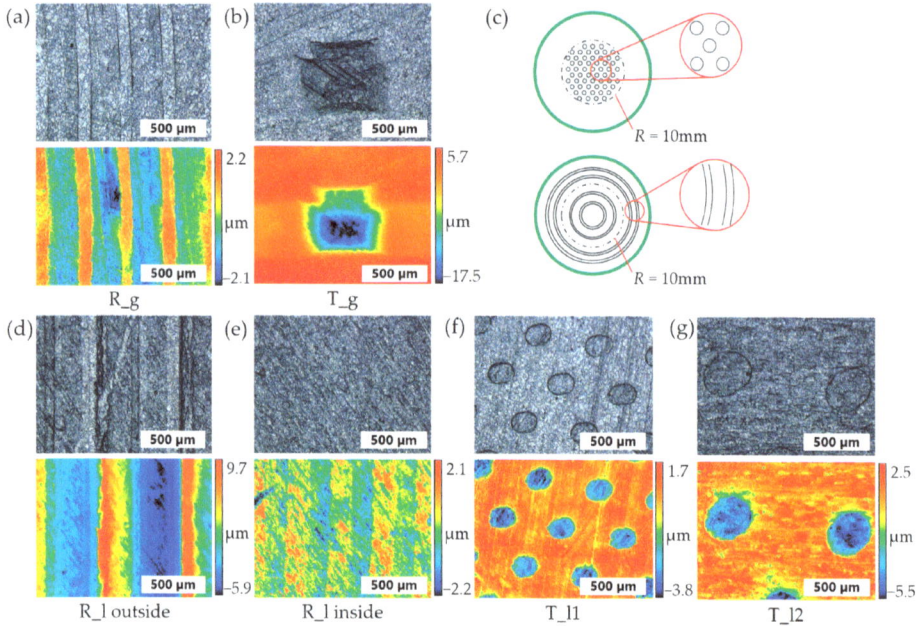

Figure 8. Light microscopic and laser scanning microscopic images of the texture patterns applied to the tappets: global textures (**a**,**b**), distribution of the local texture patterns (**c**) and their characteristics (**d**–**g**). The different height scales must be considered.

3.5. DLC-Coating Characterization

The indentation hardness H_{IT} of the coatings was 25.8 GPa with an indentation modulus E_{IT} of 199.9 GPa. The values are, therefore, within the usual range for DLC coatings. The design of the DLC coating is shown in Figure 9a in the FIB cut. The adhesive layer consists of 0.3 µm Cr, 0.2 µm CrWC and 0.1 µm WC. The thickness of the a-C:H functional layer is approx. 2 µm, whereby the transition between a-C:H:W and a-C:H is not recognizable in the FIB cut. However, this can be seen in the corresponding EDS mapping in Figure 9b.

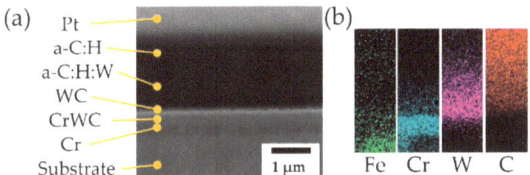

Figure 9. Coating design in the FIB cut (**a**) and the corresponding EDS mapping (**b**).

Figure 10 shows three examples of microsections through the coated texture patterns T_g and T_l2. The exact position of the microsection cannot be determined. It is noticeable that the coating growth is not influenced by the textures, as they are very wide in relation to their depth. Shadowing effects, therefore, do not occur, and even in the areas of the texture edges (see Figure 10g), a uniform coating application can be determined.

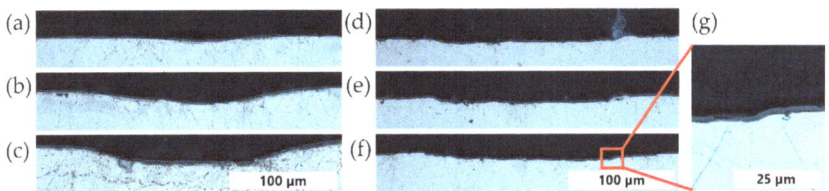

Figure 10. Coating in the textures T_g (**a–c**) and T_l2 (**d–f**) as well as exemplary enlargement in the area of the texture edges (**g**).

The measured roughness of the DLC-coated tappets is $S_a = 0.20$ μm ($\sigma = 0.043$ μm) and $S_z = 3.98$ μm ($\sigma = 0.87$ μm) and is therefore lower than the uncoated ones with $S_a = 0.35$ ($\sigma = 0.037$ μm) and $S_z = 5.25$ μm ($\sigma = 0.91$ μm). This can be explained by the fact that the surfaces are comparatively rough before coating, and the coating with a similar thickness clogs the fine structures of the surface due to the application process. The coarse shape of the surface, including the applied textures, on the other hand, does not change, which can also be seen in Figure 10 as an example.

3.6. Tribological Behavior in Model Contact

The results from the tribometer tests are shown in Figure 11 as a function of the texture variant investigated and for the untextured reference.

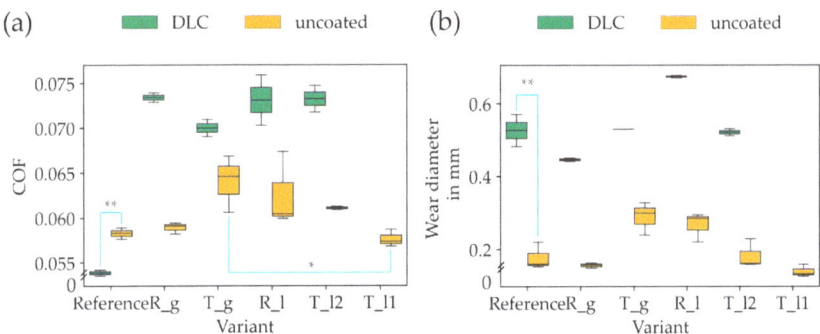

Figure 11. Mean friction values (**a**) and diameter of the wear calotte of the counter body (**b**) of the tested variants ($n = 3$ uncoated and $n = 2$ with DLC) with selected significance levels (*t*-test with $p < 0.05$ (*) and $p < 0.01$ (**)). Please note the axes cut off for reasons of presentation.

The amorphous carbon coating only led to a reduction in friction in the non-textured case, while the counter body wear increased considerably at the same time. In the case of uncoated samples, a clear difference was found between the optimized textures T_l1 and R_g compared to the textures T_g and R_l, which were assumed to be unsuitable, with negative effects presumably overriding the positive texture effect due to the shape deviations. However, only the difference in friction between T_g and T_l1 was significant, which may be due, among other things, to the sometimes-large variance in the repeated tests for tribometer tests. Additional long-term tests (see Figure 12c) at the textures T_l2 show a clear running-in behavior of the coating until the friction level of the uncoated texture variant is reached after approx. 180,000 cycles.

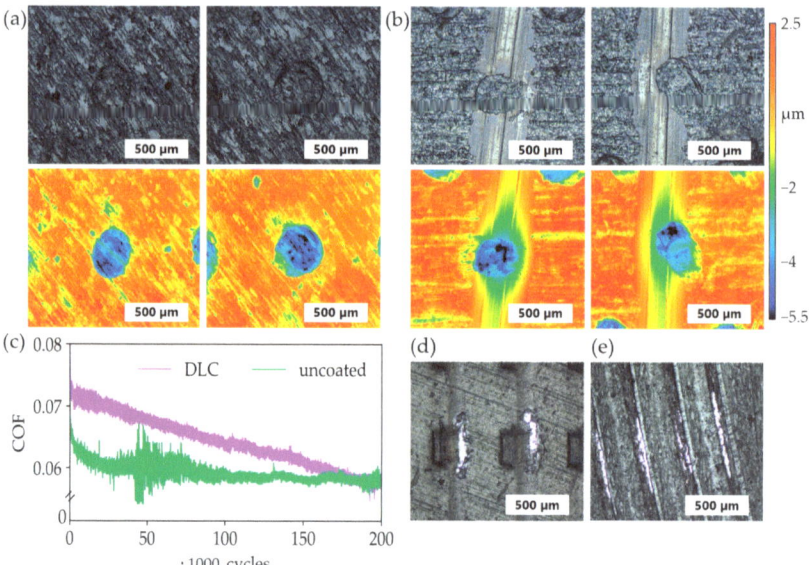

Figure 12. Wear of the coated (**a**) and uncoated (**b**) textures (two examples) with the corresponding LSM images and the friction curves (**c**) in the long-term tests with texture T_l2. Note the axis cut off for display reasons. In addition, the signs of wear on the unfavorable texture patterns T_g (**d**) and R_l (**e**), which occurred sporadically even after the short test running times.

This also explains why the COFs for the coated tappets were significantly higher in the short tests. At the same time, wear on the textures was completely prevented by the additional DLC-coating, whereas the uncoated tappets were already largely worn out after the long-term test (see Figure 12a,b). In the case of the coated tappets, the light-colored areas in Figure 12a indicate the running in of the coatings. Due to their unfavorable surface with protruding edges, however, the texture patterns T_g (see Figure 12d) and R_l (see Figure 12e) already showed coating failure at individual points in the short tests. With the textures T_l1 and R_l, however, no wear was recognizable there, as was also the case with T_l2.

4. Discussion

The numerical evaluation of textures using EHL simulation has long proven to be an effective approach, as it reduces the amount of experimental testing required. In the selected application example, the simplified model leads to similar results as the complete 3D contact model. By using the simplified model, the calculation time can be reduced considerably. For the investigated cam tappet contact, it could be shown that the optimal texture patterns strongly depend on the very different contact conditions, which makes it difficult to define a uniformly optimal texture since it would also have to be considered how long the respective conditions exist. For the load case (a), which is also considered to have similar boundary conditions and model structure, the results are similar to those from [30]. From a numerical point of view, it can also be concluded that the textures in the center of the tappet should be smaller than in the edge areas, where a larger proportion of the cam flank is in contact. In addition to simply reading out the optimum parameters, the trend in the parameter space is also important for the evaluation of the optimization model to define sufficient textures.

The tendencies from the simulations could also be recognized in tribometer tests, but the positive effects are partly overlaid by the production-related deviations of the textures so that no effective friction reduction could be proven compared to the non-textured case. We must point out that the results of the theoretical simulation and texture optimization should not be compared directly with those of the pure model test.

For the coating of steel substrates, the substrates are often pretreated by polishing to low roughness values ($R_a < 0.1$ µm). This serves to ensure good adhesion of the coating to the steel substrate and that the roughness of the applied coating is not influenced by rough substrate surfaces. In tribological contacts, a high roughness of the coating often leads to high friction and, thus, abrasive wear of the counterbody if the coating exhibits a higher hardness than the contact partner, which in the case of the a-C:H coating against steel. Here, we investigated the possibility of coating application on the microtextured surface and found that it basically achieved excellent performance. The results of the tribological tests showed no flaking, delamination or severe wear of the a-C:H coating, meaning that coating adhesion under load remained stable. However, an exception applies to texture patterns with unfavorable and sharp edges. Here, the coating failure already occurred in the short tests on individual edges. Although the additional a-C:H layer applied to the texturing did not reduce friction in the short tests, the wear on the tappets was significantly reduced in long-term tests. The initially higher friction is presumably because the unfavorable edges on the textures caused by the production process are also protected against rapid smoothing and, therefore, have a negative effect on the contact. This means that the positive effect of an additional coating only takes full effect after the contact has been run in, and it can be assumed that friction is also reduced, as well as wear. Another option for future investigations would be to polish the coating after deposition.

The production of textured metallic components should be implemented using sustainable processes. Forming processes are particularly suitable due to their high resource efficiency. Texturing can thus be integrated directly into the production of the base body. The main challenges that have so far hindered implementation in an industrial environment are texture dimensional accuracy and tool wear. In this context, the measures identified in this study are designed to meet these challenges. Regarding the application of the cup tappet, where only low texture depths are required, it was also demonstrated that the process is also suitable for large quantities. Tool wear only plays a subordinate role at this texture depth.

5. Conclusions

As part of our work, we have dealt with the numerical optimization, production and coating of textured tappets. We can summarize the following key findings:

- A numerical EHL simulation with subsequent optimization can be used to investigate textures efficiently. In addition to the numerical optimum, the prediction trend in the model is also of decisive importance for understanding the contact.
- The process combination of deep-drawing, ironing and micro-coining is suitable for applying structures that are relevant for microtexturing to metallic surfaces.
- For applications where deep textures are required, tool wear can be reduced by orienting the textures radially in the direction of the material flow.

- Tribometer tests have shown that, from a numerical point of view, better textures lead to reduced friction in comparison to unsuitable ones. At the same time, the production-related deviations in the non-post-treated state appear to outweigh the positive effects.
- Application of the coating to the microtextured surface of the tappets was effective without post-processing of the surface, as is often the case for coated steel substrates, and led to a significant reduction in wear, which was observed based on optical evaluations after the tribological tests.

Author Contributions: The following contributions are based on the Contributor Roles Taxonomy (CRediT): Conceptualization, C.O. and M.R.; methodology, C.O., M.R., S.T. and M.M.; software, C.O.; validation, C.O. and M.R.; investigation, C.O., M.R. and A.S.; resources, S.T. and M.M.; data curation, C.O. and M.R.; writing—original draft preparation, C.O., M.R. and A.S.; writing—review and editing, T.R., S.T. and M.M.; visualization, C.O. and M.R.; funding acquisition, S.T. and M.M.; Project administration, S.T. and M.M. All authors have read and agreed to the published version of the manuscript.

Funding: This research was funded by the German Research Foundation (DFG) under Project 426217784 within the Priority Program SPP 1551 "Resource-efficient construction elements". Funded by the Open Access Publishing Fund of the University of Bayreuth.

Data Availability Statement: The numerical transient simulations under consideration typically generate a very large amount of data, which is why the complete results are not made available in a public database. For further information, please contact the authors.

Acknowledgments: C. Orgeldinger, A. Seynstahl, T. Rosnitschek and S. Tremmel greatly acknowledge the continuous support of University of Bayreuth. M. Reck and M. Merklein greatly acknowledge the continuous support of the Friedrich-Alexander-Universität Erlangen-Nürnberg (FAU). The authors would especially like to thank S. Wartzack from Engineering Design at Friedrich-Alexander-Universität Erlangen-Nürnberg (FAU) for the opportunity to use resources. The authors would also like to thank M. Göken and M. Köbrich from Materials Science and Engineering I at Friedrich-Alexander-Universität Erlangen-Nürnberg (FAU) for providing access to the laboratory equipment and the performance of the FIB cross-sections and SEM-EDS measurements.

Conflicts of Interest: The authors declare no conflicts of interest.

Nomenclature

a_c	Carreau parameter
A	Texture amplitude
b_{Hertz}	Hertzian contact half-wide
C	Elasticity tensor
d_x	Texture distance in x-direction
d_y	Texture distance in y-direction
w_x	Texture width in x-direction
w_y	Texture length in y-direction
E_c	Young's modulus of the cam
E_g	Function of the cam edge geometry
E_t	Young's modulus of the tappet
F_N	Contact normal force
F_R	Friction force
$F_{R,solid}$	Solid friction force
$F_{R,fluid}$	Fluid friction force
G_c	Critical shear stress
H	Dimensionless lubricant gap height
H_0	Dimensionless distance between undeformed bodies

n	Number of test cycles
n_c	Carreau parameter
n_{cam}	Camshaft speed
p	Hydrodynamic pressure
p_{solid}	Solid contact pressure
P	Dimensionless hydrodynamic pressure
P_{solid}	Dimensionless solid contact pressure
P_{tot}	Dimensionless total contact pressure
p_{Hertz}	Hertzian contact pressure
r	Sliding radius
R	Cam radius
S	Textures geometry
t	Time
T	Dimensionless time
t_{amb}	Ambient temperature
t_{max}	End time of the simulation
t_{oil}	Oil temperature
u	Sliding speed
u_m	Mean entrainment velocity
x, y, z	Coordinates
X, Y, Z	Dimensionless coordinates
α	Oscillation angle
α_η	Pressure viscosity coefficient
$\dot{\gamma}$	Shear rate
δ	Elastic deformation
$\bar{\delta}$	Dimensionless elastic deformation
ε	Strain tensor
η	Viscosity
η_0	Base viscosity
$\bar{\eta}$	Dimensionless viscosity
η_∞	Second plateau viscosity
μ	Coefficient of friction
ν_c	Poisson's ratio of the cam
ν_t	Poisson's ratio of the tappet
ρ	Density
ρ_0	Base density
$\bar{\rho}$	Dimensionless density
σ	Stress tensor
σ_c	Combined surface roughness
φ	Cam angle
ψ	Term of the Reynolds equation
Ω	Calculation area
Ω_c	Central calculation area
∇	Nabla operator

References

1. Ciulli, E. Vastness of Tribology Research Fields and Their Contribution to Sustainable Development. *Lubricants* **2024**, *12*, 33. [CrossRef]
2. Holmberg, K.; Erdemir, A. Influence of Tribology on Global Energy Consumption, Costs and Emissions. *Friction* **2017**, *5*, 263–284. [CrossRef]
3. Shah, R.; Woydt, M.; Zhang, S. The Economic and Environmental Significance of Sustainable Lubricants. *Lubricants* **2021**, *9*, 21. [CrossRef]
4. Budinski, K.G.; Budinski, S.T. *Tribomaterials: Properties and Selection for Friction Wear and Erosion Applications*; ASM International: Novelty, OH, USA, 2021; ISBN 978-1-62708-321-8.
5. Donnet, C.; Erdemir, A. (Eds.) *Tribology of Diamond-like Carbon Films: Fundamentals and Applications*; Springer: New York, NY, USA, 2008; ISBN 978-0-387-30264-5.
6. Rosenkranz, A.; Heib, T.; Gachot, C.; Mücklich, F. Oil Film Lifetime and Wear Particle Analysis of Laser-Patterned Stainless Steel Surfaces. *Wear* **2015**, *334–335*, 1–12. [CrossRef]

7. Fowell, M.; Olver, A.V.; Gosman, A.D.; Spikes, H.A.; Pegg, I. Entrainment and Inlet Suction: Two Mechanisms of Hydrodynamic Lubrication in Textured Bearings. *J. Tribol.* **2007**, *129*, 336–347. [CrossRef]
8. Marian, M.; Weikert, T.; Tremmel, S. On Friction Reduction by Surface Modifications in the TEHL Cam/Tappet-Contact-Experimental and Numerical Studies. *Coatings* **2019**, *9*, 843. [CrossRef]
9. Lohner, T.; Ziegltrum, A.; Stemplinger, J.-P.; Stahl, K. Engineering Software Solution for Thermal Elastohydrodynamic Lubrication Using Multiphysics Software. *Adv. Tribol.* **2016**, *2016*, 6507203. [CrossRef]
10. Qiu, M.; Chyr, A.; Sanders, A.P.; Raeymaekers, B. Designing Prosthetic Knee Joints with Bio-Inspired Bearing Surfaces. *Tribol. Int.* **2014**, *77*, 106–110. [CrossRef]
11. Gao, L.; Yang, P.; Dymond, I.; Fisher, J.; Jin, Z. Effect of Surface Texturing on the Elastohydrodynamic Lubrication Analysis of Metal-on-Metal Hip Implants. *Tribol. Int.* **2010**, *43*, 1851–1860. [CrossRef]
12. Scaraggi, M.; Mezzapesa, F.P.; Carbone, G.; Ancona, A.; Tricarico, L. Friction Properties of Lubricated Laser-MicroTextured-Surfaces: An Experimental Study from Boundary- to Hydrodynamic-Lubrication. *Tribol. Lett.* **2013**, *49*, 117–125. [CrossRef]
13. Gropper, D.; Wang, L.; Harvey, T.J. Hydrodynamic Lubrication of Textured Surfaces: A Review of Modeling Techniques and Key Findings. *Tribol. Int.* **2016**, *94*, 509–529. [CrossRef]
14. Marian, M.; Almqvist, A.; Rosenkranz, A.; Fillon, M. Numerical Micro-Texture Optimization for Lubricated Contacts—A Critical Discussion. *Friction* **2022**, *10*, 1772–1809. [CrossRef]
15. Uddin, M.S.; Liu, Y.W. Design and Optimization of a New Geometric Texture Shape for the Enhancement of Hydrodynamic Lubrication Performance of Parallel Slider Surfaces. *Biosurface Biotribology* **2016**, *2*, 59–69. [CrossRef]
16. Obilor, A.F.; Pacella, M.; Wilson, A.; Silberschmidt, V.V. Micro-Texturing of Polymer Surfaces Using Lasers: A Review. *Int. J. Adv. Manuf. Technol.* **2022**, *120*, 103–135. [CrossRef]
17. Rosenkranz, A.; Szurdak, A.; Gachot, C.; Hirt, G.; Mücklich, F. Friction Reduction under Mixed and Full Film EHL Induced by Hot Micro-Coined Surface Patterns. *Tribol. Int.* **2016**, *95*, 290–297. [CrossRef]
18. Weck, M.; Fischer, S.; Vos, M. Fabrication of Microcomponents Using Ultraprecision Machine Tools. *Nanotechnology* **1997**, *8*, 145. [CrossRef]
19. Costa, H.; Hutchings, I. Some Innovative Surface Texturing Techniques for Tribological Purposes. *Proc. Inst. Mech. Eng. Part J J. Eng. Tribol.* **2015**, *229*, 429–448. [CrossRef]
20. Dietrich, J. *Praxis der Umformtechnik*; Springer Fachmedien Wiesbaden: Wiesbaden, Germany, 2018; ISBN 978-3-658-19529-8.
21. Zahner, M.; Lentz, L.; Steinlein, P.; Merklein, M. Investigation of Production Limits in Manufacturing Microstructured Surfaces Using Micro Coining. *Micromachines* **2017**, *8*, 322. [CrossRef] [PubMed]
22. Tekkaya, A.E.; Allwood, J.M.; Bariani, P.F.; Bruschi, S.; Cao, J.; Gramlich, S.; Groche, P.; Hirt, G.; Ishikawa, T.; Löbbe, C.; et al. Metal Forming beyond Shaping: Predicting and Setting Product Properties. *CIRP Ann.* **2015**, *64*, 629–653. [CrossRef]
23. Zahner, M.; Merklein, M. Analysis of Combined Extrusion Micro Coining Process to Manufacture Microstructured Tappets. *Procedia Manuf.* **2018**, *15*, 272–279. [CrossRef]
24. Zou, H.; Yan, S.; Shen, T.; Wang, H.; Li, Y.; Chen, J.; Meng, Y.; Men, S.; Zhang, Z.; Sui, T.; et al. Efficiency of Surface Texturing in the Reducing of Wear for Tests Starting with Initial Point Contact. *Wear* **2021**, *482–483*, 203957. [CrossRef]
25. Hong, Y.; Zhang, P.; Lee, K.-H.; Lee, C.-H. Friction and Wear of Textured Surfaces Produced by 3D Printing. *Sci. China Technol. Sci.* **2017**, *60*, 1400–1406. [CrossRef]
26. Hu, T.; Zhang, Y.; Hu, L. Tribological Investigation of MoS2 Coatings Deposited on the Laser Textured Surface. *Wear* **2012**, *278–279*, 77–82. [CrossRef]
27. Pettersson, U.; Jacobson, S. Influence of Surface Texture on Boundary Lubricated Sliding Contacts. *Tribol. Int.* **2003**, *36*, 857–864. [CrossRef]
28. Bellón Vallinot, I.; De La Guerra Ochoa, E.; Echávarri Otero, J.; Chacón Tanarro, E.; Fernández Martínez, I.; Santiago Varela, J.A. Individual and Combined Effects of Introducing DLC Coating and Textured Surfaces in Lubricated Contacts. *Tribol. Int.* **2020**, *151*, 106440. [CrossRef]
29. Zhang, Z.; Lu, W.; Feng, W.; Du, X.; Zuo, D. Effect of Substrate Surface Texture on Adhesion Performance of Diamond Coating. *Int. J. Refract. Met. Hard Mater.* **2021**, *95*, 105402. [CrossRef]
30. Marian, M. *Numerische Auslegung von Oberflächenmikrotexturen für Geschmierte Tribologische Kontakte*; Friedrich-Alexander-Universität Erlangen-Nürnberg (FAU): Erlangen, Germany, 2021.
31. Kalghatgi, G. Is It Really the End of Internal Combustion Engines and Petroleum in Transport? *Appl. Energy* **2018**, *225*, 965–974. [CrossRef]
32. Bae, C.; Kim, J. Alternative Fuels for Internal Combustion Engines. *Proc. Combust. Inst.* **2017**, *36*, 3389–3413. [CrossRef]
33. Weschta, M. *Untersuchungen zur Wirkungsweise von Mikrotexturen in Elastohydrodynamischen Gleit/Wälz-Kontakten*; Friedrich-Alexander-Universität Erlangen-Nürnberg (FAU): Erlangen, Germany, 2017.
34. Orgeldinger, C.; Tremmel, S. Understanding Friction in Cam–Tappet Contacts—An Application-Oriented Time-Dependent Simulation Approach Considering Surface Asperities and Edge Effects. *Lubricants* **2021**, *9*, 106. [CrossRef]
35. Habchi, W. *Finite Element Modeling of Elastohydrodynamic Lubrication Problems*; John Wiley & Sons: Hoboken, NJ, USA, 2018; ISBN 978-1-119-22514-0.
36. Hertz, H. Ueber die Berührung fester elastischer Körper. *J. Die Reine Angew. Math.* **1882**, *1882*, 156–171. [CrossRef]

37. Reynolds, O. On the Theory of Lubrication and Its Application to Mr. Beauchamp Tower's Experiments, Including an Experimental Determination of the Viscosity of Olive Oil. *Philos. Trans. R. Soc. Lond.* **1886**, *177*, 157–234. [CrossRef]
38. Tan, X.; Goodyer, C.E.; Jimack, P.K.; Taylor, R.I.; Walkley, M.A. Computational Approaches for Modelling Elastohydrodynamic Lubrication Using Multiphysics Software. *Proc. Inst. Mech. Eng. Part J J. Eng. Tribol.* **2012**, *226*, 463–480. [CrossRef]
39. Dowson, D.; Higginson, G.R. *CHAPTER 1-INTRODUCTION*; Dowson, D., Higginson, G.R., Eds.; International Series on Materials Science and Technology; Pergamon Press Ltd.: Oxford, UK, 1977; ISBN 978-0-08-021302-6.
40. Roelands, C.J.A. *Correlational Aspects of the Viscosity-Temperature-Pressure Relationship of Lubricating Oils*; TU Delft: Delft, The Netherlands, 1966.
41. Carreau, P.J. Rheological Equations from Molecular Network Theories. *Trans. Soc. Rheol.* **1972**, *16*, 99–127. [CrossRef]
42. Bair, S. A Rough Shear-Thinning Correction for EHD Film Thickness. *Tribol. Trans.* **2004**, *47*, 361–365. [CrossRef]
43. Marian, M.; Weschta, M.; Tremmel, S.; Wartzack, S. Simulation of Microtextured Surfaces in Starved EHL Contacts Using Commercial FE Software. *Matls. Perf. Charact.* **2017**, *6*, MPC20160010. [CrossRef]
44. Masjedi, M.; Khonsari, M.M. Film Thickness and Asperity Load Formulas for Line-Contact Elastohydrodynamic Lubrication with Provision for Surface Roughness. *J. Tribol.* **2012**, *134*, 011503. [CrossRef]
45. Patir, N.; Cheng, H.S. An Average Flow Model for Determining Effects of Three-Dimensional Roughness on Partial Hydrodynamic Lubrication. *J. Lubr. Technol.* **1978**, *100*, 12–17. [CrossRef]
46. Winkler, A.; Marian, M.; Tremmel, S.; Wartzack, S. Numerical Modeling of Wear in a Thrust Roller Bearing under Mixed Elastohydrodynamic Lubrication. *Lubricants* **2020**, *8*, 58. [CrossRef]
47. Orgeldinger, C.; Rosnitscheck, T.; Tremmel, S. Numerical Optimization of Highly Loaded Microtextured Contacts: Understanding and Mastering Complexity. *ILT* **2023**, *75*, 741–747. [CrossRef]
48. Rasmussen, C.E.; Williams, C.K.I. *Gaussian Processes for Machine Learning*; Adaptive Computation and Machine Learning; MIT Press: Cambridge, MA, USA, 2006; ISBN 978-0-262-18253-9.
49. Gnanavel, C.; Saravanan, R.; Chandrasekaran, M.; Pugazhenthi, R. Restructured Review on Electrical Discharge Machining—A State of the Art. *IOP Conf. Ser. Mater. Sci. Eng.* **2017**, *183*, 012015. [CrossRef]
50. Singh, A.K.; Mahajan, R.; Tiwari, A.; Kumar, D.; Ghadai, R.K. Effect of Dielectric on Electrical Discharge Machining: A Review. *IOP Conf. Ser. Mater. Sci. Eng.* **2018**, *377*, 012184. [CrossRef]
51. Oliver, W.C.; Pharr, G.M. An Improved Technique for Determining Hardness and Elastic Modulus Using Load and Displacement Sensing Indentation Experiments. *J. Mater. Res.* **1992**, *7*, 1564–1583. [CrossRef]
52. Oliver, W.C.; Pharr, G.M. Measurement of Hardness and Elastic Modulus by Instrumented Indentation: Advances in Understanding and Refinements to Methodology. *J. Mater. Res.* **2004**, *19*, 3–20. [CrossRef]
53. Bückle, H. *Mikrohärteprüfung Ihre Anwendung*; Berliner Union: Stuttgart, Germany, 1965.
54. Richard, H.A.; Sander, M. *Ermüdungsrisse: Erkennen, Sicher Beurteilen, Vermeiden*; Vieweg+Teubner Verlag: Wiesbaden, Germany, 2012; ISBN 978-3-8348-1594-1.
55. Reck, M.; Merklein, M. Wear Analysis of Micro Textured Coining Tools. In Proceedings of the XLI. Colloquium on Metal Forming; Friedrich-Alexander-Universität Erlangen-Nürnberg: Erlangen, Germany, 2023; pp. 104–109.
56. Sauer, B. (Ed.) *Konstruktionselemente des Maschinenbaus 2: Grundlagen von Maschinenelementen für Antriebsaufgaben*; Springer Berlin Heidelberg: Berlin/Heidelberg, Germany, 2018; ISBN 978-3-642-39502-4.
57. Meidert, M. Beitrag zur Deterministischen Lebensdauerabschätzung von Werkzeugen der Kaltmassivumformung. Ph.D. Thesis, Friedrich-Alexander-Universität Erlangen-Nürnberg (FAU), Erlangen, Germany, 2006.

Disclaimer/Publisher's Note: The statements, opinions and data contained in all publications are solely those of the individual author(s) and contributor(s) and not of MDPI and/or the editor(s). MDPI and/or the editor(s) disclaim responsibility for any injury to people or property resulting from any ideas, methods, instructions or products referred to in the content.

 *lubricants*

Article

Machine-Learning-Based Wear Prediction in Journal Bearings under Start–Stop Conditions

Florian König *, Florian Wirsing, Ankit Singh and Georg Jacobs

Institute for Machine Elements and Systems Engineering, RWTH Aachen University, Schinkelstrasse 10, 52062 Aachen, Germany; georg.jacobs@imse.rwth-aachen.de (G.J.)
* Correspondence: florian.koenig@imse.rwth-aachen.de

Abstract: The present study aims to efficiently predict the wear volume of a journal bearing under start–stop operating conditions. For this purpose, the wear data generated with coupled mixed-elasto-hydrodynamic lubrication (mixed-EHL) and a wear simulation model of a journal bearing are used to develop a neural network (NN)-based surrogate model that is able to predict the wear volume based on the operational parameters. The suitability of different time series forecasting NN architectures, such as Long Short-Term Memory (LSTM), Gated Recurrent Unit (GRU), and Nonlinear Autoregressive with Exogenous Inputs (NARX), is studied. The highest accuracy is achieved using the NARX network architectures.

Keywords: journal bearing; wear prognosis; machine learning; surrogate model

Citation: König, F.; Wirsing, F.; Singh, A.; Jacobs, G. Machine-Learning-Based Wear Prediction in Journal Bearings under Start–Stop Conditions. *Lubricants* **2024**, *12*, 290. https://doi.org/10.3390/lubricants12080290

Received: 17 May 2024
Revised: 6 August 2024
Accepted: 9 August 2024
Published: 15 August 2024

Copyright: © 2024 by the authors. Licensee MDPI, Basel, Switzerland. This article is an open access article distributed under the terms and conditions of the Creative Commons Attribution (CC BY) license (https:// creativecommons.org/licenses/by/ 4.0/).

1. Introduction

Preventing wear-induced failures of machines is one important measure to become more sustainable and reduce CO_2 emissions, as maintenance and replacement efforts due to wear-induced failures share about 3% of global primary energy consumption [1]. Journal bearings are regarded as one wear-critical component in a machine. Hydrodynamic journal bearings can operate free of wear during nominal operation, but they are prone to abrasive wear during start–stop operation [2–4]. Advancing wear losses can lead to a loss of conformity, which may cause irregular vibrations or even a loss of load-carrying capacity. The latter can cause severe consequential damage to the machine. Therefore, potential methods to predict the wear of a journal bearing during mixed-friction operation are needed. Various studies have tackled the damage caused by mixed-friction operation, and promising physical approaches have been developed and validated [2,5,6]. These models can provide accurate results if the operating conditions of a specific machine in terms of load, speeds, and temperature as well as the physical properties of the specific bearing material–lubricant combinations are predefined. However, in many cases, these numerical approaches are still very time-consuming, making their effective utilization in design and direct integration into condition monitoring impractical.

In recent years, in the age of artificial intelligence (AI) and machine learning (ML), so-called surrogate or meta models have been developed in the field of tribology [7,8]. These models can replace computationally expensive physical and empirical models for different purposes, including wear prediction [9–11]. After successful training, these models can predict the desired output with high accuracy within much shorter calculation times than the original physical simulation approaches. Hence, ML-based models can handle a large variety of operating conditions and parameters in real time or near real time. This capability allows for predicting wear under dynamic conditions, such as start–stop conditions with sequential mixed-friction operation, which are often observed in real machinery. From a practical point of view, ML-based models of tribological systems can be applied in the design and the use phases of these systems. Specifically, the designer can use ML-based models to efficiently assess the impact of the design as well as operational conditions on wear. In the use phase, wear monitoring can benefit from a correlation between sensor data, i.e., measurable bearing condition indicators, and the corresponding wear losses, which can typically not be directly measured.

The focus of this study is use phase wear monitoring. For this purpose, it is hypothesized that the knowledge gained through experimentally validated physical simulation methods and the sensor data obtained through condition monitoring with tribology-related signals in the real system, e.g., temperature, load, torque, or Acoustic Emission, are complementary and can be used to develop novel methods for accurate and robust ML-based wear monitoring during the use phase of a particular machine.

As a first step towards this vision, the aim of this study is to evaluate the capabilities of ML for the prediction of wear in journal bearings under start–stop conditions. For this purpose, the wear data obtained with an existing, experimentally validated coupled mixed-elasto-hydrodynamic (mixed-EHL) and wear simulation model are used to train and evaluate an ML-based surrogate model. Three similar start–stop cycles in terms of time and speed with different bearing loads and temperatures serve as the input, whereas the corresponding wear volume serves as the output.

2. Methods and Parameters

The methodology of this work is shown in Figure 1. Firstly, the mixed-EHL and wear simulation model was used to generate wear data. Following the pre-processing of the generated data, multiple variants of the three time series forecasting recurrent neural network (RNN) architectures, namely Long Short-Term Memory (LSTM), Gated Recurrent Unit (GRU), and Nonlinear Autoregressive with Exogenous Inputs (NARX), were built. The aim was to explore the suitability of the chosen architectures as well as the impact of the number of hidden layers (HL) and neurons per layer on the prediction accuracy. To compare RNN architectures, each network architecture was simulated using the same training and test data sets. At the beginning of the training, the weights and biases of the neural network must be set to specific initial values. The initial determination of the values is called the "initial guess". The training algorithm does this by means of random initialization. As the choice of initial values for weights and biases influences the training, the training was repeated several times, and the best result was saved. Subsequently, the aim was to elucidate whether the best-performing RNN can predict wear if the sampling rate of the data sets is reduced. The reason for this is twofold. On the one hand, training with large data sets is computationally expensive and time-consuming. By using lower sampling rates, not only can the time be reduced but also the necessary hardware costs. On the other hand, the acceleration and deacceleration rates as well as the operating time of dynamically operating machines may vary, which affects the number of sampling points in a whole cycle and thereby the seasonality of the data. From a data science point of view, this different seasonality of the data can also be achieved by varying the sampling rate of the data sets, as it is conducted in this study. It should be noted that the stochastic operation is far more complex in terms of loads, speeds, and temperatures. Hence, this can only be seen as a first step towards wear prediction under dynamic conditions. In this study, the best-performing model architecture from our initial screening of LSTM, GRU, and NARX was trained again using training data with different sampling rates. Specifically, the sample rate of the training data was heuristically changed by a downsampling factor of 50, 100, and 400 using the function downsample within MATLAB. The downsampling factor determines the data points that are maintained from the original training data. With a downsampling factor of 50, only every 50th value is maintained. Subsequently, the model was tested with testing data at a downsampling factor of 200 to obtain a first idea of the model's adaptability to condition-monitoring systems with lower sampling rates as well as dynamic operating conditions.

Finally, the best-performing RNN model from our initial screening was trained on series A and B and then tested on series C to show the transferability of the model to different operational conditions.

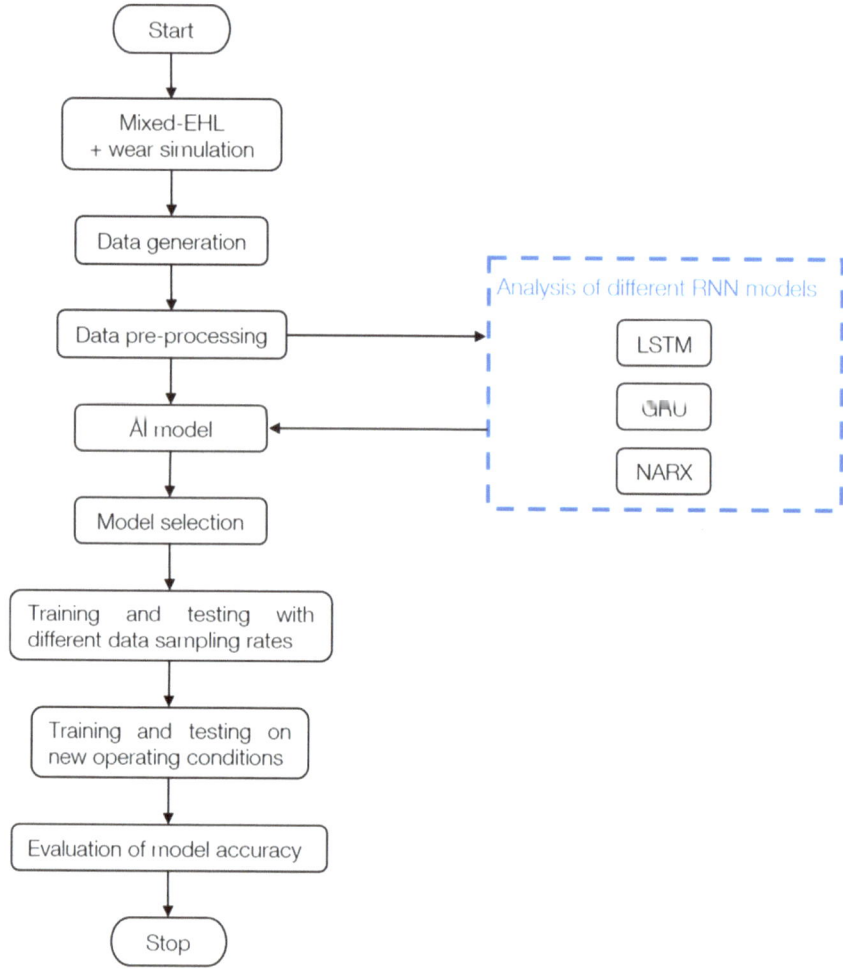

Figure 1. Process chart depicting the methodology of the present work.

2.1. Coupled Mixed-EHL and Wear Simulation Model

In this study, a coupled transient mixed-elasto-hydrodynamic (mixed-EHL) and wear simulation model is used to simulate the wear behavior during multiple thousand start–stop cycles. The basic structure of the physical simulation model and its combination with a wear simulation with empirical wear laws is shown in Figure 2.

In this model, which is built in AVL Excite PowerUnit, the initial contact geometry, the surface topography of the bearing and shaft, and the transient operating conditions are used as input parameters. Furthermore, the temperature, boundary friction, and wear coefficient are used as inputs. The variation in temperature affects the lubricant's viscosity and therefore the hydrodynamic film formation. The relationship between temperature, pressure, and lubricant viscosity, which was used in this study, is shown in Figure A1 in Appendix A. The friction coefficient was modeled with the boundary friction model according to Knauss and Offner [12]. Changing the lubricant type, the base oil viscosity, and the additives will certainly have an influence on the friction coefficient.

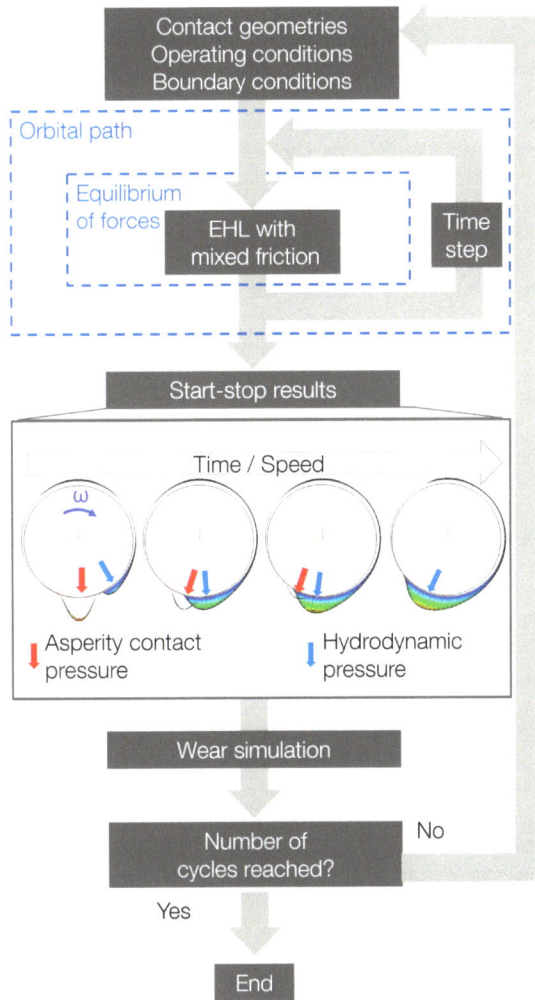

Figure 2. Structure of the mixed-EHL and wear simulation.

In transient mixed-elasto-hydrodynamic simulations (EHL with mixed friction), the local, transient wear-critical pressures are calculated. After each simulation step, which corresponds to one start–stop cycle, the wear pattern is calculated using the energetic approach according to Fleischer [13,14], which calculates the local wear depth as a function of the locally resolved coefficient of friction (CoF), the asperity contact pressure, as well as a predefined wear coefficient. Subsequently, the maximum wear per simulation step is extrapolated to 0.25 µm for the maximum wear depth to calculate the wear loss caused by multiple start–stop cycles within one simulation step. In the next step, the wear volume is calculated via integration. After the wear calculation, a new simulation model is built with an updated contour and bearing roughness. This process is repeated with updated wear profiles and surface topographies of the bearing in each simulation step until a predefined simulation time is reached.

2.2. Data Generation from the Simulation Model

The coupled, transient mixed-EHL + wear simulation model can be used to predict the contact conditions inside of the journal bearing as well as volumetric wear losses for different operational conditions, i.e., loads, temperatures, and speeds. The data used in this study are from three different start–stop time series, namely series A, B, and C. In all cases, the modeled bearing has a diameter of 30 mm, a width of 15 mm, and a radial clearance of 25 μm. For Series A, a radial load of 900 N and a temperature of 80 °C were chosen, while a load of 900 N and a temperature of 90 °C and a load of 1350 N and a temperature of 80 °C were chosen for series B and C, respectively. With an increase from 80 °C to 90 °C, the base viscosity at 40 °C decreases from 7.2 to 5.7 mPas. In all series, the start–stop cycle is fixed to a 2 s acceleration from 0 to 600 min^{-1}, 2 s at a constant speed for 600 min^{-1}, and a 3 s deceleration from 600 to 0 min^{-1}. Each start–stop cycle lasted 7 s, totaling 45 shaft revolutions per start–stop cycle (10 + 20 + 15 for starting, constant, and stopping mode, respectively). The coupled mixed-EHL + wear simulation was conducted for a total of 2500 start–stop cycles. With 4884 time steps per start–stop cycle, the overall time series contains more than 12 million time steps. The three different load–temperature combinations generate three distinct time series for the simulated contact conditions in the bearing as well as the resulting wear volume.

2.3. Data Pre-Processing

The three distinct data sets for series A, B, and C contain constant input data (load, temperature), transient input data (time, speed), and transient mixed-EHL simulation output data (CoF), alongside the transient output data of the wear simulation (wear volume, wear depth, and wear rate). The input and output variables are referred to as features. From all of these features, a few features were carefully chosen with the idea that only certain measured values are suitable to be measured in real machinery, such as the temperature and speed. The feature temperature was excluded due to the isothermal modeling as a first attempt and could be integrated at a later stage. In contrast, other features, such as the radial bearing load and friction, are difficult to measure and therefore commonly not available. The bearing load can be assumed from the overall torque or energy consumption of a machine. In this study, the radial bearing load remained constant within each simulation run, so it was also excluded from the model [15]. Friction cannot be measured directly, but it is an important parameter. Hence, various studies have shown that friction phenomena can instead be measured with Acoustic Emission technology. Because this is not within the scope of this study, the integral friction coefficient (CoF) was used.

However, it should be noted that the temperature and bearing load are implicitly considered. First, variations in the predefined isothermal temperature and the external load impact the oil viscosity due to temperature– and pressure–viscosity relationships, which affect the hydrodynamic film's load-carrying capacity. Consequently, the ratio between oil and solid contact pressure is affected, influencing the global CoF, which contains both solid body and fluid friction via the integration of local shear stresses. Second, to incorporate the load's influence, which also remained constant throughout the simulation, frictional torque was used instead of CoF, as it also contains the external load as a variable, as shown in Equation (1),

$$\text{Frictional torque} = \text{CoF} \times \text{load} \times r \tag{1}$$

where r is the radius of the bearing. Hence, the data were significantly reduced to only contain frictional torque and speed as input parameters for the ML models. The wear prediction model needs to anticipate short-term fluctuations, like changes in operational conditions (such as start–stop cycles), termed seasonality, and long-term wear patterns, known as trend. Hence, "wear volume" was chosen as the output feature instead of transient measures like wear rate, as it effectively encapsulates both seasonality and trend aspects.

In the future, the aim is to replace the simulation data with measured data. Using the measured speed, temperature, and friction torque can be the first step to demonstrate the transferability from simulation to reality.

2.4. Machine Learning Models

The transient data obtained from the simulation model can be treated as a time series for ML-based surrogate modeling. Pragmatically speaking, wear monitoring and prognosis are a combination of regression modeling between input and output data and time series forecasting, commonly performed with RNNs, which are used to predict posterior data while considering the previous and current state of the system. RNNs are able to retain past sequential data in memory, making them exceptionally well-suited for tasks involving sequential data, such as time series analysis. Following the pre-processing of the data obtained from mixed EHL and wear simulations, the different RNN architectures LSTM, GRU, and NARX were trained and compared. These architectures are further introduced in Sections 2.4.1–2.4.3.

2.4.1. LSTM

Long Short-Term Memory (LSTM) is an RNN that avoids the vanishing gradient problem, which occurs during the training of long time series. As shown in Figure 3, LSTMs feature a unique architecture comprising an input gate, a forget gate, and an output gate, enabling them to retain and utilize information over extended sequences selectively. The input gate regulates the flow of new inputs (labeled x) into the cell state (labeled c). Conversely, the forget gate determines which information from the previous cell state is discarded, thereby facilitating the preservation of relevant long-term dependencies. Finally, the output gate controls the flow of information from the cell state to the output (labeled y). This gating mechanism allows LSTMs to capture and retain important temporal dependencies, making them particularly interesting for using sequential data [16–18].

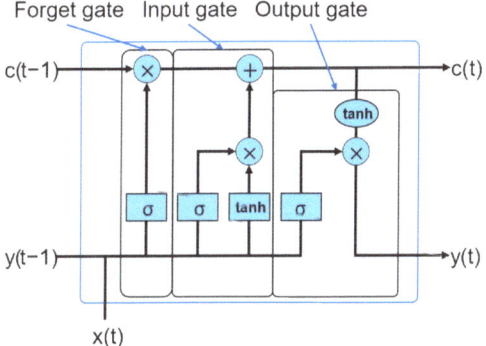

Figure 3. Schematic representation of the LSTM model.

2.4.2. GRU

Gated Recurrent Unit (GRU) networks are a specialized variant of RNNs, similar to LSTM networks, designed to overcome the vanishing gradient problem inherent in traditional RNN architectures. Compared to LSTMs, GRUs employ a simpler architecture, featuring fewer gating mechanisms while maintaining competitive performance [17]. Within the GRU architecture, depicted in Figure 4, two primary gates regulate the flow of information: the update gate and the reset gate. The update gate in GRUs governs the flow of new inputs (labeled x) into the current state (labeled c), deciding how much past information to retain and how much new information to integrate. Similarly, the reset gate adjusts the contribution of past information when computing the current state, facilitating selective forgetting of irrelevant information. GRU models dynamically adjust these gates to capture and retain relevant information over extended sequences. Compared to LSTMs, GRUs are easier to train and faster to execute at the cost of modeling accuracy [19,20].

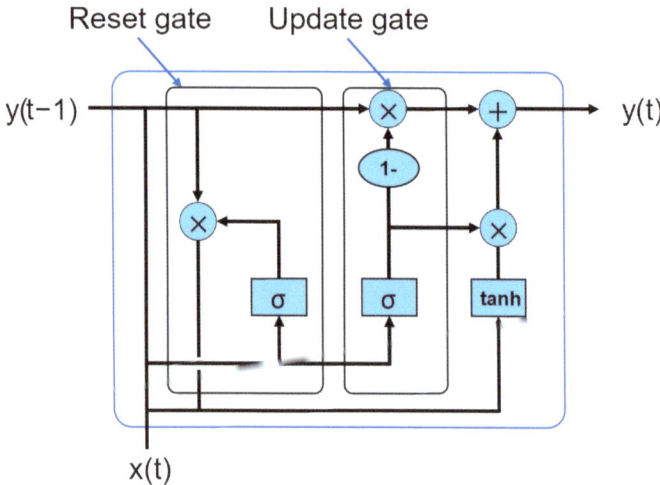

Figure 4. Schematic representation of the GRU model.

2.4.3. NARX

NARX models, or Nonlinear Autoregressive models with exogenous inputs, are specialized tools for time series forecasting. NARX models capture complex patterns and dynamics due to their capability to integrate speed and frictional torque as input signals into the forecasting process. Unlike traditional autoregressive models, NARX models incorporate exogenous inputs alongside the autoregressive terms, allowing them to account for external influences on the system's behavior [21–25].

The NARX model equation is shown in Equation (2), where y(t), the predicted value of the model, is predicted using its previous values, $y(t-1)$ to $y(t-n_y)$, and the corresponding previous values of an input signal, $x(t-1)$ to $x(t-n_x)$. There are two forms of NARX architecture: series-parallel and parallel architecture. The series-parallel NARX architecture was chosen for this study as the availability of the previous values of y(t) makes the network more accurate [26].

$$y(t) = f(y(t-1), y(t-2), \ldots, y(t-n_y), x(t-1), x(t-2), \ldots, x(t-n_x)) \quad (2)$$

Here y(t) is the wear volume. The Neural Net Time Series App in MATLAB version R2020b was used for training and testing the NARX models. A schematic representation of the NARX network is shown in Figure 5.

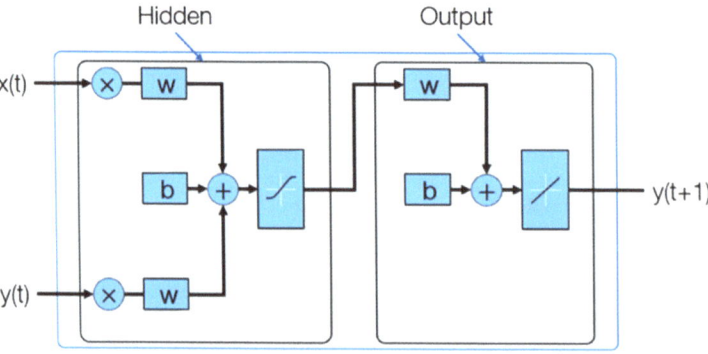

Figure 5. Schematic representation of the series-parallel NARX model.

3. Results and Discussion

The aim of this project was to train an RNN model that predicts the wear volume data generated from the physics-based simulation model. Three time series, A, B, and C, were generated with different operating conditions, as described in the methods in Section 2.2.

3.1. Mixed-EHL and Wear Simulation Results

The speed and the coefficient of friction (CoF) predicted by the mixed-EHL model for the first three and last three start–stop cycles of series A are exemplarily shown in Figure 6a. It can be observed that the CoF is zero at the initial start-up and stopping time due to standstill. After a certain number of start–stop cycles, the CoF is significantly reduced due to the running in process. During the stationary phase, hydrodynamic lubrication is present, which leads to very low viscous friction losses, as it can be seen in Figure 6b.

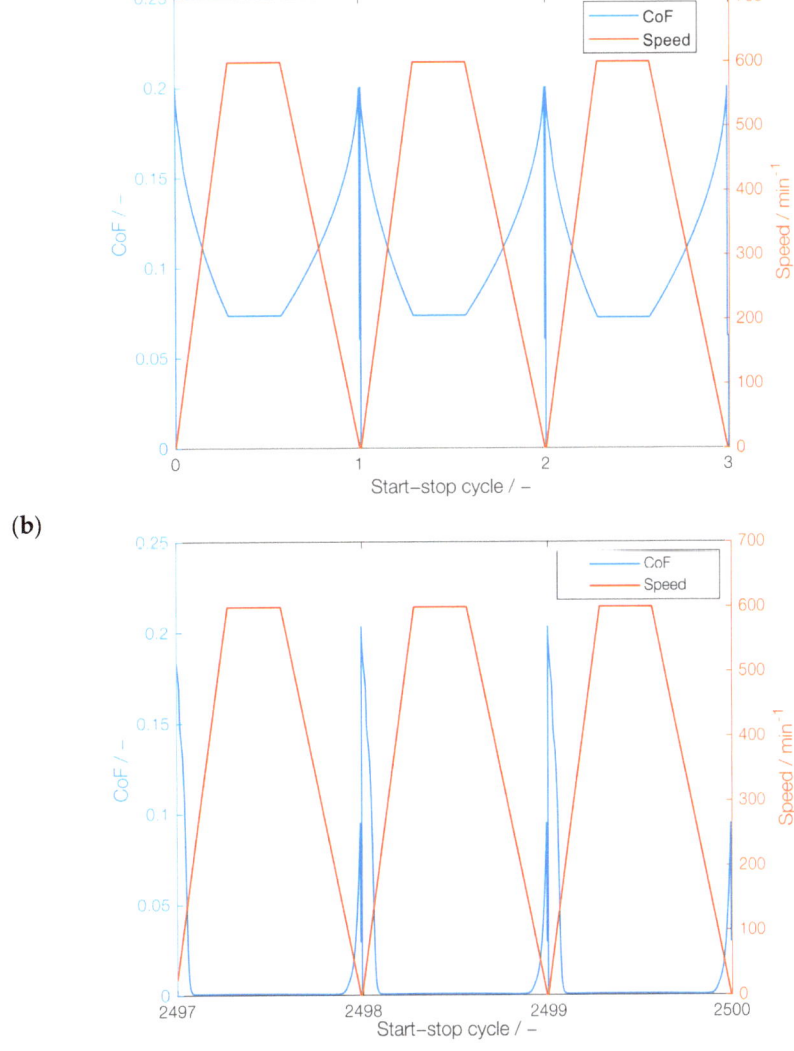

Figure 6. CoF and speed obtained from physics-based simulation model: (**a**) cycles 1–3, (**b**) cycles 2498–2500.

The results of the coupled wear simulation are exemplarily shown in Figure 7. Figure 7a shows the long-term trend of the wear volume. As it can be seen, the wear rate significantly decreases over the number of start–stop cycles after a pronounced wearing-in within the first few hundred start–stop cycles. Turning our attention to the seasonality, the wear volume generated within the first three cycles (1–3) is shown in Figure 7b, while the wear volume originated from the last three cycles (2498–2500) is shown in Figure 7c.

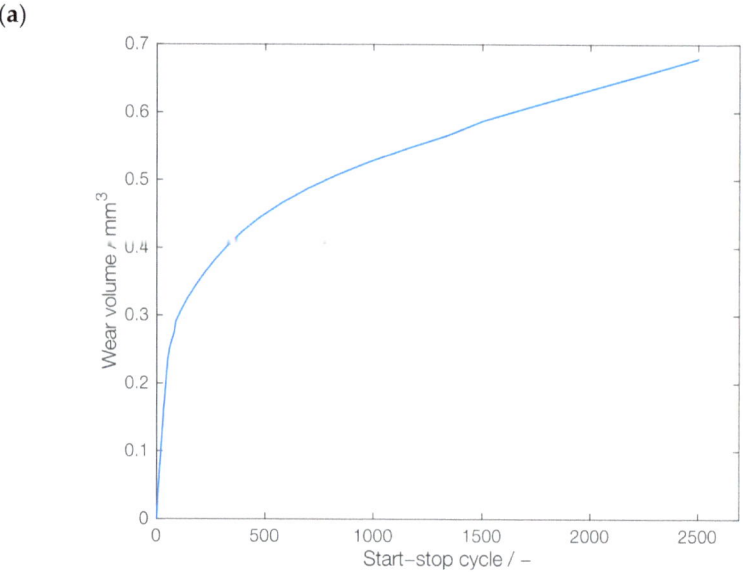

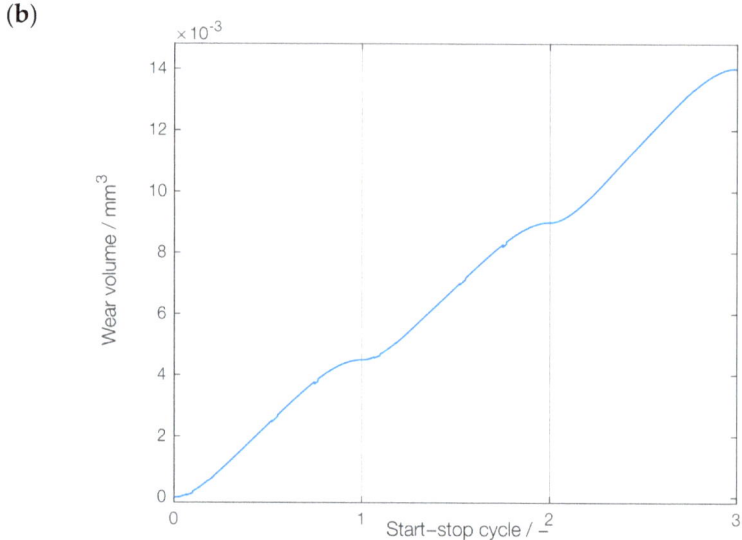

Figure 7. *Cont.*

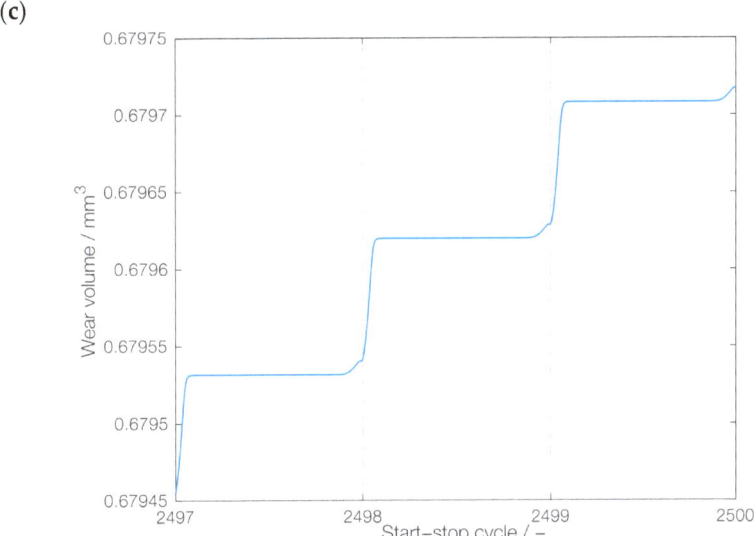

Figure 7. Wear volume obtained from physics-based simulation model: (**a**) total number of start–stop cycles, (**b**) cycles 1–3, (**c**) cycles 2498–2500.

Similarly to the observed trend, the generated wear volume loss within the first three cycles (0.015 mm^3) is significantly higher than the loss within the last three cycles of the data set (0.0003 mm^3), as the wear during the first three cycles was generated during the running-in period. In contrast, the wear during the last three cycles was predominantly caused by the mixed-friction operation during start-up and only slightly increased during stopping. The challenge is to incorporate the overall trend and seasonality effects into an efficient, predictive RNN model.

3.2. ML Models and Architecture Comparison

A detailed comparison of LSTM, GRU, and NARX was conducted to evaluate the performance of these RNN architectures for predicting wear in journal bearings during start–stop cycles. The number of layers and the number of neurons per layer were varied for each of the three different RNN architectures to evaluate the performance under various configurations.

The data set was initially downsampled by a downsampling factor of 200 to accelerate convergence during training. An 80–20 data split was used, with 80% allocated to training and 20% reserved for testing. This ratio was chosen based on previous studies. Due to the early pre-screening of these models, the validation data were included for both LSTM and GRU models. Overfitting is prevented by using gradient clipping, limiting the absolute value of the gradients to 1. In the NARX model, the 80% training data set is subdivided, with 60% used for training and 20% as a validation data set. Tables A2 and A3 in the Appendix A provide detailed information on the training parameter for LSTM, GRU, and NARX. The evaluation was based on the Root Mean Square Error (RMSE) metric for both training as well as testing error analysis. Figure 8 shows the comparison of the three RNN architectures; the complete list of RMSE values can be found in Table A1 in the Appendix A. Variants for each RNN architecture include models with a single hidden layer and two hidden layers for varying numbers of neurons. The test case results for each network are shown in Figure 8. The best results for LSTM were obtained for two hidden layers with 50 neurons in each layer. For GRU, the best results were obtained for a single hidden layer with 100 neurons, whereas for NARX, a single hidden layer with 50 neurons gave the best result. The RMSE value of the NARX model is much lower compared to LSTM and GRU, which can also be observed in the deviations, which are shown in Figure 9. The predictions from LSTM and GRU show local fluctuation, making them unsuitable for prediction of the seasonality. Additionally, the training process of the NARX network was considerably faster than the other two networks. However, due to the uncertain outcome of the comparison between different networks, the training time was not our focus and therefore not measured. Therefore, the NARX model was selected as the most promising model for further training and testing.

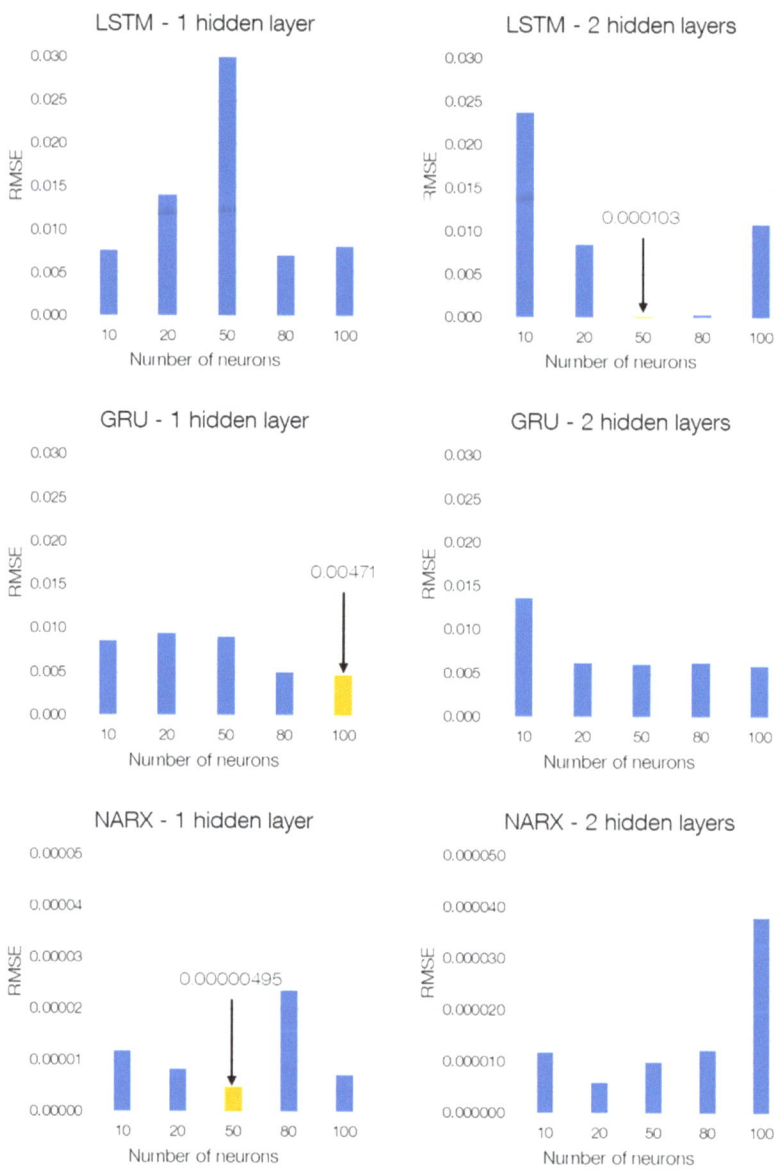

Figure 8. ML model and architecture comparison for the test case.

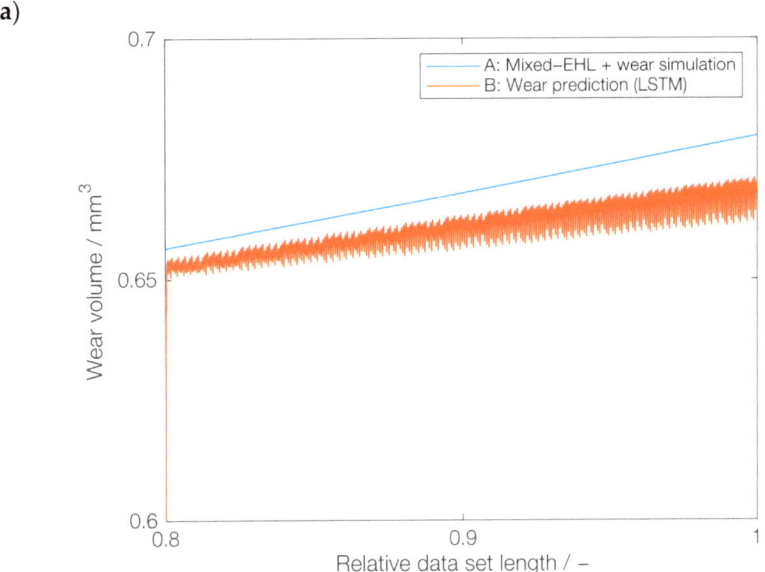

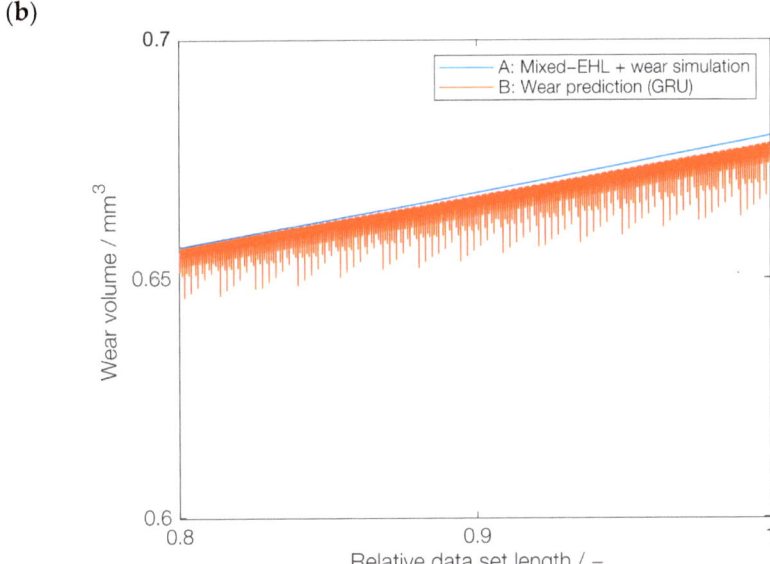

Figure 9. *Cont.*

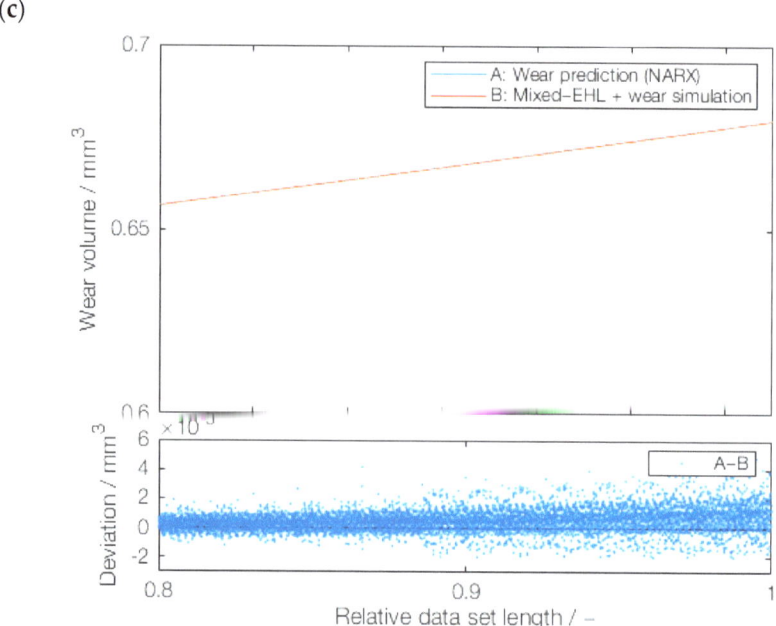

Figure 9. ML model comparison: (**a**) LSTM (2 HL, 50 neurons), (**b**) GRU (1 HL, 100 neurons), and (**c**) NARX (1 HL, 50 neurons), including deviation from the physics-based wear volume values.

3.3. Exploring the Adaptability of the NARX Model to Varying Data Resolutions

During dynamic operation, the acceleration and deceleration may be subjected to change. Under the assumption of a sampling rate of a condition-monitoring system, this leads to a different number of data points for a given start–stop cycle. The question is whether this change affects the outcome of the model. Therefore, this section focuses on the prediction of multiple downsampled versions of the Series A data set. These downsampled versions were produced by applying downsampling factors of 50, 100, and 400. The NARX model was then trained in three different orders, specifically, 50-400-100, 400-100-50, and 100-50-400. Following the training process, the NARX model with a single hidden layer of 50 neurons was tested on series A with a downsampling factor of 200. The idea to train on different downsampling factors (50, 100, and 400) and to test it on a new downsampling factor of 200 was to artificially check the NARX model's adaptability to varied sampling rates as well as varied acceleration and deacceleration rates. Although downsampling does not directly influence acceleration and deceleration rates, the concept of the sampling rate offers a parallel understanding: just as a higher sampling rate captures finer details in signal processing, faster changes in velocity signify more rapid acceleration or deceleration. Therefore, if NARX predicts the wear successfully for downsampled data, it is assumed that a similar principle will work for dynamic cycle lengths. The best test results (data set of downsampling factor 200) were achieved with a training sequence of 50-400-100, shown in Figure 10. As it can be seen in Figure 10a, the predicted wear volume of the NARX model is very close to the wear volume of the simulation model. The NARX model was successful in predicting the trend but was unable to capture the local seasonality with high accuracy, as it can be seen in Figure 10b. Even though there are some fluctuations in the prediction, the error is within the range of 0.01.

From the comparison between the best test results in Figure 10 and the other two results, which are shown in Appendix A, Figures A2 and A3, it can be observed that both the sequential training with 50-400-100 (Figure 10) and 100-50-400 (Figure A3) lead to comparable results, with a slightly better prognosis result using 50-400-100. In contrast, the sequential training 400-50-100 (Figure A2) leads to larger deviations, particularly after multiple hundred start–stop cycles. One potential reason is that the initial training of NARX with a downsampling factor of 400 can lead to a wrong direction of weights and biases, which ultimately affects the results. This would agree with the results for 50-400-100 (Figure 10) and 100-50-400 (Figure A3), which leads to slightly better results when finishing the training with a downsampling factor of 100 instead of 400. Based on these results, it can be assumed that the downsampling factor of 400 is not beneficial for training, which should be further studied in the future. Nevertheless, based on the positive results with 50-400-100 (Figure 10) and 100-50-400 (Figure A3), the NARX model can be suitable for wear trend prediction under varying dynamic conditions and sampling rates. The suitability of NARX in terms of seasonality is also demonstrated by the minor deviations between its wear prediction and the physics-based wear simulation.

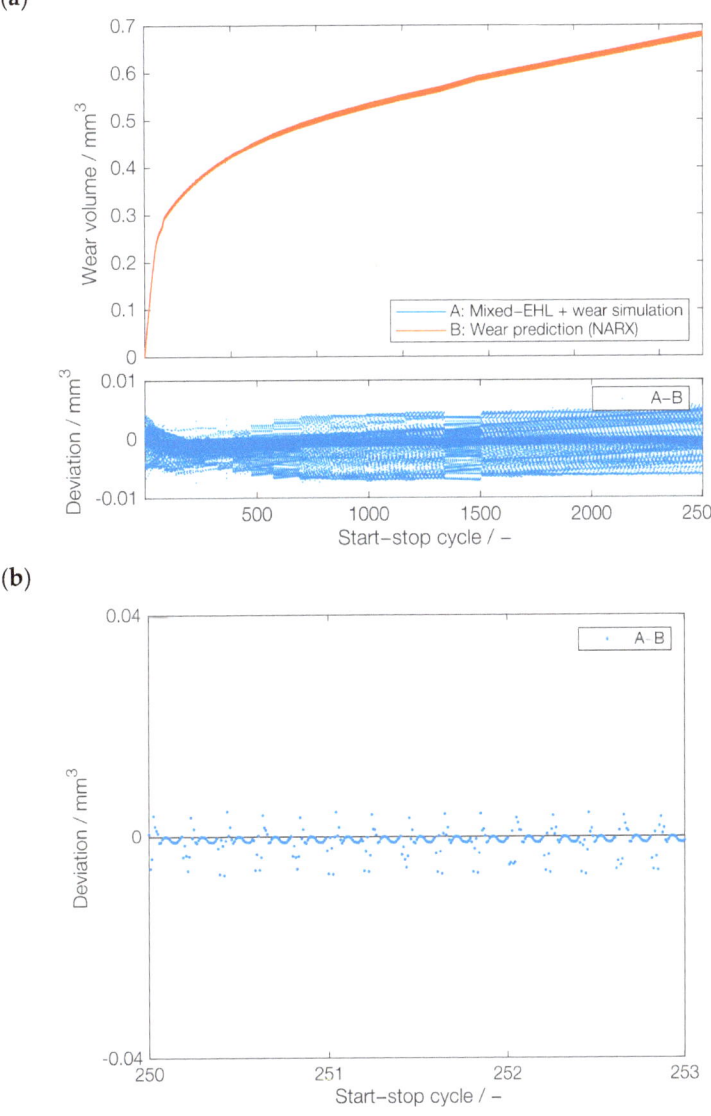

Figure 10. Prediction of a novel, downsampled series: (**a**) test results with a downsampling factor of 200 by NARX model trained on downsampling factors of 50, 400, and 100, and (**b**) deviations from the actual wear volume values (cycles 251–253).

3.4. Utilizing the Trained NARX Model for Forecasting a Novel Series under New Operating Conditions

The NARX model with a single hidden layer of 50 neurons underwent training using two distinct series from the data sets Series A and Series B, featuring loads of 900 N and temperatures of 80 °C and 90 °C, respectively. The training incorporated different downsampling factors, specifically 50 and 100. Afterwards, the trained NARX model was tested with an entirely new series, Series C, without downsampling. In this case, low RMSE values were observed (0.000571). As it can be seen from Figure 11a, the prediction curve is almost identical to the wear volume curve obtained in physics-based simulations. The deviations between NARX and the coupled mixed-EHL and wear simulation are within 0.005, as shown in Figure 11b. This underscores NARX's remarkable predictive capability in accurately forecasting unknown time series data, highlighting its potential for predictive maintenance applications in dynamic systems.

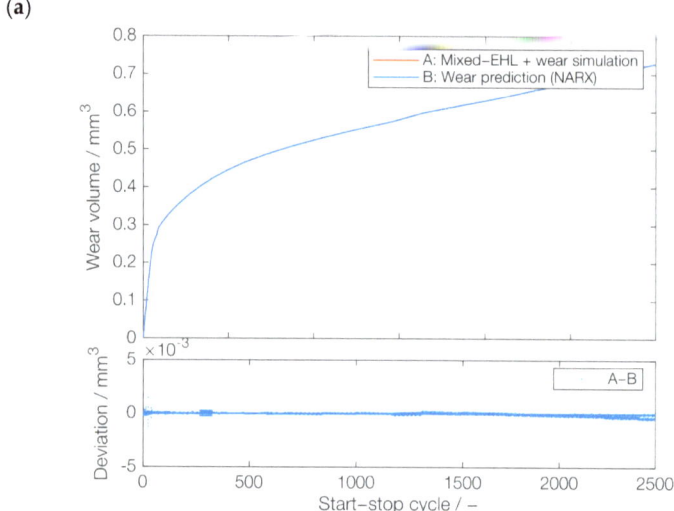

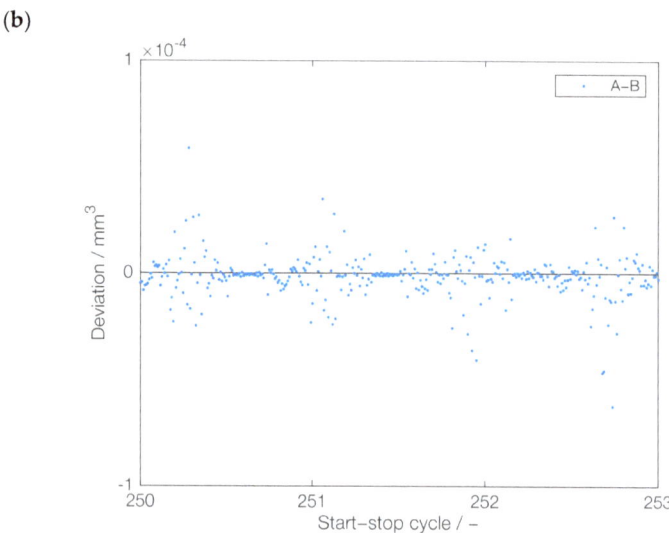

Figure 11. Prediction of series C (no downsampling): (**a**) wear volume by the NARX model for the test case trained on series A and B and (**b**) deviations from the actual wear volume values (cycles 251–253).

4. Conclusions

Journal bearings' start–stop operation increases the risk of failure due to abrasive wear. Considering the high computational cost of mixed-EHL and wear simulations to predict abrasive wear under dynamic conditions, this work aimed to study the potential of machine learning with recurrent neural networks (NNs) to predict wear after 2500 start–stop cycles.

Based on the results of this work, the following conclusions can be drawn:

- The NARX model exhibited superior predictive performance in wear prediction compared to the LSTM and GRU models, showcasing a lower RMSE.
- The trained NARX model demonstrated robust predictive capabilities, accurately forecasting all time series, irrespective of varying downsampling requirements, underscoring its adaptivity and generalization ability.
- The NARX's demonstrated effectiveness across diverse scenarios highlights its potential utility in various applications, accentuating its generalization capabilities to handle varying input parameters and data configurations.

Following the original hypothesis, the knowledge gained through experimentally validated physical simulation methods can be used for accurate and robust ML-based wear monitoring. In the future, the input data should be replaced by sensor data obtained through condition monitoring with tribology-related signals in the real system, e.g., temperature, torque, vibration, or Acoustic Emission [3,27]. Furthermore, for the condition monitoring of dynamic operating machinery under real-life, stochastic conditions, which was originally formulated within the hypothesis, this work should be extended towards more complex load cases in terms of speed profiles, temperatures, and loads. In addition, the deviations between the surrogate model and the real data from the experiment should be incorporated using statistical methods [28]. Furthermore, different failure modes should be addressed in the future by applying ML for classification [29].

Author Contributions: Conceptualization, F.K.; methodology, F.K. and A.S.; software, F.K., A.S. and F.W.; validation, F.K., A.S. and F.W.; formal analysis, F.W. and A.S.; investigation, F.K., A.S. and F.W.; resources, G.J.; data curation, F.K., A.S. and F.W.; writing—original draft preparation, F.K., A.S. and F.W.; writing—review and editing, F.K., F.W. and G.J.; visualization, F.K. and F.W.; supervision, F.K.; project administration, F.K.; funding acquisition, F.K. All authors have read and agreed to the published version of the manuscript.

Funding: Funded by the Federal Ministry of Education and Research (BMBF) and the Ministry of Culture and Science of the German State of North Rhine-Westphalia (MKW) under the Excellence Strategy of the Federal Government and the Länder.

Data Availability Statement: The data sets presented in this article are not readily available because the data are part of an ongoing study. Requests to access the data sets should be directed to F.K.

Acknowledgments: F.K. acknowledges the funding provided within the RWTH ERS SeedFund start-up project 'beArIngs'.

Conflicts of Interest: The authors declare no conflicts of interest.

Appendix A

Table A1. Test results of all trained RNN architectures in Section 3. The best-performing RNN of each architecture is highlighted by bold and underscored letters.

Model	Hidden Layer	Neurons	RMSE-Test
LSTM	1	10	7.72×10^{-3}
LSTM	1	20	1.41×10^{-2}
LSTM	1	50	2.19×10^{-1}

Table A1. Cont.

Model	Hidden Layer	Neurons	RMSE-Test
LSTM	1	80	7.10×10^{-3}
LSTM	1	100	8.15×10^{-3}
LSTM	2	10	2.38×10^{-2}
LSTM	2	20	8.53×10^{-3}
LSTM	**2**	**50**	**1.03×10^{-4}**
LSTM	2	80	1.26×10^{-4}
LSTM	2	100	1.09×10^{-2}
GRU	1	10	8.59×10^{-3}
GRU	1	20	9.45×10^{-3}
GRU	1	50	9.08×10^{-3}
GRU	1	80	5.05×10^{-3}
GRU	**1**	**100**	**4.71×10^{-3}**
GRU	2	10	1.37×10^{-2}
GRU	2	20	6.25×10^{-3}
GRU	2	50	6.11×10^{-3}
GRU	2	80	6.26×10^{-3}
GRU	2	100	5.91×10^{-3}
NARX	1	10	1.20×10^{-5}
NARX	1	20	8.51×10^{-6}
NARX	**1**	**50**	**4.95×10^{-6}**
NARX	1	80	2.37×10^{-5}
NARX	1	100	7.33×10^{-6}
NARX	2	10	1.20×10^{-5}
NARX	2	20	6.07×10^{-6}
NARX	2	50	1.00×10^{-5}
NARX	2	80	1.25×10^{-5}
NARX	2	100	3.82×10^{-5}

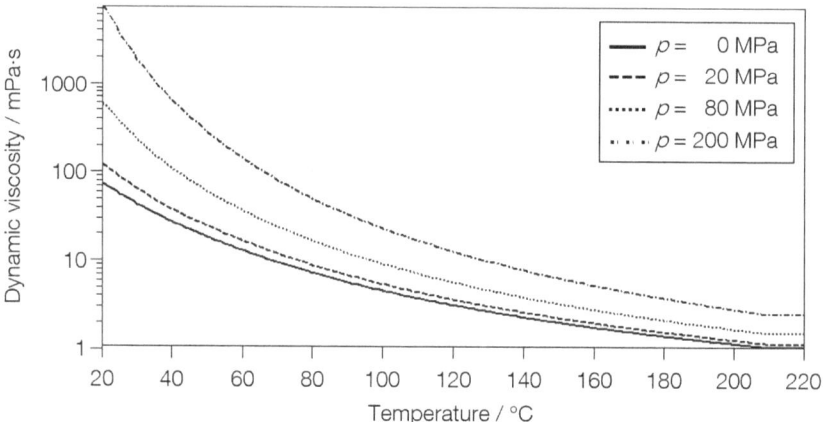

Figure A1. Relationship between temperature, pressure, and lubricant viscosity.

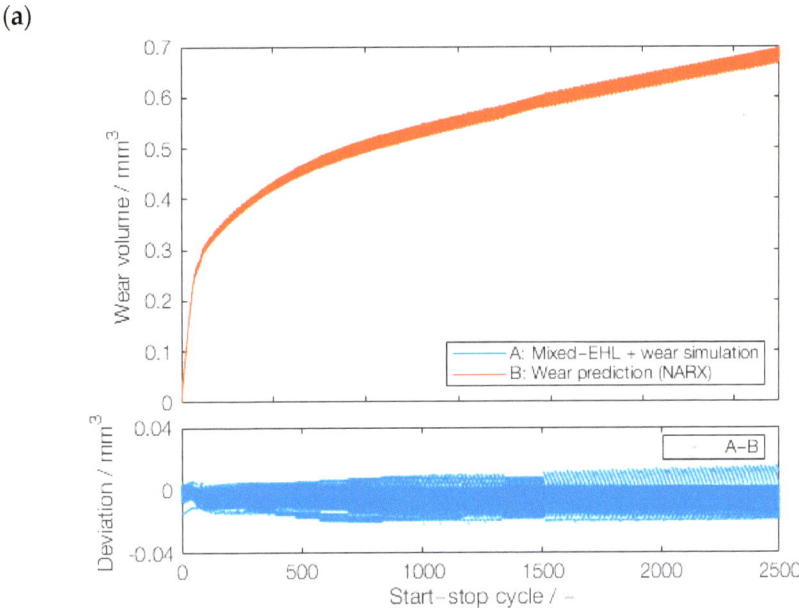
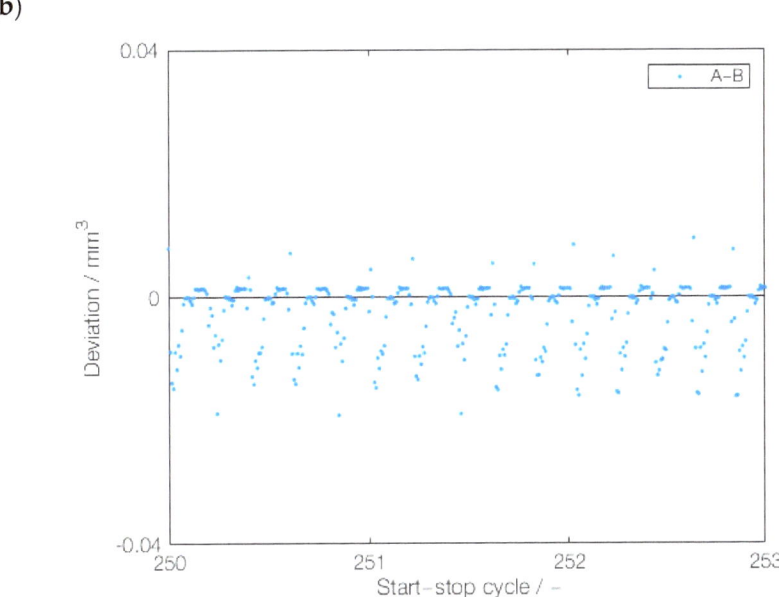

Figure A2. Prediction of a novel, downsampled series: (**a**) test results with a downsampling factor of 200 by NARX model trained on downsampling factors of 400, 100, and 50, and (**b**) deviations from the actual wear volume values (cycles 251–253).

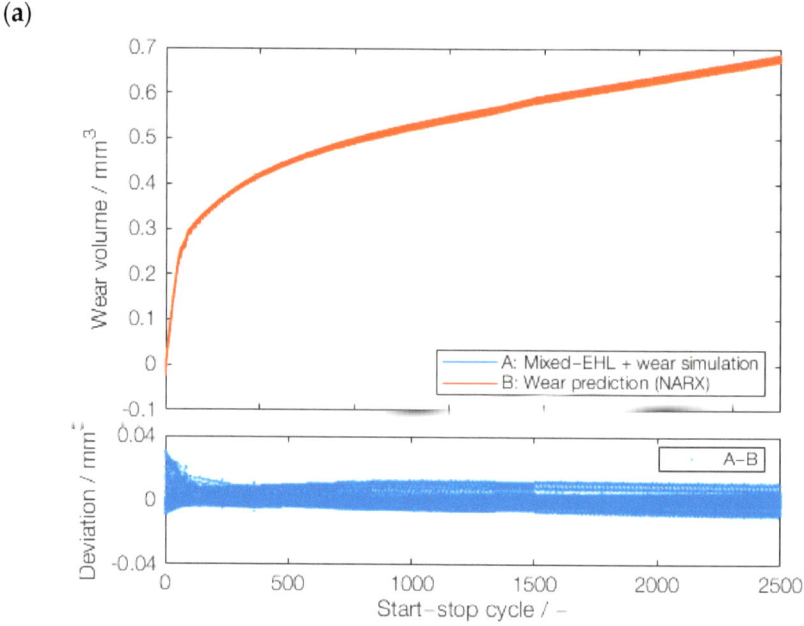
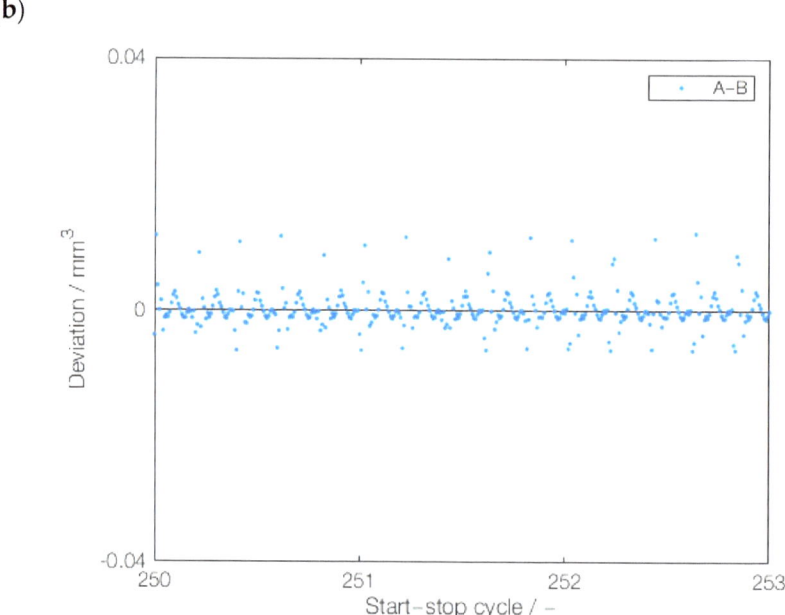

Figure A3. Prediction of a novel, downsampled series: (**a**) test results with a downsampling factor of 200 by NARX model trained on downsampling factors of 100, 50, and 400, and (**b**) deviations from the actual wear volume values (cycles 251–253).

Table A2. Training parameters of LSTM and GRU.

Training Parameter	Description
Normalization	• Scaling of the data in the value range $[-1, 1]$ using the MATLAB command mapminmax.
Learning rate	• The initial learning rate is set to 0.005. • The learning rate is gradually reduced every 50 epochs. • With each reduction, the learning rate is reduced to 20% of its previous value.
Optimization methods	• Use of the Stochastic Gradient Descent with Momentum (SGDM) as a training algorithm.
Regularization techniques	• Gradient clipping. ○ Limiting the absolute value of the gradients to 1 to prevent unstable training conditions.

Table A3. Training parameters of NARX.

Training Parameter	Description
Normalization	• Scaling of the data in the value range $[-1, 1]$ using the MATLAB command mapminmax.
Learning rate	• The learning rate is dynamically and internally adjusted by the training algorithm used; a specific learning rate is not specified.

Optimization methods	• Use of the Levenberg–Marquardt algorithm as a training algorithm.
	    ○ Special method for solving non-linear least squares problems.
	    ○ Combination of the gradient descent method with the Gauss–Newton method.
	    ○ The Levenberg–Marquardt algorithm switches adaptively between both methods to accelerate convergence.

Regularization techniques	• No application.

References

1. Holmberg, K.; Erdemir, A. Influence of tribology on global energy consumption, costs and emissions. *Friction* **2017**, *5*, 263–284. [CrossRef]
2. König, F.; Chaib, A.O.; Jacobs, G.; Sous, C. A multiscale-approach for wear prediction in journal bearing systems—From wearing-in towards steady-state wear. *Wear* **2019**, *426–427*, 1203–1211. [CrossRef]
3. Zhang, H.; Ma, J.; Li, X.; Xiao, S.; Gu, F.; Ball, A. Fluid-asperity interaction induced random vibration of hydro-dynamic journal bearings towards early fault diagnosis of abrasive wear. *Tribol. Int.* **2021**, *160*, 107028. [CrossRef]
4. Vencl, A.; Rac, A. Diesel engine crankshaft journal bearings failures: Case study. *Eng. Fail. Anal.* **2014**, *44*, 217–228. [CrossRef]
5. Maier, M.; Pusterhofer, M.; Grün, F. Modelling Approaches of Wear-Based Surface Development and Their Experimental Validation. *Lubricants* **2022**, *10*, 335. [CrossRef]
6. Maier, M.; Pusterhofer, M.; Grün, F. Wear simulation in lubricated contacts considering wear-dependent surface topography changes. *Mater. Today Proc.* **2023**, *93*, 563–570. [CrossRef]
7. Marian, M.; Tremmel, S. Current Trends and Applications of Machine Learning in Tribology—A Review. *Lubricants* **2021**, *9*, 86. [CrossRef]
8. Yin, N.; Yang, P.; Liu, S.; Pan, S.; Zhang, Z. AI for tribology: Present and future. *Friction* **2024**, *12*, 1060–1097. [CrossRef]
9. He, Z.; Shi, T.; Xuan, J.; Li, T. Research on tool wear prediction based on temperature signals and deep learning. *Wear* **2021**, *478–479*, 203902. [CrossRef]
10. Bote-Garcia, J.-L.; Gühmann, C. Wear volume estimation for a journal bearing dataset. *tm-Tech. Mess.* **2022**, *89*, 534–543. [CrossRef]

11. Ates, C.; Höfchen, T.; Witt, M.; Koch, R.; Bauer, H.-J. Vibration-Based Wear Condition Estimation of Journal Bearings Using Convolutional Autoencoders. *Sensors* **2023**, *23*, 9212. [CrossRef] [PubMed]
12. Offner, G.; Knaus, O. A Generic Friction Model for Radial Slider Bearing Simulation Considering Elastic and Plastic Deformation. *Lubricants* **2015**, *3*, 522–538. [CrossRef]
13. Fleischer, G. Zur Energetik der Reibung. *Wiss. Z. Tech. Univ. Magdebg.* **1990**, *34*, 55–66.
14. Bartel, D.; Bobach, L.; Illner, T.; Deters, L. Simulating transient wear characteristics of journal bearings subjected to mixed friction. *Proc. Inst. Mech. Eng. Part J J. Eng. Tribol.* **2012**, *226*, 1095–1108. [CrossRef]
15. Moder, J.; Bergmann, P.; Grün, F. Lubrication Regime Classification of Hydrodynamic Journal Bearings by Machine Learning Using Torque Data. *Lubricants* **2018**, *6*, 108. [CrossRef]
16. Lindemann, B.; Müller, T.; Vietz, H.; Jazdi, N.; Weyrich, M. A survey on long short-term memory networks for time series prediction. *Procedia CIRP* **2021**, *99*, 650–655. [CrossRef]
17. Mateus, B.C.; Mendes, M.; Farinha, J.T.; Assis, R.; Cardoso, A.M. Comparing LSTM and GRU Models to Predict the Condition of a Pulp Paper Press. *Energies* **2021**, *14*, 6958. [CrossRef]
18. van Houdt, G.; Mosquera, C.; Nápoles, G. A review on the long short-term memory model. *Artif. Intell. Rev.* **2020**, *53*, 5929–5955. [CrossRef]
19. Yang, S.; Yu, X.; Zhou, Y. LSTM and GRU Neural Network Performance Comparison Study: Taking Yelp Re-view Dataset as an Example. In Proceedings of the 2020 International Workshop on Electronic Communication and Artificial Intelligence (IWECAI), Shanghai, China, 12–14 June 2020; pp. 98–101. [CrossRef]
20. Gao, S.; Huang, Y.; Zhang, S.; Han, J.; Wang, G.; Zhang, M.; Lin, Q. Short-term runoff prediction with GRU and LSTM networks without requiring time step optimization during sample generation. *J. Hydrol.* **2020**, *589*, 125188. [CrossRef]
21. Chen, S.; Billings, S.A.; Grant, P.M. *Non-Linear Systems Identification Using Neural Networks*; Acse Report 370; Department of Automatic Control and System Engineering, University of Sheffield: Sheffield, UK, 1989; Available online: https://eprints.whiterose.ac.uk/78225/ (accessed on 5 January 2024).
22. Narendra, K.S.; Parthasarathy, K. Learning automata approach to hierarchical multiobjective analysis. *IEEE Trans. Syst. Man Cybern.* **1991**, *21*, 263–272. [CrossRef]
23. Ouyang, H.-T. Nonlinear autoregressive neural networks with external inputs for forecasting of typhoon inundation level. *Environ. Monit. Assess.* **2017**, *189*, 376. [CrossRef] [PubMed]
24. Kotu, V.; Deshpande, B. Time Series Forecasting. In *Data Science*; Elsevier: Amsterdam, The Netherlands, 2019; pp. 395–445.
25. Siegel, A.F.; Wagner, M.R. Time Series. In *Practical Business Statistics*; Elsevier: Amsterdam, The Netherlands, 2022; pp. 445–482.
26. Olorunlambe, K.A.; Eckold, D.G.; Shepherd, D.; Dearn, K.D. Bio-Tribo-Acoustic Emissions: Condition Monitoring of a Simulated Joint Articulation. *Biotribology* **2022**, *32*, 100217. [CrossRef]
27. Poddar, S.; Tandon, N. Detection of particle contamination in journal bearing using acoustic emission and vibration monitoring techniques. *Tribol. Int.* **2019**, *134*, 154–164. [CrossRef]
28. König, F.; Wirsing, F.; Jacobs, G.; He, R.; Tian, Z.; Zuo, M.J. Bayesian inference-based wear prediction method for plain bearings under stationary mixed-friction conditions. *Friction* **2023**, *12*, 1272–1282. [CrossRef]
29. König, F.; Sous, C.; Chaib, A.O.; Jacobs, G. Machine learning based anomaly detection and classification of acoustic emission events for wear monitoring in sliding bearing systems. *Tribol. Int.* **2021**, *155*, 106811. [CrossRef]

Disclaimer/Publisher's Note: The statements, opinions and data contained in all publications are solely those of the individual author(s) and contributor(s) and not of MDPI and/or the editor(s). MDPI and/or the editor(s) disclaim responsibility for any injury to people or property resulting from any ideas, methods, instructions or products referred to in the content.

Article

Influence of Lubrication Cycle Parameters on Hydrodynamic Linear Guides through Simultaneous Monitoring of Oil Film Pressure and Floating Heights

Burhan Ibrar [1,*], Volker Wittstock [1], Joachim Regel [1] and Martin Dix [1,2]

[1] Professorship for Machine Tools and Production Processes, Chemnitz University of Technology, 09126 Chemnitz, Germany; psp@mb.tu-chemnitz.de
[2] Fraunhofer Institute for Machine Tools and Forming Technology IWU, 09126 Chemnitz, Germany
* Correspondence: burhan.ibrar@mb.tu-chemnitz.de

Abstract: Hydrodynamic linear guides in machine tools offer a high load capacity and excellent damping characteristics, improving stability, precision, and vibration reduction. This study builds on previous research where floating heights were verified with a simulation model limited to measured floating heights. Advancements include incorporating pressure sensors into a fixed steel rail, enabling simultaneous measurement of oil film pressure and floating heights for a comprehensive understanding of lubrication conditions within the lubrication gap. The experimental results explore the effects of different lubrication methods, providing valuable insights into cavitation and lubrication adequacy. The results demonstrate the feasibility of utilizing pressure sensors to measure oil film pressure within the lubrication gap, providing a nuanced understanding of lubrication dynamics. By measuring both floating heights and pressure measurement, distinctions between hydrodynamic lubrication, mixed friction regions, and instances of lubricant deficiency become readily discernible. The variations in real-time oil film pressure and floating heights help to optimize the lubrication cycle for hydrodynamic linear guides, enhancing system performance and longevity.

Keywords: oil film pressure measurement; miniature strain gauge pressure sensors; lubrication improvements; hydrodynamic linear guides; Stribeck curve

1. Introduction

Hydrodynamic linear guides (HDGs) are essential components in various engineering applications, facilitating precise and smooth linear motion due to good damping properties. They are successfully used not only in large machines but also in medium-sized machines [1]. They have a good damping coefficient, are low cost [2], and lead to high-quality machined surfaces. Therefore, this type of guideway remains crucial for the feed axes in numerous machine tools [2]. Contrary to the constant supply of lubricant in hydrostatic bearing, the efficiency is better. The constant supply of lubricant in hydrostatic bearings increases the running cost, which reduces the efficiency; whereas, roller bearing leads to low-quality machine surfaces due to the dynamic excitation of rolling contact. The wear caused by friction is considered the main reason for the failure of mechanical systems and the major source of energy loss [3]. Depending on the application, a static surface pressure of approximately p_{stat} = 0.5 MPa, corresponding to the weight load (F_{stat}), is recommended for the bearing surfaces of the guideways [4]. However, significantly lower values are used for higher dynamics of the feed axis, such as in grinding machines [5,6]. The hydrodynamic (HD) pressure is generated due to wedge-shaped film and relative movement between the guide surfaces. The initial moment of floating and forming a lubrication wedge is due to oil inlet pressure [6]. One of the key parameters in hydrodynamic guides is the oil film pressure, which influences the operation of guides [2,7]. The accuracy of the straight-line motion of machine tool tables with a hydrodynamic linear guide is influenced not only

by the geometry of the guideway, but also the thickness of the lubricant films [8]. The floating heights of a hydrodynamically lubricated guide are inconsistent depending on the speed and load, assuming a sufficient oil supply and constant viscosity of the lubricant. In machine tools, however, the change of state of friction is undesirable [8]. Additionally, the presence of mixed friction regimes during start-up and deceleration phases introduces instability and potential wear, adversely affecting system accuracy and efficiency. Furthermore, issues such as air entrapment within the lubrication gap and variations in oil film pressure further compound the complexities of lubrication dynamics, leading to suboptimal performance and potential mechanical failure. Understanding lubrication conditions is crucial for mitigating challenges (adverse effects of dynamic changes in floating heights and mixed friction regimes) in hydrodynamic linear guides. Therefore, the primary objective of this study is to gain insight into the lubrication conditions (with the help of pressure sensors) within the lubrication gap and propose solutions to improve the lubrication conditions, which enhance the performance and reliability of hydrodynamic linear guides in engineering applications.

2. Research Objective

Through systematic experiments and analysis, the aim of this novel study is to gain insights into the lubrication condition inside the lubrication gap. It also seeks to propose solutions that enhance the overall performance, stability, and longevity of hydrodynamic linear guides across various operation modes of the machine tools (rapid feed rate versus process feed rate). It will investigate how different factors in the lubrication cycle influence pressure sensor readings and access the impact of the number of oil-supplied lubrication grooves in the slide bar on oil film pressure. The effectiveness of existing pressure integration methods will be evaluated and assess how the amount of oil (or air entrapment) affects sensor performance. The study also seeks to understand the nuances of oil film pressure measurements and their implications for improving the lubrication cycle.

3. Literature Review

In the domain of hydrodynamic linear guides, this study pioneers the integration of sensors to measure oil film pressure within the lubrication gap, filling a significant gap in existing research. Despite decades of research in this area, there has been a notable absence of methods to measure oil film pressure in hydrodynamic linear guides. While experimental efforts dating back to 1969 have focused on measuring dynamic friction and floating heights/film thicknesses, the absence of a developed method for oil film pressure measurement underscores the novelty and importance of our research endeavor. Our research addresses this gap by introducing an innovative approach, aiming to deepen the understanding of lubrication dynamics in hydrodynamic linear guides. However, the existing literature was reviewed to examine variable dependencies, the design of lubrication grooves, pressure measurement for other applications, and types of lubrication.

In their experimental research, Shiozaki et al. [5] investigated the effects of lubricant viscosity on film thickness and frictional resistance in models and surface grinding machines through an oil bath lubrication method. The study demonstrated that film thickness increases at the leading edge and decreases at the trailing edge as the table moves, with a minimum thickness exceeding 10 μm. A direct relationship between the table speed and the inclination was observed, indicating that speed influences system dynamics. Additionally, increased load slightly reduced film thickness, pointing to a complex relationship between lubricant efficiency and load. Remarkably, higher speeds ensured consistent frictional resistance, even with changes in slider inclination, and higher viscosity reduced the speeds at which frictional resistance was minimized. The findings emphasize the importance of selecting appropriate lubricants and operational parameters to minimize friction and enhance efficiency in surface grinding processes. The study, however, did not address oil film pressure in the lubrication gap, a crucial aspect for understanding lubrication conditions.

Kortendieck et al. [8] explored the complexity of lubrication in hydrodynamically lubricated linear guides in machine tools, highlighting the importance of effective oil distribution for minimizing friction and ensuring smooth operations. Through an experimental setup involving a 900 mm long table with a maximum stroke of 150 mm and a feed rate of 3.8 m/min, the study underlines the necessity of optimizing lubricant distribution across sliding surfaces, considering surface smoothness, gap size, and viscosity. This study presents design recommendations for lubrication grooves, emphasizing perpendicular orientation, comprehensive coverage, and strategic placement. These recommendations aim to enhance oil supply and maintain lubrication film integrity, which are crucial for reducing losses and ensuring precision in machining applications. The research also examines the impact of motion sequences on lubricant behavior and the importance of accommodating gap height variations, concluding that a thorough understanding of lubrication dynamics and strategic design is vital for improving the performance and durability of machine tool sliding guides.

Gläser et al. [9] developed a simulation model by applying the finite difference method to the Reynolds equation [10] to estimate the floating angle. The conformity of the floating angle between the numerical and simulated results was good at velocities up to 40 m/min. On the one hand, they measured friction coefficients of sliding surfaces with very low variations and on the other hand, with a slight concavity deviation of the flatness. It has been realized that with concave sliding surfaces there is relatively 60% more friction and the floating angle showed irregular behavior.

Zhang et al. [11] continued the work of Gläser et al. [9] and estimated the oil film pressure and floating heights of HDGs using a 2D simulation model (single-phase) based on the Reynolds equation [10]. However, due to the unavailability of a pressure measurement system, the simulation model was calibrated using only experimental floating heights (measured using eddy current sensors). The lubrication was adjusted subjectively by visual inspection. Both Gläser et. al. [9] and Zhang et. al. [11] used only two lubrication grooves close to the ends of the carriage. Neugebauer et al. [12] have also studied various shapes of the lubrication gap and their effects on oil film pressure.

Optical sensors can be used as an alternative to eddy current sensors for measuring oil film thickness in hydrodynamically lubricated bearings, as demonstrated in [13,14]. These sensors work by transmitting light through the lubricant film and reflecting it from the shaft surface back to the sensor. However, they tend to be more expensive due to their use of advanced technologies such as Fiber Bragg Grating [15] or Fabry–Perot interferometry [16].

Previously, 16 manometer tubes were used by Sinanoglu et al. [17] to study the pressure development in journal bearings under different velocity variations and various shaft surface textures [17]. Additionally, Ichikawa et al. [18], Mihara et al. [19,20], Mihara and Someya [21], and Someya and Mihara [22] used thin film sensors, consisting of thin material layers with a total thickness of 6 µm, on the sliding surface of bearings to measure the oil film pressure in journal bearings. Mihara and Someya [21] conducted oil film pressure measurements, along with temperature and strain measurements, on the bearing surface in an engine test. Ronkanien et al. [23] used optical pressure sensors to measure the oil film pressure in journal bearings. Iwata et al. [24] displayed the successful implementation of a recently created thin film sensor for gauging the distribution of oil film pressure in the main bearings of high-speed motorcycle engines. The results of the study emphasize the significance of accurate oil film pressure measurement in optimizing engine performance, thereby offering valuable insights for further exploration and improvement in engine design and lubrication systems.

Previous research efforts have laid the groundwork for the current investigation, emphasizing the critical role of proper lubrication in achieving optimal performance and longevity in linear guide systems. Ibrar et al. [25] and Wittstock et al. [26] developed an integration method for strain gauge miniature pressure sensors into the rail to measure hydrodynamic pressure generated inside a lubrication gap. This method is further used in this study to understand the lubrication conditions and influencing parameters. These findings serve as a pertinent reference point for the advancement of similar technology in diverse applications. Firstly, based on [27,28], the pressure sensors were installed in an identical rail made of acrylic glass (Plexiglas) for easier fabrication. Different integration methods were tested in the Plexiglas rail to measure the pressure and observe the oil or trapped air [25]. Subsequently, the finalized integration method, which showed a significant improvement in pressure measurement, was used to integrate pressure sensors into the steel rail [25]. The author also calibrated the pressure sensors statically and dynamically before and after integrating them into the steel rail [25]. This novel approach enables a comprehensive evaluation of lubrication dynamics and system behavior under dynamic operating conditions, paving the way for informed decisions in lubrication strategy optimization and maintenance practices.

4. Experimental Setup

Figure 1 depicts the hydrodynamic linear guide system used in this study. The system features a carriage with two sliding bars that move over two parallel steel rails. Distance eddy current sensors are mounted at each of the four ends of the carriage to measure the lubrication gap, as well as the roll and pitch of the carriage. Shell Tonna S-68 lubricant is used, with a density of 879 kg/m^3 and a viscosity of 68 mm^2/s at 40 °C (8.6 mm^2/s at 100 °C). Pressure sensors have a round sensing surface with a diameter of 6 mm and a thickness of 0.6 mm. Seven strain gauge pressure sensors by Kyowa [29] named "PS-5KD", calibrated, are integrated into a steel rail (integration method is shown in Figure 2c) to provide a detailed examination of the pressure within the lubricating gap. The pressure sensor comprises miniature strain gauge pressure sensors that can safely operate within a temperature range of −20 to 70 °C [29], allowing for the measurement of oil film pressure inside the lubrication gap. The pressure rise rate is 213 bar/s.

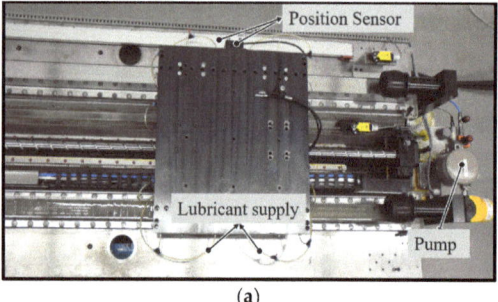
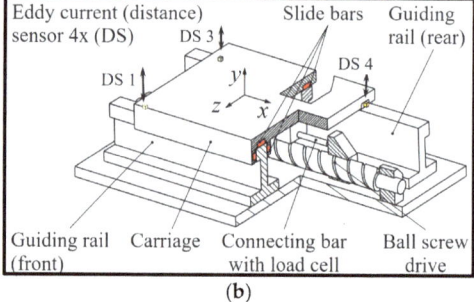

Figure 1. (**a**) Top view of the test stand for hydrodynamic linear guides with pressure sensors integrated into one steel rail, and (**b**) schematic configuration of the hydrodynamic linear guide test stand.

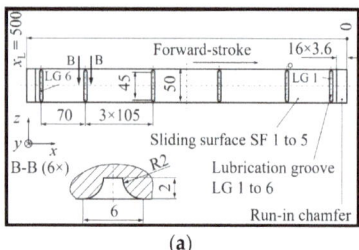
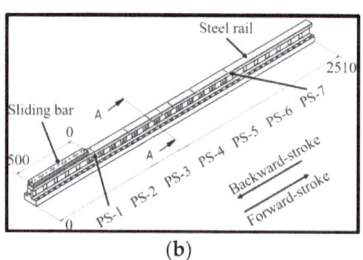
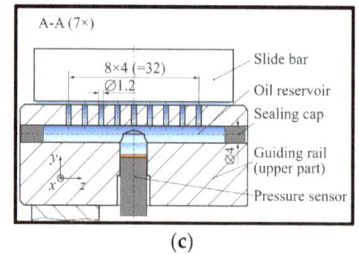

Figure 2. (**a**) Sliding surface geometry with six lubrication grooves, (**b**) steel rail with seven integrated pressure sensors, and (**c**) cross-sectional view (A-A) of the pressure sensors integration method with a strip including small holes. All units are in millimeters (mm).

Figure 2a demonstrates the inclusion of six lubrication grooves, LG 1 to 6, in each sliding bar accompanied by four holes with a diameter of 1.8 mm in each groove to supply lubricant in the lubrication gap. The sliding bars have undergone precise machining to guarantee an optimal and seamless contact with the steel rails if no lubrication oil is supplied. This was assessed using contact paper. With the help of pressure sensors, it has been realized that two lubrication grooves (used in [9,10]) are not sufficient to lubricate the 500 mm long lubrication gap irrespective of the lubrication time. Contrary to [9,10], a combination of two, four, and six oil-supplied lubrication grooves have been compared using integrated pressure sensors to test which combination improves the lubrication conditions. Figure 2b,c illustrates the inclusion of the pressure sensors PS 1 to PS 7 in the steel rail, allowing for accurate monitoring of the oil film pressure in a longitudinal manner along the length of the carriage. However, it is important to note that these sensor systems are unable to detect pressure variations across the width of the rail (average pressure in transverse direction). The distances between the pressure sensors are as follows: 5 mm from the start position of the sliding bar to PS1, 300 mm from PS1 to PS2, 200 mm from PS2 to PS3, 100 mm from PS3 to PS4 and from PS4 to PS5, 200 mm from PS5 to PS6, and 300 mm from PS6 to PS7.

Integrating sensors into the steel rail enables the full utilization of the 1.7 m stroke length, allowing for a maximum feed rate of 100 m/min in both forward and backward directions. The pressure data are obtained throughout the extensive range of constant velocity, acceleration, and deceleration of the carriage. Nevertheless, limitations such as the stick-slip effect and the necessity to minimize abrupt motion of the carriage restrict the attainable range of both acceleration and deceleration (3 m/s^2).

The advantage of integrating pressure sensors in a steel rail is that both oil film pressure and floating heights can be measured simultaneously. The steel rail, composed of P20/1.2311 (40-CrMnMo-7) grade alloy tool steel, has undergone quenching and tempering procedures to achieve an impressive strength of 1000 N/mm^2. Furthermore, gas-nitriding techniques have been applied, achieving a hardening depth of 0.3 mm and a hardness level of 600 HV.

Utilizing a screw-drive mechanism, the system operates with a force sensor strategically placed between the ball screw drive and carriage to gauge the applied force precisely. To eliminate any horizontal shifting, roller bearing side guiding is incorporated on rear side of the carriage, utilizing a steel rail without pressure sensors. In contrast, the carriage exclusively interfaces with one side of the front steel rail, which includes integrated sensors, ensuring optimal performance through hydrodynamic lubrication.

To obtain the rail surface profile, a dry run of the carriage was conducted without the application of lubricant, operating at a speed of 0.1 m/min, which is one-tenth of the minimum speed used for measurements. The detected profile (as shown in Figure 3) is then subtracted from the actual measured floating heights and yields the absolute and realistic floating height values.

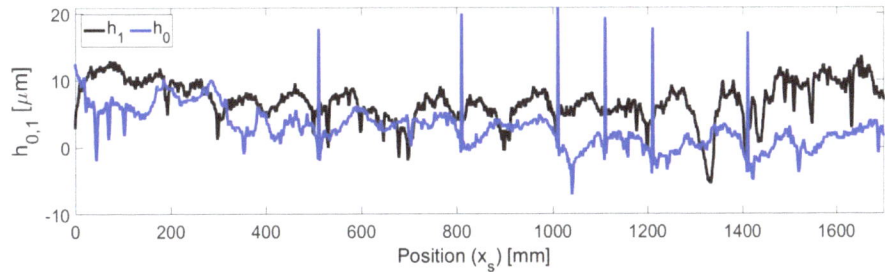

Figure 3. Surface profile of the rail (sensor side) measured through a dry run of the carriage at a speed of 0.1 m/min.

Before using the slide bars with six oil-supplied lubrication grooves to measure the pressure, it was necessary to evaluate the contact pattern between the sliding bar and the steel rail with integrated sensors [25]. Ink spots were applied to the steel rail, and the bars were meticulously cleaned to remove any oil from the lubrication gap. The carriage was then placed on both steel rails and moved back and forth to examine the surface contact. This analysis uncovered an irregular and nonuniform contact pattern between the sliding bar and the steel rail, leading to low or negative pressure readings in areas with insufficient contact. Nevertheless, this situation accurately reflects the real-world conditions experienced in machine tools.

5. Experimental Design

In this research study, a series of experiments was meticulously designed as shown in Table 1 to investigate the impact of various factors on pressure measurements within the lubrication system. Specifically, the experiments were structured to analyze the influence of different parameters, including feed rate variations (10, 20, 30, 40, and 50 m/min), lubrication time (duration of lubricant supply), pause time (interval between lubrication supply and measurement stroke), static pressure p_{stat} (ratio of the weight of the carriage to the contact surface area between the rail and the carriage), and the number of oil-supplied lubrication grooves present. By systematically altering these variables across multiple experimental setups, the study aimed to understand how each factor contributes to variations in pressure measurements. This approach allowed for a detailed examination of the intricate dynamics within the lubrication system, providing valuable insights into its behavior under diverse operating conditions.

Due to unidentified issues, the pressure sensor PS 3 displayed errors throughout from Exp. No. 10 to Exp. No. 18. As a result, the data collected from PS 3 during these experiments have been deemed unreliable and has been omitted from the analysis.

The lubrication cycle employed in the study consisted of several key phases as shown in Figure 4. Initially, lubrication was applied to the system, followed by a variable pause time to allow for proper distribution and settling of the lubricant. Subsequently, measurements were taken during the forward stroke of the system to assess its performance under lubricated conditions. After this, the lubrication process was repeated, followed by another variable pause period, before measurements were taken during the system's backward stroke. Each stroke is separated by the addition of the lubrication interval (t_L) and the pause time (t_P). This cycle was repeated five times to ensure thorough data collection and consistency in the experimental procedure.

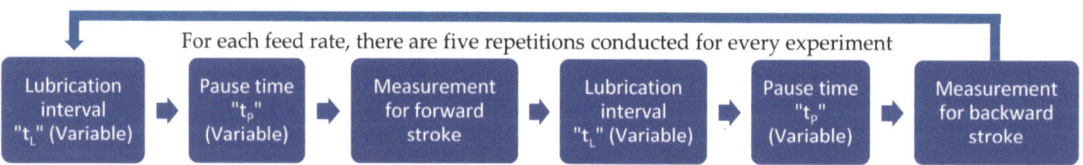

Figure 4. Lubrication cycle with variable parameters used in this study.

In this study, the analysis of the influence of different static pressures on pressure measurement and system performance was not discussed, because this aspect was already covered by Ibrar et al. [25] during the development of the integration method for pressure sensors.

Table 1. Experimental design at different variable lubrication cycle's parameters and highlighted are the experiments discussed in Section 6 accordingly.

Experiment (Exp.) No.	Number of Oil-Supplied Lubrication Grooves (see Figure 2a)	Static Pressure p_{stat} [bar]	Feed Rate v_s [m/min]	Pause Time t_P [s]	Lubrication Interval/Time t_L [s]
1	6 (LG—1, 2, 3, 4, 5, and 6)	0.24	10, 20, 30, 40, and 50	5	8
2	6 (LG—1, 2, 3, 4, 5, and 6)	0.454	10, 20, 30, 40, and 50	5	8
3	6 (LG—1, 2, 3, 4, 5, and 6)	0.7	10, 20, 30, 40, and 50	5	8
4	6 (LG—1, 2, 3, 4, 5, and 6)	0.24	10, 20, 30, 40, and 50	30	8
5	6 (LG—1, 2, 3, 4, 5, and 6)	0.454	10, 20, 30, 40, and 50	30	8
6	6 (LG—1, 2, 3, 4, 5, and 6)	0.7	10, 20, 30, 40, and 50	30	8
7	6 (LG—1, 2, 3, 4, 5, and 6)	0.24	10, 20, 30, 40, and 50	60	8
8	6 (LG—1, 2, 3, 4, 5, and 6)	0.454	10, 20, 30, 40, and 50	60	8
9	6 (LG—1, 2, 3, 4, 5, and 6)	0.7	10, 20, 30, 40, and 50	60	8
10	6 (LG—1, 2, 3, 4, 5, and 6)	0.24	10, 20, 30, 40, and 50	5	5
11	6 (LG—1, 2, 3, 4, 5, and 6)	0.454	10, 20, 30, 40, and 50	5	5
12	6 (LG—1, 2, 3, 4, 5, and 6)	0.7	10, 20, 30, 40, and 50	5	5
13	4 (LG—1, 2, 5, and 6)	0.24	10, 20, 30, 40, and 50	5	8
14	4 (LG—1, 2, 5, and 6)	0.454	10, 20, 30, 40, and 50	5	8
15	4 (LG—1, 2, 5, and 6)	0.7	10, 20, 30, 40, and 50	5	8
16	2 (LG—1, and 6)	0.24	10, 20, 30, 40, and 50	5	8
17	2 (LG—1, and 6)	0.454	10, 20, 30, 40, and 50	5	8
18	2 (LG—1, and 6)	0.7	10, 20, 30, 40, and 50	5	8

6. Analysis and Findings

Analysis methodology of pressure and floating: Floating thicknesses of the oil film (h_0 and h_1) depend directly on lubrication cycle parameters. As the lubricant enters the gap between the slide bar and rail (see Figures 1 and 2), the floating heights increase; when lubrication stops, the height decreases due to the squeezing effect. Initial floating heights significantly influence the pressure within the lubrication gap. Pressure sensors provide critical real-time data on lubrication conditions, essential for improving lubrication and preventing failure.

To grasp the pressure curves and floating heights measured, the data from Exp. No. 3 are presented in Figure 5, where the pressure curves closely resemble the ideal scenario of an excellent floating behavior. The floating heights and pressure were measured during an exemplary forward stroke with relatively high feed rates of 30 m/min, with a static pressure of 0.7 bar and a 5 s pause time. Figure 5a shows floating heights at both the trailing edge (h_0) and the leading edge (h_1) and the inclination angle (α) of the slide. It represents that at the start of the slide bar's movement, the slide bar is lifted and runs relatively smoothly afterward. Figure 5b shows the pressure curves measured through pressure sensors integrated into the steel rail. In Figure 5c, the slide with lubrication grooves and the rail with the position of pressure sensors are schematically illustrated. Vertical dashed lines show the position of all seven pressure sensors; therefore, they represent the starting point of each pressure curve of PS 1 to PS 7.

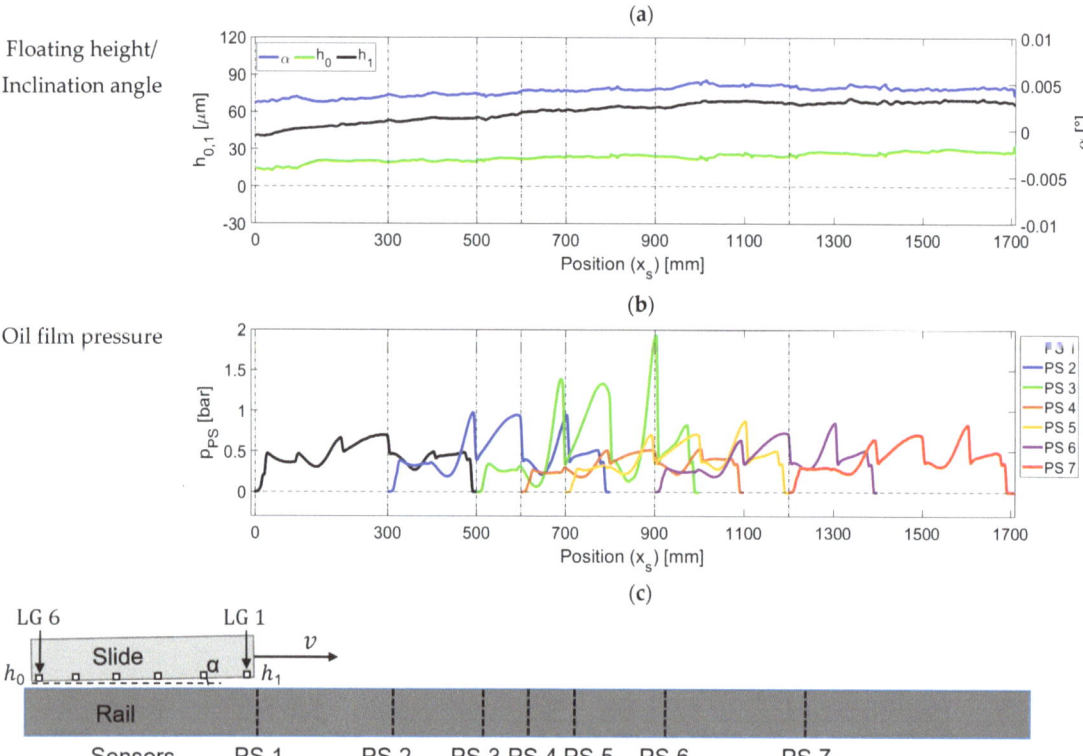

Figure 5. (Exp. No. 3) The corresponding measured floating heights $h_{0,1}$, along with the lubrication floating angle α during the forward stroke, are presented in (**a**). The pressure curves of sensors PS 1 to PS 7, related to slide bar movement, are plotted against the stroke position (x_s) in (**b**). Sketch of a slide featuring lubrication grooves and rail with the position of pressure sensors integrated in (**c**).

The first pressure sensor PS 1 (as shown in Figure 2) sits right outside the lubrication gap and beside the inlet floating height h_1 when the slide bar is at the rest position. As soon as the slide bar moves, the first sensor PS 1 comes inside the lubrication gap and the pressure sensor records the first pressure curve (shown by the black curve in Figure 5 from the 0 to 500 mm position). The second pressure sensor PS 2 sits at a distance of 300 mm from the first sensor; therefore, the blue pressure curve starts from 300 mm and completes at 800 mm and so on. Six pressure maxima corresponding to six oil-supplied lubrication grooves can be clearly seen in the third pressure curve (from 500 to 1000 mm position) measured from the third pressure sensor PS 3. The variability of oil film pressure along the length of the slide bar (due to the difference between the floating thicknesses h_0 and h_1) can be validated with the help of the integrated pressure sensors as it passes over them. The pressure curve measured through PS 3 shows that at the leading edge, the pressure is minimal as the floating height h_1 is relatively large whereas the maximum pressure generated between the LG 5 and LG 6 is close to the trailing floating height (h_0), which is relatively small. The pressure curves measured by pressure sensors from PS 4 to PS 7 show similar curves with the same pressure values. This is because, after reaching maximum thickness, the pressure decreases, causing the slide bar to start sinking.

During the forward strokes, the floating height sensor h_0 at the trailing edge runs over one of the small holes in the rail (which connects the pressure sensor and lubrication gap). The floating height h_0 has seven peaks (small fluctuations) in all experimental results because when it runs over the small holes of the pressure measurement then it records the larger values as compared to the lubrication gap. Due to the assembly and disassembly of experimental setup during different experiments, perfect synchronization of position (x_s) could not be achieved. This issue should be addressed by subtracting the surface profile of the steel rail (measured with same floating heights sensor before experimental run) from the actual floating height measurement. However, minor fluctuations in h_0 are observed due to slight discrepancies of the position between the dry run measurement and the actual pressure measurements.

Variation of Feed Rate: v_s. Comparing the pressure curves presented in Figures 5 and 6, for feed rates of 30 m/min and 10 m/min, respectively, for a surface pressure of 0.7 bar, reveals slight differences. The pressure curves recorded by the sensors indicate slightly higher values at lower speeds, contrary to expectations. However, as the speed increases, there is a greater likelihood of air bubble formation within the oil reservoir (see Figure 2c). However, the actual increase in pressure due to the increment of the feed rate cannot be measured through this setup. This is because of a very small sensor area as compared to the total lubrication gap and when the pressure inside the lubrication gap increases, then the floating heights will increase. The change of floating heights will neutralize the pressure generated and the pressure inside the lubrication gap becomes equal to the weight of the carriage. In addition, if the pressure reduces inside the lubrication gap, then floating heights are reduced until the pressure becomes equal to the weight of the carriage. Therefore, to capture the actual increase in pressure, the floating heights should be kept constant, which is not realistically possible. However, if the slide bar moves at higher speed, then the floating heights have higher values as compared to lower feed rates, which proves that at higher speed, the pressure generated is also higher but cannot be measured with the pressure sensors. In addition, the minimum lubrication gap is influenced by both the feed rate and the surface pressure or load that the system must withstand.

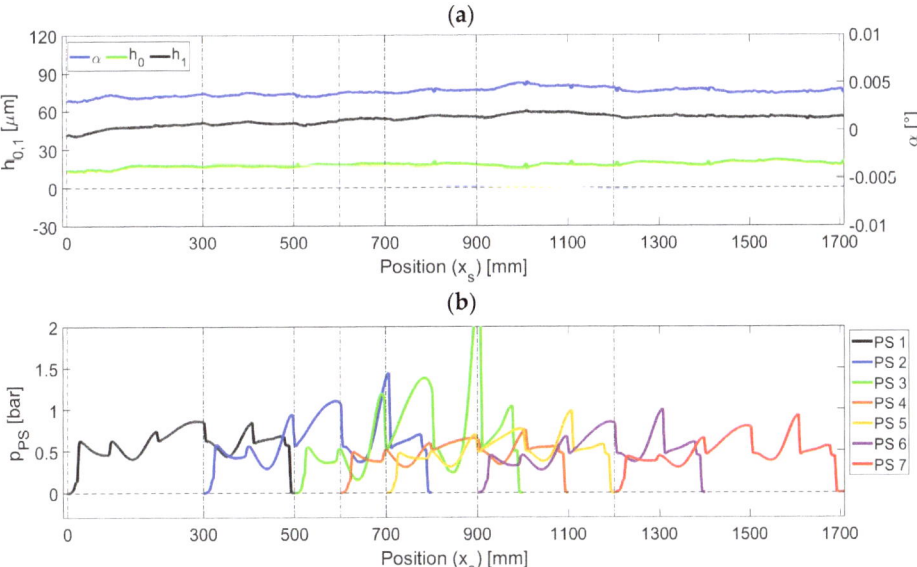

Figure 6. (Exp. No. 3) The pressure curves of sensors PS 1 to PS 7 are shown in (**b**), and the measured floating heights $h_{0,1}$, along with the lubrication floating angle α, are shown in (**a**) during the forward stroke at a feed rate of 10 m/min.

Variation of the Number of Oil-Supplied Lubrication Grooves: (LG) To compare the different number of oil-supplied lubrication grooves, a slide bar with six lubrication grooves was used. However, oil was supplied through two, four, and six lubrication grooves for different tests. With the help of pressure sensors, it has been realized that two lubrication oil–supplied grooves are not sufficient to lubricate the 500 mm long lubrication gap irrespective of the lubrication interval. It is apparent that having fewer than six oil-supplied grooves was not sufficient to lubricate the lubrication gap properly, leading to poor lubrication during operation. This leads to irregular floating heights and an inappropriate pressure measurement, which is an undesirable outcome.

The different oil film pressures measured at a feed rate of 30 m/min with six, four, and two oil-supplied lubrication grooves, respectively, are shown in Figure 5 (Exp. No. 3), Figure 7 (Exp. No. 15), and Figure 8 (Exp. No. 18). It is evident from the results that it is important to have six oil-supplied lubrication grooves because it shows promising results (pressure curves) and because it can lubricate the whole lubrication gap before each stroke, while the pressures measured with two and four oil-supplied lubrication grooves have zero pressure regions due to poor lubrication between the lubrication grooves. However, the maximum values of pressure peaks corresponding to the lubrication grooves also increase with fewer grooves. This is understandable, as with fewer oil-supplied lubrication grooves, the guiding surfaces only separate in regions where the lubricant is provided. In other words, when lubricating with fewer grooves, the weight of the entire carriage is distributed over a smaller area, which leads to higher maximum pressure values over a relatively small area. The results clearly show that the floating thickness (h_0) decreases with fewer oil-supplied lubrication grooves; h_0 rises with the number of oil-supplied lubrication grooves from two to four and six. The differences in pressure and floating height measurements due to varying numbers of oil-supplied lubrication grooves are almost independent of feed rate, surface static pressure, pause time, and lubrication interval.

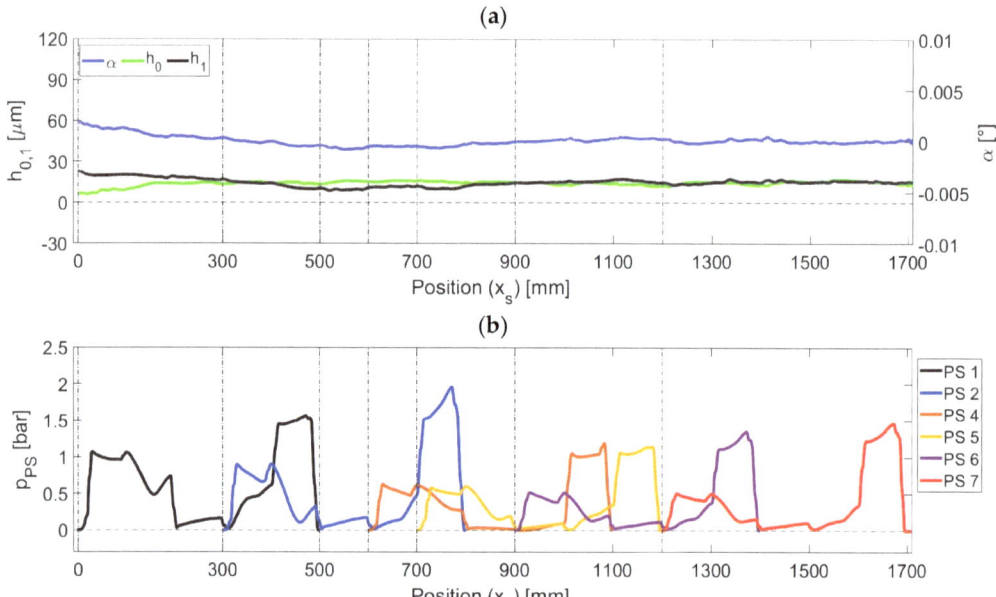

Figure 7. (**a**) Floating heights ($h_{0,1}$) and floating angle (α) measured during Exp. No. 15 at a feed rate of 30 m/min and 0.7 bar static pressure, using a slide with four oil-supplied grooves. (**b**) Oil film pressure measured under the same conditions.

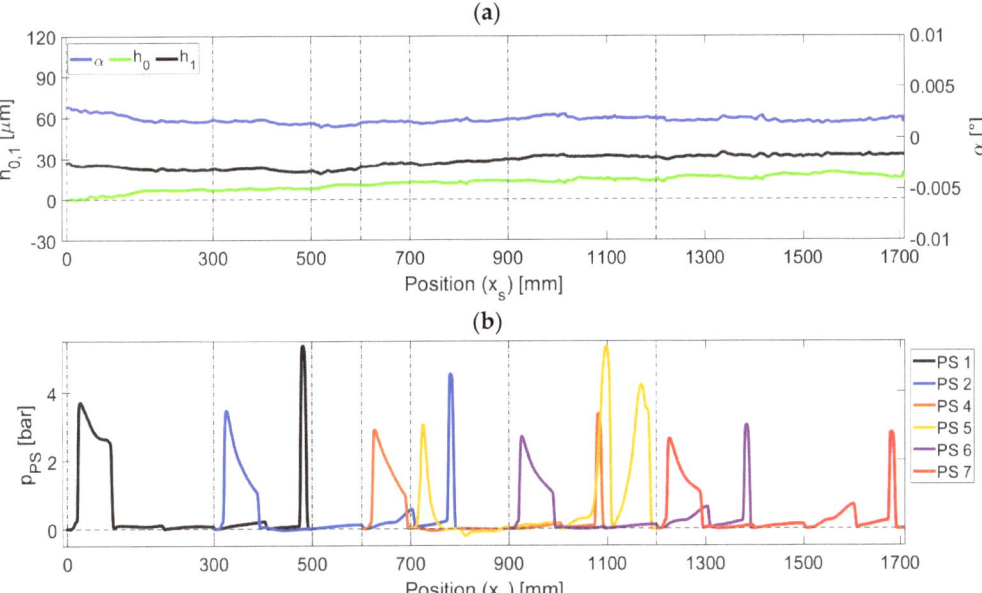

Figure 8. (**a**) Floating heights ($h_{0,1}$) and floating angle (α) measured during Exp. No. 18 at a feed rate of 30 m/min and 0.7 bar static pressure, using a slide with two oil-supplied grooves. (**b**) Oil film pressure measured under the same conditions.

Variation of Lubrication Interval: Different lubrication intervals (t_L) (Exp. No. 1 with 8 s t_L and Exp. No. 10 with 5 s t_L for static pressure of 0.24 bar) were tested to explore their effects on pressure distribution inside the lubrication gap and lubrication effectiveness. The difference between different pause times for other static pressures (0.454 and 0.7 bar) and a pause time of 5 s also show similar results. In other words, the pressure curves do not change their shape due to different static pressures, but only the maximum values in response to the weight of the carriage. This parameter is adjusted to thoroughly lubricate the lubrication gap and assess its influence on the system performance during operation. It is important to consider the number of oil-supplied lubrication grooves along with the lubrication interval time because the main goal is to lubricate the lubrication gap completely to achieve low friction and better lubrication conditions during operation. In Figure 8, it can be seen that the pressure sensors showed two pressure peaks where the lubrication grooves meet the pressure sensors and in between two lubrication grooves, the sensors did not measure anything. This is due to poor lubrication inside the lubrication gap between both lubrication grooves. In Figure 9, the pressure measured inside the lubrication gap is shown for a feed rate of 10 m/min and a static pressure of 0.24 bar for two different lubrication interval times of 5 and 8 s. It can be clearly seen that the pressure values recorded with sensors from PS 4 to PS 7 are relatively higher with a lubrication interval of 8 s. This is due to better lubrication conditions inside the lubrication gap before each stroke and more oil stays inside the lubrication gap until the end of the stroke.

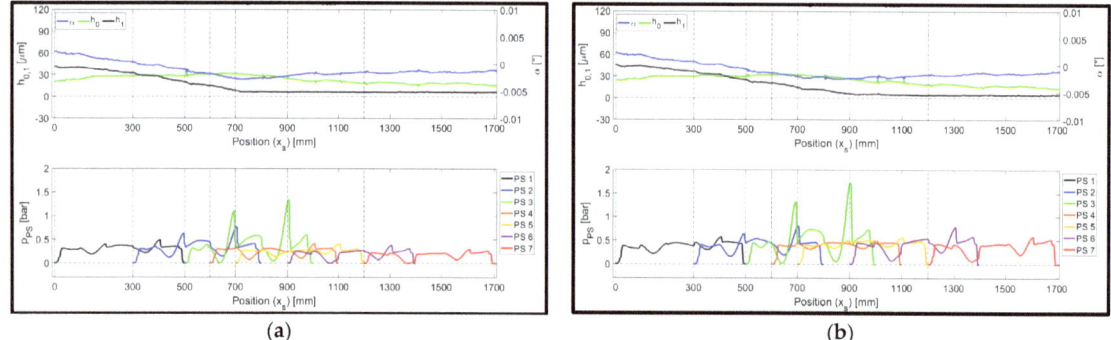

Figure 9. Comparison between different lubrication intervals (t_L). (**a**) Exp. No. 10 with a 5 s lubrication interval and (**b**) Exp. No. 1 with an 8 s lubrication interval.

The cavitation occurs inside the lubrication gap due to poor lubrication and relative motion between guiding surfaces, which draws the oil out from the oil reservoir. Initially, the integration method of pressure sensors was developed using a Plexiglas rail and carriage, enabling visual observation of flow and air entrapment in the lubrication gap. Later, when testing different lubrication grooves, the pressure sensors showed negative values between grooves due to increased air from poor lubrication conditions. This phenomenon, where pressure sensors indicated negative values in certain regions and increased with longer pause times, was also reported in [19].

The presence of air bubbles inside the oil reservoir can reduce the pressure values measured due to the compressibility of air, which can be confirmed from the comparison of different lubrication intervals as shown in Figure 9. To avoid these losses, it is important to keep the oil reservoir always filled with oil. In this experimental study, the oil reservoir was filled only before each experiment run, because it is not feasible or safe to refill it after every stroke. However, there is a possibility to avoid these losses by providing the oil to the reservoir before each stroke with the help of a pump in future experiments.

To avoid or reduce cavitation within the lubrication gap, it is important to optimize lubrication parameters. Increasing the number of lubrication grooves ensures that the entire lubrication gap is thoroughly filled with oil. Additionally, a longer lubrication time and a shorter pause time help maintain proper lubrication during strokes. Furthermore, the feed rate and floating heights are directly proportional to the generation of cavitation, meaning that higher feed rates and increased floating heights can increase cavitation.

Variation of the Pause Time: t_P In real application, there is no specific pause time and the lubricant is supplied usually before or during strokes. It is not optimal to supply lubricant during machining or operation and if the lubricant is supplied before strokes then there will be a pause, which can be very short or long depending on the requirement. Therefore, it is important to understand the effects of pause duration on pressure measurement, and various pause durations (5 s (Exp. No. 3), 30 s (Exp. No. 6) and 60 s (Exp. No. 9)) are tested with a 0.7 bar static surface pressure to investigate their impact on the lubricant settling and its distribution within the lubrication gap.

To enhance the comprehension of oil film pressure across different scenarios, the average pressure ($P_{avg} = \frac{1}{L}\int_0^L P(x)dx$) by integrating pressure over the length (L) of the lubrication gap (sliding bar) has been derived as shown in equation 1. This aggregated value provides a useful benchmark for comparison, shedding light on lubrication conditions and potential enhancements.

In Figure 10, the oil film pressure along with floating heights are shown for pause times of 5, 30, and 60 s for different feed rates (10, 30, and 50 m/min). It can be seen that the pressure measured after 60 s of pause time is comparatively small from all seven sensors and pressure curves go in a negative direction in some regions, which is not optimal.

Whereas, the pressure with a 5 s pause time shows promising results (also shown in Figure 5) because the pressure curves are in the positive region and the pressure measured through all seven sensors is also significant. However, with the 30 s pause time, the pressure sensors measured pressure curves with values between a 50 and 60 s pause time. All seven sensors measure similar pressure curves qualitatively, but quantitatively they are different due to speed variation, pause time, lubrication interval, and surface pressure. A short pause time causes the slide bar to descend during operation, leading to more irregular floating heights during operation and high friction. Conversely, a longer pause slows down the process, resulting in a decrease in lubricant volume within the gap. This can lead to increased friction due to a mixed lubrication regime and an increased angle of attack due to pressure variations along the length of the slide bar.

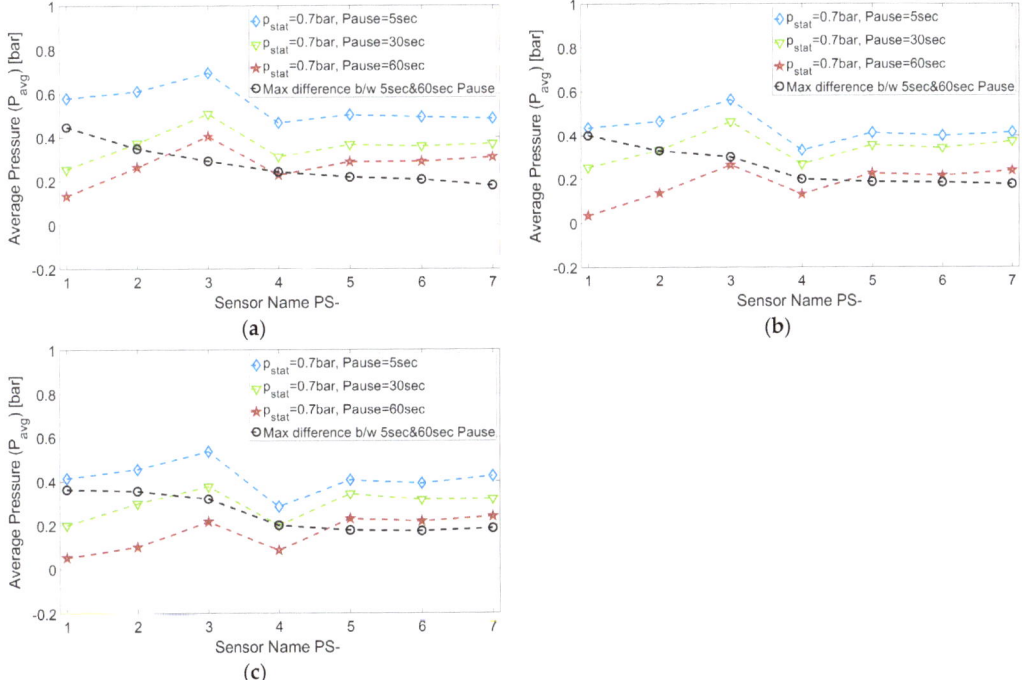

Figure 10. Comparison of average oil film pressure between different pause times t_P (5 s (Exp. No. 3), 30 s (Exp. No. 6) and 60 s (Exp. No. 9)) for all seven pressure sensors. (**a**) v_s = 10 m/min, (**b**) v_s = 30 m/min, and (**c**) v_s = 50 m/min.

It is also evident that the sensors generate reproducible results as long as the floating heights or inclination angles are similar. From the results, it can also be observed that the pressure curves are highly sensitive to the slide bar's inclination angle and the reason for instable floating heights is due to the varying pressure along the length of the slide bar.

7. Discussion

In the quest for improving the lubrication system of hydrodynamic linear guides, the integration of pressure sensors presents a promising opportunity to enhance functionality and efficiency. By utilizing pressure sensor technology, various significant advancements can be made as follows, with a specific focus on optimizing the lubrication system:

1. Real-Time Monitoring: Pressure sensors enable the continuous monitoring of oil film pressure within the lubrication gap. This constant feedback allows for dynamic adjustments to lubrication system parameters, such as flow rate and timing. As conditions change during

operation, this real-time monitoring optimizes the distribution of the lubricant and ensures the ideal film thickness.

2. Detection of Lubrication Issues: With the presence of pressure sensors, it becomes possible to quickly identify and address any lubrication deficiencies or irregularities. Whether it is insufficient lubrication or the presence of air pockets, the data from pressure sensors empowers proactive interventions to ensure optimal lubrication performance.

3. Prevention of Failures: Pressure sensors have the capability to detect changes in pressure that indicate potential failures in the lubrication system, such as oil film breakdown or bearing overload. By promptly identifying these issues, pressure sensors facilitate preventive maintenance measures, reducing the risk of catastrophic failures and prolonging the lifespan of linear guides.

4. Enhanced Performance: Through precise monitoring and control of the lubrication process facilitated by pressure sensors, linear guides can achieve superior performance. Optimized lubrication parameters minimize friction, wear, and heat generation, resulting in smoother operation, reduced vibration, and improved accuracy and repeatability

5. Optimizing Sensor Placement: One challenge is to minimize the number of sensors required while still obtaining accurate and comprehensive data. Reducing the sensor count can help lower costs and simplify the system without compromising performance.

6. Machine Learning Approaches: Implementing machine learning algorithms can significantly improve the analysis and interpretation of the data collected by pressure sensors and four floating height sensors (to measure the thickness of lubrication gap and machine table's orientation). Additionally, the data from the current measurement of the motor to assess friction conditions can also enhance the overall monitoring capability. These algorithms can predict potential failures, optimize lubrication schedules, and adapt to changing operational conditions, thereby improving the efficiency and reliability of the lubrication system.

In brief, while the integration of pressure sensors in the lubrication system of hydrodynamic linear guides offers substantial benefits in monitoring, detecting issues, preventing failures, and enhancing performance, addressing challenges such as reducing the number of sensors and leveraging advanced technologies like multisensor systems and machine learning will be crucial for future advancements.

8. Conclusions

Using a newly developed pressure integration method, for the first time, this study provides an insight into the oil film pressure behavior of a hydrodynamic linear guide at various feed rates, different surface pressures, and various numbers of lubrication grooves, and different parameters of lubrication cycles have been measured. Pressure sensors are used in this study to comprehend the lubrication conditions inside the lubrication gap, which helps in taking appropriate actions to prevent failures. The measured transient pressure curve gives very important information regarding the lubricant conditions, the contact built between the sliding surfaces, the table inclination angle, the sliding surface conditions, and the presence of air pockets.

The study underscores the importance of minimizing the pause time for accurate pressure measurements, while acknowledging its impact on slide bar orientation. Pressure curves obtained with a 5 s pause time and longer lubrication intervals closely resemble the ideal scenario. However, longer pause times, coupled with an additional lubrication system to maintain a filled oil reservoir, could improve pressure measurements. Careful consideration of lubrication intervals and the number of lubrication grooves is crucial to ensure thorough lubrication before each stroke. In future research, it will be beneficial to compare three different slides featuring two, four, and six physical lubrication grooves.

Regarding feed rate variations, it is essential to note that pressure measurements may not precisely align with simulation results due to rapid changes in floating heights caused by pressure build-up in the lubrication gap. To obtain comparable simulation results, the instantaneous floating heights from the experimental results should be used in the simulation to calculate the pressure. The study emphasizes that the pressure sensors' limited area may not capture instantaneous pressure rises within the lubrication gap. Ideally, the average pressure value of any pressure curve should be above or closely match the carriage's static pressure.

The results presented in this paper demonstrate that the pressure sensors effectively measure the average oil film pressure, which closely approximates the carriage's average static pressure. However, if lubrication parameters are not properly selected, the average oil film pressure can be significantly reduced, leading to less favorable outcomes. Additionally, the number of lubrication grooves influences the accuracy of oil film pressure measurements. To ensure the precise measurement of the oil film pressure, lubrication parameters are optimized to ensure that the entire lubrication gap is fully lubricated before each stroke.

The sensors measure average pressure across the width of the slide; however, the sensors still provide valuable information about the lubrication conditions of the hydrodynamic slide. While the results demonstrate promising potential for monitoring lubrication conditions, challenges such as air entrapment within the oil reservoir have been identified. To address this issue, future research will focus on further improvement of the existing lubrication system to ensure a reliable oil supply to the oil reservoir during measurements. There is a possibility if the reservoir is attached to a pump (which is used for lubrication through lubrication grooves) to obtain oil supply that fills the oil reservoir after each stroke. However, a nonreturning valve should be installed between the pump and the oil reservoir to prevent backflow toward the pump/tank during pressure measurement. The pressure requirement to keep the oil reservoir filled is very small (1/10 of the small pressure measured during operation) to avoid a large change in pressure measurement during operation.

The results outlined in this research are based on controlled laboratory experiments that were conducted with adequate lubrication. Furthermore, future research will concentrate on optimizing the deployment of sensors, potentially reducing their quantity. The objective is to utilize pressure sensors for the ongoing monitoring of linear guide lubrication in machine tools, ultimately enhancing efficiency and dependability without depending on float measurements.

Author Contributions: Conceptualization, B.I.; methodology, B.I.; software, B.I.; validation, B.I.; formal analysis, B.I.; investigation, B.I.; resources, B.I.; data curation, B.I.; writing—original draft preparation, B.I.; writing—review and editing, B.I., V.W., J.R. and M.D.; visualization, V.W. and B.I.; supervision, V.W. and J.R.; project administration, J.R.; funding acquisition, V.W. All authors have read and agreed to the published version of the manuscript.

Funding: This work is funded by the Deutsche Forschungsgemeinschaft (DFG, German Research Foundation)—Project-ID 285064832 (WI 4053/9-2). The authors thank them for their support.

Data Availability Statement: The original contributions presented in the study are included in the article, further inquiries can be directed to the corresponding author.

Conflicts of Interest: I "Burhan Ibrar" and all other authors declare that the research was conducted in absence of any commercial or financial relationship that could be construed as a potential conflict of interest.

References

1. Uriarte, L.; Zatarain, M.; Axinte, D.; Yagué-Fabra, J.; Ihlenfeldt, S.; Eguia, J.; Olarra, A. Machine tools for large parts. *CIRP Ann.* **2013**, *62*, 731–750. [CrossRef]
2. Hirsch, A.; Zhu, B. Hydrodynamische Gleitführungen in Werkzeugmaschinen—Ein Auslaufmodell? (Hydrodynamic Sliding Guideways in Machine Tools—A Discontinued Model?). *Konstruktion* **2011**, *63*, 67–72.
3. Bhushan, B. *Introduction to Tribology*; Wiley: New York, NY, USA, 2013. [CrossRef]
4. Brecher, C.; Weck, M. *Werkzeugmaschinen Fertigungssysteme: Konstruktion, Berechnung und Messtechnische Beurteilung*; Springer Vieweg: Berlin, Germany, 2017.
5. Shiozaki, S.; Nakano, Y. Fluctuation of lubricant film thickness at table slideways of surface grinding machines. *JSME* **1969**, *12*, 1212–1222. [CrossRef]
6. Lakawathana, P.; Matsubara, T.; Nakamura, T. Mechanism of hydrodynamic load capacity generation on a slideway. *JSME Ser. C* **1998**, *41*, 125–133.

7. Brewe, D.E. Slider Bearings. In *Modern Tribology Handbook*; Bhushan, B., Ed.; CRC Press: Boca Raton, FL, USA, 2001; Volume 2, pp. 969–1039.
8. Kortendieck, W. Ölverteilung in Werkzeugmaschinen-Gleitführungen. In *Aufbau des Schmierfilms in Hydrodynamischen Gleitführungen und Einfluß der Schlittenbewegungen auf den Schmierzustand*; Machinenmarkt: Würzburg, Germany, 1969; Volume 75, pp. 2095–2100.
9. Gläser, M.; Wittstock, V.; Hirsch, A.; Putz, M. Simulation method for the floating of hydrodynamic guides. *Procedia CIRP* **2017**, *62*, 346–350. [CrossRef]
10. Grote, K.; Antonsson, E.K. *Springer Handbook of Mechanical Engineering*; Springer: New York, NY, USA, 2009.
11. Zhang, Y.; Wittstock, V.; Putz, M. Simulation for instable floating of hydrodynamic guides during acceleration and at constant velocity. *J. Mach. Eng.* **2018**, *18*, 5–15. [CrossRef]
12. Neugebauer, R.; Hirsch, A.; Kolouch, M.; Zhu, B. Optimierung der Schmiernutenform in hydrodynamischen Gleitführungen (Optimization of the Oil Groove in Hydrodynamic Slide Guideways). *Konstruktion* **2012**, *63*, 71–74.
13. Jaloszynski, T.M.; Evers, L.W. Dynamic Film Measurements in Journal Bearings Using an Optical Sensor. *SAE Tech. Pap. Ser.* **1997**, *970846*, 95–103.
14. Glavatskikh, S.; Larsson, R. Oil Film Thickness Measurement by Means of an Optical Lever Technique. *Lubr. Sci.* **2000**, *13*, 23–35. [CrossRef]
15. Valkonen, A.; Kuosmanen, P.; Juhanko, J. Measurement of Oil Film Pressure in Hydrodynamic Journal Bearings. In Proceedings of the 7th International DAAAM Baltic Conference—Industrial Engineering, Tallin, Estonia, 22–24 April 2010.
16. Binu, K.G.; Yathish, K.; Mallya, R.; Shenoy, B.S.; Rao, D.S.; Pai, R. Experimental Study of Hydrodynamic Pressure Distribution in Oil-Lubricated Two-Axial Groove Journal Bearing. *Mater. Today Proc.* **2015**, *2*, 3453–3462. [CrossRef]
17. Sinanoglu, C.; Nair, F.; Karamis, M.B. Effects of shaft surface texture on journal bearing pressure distribution. *J. Mater. Process. Technol.* **2005**, *168*, 344–353. [CrossRef]
18. Ichikawa, S.; Mihara, Y.; Someya, T. Study on main bearing load and deformation of multi-cylinder internal combustion engine: Relative inclination between main shaft and bearing. *JSAE Rev.* **1995**, *16*, 383–386. [CrossRef]
19. Mihara, Y.; Hayashi, T.; Nakamura, M.; Someya, T. Development of measuring method for oil-film pressure of engine main bearing by thin film sensor. *JSA Rev.* **1995**, *16*, 125–130. [CrossRef]
20. Mihara, Y.; Kajiwara, M.; Fukamatsu, T.; Someya, T. Study on the measurement of oil-film pressure of engine main bearing by thin-film sensor—The influence of bearing deformation on pressure sensor output under engine operation. *JSA Rev.* **1996**, *17*, 281–286. [CrossRef]
21. Mihara, Y.; Someya, T. Measurement of oil-film pressure in engine bearings using a thin-film sensor. *Tribol. Trans.* **2002**, *45*, 11–20. [CrossRef]
22. Someya, T.; Mihara, Y. New thin film sensors for engine bearings. In Proceedings of the CIMAC Congress, Kyoto, Japan, 7–11 June 2004. Paper No. 91, 16.
23. Ronkainen, H.; Hokkanen, A.; Kapulainen, M.; Lehto, A.; Martikainen, J.; Stuns, I.; Valkonen, A.; Varjus, S.; Virtanen, J. Optical Sensor for Oil Film Pressure Measurement in Journal Bearings. In Proceedings of the 13th Nordic Symposium of Tribology, Nordtrib 2008, Tampere, Finland, 10–13 June 2008; p. 12.
24. Iwata, T.; Oikawa, M.; Owashi, M.; Mihara, Y.; Ito, K.; Ninomiya, Y.; Kato, Y.; Kubota, S. The Verification of Engine Analysis Model Accuracy by Measuring Oil-film Pressure in the Main Bearings of a Motorcycle High-Speed Engine Using a Thin-Film Sensor. *Lubricants* **2022**, *10*, 314. [CrossRef]
25. Ibrar, B.; Wittstock, V.; Regel, J.; Dix, M. Integration and verification of miniature fluid film pressure sensors in hydrodynamic linear guides. *J. Mach. Eng.* **2023**, *23*, 38–55. [CrossRef]
26. Wittstock, V.; Ibrar, B.; Dix, M. Experimentelle Verifizierung des hydrodynamischen Drucks in Gleitführungen zur Kontrolle der Schmierung bei hohen Geschwindigkeiten. 15. VDI Fachtagung, Gleit- und Wälzlagerungen, 13.-14.06.2023, Schweinfurt. In *VDI-Berichte 2415*; VDI Verlag GmbH: Düsseldorf, Germany, 2023; pp. 211–220. [CrossRef]
27. Drews, K. Über Untersuchungen der Reibungs- und Bewegungsverhältnisse an Werkzeugmaschinen-Flachführungen im Grenzreibungsgebiet. Ph.D. Thesis, TH Darmstadt, Darmstadt, Germany, 1967.
28. Piwowarsky, E. Anforderungen an Lineargleitführungen und Einfluss der Schmierstoffe. *Tribol. Schmier.* **1999**, *1999*, 33–37.
29. Kyowa. Available online: https://www.kyowa-ei.com/eng/product/category/sensors/ps-d/index.html (accessed on 8 February 2024).

Disclaimer/Publisher's Note: The statements, opinions and data contained in all publications are solely those of the individual author(s) and contributor(s) and not of MDPI and/or the editor(s). MDPI and/or the editor(s) disclaim responsibility for any injury to people or property resulting from any ideas, methods, instructions or products referred to in the content.

Article

Investigation of Failure Mechanisms in Oil-Lubricated Rolling Bearings under Small Oscillating Movements: Experimental Results, Analysis and Comparison with Theoretical Models

Fabian Halmos *, Sandro Wartzack and Marcel Bartz

Department of Mechanical Engineering, Friedrich-Alexander-Universität Erlangen-Nürnberg (FAU), Engineering Design, Martensstraße 9, 91058 Erlangen, Germany; wartzack@mfk.fau.de (S.W.); bartz@mfk.fau.de (M.B.)
* Correspondence: halmos@mfk.fau.de

Abstract: Bearing life calculation is a well-researched and standardized topic for rotating operation conditions. However, there is still no validated and standardized calculation for oscillating operation, only different calculation approaches. Due to the increasing number of oscillating rolling bearings, for example, in wind turbines, industrial robots, or 3D printers, it is becoming more and more important to validate one of these approaches or to formulate a new one. In order to achieve this goal, the damage mechanisms for oscillating operating conditions must first be analyzed in more detail by means of experimental investigations. The open question is whether fatigue is the relevant damage mechanism or whether wear damage, such as fretting corrosion or false brinelling, dominates. The present work therefore shows under which oscillation angle and frequency fatigue occur in oil-lubricated cylindrical roller bearings.

Keywords: bearings; oscillation; oscillating movements; wear; fatigue; rating life

1. Introduction

1.1. Motivation and State-of-the-Art

There are many applications where rolling bearings are used in oscillating movements. The most prominent examples are pitch bearings in wind turbines [1,2] and joints in industrial robots. However, bearings are also used in oscillating movements on a smaller scale, for example, in 3D printers.

Although there have been more than 1000 tests with oscillating movements since 1964, there is still no standardized method for calculating the life of rolling bearings under oscillating movements, only different calculation approaches [3,4]. The best-known current approaches are those of Harris [5,6] and Houpert [7]. Another possibility is to convert the oscillating movement into a continuous rotary motion and use the basic rating life equation according to DIN ISO 281 [8]. The different approaches are briefly described below. For a deeper insight into the various calculation approaches, the work of Menck is recommended [9].

All calculation approaches are based on the standardized rating life calculation for continuous rotary motion according to DIN ISO 281 (see Equation (1)). The relationship between dynamic capacity C and equivalent dynamic load P is maintained, but the oscillating movement is taken into account in different ways. All approaches use the oscillation angle φ, which is defined as the angle between the starting point and the return point of the oscillatory movement. In addition, all approaches use the life exponent p in their respective rating life equation L_{Osc}, where $p = 3$ for ball bearings and $p = 10/3$ for roller bearings, as defined in DIN ISO 281:

$$L_{10} = \left(\frac{C}{P}\right)^p \tag{1}$$

1.1.1. Approach by Harris—Method 1

Harris introduces a modified load P_{RE} that depends on the oscillation angle φ and the life exponent p_H, with $p_H = 3$ for point contact (ball bearings) and $p_H = 4$ for line contact (roller bearings) [5]:

$$P_{RE} = \left(\frac{\varphi}{180°}\right)^{\frac{1}{p_H}} \cdot P \quad (2)$$

P_{RE} can be inserted into the basic rating life equation L_{Osc} and gives the number of oscillations in 10^6 cycles:

$$L_{Osc} = \left(\frac{C}{P_{RE}}\right)^p \quad (3)$$

1.1.2. Approach by Harris and Rumbarger—Method 2

A more recent approach by Harris and Rumbarger, specifically aimed at wind turbines, introduces a modified load rating C_{Osc} and a critical oscillation angle φ_{CRIT} [6]. In this context, φ_{CRIT} is the oscillation angle at which the loaded raceway areas touch but do not intersect. As the following equations Equations (4) and (5) indicate, φ_{CRIT} depends on the number of rolling elements Z, the contact angle α, the rolling element diameter D_w, and the pitch diameter D_{pw}. In Equation (4), the upper sign (+) refers to the inner raceway and the lower sign (−) refers to the outer raceway:

$$\varphi_{crit} = \frac{720°}{Z \cdot (1 \pm \gamma)} \quad (4)$$

$$\gamma = \frac{D_w \cdot \cos(\alpha)}{D_{pw}} \quad (5)$$

For oscillation angles $\varphi > \varphi_{CRIT}$, the calculation of C_{Osc} is expressed as follows:

$$C_{Osc} = \left(\frac{180°}{\varphi}\right)^{\frac{1}{p_H}} \cdot C \quad (6)$$

In Equation (6), the life exponent p_H is, again, $p_H = 4$ for roller bearings and $p_H = 3$ for ball bearings.

For $\varphi < \varphi_{CRIT}$, the calculation of C_{Osc} is differentiated for ball bearings (see Equation (7)) and roller bearings (see Equation (8)):

$$C_{Osc} = \left(\frac{180°}{\varphi}\right)^{\frac{3}{10}} \cdot Z^{0.033} \cdot C \quad (7)$$

$$C_{Osc} = \left(\frac{180°}{\varphi}\right)^{\frac{2}{9}} \cdot Z^{0.028} \cdot C \quad (8)$$

Again, C_{Osc} can be inserted into the basic rating life equation, which gives the number of oscillations L_{Osc} in 10^6 cycles:

$$L_{Osc} = \left(\frac{C_{Osc}}{P}\right)^p \quad (9)$$

1.1.3. Approach by Houpert

Houpert changes neither the load rating C nor the load P, but introduces the oscillation factor A_{Osc} [7]:

$$L_{Osc} = A_{Osc} \cdot \left(\frac{C}{P}\right)^p \quad (10)$$

The factor A_{Osc} depends on the load zone parameter e and the oscillation angle φ. However, the calculation is only valid for angles greater than $2\pi/Z$ and is complex due to the calculation of the load zone parameter. For more information on calculating the load zone parameter e and values for A_{Osc} for oscillation angles between $10°$ and $90°$, see [7]. Recently, Houpert and Menck [10] corrected an error in the original calculation approach that had come simultaneously to the attention of Menck as well as Breslau and Schlecht [11] in 2020.

1.1.4. Approach with an Equivalent Speed n

Another approach, which can also be found in the catalogs of various rolling bearing manufacturers, uses the oscillation angle φ and the oscillation frequency n_{Osc} to form an equivalent speed n that can be used in the basic rating life calculation according to DIN ISO 281:

$$n = n_{Osc} \cdot \frac{\varphi}{180°} \tag{11}$$

All calculation approaches based on the basic rating life calculation according to DIN ISO 281 can be modified using the coefficient a_{ISO} to calculate the extended rating life. However, the use of this calculation for oscillating movements is controversial because lubrication cannot be guaranteed, especially with small oscillating movements and the use of grease. As a result, the parameter for the lubrication condition (viscosity ratio) κ is less than 0.1 under the usual operating conditions for oscillating movements [6]. According to DIN ISO 281, this means that the calculation of the a_{ISO} factor is no longer valid [8].

1.2. Research Objective of This Paper

It has been shown that there are a large number of applications in which rolling bearings are used under oscillating movements. However, there is no validated calculation method, only various approaches, some of which differ considerably in their results. Most of the current work focuses on the occurrence of wear during small oscillating movements or vibrations. Particular attention is paid to the occurrence of false brinelling and stall marks and their prevention by the use of suitable lubricants [12,13]. Studies on fatigue and oscillating movements are rare and are limited to relatively large oscillation angles [14].

The research objective of this paper is therefore to investigate whether fatigue is possible with small oscillating movements and whether the calculation approaches presented above are valid at all. For this reason, rating life tests are carried out for cylindrical roller bearings under oil lubrication to determine if fatigue occurs as a failure mechanism.

2. Materials and Methods

2.1. Experimental Setup

The Chair of Engineering Design has built a test rig as part of an FVA project to investigate the fatigue life of rolling bearings under small oscillating movements [15]. The WSBP (Figure 1) is a custom test rig that allows four radial bearings to be tested simultaneously. Cylindrical roller bearings (NU210) or deep groove ball bearings (6210) can be used as test bearings.

The test shaft (see Figure 2a) is directly driven by a synchronous motor. Continuous rotation or oscillation angles ranging from $0.1°$ to $360°$ are possible with an oscillation frequency of up to 20 Hz. A worm gear screw jack generates the radial test load up to a maximum of 60 kN. A recirculating oil lubrication system continuously supplies the bearings with oil. At eight lubrication points between the bearings, each of the nozzle's four holes simultaneously lubricates the contacts between the outer ring and rolling element, as well as between the inner ring and rolling element.

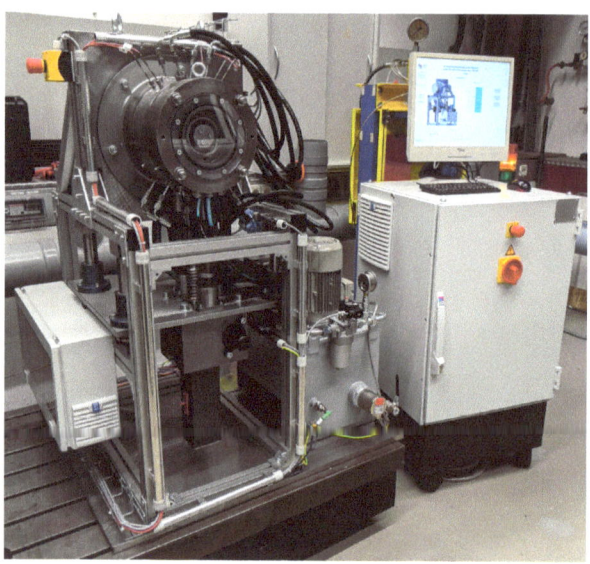

Figure 1. WSBP test rig at the Chair of Engineering Design at FAU Erlangen-Nürnberg.

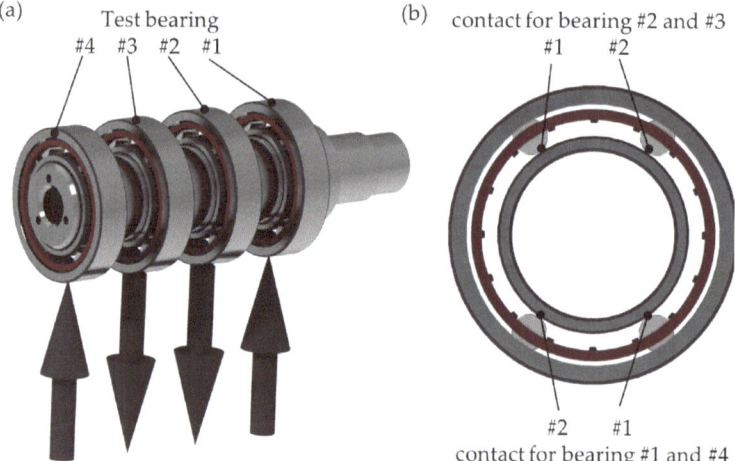

Figure 2. Test shaft with the position of the four test bearings (**a**), test bearing with the reduced number of four rolling elements, and with the position of the contacts for the test bearings 1 & 4 and 2 & 3 (**b**).

Bearing and oil temperature, oil flow, and radial force can be measured and recorded. In addition, an optical measuring system can measure slippage that might occur in the outer test bearing #4.

The number of rolling elements in each test bearing is reduced to four (see Figure 2b). They are arranged symmetrically, whereby only the two rolling elements at the bottom or the top are loaded according to the radial load for each test bearing. The remaining two rolling elements serve purely to guide the cage. This setup has several advantages: The force that is required to achieve the intended contact pressure is reduced; moreover, even at larger oscillation angles, the contact areas of adjacent contacts do not intersect, which significantly improves the possibility of evaluating and analyzing the damage.

2.2. Test Procedure, Plan, and Analysis

Screening tests with 100,000 oscillation cycles have shown that small oscillation angles in the range of 2° to 5° at oscillation frequencies of 5 Hz to 15 Hz show little wear, so fatigue is possible at a later stage [16]. For this reason, the oscillation angle φ was set to 3° and the oscillation frequency to 10 Hz for the rating life tests performed in this work. Cylindrical roller bearings (NU 210) by two different manufacturers were used. The Hertzian contact pressure at the inner ring-roller contact was set to 3000 MPa in order to minimize test run times. The shutdown criteria for the fatigue life test is the formation of a spalling. These can be detected from a certain size via the deviation of the actual oscillation movement.

Prior to testing, unneeded rolling elements are removed and the bearings are cleaned by hand with isopropanol. All tests are performed with FVA 3 oil at 30 °C at a total flow rate of 4 L/min (0.5 L/min for each of the eight lubrication points). After testing, a laser scanning microscope (Keyence VK-X200, Keyence Deutschland GmbH, Neu-Isenburg, Germany) is used to evaluate and qualitatively assess the contact zone/track (see Figure 3c).

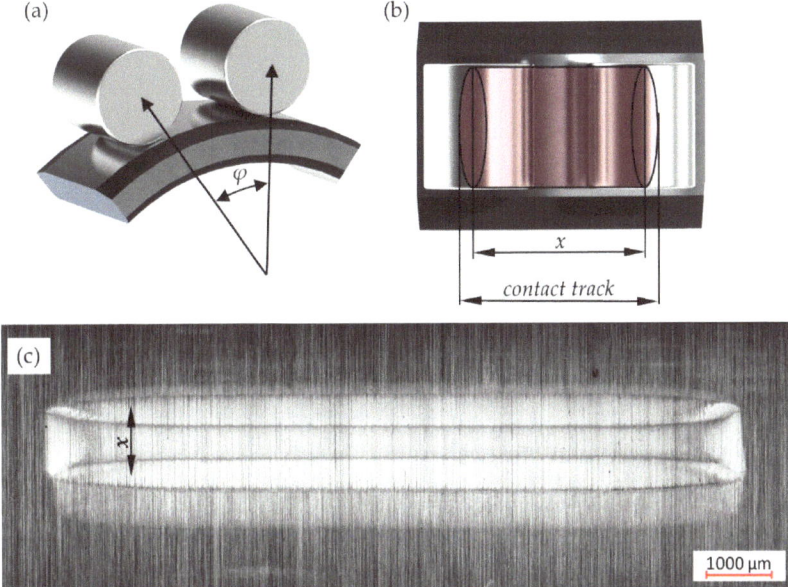

Figure 3. Oscillation angle φ (**a**), oscillation path x and contact track (**b**), and typical contact track with its two reversal points on the raceway (**c**).

The aim was to generate as many failures as possible due to fatigue and to compare the experimental rating life with the theoretical rating life. In addition, it should be clarified whether "classical" sub-surface or surface-initiated fatigue occurs.

Table 1 shows the characteristics of the test bearings and the test parameters. For the purpose of life calculation, each contact is considered as an individual bearing. This is possible because the individual contacts do not intersect during the small oscillating movement. The advantage of this is that with the four test bearings, eight contacts can be considered individually in the life calculation. In addition, the dimensionless $x/2b$ ratio is specified, which indicates the oscillation path x in relation to the semi-minor half-axis b of the contact ellipse. The ratio facilitates the comparison of different bearing types and sizes.

Table 1. Overview of NU210 test bearing and test parameters.

Pitch diameter D_{pw}	70.48 mm
Rolling element diameter D_{we}	10.98 mm
Rolling element length l	11.4 mm
Number of rolling elements Z	1 (for calculation)
Dynamic load capacity C_r	8.46 kN
Oscillation angle φ	3°
Oscillation frequency	10 Hz
Oscillation path x IR	0.779 mm
$x/2b$ IR	1.91

Figure 3 shows the oscillation angle φ and the oscillation path x on the raceway resulting from the oscillation movement as a rendering. Figure 3c shows a typical contact track on the actual raceway. It consists of the two reversal points of the oscillation movements, which correspond to the shape of the Hertzian contact ellipse.

3. Results

A total of eight rating life tests were performed, in which a total of over 135 million oscillation cycles were completed. This corresponds to a pure test duration of 157 days. In addition, further tests were carried out with cylindrical roller bearings at a lower contact pressure and with deep groove ball bearings to validate the results. In total, more than 289 million oscillation cycles were performed.

Table 2 lists all the test runs, including the bearing type (cylindrical roller bearing, CRB, or deep groove ball bearing, DGB), the Hertzian pressure at the inner ring-roller contact, and the number of oscillation cycles until failure. It is also indicated on which bearing inner or outer ring the damage has occurred.

Table 2. Overview for all tests with bearing type, Hertzian pressure, number of oscillation cycles, position of failure at inner-ring (IR) or outer-ring (OR), which bearing and contact failed, and which manufacturer was used.

Name	ZRL/DGB	Hertzian Pressure in MPa	Number of Oscillation Cycles	Failure at IR or OR	Manufacturer
RV1	CRB	3000	8,968,438	OR 3_1	A
RV2	CRB	3000	21,893,705	OR 3_1	A
RV3	CRB	3000	16,436,337	OR 3_1	A
RV4	CRB	3000	15,796,133	OR 4_1	A
RV5	CRB	3000	19,427,977	OR 3_1	A
SV1	CRB	3000	34,209,108	IR 2_2	B
SV2	CRB	3000	11,170,668	IR 3_1	B
SV3	CRB	3000	7,653,117	IR 2_2	B
AV1	CRB	2500	24,300,000	-	B
AV2	CRB	2500	23,900,000	OR 1_1	B
AV3	CRB	2500	43,000,000	-	B
AV4	CRB	2500	47,000,000	-	B
RV1	DGB	3000	15,280,000	IR 2_1	B

It should be noted that the test run was shut down as soon as damage was detected. Damage is detected via the drive motor control: As soon as the actual oscillation angle deviates from the specified oscillation angle and an increase in the frictional torque occurs at the same time, damage has occurred. While in theory, all eight contacts could fail at the same time, in reality, damage always occurs on one of the eight contacts first. Due to the subsequent shutdown, only one of the eight contacts failed in the following tests.

3.1. Rating Life Tests RV 1 to SV 3

Rolling bearings from two different manufacturers were used for the rating life tests. In principle, the observed failures are very similar for both manufacturers, though the location of the failures differs. While the failure is mainly localized on the outer ring for manufacturer A, it is mainly located on the inner ring for manufacturer B.

In tests RV 1, RV 2, RV 3, and RV 5, bearing 3 failed due to a spalling at contact 1 of the outer ring. In test RV 4, bearing 4 failed at contact 1 of the outer ring due to a spalling.

In tests SV 1 and SV 3, a spalling occurred at the inner ring contact 2 of bearing 2. Test SV 2 failed due to a spalling at the inner ring contact 1 of bearing 3.

The location and shape of the observed spalling are similar for all tests. The spalling is located at the outer end of the contact track in all tests (see Figure 4). The spalling runs at an angle of approx. 45° from the surface into the depth (orthogonal to the oscillation direction). The deepest point is characterized by a course of almost 90° to the surface, which can be observed in particular in the spalling on the inner ring. In addition, a kind of elevation can be observed in the center of the spalling, which corresponds to the center of the oscillation movement. This elevation can be measured, but is also very easy to recognize optically (see Figure 5).

Figure 4. Contact trace of the RV 3 bearing 3 contact 1 OR with a spalling.

The remaining contact track shows a strong reddish discoloration due to hematite formation as well as a bluish tribological layer formation due to metal particles and higher local pressure. It should be noted here that the spalling is not usually detected immediately, but only shuts down approx. 500,000 cycles later.

The shape and location of the spalling can be explained by a surface-initiated crack formation that occurs at the outer end of the contact track. Cyclic loading causes the crack to grow in depth, resulting in the observed spalling. This can also be confirmed by crack formation in this and other tests in which cracks were observed at the transition from the contact track to the unloaded raceway (see Figure 6).

Looking at tests RV 1, RV 2, RV 3, and RV 5, it is noticeable that the fault occurs on the same contact, in bearing 3 on contact 1. Basically, the probability of a particular contact failing is 1/8, as there are a total of 8 contacts that can fail, and the first failure is followed by a shutdown. At first sight, it seems very unlikely for all four attempts to fail on the same contact, though possible explanations can be given. One of them is the design of the test rig. As bearings 2 and 3 are used to apply the force, those will definitely receive the full test load. Bearing 1, on the other hand, is positioned next to the motor bearing, raising the possibility that the motor bearing receives a fraction of the test load. In addition, the contacts of bearings 2 and 3 are at the upper position, so that they are lubricated only by the injected oil, while the contacts of bearings 1 and 4 are located at the bottom position, where an oil bath is formed.

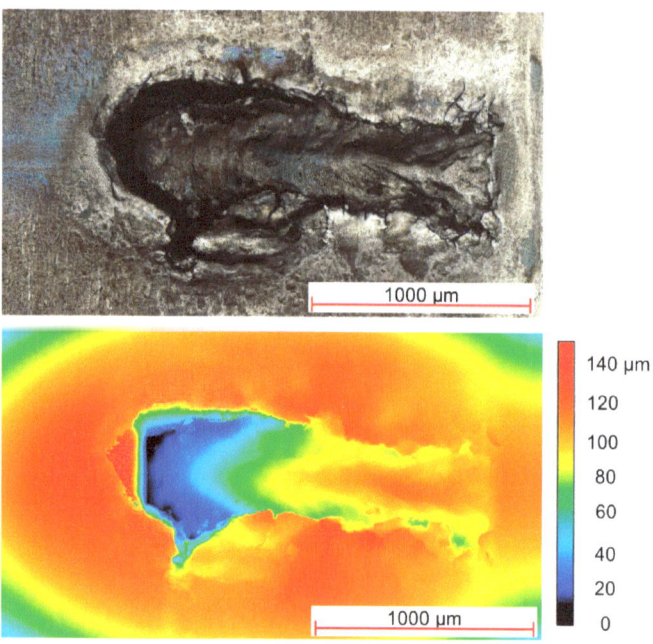

Figure 5. Right end of the contact trace of the SV 2 bearing 3 contact 1 IR with a spalling.

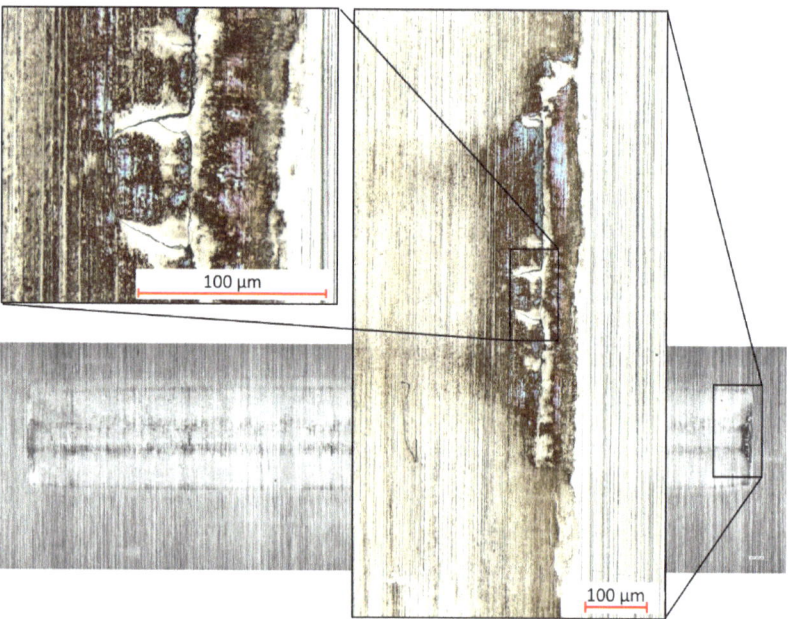

Figure 6. Contact trace of the SV 3 bearing 3 contact 1 IR with cracks in the surface.

Additional tests must therefore be carried out to check and verify that the occurring failure mechanism is indeed fatigue damage and not a localized exceeding of the yield point due to tilting of the rolling elements or deflection of the shaft. For this purpose, tests with deep groove ball bearings and tests with cylindrical roller bearings were carried out at lower Hertzian pressures.

3.2. Check #1 with Deep Groove Ball Bearings: Is Tilting the Cause of the Damage?

A simple verification was carried out to rule out possible tilting of the rolling elements. Deep groove ball bearings were used as test bearings, where tilting of the rolling elements is not possible due to their design. The tests were carried out with identical test parameters (oscillation frequency, oscillation angle, and Hertzian pressure). Several tests were carried out, whereby a random shutdown occurred in one test after approx. 15 million oscillation cycles. This shutdown proved to be a stroke of luck, as crack formation and the start of crack propagation could be observed in its initial stage (see Figure 7).

Figure 7. Crack formation and beginning of a spalling at the edge of the contact track.

Of particular interest is the location of the crack formation, which is located directly at the transition from the contact track to the unloaded raceway.

This suggests that a similar damage mechanism can also be observed in deep groove ball bearings, as has already been observed in cylindrical roller bearings. As the failure was again on one of the two center bearings, it is necessary to ascertain whether the radial bearing test rig design results in a significantly increased test load on the center bearings, which may lead to premature non-fatigue damage. Additional tests were performed with a reduced contact pressure to address this concern.

3.3. Check #2: Is the Damage Still Present at Lower Hertzian Pressures?

In these tests, the Hertzian pressure in the inner ring-rolling element contact was set at 2500 MPa. The lower contact pressure results in a significantly longer theoretical rating life and also a significantly longer actual test rating life. A total of four tests were performed. Unfortunately, three out of four tests had to be canceled before any damage occurred, due to a failure of the motor bearing or a power failure at the building as a result of unannounced maintenance work. Regrettably, a restart at the identical contact point is not possible on the test bench due to its design.

As a result, there was only one test, which ran until it failed due to a spalling at bearing 1 contact 1 on the outer ring. This is particularly interesting, as the bearings from this manufacturer have only ever failed on the inner ring and there is a lower pressure of approx. 2150 MPa on the outer ring. The combination of the lower pressure and the position on the outer ring supports the assumption that this is a fatigue failure, rather than a different type of failure due to a localized increase in stress.

4. Discussion

The results show that fatigue can occur as a failure mode for rolling bearings under small oscillating movements. This was demonstrated for both line and point contacts at a pressure of 3000 MPa and an oscillation angle of 3° at an oscillation frequency of 10 Hz. It should be noted that these are oil-lubricated bearings.

To compare the theoretically calculated rating life with the actual rating life, the actual rating life must first be determined. For this purpose, the experimental rating life B_{10} is calculated for a 10% probability of failure using the failure times of the eight rating life tests (RV 1-5 and SV 1-3) and a 2-parametric Weibull distribution. This results in an experimental rating life of $B_{10} = 16.99 \times 10^6$ oscillation cycles (see Figure 8).

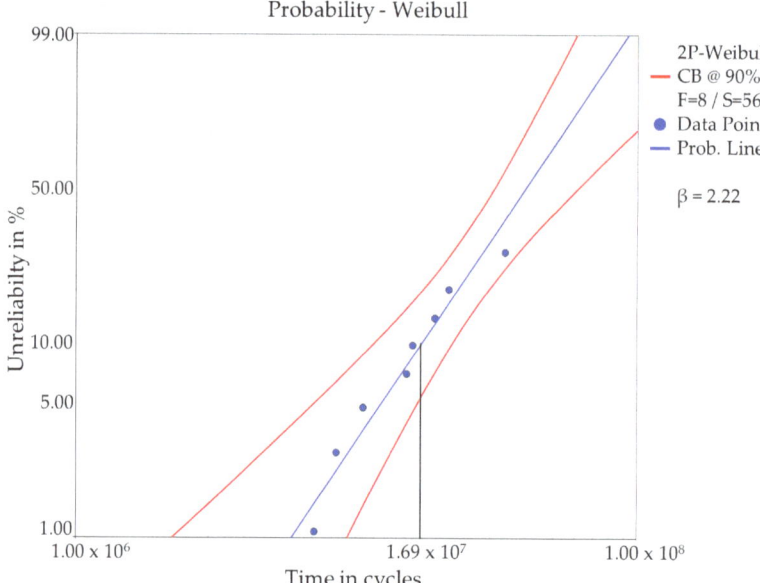

Figure 8. Two parametric Weibull plots for the tests RV 1-5 and SV 1-3 with 8 failures (data points) and 90% confidence boundaries.

For a conservative estimate of the theoretical rating life, the calculation model that produces the highest rating life is used, in this case, the *Approach with an equivalent speed n*. This results in a rating life of $L_{10} = 14.03 \times 10^6$ oscillation cycles for an oscillation angle of 3°. As a further part of the conservative estimate, only the nominal rating life is calculated, as the a_{ISO} factor ($a_{ISO} = 0.2$ for the operating conditions present here) would only lead to a further reduction in the theoretical rating life.

The results show that the experimental rating life B_{10} determined in the test exceeds the theoretical rating life L_{10}. Thus, with good lubrication conditions and a sufficiently large oscillation angle, a conservative estimation of the service life is possible using the calculation of the theoretical service life with the existing calculation approaches for cylindrical rolling bearings.

The accumulation of failures in test bearing 3 is striking. However, the validation tests demonstrated that these failures were always due to fatigue and not a localized exceeding of the yield point. One potential explanation for the number of failures is that the load on test bearing 3 is greater than on the other test bearings, which increases the probability of failure due to fatigue damage. This is not a critical issue, as the theoretical rating life would be reduced if the higher load were taken into account. Consequently, the experimentally determined rating life is still higher than the theoretical rating life.

The tests have therefore shown that fatigue is indeed occurring. However, it has not yet been conclusively clarified whether it is surface-initiated or classic sub-surface fatigue. The results indicate that the cracks are surface-induced and originate directly in the transition from the contact track to the unaffected part of the track. Unfortunately, an evaluation of various microsection analyses has not yet been successful, so this is still a hypothesis that needs to be finally verified with FIB sections or other methods. This should be the focus of further work.

Author Contributions: Conceptualization, methodology, validation, writing—original draft preparation, F.H.; writing—review and editing, M.B.; supervision, M.B. and S.W. All authors have read and agreed to the published version of the manuscript.

Funding: The test rig WSBP was funded by the Forschungsvereinigung Antriebstechnik e.V., project 824 I (IGF 19915 N). The remaining research received no external funding.

Data Availability Statement: For insights into in-depth data, e.g., test data, etc., please contact the corresponding author.

Acknowledgments: We acknowledge the support of the students Alexandra Neumann, Louis Römert, Jonas Ruhl, and Luzia Scheper for the conduct and evaluation of the experiments.

Conflicts of Interest: The authors declare no conflicts of interest.

References

1. Schwack, F.; Halmos, F.; Stammler, M.; Poll, G.; Glavatskih, S. Wear in wind turbine pitch bearings—A comparative design study. *Wind Energy* **2021**, *25*, 700–718. [CrossRef]
2. Behnke, K.; Schleich, F. Exploring limiting factors of wear in pitch bearings of wind turbines with real-scale tests. *Wind Energy Sci.* **2023**, *8*, 289–301. [CrossRef]
3. Tawresey, J.S.; Shugarts, W.W. An Experimental Investigation of the Fatigue Life and Limit Load Characteristics of Needle Roller Bearings Under Oscillating Load Conditions. Available online: https://apps.dtic.mil/sti/citations/AD0437467 (accessed on 30 April 2024).
4. Rumbarger, J.H.; Jones, A.B. Dynamic Capacity of Oscillating Rolling Element Bearings. *J. Lubr. Tech.* **1968**, *90*, 130–138. [CrossRef]
5. Harris, T.A.; Kotzalas, M.N. Rolling Bearing Analysis. In *Essential Concepts of Bearing Technology*, 5th ed.; CRC Press: Boca Raton, FL, USA, 2007.
6. Harris, T.A.; Rumbarger, J.H.; Butterfield, C.P. *Wind Turbine Design Guideline DG03: Yaw and Pitch Rolling Bearing Life*; Technical Report NREL/TP-500-42362; NREL: Golden, CO, USA, 2009. [CrossRef]
7. Houpert, L. Bearing Life Calculation in Oscillatory Applications. *Tribol. Trans.* **1999**, *42*, 136–143.
8. *DIN ISO 281*; Wälzlager—Dynamische Tragzahlen und nominelle Lebensdauer. Beuth: Berlin, Germany, 2010.
9. Menck, O.; Stammler, M. Review of rolling contact fatigue life calculation for oscillating bearings and application-dependent recommendations for use. *WES* **2024**, *9*, 777–789. [CrossRef]
10. Houpert, L.; Menck, O. Bearing Life Calculations in Rotating and Oscillating Applications. *J. Tribol.* **2022**, *144*, 071601-1–071601-18. [CrossRef]
11. Breslau, G.; Schlecht, B. A Fatigue Life Model for Roller Bearings in Oscillatory Applications. In *FVA (Hrsg.): Bearing World Journal Nr. 5*, 1st ed.; VDMA: Frankfurt am Main, Germany, 2020; pp. 65–80.
12. Schwack, F.; Bader, N.; Leckner, J.; Demaille, C.; Poll, G. A study of grease lubricants under turbine pitch bearing conditions. *Wear* **2020**, *454–455*, 203335. [CrossRef]
13. Wandel, S.; Bader, N.; Glodowski, J.; Lehnhardt, B.; Leckner, J.; Schwack, F.; Poll, G. Starvation and Re-lubrication in Oscillating Bearings: Influence of Grease Parameters. *Triology Lett.* **2022**, *70*, 114. [CrossRef]
14. Cavacece, F.; Frache, L.; Tonazzi, D.; Bouscharain, N.; Philippon, D.; Le Jeune, G.; Maheo, Y.; Massi, F. Roller bearing under high loaded oscillations: Life evolution and accommodation mechanisms. *Tribol. Int.* **2020**, *147*, 106278. [CrossRef]

15. *FVA AB 824 I*; Ermüdungslebensdauer von Wälzlagern unter Kleinen Schwenkbewegungen. FVA: Frankfurt, Germany, 2021.
16. Halmos, F.; Bartz, M.; Wartzack, S. Schadens-und Ausfallmechanismen bei Wälzlagern unter Schwenkbewegungen. In *Gleit-und Wälzlagerungen 2023*; VDI Verlag: Düsseldorf, Germany, 2023; Volume 2415, pp. 321–332. [CrossRef]

Disclaimer/Publisher's Note: The statements, opinions and data contained in all publications are solely those of the individual author(s) and contributor(s) and not of MDPI and/or the editor(s). MDPI and/or the editor(s) disclaim responsibility for any injury to people or property resulting from any ideas, methods, instructions or products referred to in the content.

Article

Improved Operating Behavior of Self-Lubricating Rolling-Sliding Contacts under High Load with Oil-Impregnated Porous Sinter Material

Nicolai Sprogies *, Thomas Lohner and Karsten Stahl

Gear Research Center (FZG), Department of Mechanical Engineering, School of Engineering and Design, Technical University of Munich, Boltzmannstraße 15, D-85748 Garching, Germany
* Correspondence: nicolai.sprogies@tum.de

Abstract: Resource and energy efficiency are of high importance in gearbox applications. To reduce friction and wear, an external lubricant supply like dip or injection lubrication is used to lubricate tribosystems in machine elements. This leads to the need for large lubricant volumes and elaborate sealing requirements. One potential method of minimizing the amount of lubricant and simplifying sealing in gearboxes is the self-lubrication of tribosystems using oil-impregnation of porous materials. Although well established in low-loaded journal bearings, self-lubrication of rolling-sliding contacts in gears is poorly understood. This study presents the self-lubrication method using oil-impregnated porous sinter material variants. For this, the tribosystem of gear contacts is transferred to model contacts, which are analyzed for friction and temperature behavior using a twin-disk tribometer. High-resolution surface images are used to record the surface changes. The test results show a significant increase in self-lubrication functionality of tribosystems by oil-impregnated porous sinter material and a tribo-performance comparable to injection-lubricated tribosystems of a sinter material with additionally solid lubricant added to the sinter material powder before sintering. Furthermore, the analyses highlight a significant influence of the surface finish, and in particular the surface porosity, on the overall tribosystem behavior through significantly improved friction and wear behavior transferable to gear applications.

Keywords: gearbox; sinter material; oil-impregnation; self-lubrication; EHL; tribology; friction

Citation: Sprogies, N.; Lohner, T.; Stahl, K. Improved Operating Behavior of Self-Lubricating Rolling-Sliding Contacts under High Load with Oil-Impregnated Porous Sinter Material. *Lubricants* **2024**, *12*, 259. https://doi.org/10.3390/lubricants12070259

Received: 19 May 2024
Revised: 21 June 2024
Accepted: 9 July 2024
Published: 21 July 2024

Copyright: © 2024 by the authors. Licensee MDPI, Basel, Switzerland. This article is an open access article distributed under the terms and conditions of the Creative Commons Attribution (CC BY) license (https://creativecommons.org/licenses/by/4.0/).

1. Introduction

Gearboxes are widely used in mechanical engineering due to their versatility in terms of performance class, transmission ratio, and axis positions. They offer a simple yet reliable and efficient design, making them suitable for various applications, including industry, automotive engineering, aerospace, or wind turbines. A liquid-lubricated gearbox consists of machine elements, including gears, bearings, and seals, as well as lubricants and the housing. The use of liquid lubricants reduces power losses during operation and increases lifetime by separating the surfaces in tribological contacts by a hydrodynamic lubricant film, enabling low friction and good heat transfer. Integrating the required lubricant volume for contact lubrication into the pores of a sintered component can significantly simplify or even eliminate the lubrication and sealing system, thereby reducing no-load gearbox power losses, weight, and the risk of oil leakage [1].

Commonly used gearbox lubrication methods are dip lubrication, such as in grease or oil sumps, and injection lubrication, including interval or continuous, circulating, or loss lubrication. The selection of a lubrication method for a gearbox depends, among various factors, on the type of gear, the power loss, the prevailing speed, the type of lubricant, the lubricant supply to tribological contacts, and the lifetime requirements. The lubrication methods presented are used in applications from low to high loads and speeds in continuous and intermittent operation [2,3]. The back-coupling of hydrodynamics with elastic deformation of rolling-sliding elements results in elasto-hydrodynamically lubricated (EHL) contacts [4]. In self-lubricating EHL contacts with oil-impregnated sintered rolling-sliding components, similar to the human knee joint [5], compression of the pore structure causes oil to extrude out, lubricating the contact.

Self-lubrication is already used in low-load applications, such as sintered journal bearings or roller-bearing cages, whose porous structures serve as oil reservoirs [6]. For example, the use of sintered journal bearings is standardized by ISO 2795/DIN 1850 III [7]. The formation of an oil film in such self-lubricating contacts is affected by the flow of oil from the pore reservoir in the component and its exposed surface and vice versa. The flow is determined by a number of factors, including centrifugal force as a function of rotational speed, oil viscosity [8], surface roughness [9], diffusion due to concentration gradients [10], capillary flow, and the different thermal expansion of porous solid material and oil. Additionally, the compression or deformation behavior of the material [11,12] and the operation-induced pressure fluctuations [13] must be considered. In their studies of oil-impregnated sintered bearings, Roberts [8] and Fote et al. [9] identified surface porosity and oil viscosity as significant factors affecting journal-bearing lubrication. For example, surface treatments such as grinding can cause covering of the pores necessary for oil flow, thereby reducing the porosity-dependent permeability. Also, the polarity and viscosity of the oil can affect the surface tension, thereby influencing the flow of oil into and out of the porous surface. The studies of centrifugal force by Marchetti et al. [11,12] also show that no oil is extruded below a pore-structure-dependent threshold. For sintered journal bearings, the lubricant flow in the porous material is described by the effect of self-circulation: Oil is intruded into the porous material at the contact area under high pressure, while oil can extrude from the material in unloaded areas. Also, concentration differences result in compensating flows within the porous structure [14,15]. In porous journal bearings, oil extrusion is particularly affected by the surface porosity. Dizdar [16] and Lipp [17] define the transition zone from open to closed porosity at a density of the porous matrix of ρ = 7.1–7.2 g/cm^3. In open porosity, the surface pores are mostly directly connected to the inner porous body, allowing the resulting channels to transport oil into and out of the porous material [18]. Oil-carrying pores of various sizes are randomly distributed on the surface, resulting in locally inhomogeneous hydrodynamic oil film formation and load distribution in the tribological contact. Li and Olofsson's [19] investigations of externally lubricated sliding contacts with sintered bodies demonstrate the correlation between the measured coefficient of friction and wear and pore size. According to Balasoiu et al. [20], increased oil permeability leads to increased oil extrusion but decreased surface load-carrying capacity.

The use of powder metallurgical sinter materials offers the possibility of economically producing near-net-shape parts with high durability. The thermomechanical properties are mainly determined by the density and pore structure of the sintered component [21], which are adjusted by the powder raw material and process conditions. In direct comparison with solid steel, sinter materials often exhibit poorer mechanical properties such as lower tensile strength, fatigue strength, ductility, and stiffness with increasing porosity [22]. However, Ebner et al. [23,24] transferred in their pioneering works the known self-lubricating tribosystem of oil-impregnated sinter materials to highly-loaded EHL rolling-sliding contacts, prevailing in gears and roller bearings. When considering highly loaded contacts, the lubricant flow driving causes are analogous to those reported for journal bearings. These include centrifugal force, elastic deformation of the contact partners, and thermal expansion of the oil due to frictional heat production during operation [25]. Transferring the knowledge of the self-lubrication mechanism, the general functionality of the system with porous sinter materials is confirmed using an optical tribometer and a twin-disk tribometer, as well as accompanying tribosimulation. The best functionality of self-lubrication is demonstrated with case-hardened sinter material density of ρ_{sinter} = 7 g/cm^3 [26], whereas the operating behavior is characterized in stable, metastable, and unstable conditions by a stability coefficient Ξ determined by the in-/decreasing or stationary trends of the measured friction and temperature [25]. Further, Ebner et al. correlated the operating behavior with the global and local surface porosity of the test specimens and the impregnated oil. In most cases, the tribosystem runs in severe mixed lubrication regimes, determined with contact resistance measurements. The investigations reveal cases of starved lubrication, where the amount of oil is not sufficient to form a lubricant film thickness as high as in the case of external lubrication [27]. Further, investigations using optical interferometry by Omasta et al. [28] show the local formation of a lubricant film on porous surfaces, which has a thickness of up to 40% compared to external lubrication. An ISO viscosity grade (VG) of 100 with friction modifier additives such as plastic deformation additives has proven to be the most suitable for self-lubricating tribosystems with severe mixed lubrication [29]. Nevertheless, the EHL tribosystems tend toward failure within specific load and speed limits due to reaching their thermal limits, in particular because of the avoidance of the heat convection due to external oil volumes.

Zhang et al. [30], Rabaso et al. [31], and others ([32,33]) investigated the friction and wear-reducing effects of molybdenum (Mo) and tungsten (W) nanoparticles as oil additives in sliding contacts in a vacuum atmosphere. In addition, Zhu et al. [34] reviewed the friction and wear resistance benefits. Vazirisereshk et al. [35] showed in their review limitations of coatings and composite materials, including MoS_2 and WS_2, and identified current applications such as journal bearings or sealings especially suited for operations in aerospace, with extremely low or high temperature or radioactive surrounding, and also highlighted the oxidation tendency in humid oxygenic environments. Among others (e.g., Kovalchenko et al. [36]), Dhanasekaran et al. [37] mixed solid lubricant MoS_2 powder into the sinter material powder mixture (here: Fe-C-Cu) during the sintering process for a dry-lubricating sintered structure, achieving with an addition of <5% of MoS_2 in the reference material decreased friction in pin-on-disk tests compared to non-compound dry-running material variants, by showing similar to better mechanical properties in terms of strength and wear resistance.

The literature on self-lubricating oil-impregnated sinter materials indicates that it is necessary to properly design the porosity and its distribution as well as the oil properties. For self-lubricating oil-impregnated machine elements with low-load applications like in journal bearings, the compromise between a supportively higher porosity of the tribosystem partners for oil extrusion for lubricant film formation within the tribocontact and a lower porosity for oil intrusion out of the tribocontact back into the porous structure in order to achieve the required stable operating conditions within a broad transmittable power is already found. To increase the power density of self-lubricating oil-impregnated machine elements for high-load applications like cylindrical gears, suitable material pairings with sufficient mechanical properties as well as frictional and thermal behavior are determined in this study by, e.g., varying the amount of oil-supplying oil-impregnated EHL tribosystem partners. Further, as self-lubricating tribosystems primarily operate in severe mixed lubrication regimes with comparable smaller lubricant film thicknesses, the oil intrusion inhibiting and the relative film thickness increasing effect of the surface needs to be controlled by adjusting, next to the surface roughness, the surface porosity due to surface finishing, which is evaluated in this study as well. Because of the thermal limitations due to the limited heat removal from the tribological contact, further tribological performance increasing effects such as the solid-friction reducing effect of solid lubricants have to be implemented in the self-lubricating EHL tribosystem, which are investigated first within this study. To finally increase the power density of the self-lubricating EHL tribosystem with oil-impregnated sinter materials, material–surface variants are systematically investigated in this study regarding their influence on the frictional performance at various operating conditions. Also, the stability of the operating behavior as well as of the material and surface condition alteration with further operation progress are analyzed to find suitable specifications of the EHL tribosystem with reduced frictional power loss and proper durability to enhance the performance of self-lubricating materials at higher loads and speeds. To summarize, this study investigates, in an experimental approach, the influence of material pairing and surface finish on frictional and thermal behavior.

2. Materials and Methods

The section below describes the materials and methods to generate the results of this study. First, the material and surface variants of the test specimens are introduced. Second, its testing on the twin-disk tribometer is described. Third, the measurement procedure and the analysis of the results are outlined.

2.1. Specimen

The materials for the self-lubricating EHL tribosystems investigated in this study are 16MnCr5, a conventional case-hardening steel (hereafter referred to steel), and variants according to DIN-30910 [38] open-pore sintered metal, Sint-D31 (hereafter referred to as sinter). The mechanical material properties are given in Table 1. The sintered test specimens were produced in three material variants according to DIN EN ISO 5755 [39] and consist of conventional case-hardened P-FL-05M1 sinter material (sinter$_{ref}$), solid lubricant-alloyed case-hardened P-FL-05M1 sinter material (sinter$_{sl}$) with an addition of 2% of molybdenum sulfide MoS_2 and tungsten sulfide WS_2 solid lubricant additives in the powder mixture, which does not affect the mechanical properties regarding, e.g., Young's Modulus, and plasma-nitrided P-FL-05Cr3M sinter material (sinter$_{pni}$). Nearly homogeneous pore distributions and isotropic material properties can be achieved through standard-compliant production in the axial press sintering process. The sintered metal test specimens have a density of $\rho = 7$ g/cm^3, which Ebner [25] classified as the most functional for the self-lubricating EHL tribosystems, resulting in an open porosity ($\Phi = 10\%$). During the manufacturing process, each material underwent a quality assurance measurement of density to ensure repeatability and consistency in global porosity. The porosity requirement is derived from a trade-off between material properties (particularly strength and hardness) and a maximum pore volume for oil storage.

Table 1. Mechanical properties of the considered materials.

Material	Young's Modulus E in GPa	Poisson's Ratio ν	Surface Hardness	Density ρ in g/cm^3
steel	210	0.30	745 HV5	7.8
sinter$_{ref/sl}$	130	0.27	755 HV5	7.0
sinter$_{pni}$	140	0.27	785 HV5	7.0

The specimen geometry is shown in Figure 1. The materials used are machined from a cylindrical blank to the target geometry for test purposes. The rolling-sliding contact is formed by the contact of two specially machined cylindrical disks of width $l_{eff} = 5$ mm and radius $R = 40$ mm (see Sections 2.2 and 2.3). Based on reference surfaces in a longitudinally ground state with a parallel relative orientation to the running direction, various surface finishes (see surface images made with a Leica M205A from Leica Microsystems GmbH (Wetzlar, Germany) encoded stereo microscope in Table 2) are tested to investigate the influence of the surface on the self-lubricating EHL tribosystem. To consider the transfer to gear contacts in particular, axially ground disks are considered, as they represent the characteristic surface structure of a gear contact with conventionally ground tooth flanks with a surface structure perpendicular to the rolling-sliding direction. The arithmetic mean roughness Ra of the surface variants used is ≤ 0.25 μm for longitudinally and axially ground and non-finished at sinter$_{pni}$, ≤ 0.10 μm for superfinished, and ≤ 0.01 μm for polished. Surface roughness measurements were performed using the profile method according to DIN EN ISO 13565 1-3 [40], measured with a length of $L_t = 4.0$ mm transversely to the current surface structure orientation and a cut-off wavelength of $\lambda_c = 0.8$ mm.

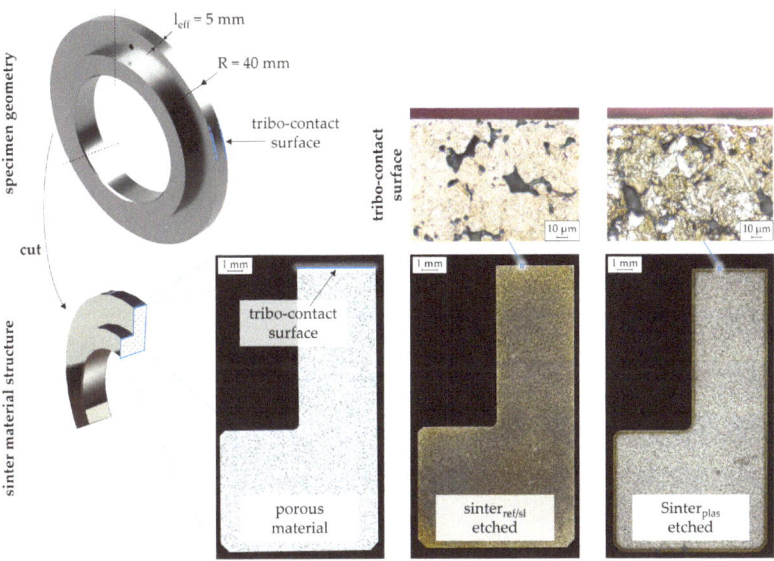

Figure 1. Specimen geometry and section images of the porous sinter material variants.

Table 2. Surfaces of the considered specimen.

Longitudinally Ground (lgr)	Axially Ground (agr)	Superfinished (suf)	Mechanically Polished (mpo)	Plasma Nitride (pni)

For self-lubrication tests, the sintered specimens were impregnated with oil using a pressure impregnation device (see [25]). The theoretical maximum impregnation quantity of a component is $m_{max} \approx 2$ g and can be determined according to [25] from the porosity volume V_{por} and the density of the designated oil $\rho_{oil} = 0.849$ g/cm³. V_{por} is set at approximately 10% of the specimen volume $V_{spe} = 24.2$ cm³ through a validated manufacturing process (see Equations (1) and (2)). The impregnability of sintered components depends mainly on the porosity distribution inside the material and on the surface, the radius of the pore channels (equivalent to capillaries), the oil properties like viscosity, and the environmental conditions, e.g., process pressure during impregnation and temperature [41,42]. Since the pore network is randomly distributed and, according to Lados et al. [43], there may be isolated or hard-to-reach pores, the measured impregnation quantity m_{oil} is lower than m_{max}. The impregnation amount m_{oil} of all specimens is consistently lower than the theoretically possible impregnation amount m_{max}, ranging from $m_{oil} = 1.501$ g to $= 1.863$ g. The average oil impregnation amount across all tests is $\overline{m}_{oil} = 1.671$ g, corresponding to an impregnation degree of $\chi_{imp} \approx 75\%$. Prior to weighing, the oil-impregnated specimens are cleaned with a compressed air pistol and dry tissues for the surfaces, as well as cotton swabs for the small bore of the Pt100 and for sharp edges.

$$V_{por} = 0.1 \cdot V_{spe} \tag{1}$$

$$m_{max} = V_{por} \cdot \rho_{oil} \tag{2}$$

The test oil PAO100+PD used is a synthetic oil based on polyalphaolefin. It is a fully formulated oil of the ISO VG 100 class with a plastic deformation (PD) additive. The PD additives belong to the class of friction modifiers, which provide solid friction-reducing properties [44] through the tribo-induced reaction of additives with metallic surfaces. The oil data can be found in Table 3.

Table 3. Data of the considered oil PAO100+PD [45].

ISO VG	Density ρ_{oil} at 15 °C in kg/m^3	Kin. Viscosity $\eta_{40°C}$ at 40 °C in mm^2/s	Kin. Viscosity $\eta_{100°C}$ at 100 °C in mm^2/s
100	849	105.0	15.7

2.2. Test Rig

The experimental tests in this study were conducted using a twin-disk tribometer, which was also used by Ebner et al. (e.g., [23,25]) to investigate self-lubricating rolling-sliding contacts. Figure 2 shows a schematic of the mechanical configuration of the twin-disk tribometer. The description and formulations presented here are based on the publications of Ebner et al. [23] and Lohner et al. [44].

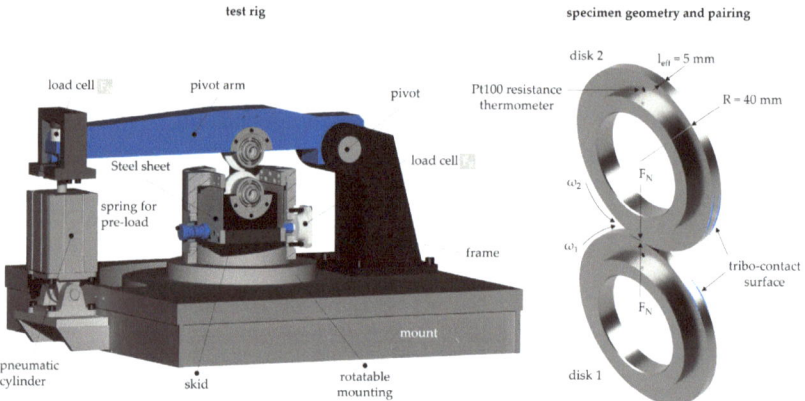

Figure 2. FZG twin-disk tribometer according to Ebner [23] and disk pairs used for the tests of this study.

A normal force F_N is applied to the disk contact by a pneumatic cylinder mounted at the end of a pivot arm. The lower disk is mounted in a skid that is attached to the frame by thin steel sheets. The skid is laterally supported by a load cell, which measures the frictional force F_R in the disk contact as a reaction force of the operating conditions in the contact. The maximum load that can be applied on the disk contact is F_N = 16 kN, resulting in individual pressure depending on the disk geometry and material. The maximum sum velocity is v_Σ = 12 m/s. In this study, the upper disk (v2) rotates at a speed that is either the same or faster than the lower disk (v1), both of which rotate with a speed between 2 m/s and 8 m/s, depending on the test method (see Section 2.3). Based on the considered normal loads, the calculated Hertzian pressure p_H in the disk contact is between 600 N/mm^2 and 1.043 N/mm^2. Within this study, the upper disk (v_2) rotates at a faster speed than the lower disk (v_1). The measurands are the normal force (F_N), the frictional force (F_R), the surface velocities of the lower and upper disk (v_1 and v_2) derived from the measured shaft speeds, and the bulk temperature (ϑ_M) of the upper test disk. ϑ_M is measured 5 mm below the disk surface using a Pt100 resistance temperature sensor, whose signal is transmitted from the rotating shafts via a mercury rotary signal transmitter and whose signal is transformed to a voltage signal in a Wheatstone Bridge after the transmitter. The location of the Pt100 sensor is selected as a compromise in order to obtain the temperature of the surface region as closely as possible, but also to ensure that there is sufficient distance between the sensor and the surface to prevent any interference with the contact stiffness and material strength when the disk is loaded. For evaluation, the bulk temperature measurement can be utilized to determine the level of dissipated energy and, thus, the frictional behavior of the tribosystem. Before each test, the contact pattern is carefully evaluated to ensure an evenly distributed load during the line contact of the disks, and any misalignment is mechanically corrected. To analyze the frictional behavior, the coefficient of friction (CoF) µ is determined from the measured friction force F_R and normal force F_N at the disk contact. According to Vojacek [46], the considered twin-disk tribometer has an accuracy for measuring µ of $\Delta\mu = 0.0025$.

$$v_\Sigma = v_1 + v_2 \tag{3}$$

$$s = ((v_1 - v_2)/v_1) \cdot 100\% \text{ with } v_1 < v_2 \tag{4}$$

$$\mu = F_R/F_N \tag{5}$$

For a comparative test with external lubrication, an injection-lubrication unit is used. Thereby, the PAO100+PD oil is injected directly into the contact inlet with a volume flow rate of $\dot{V} = 1.5$ L/min at ambient oil temperature ($\vartheta_{oil} = 25\ °C$) without any temperature regulation. For all measurements, the disk pair of a longitudinally ground, oil-impregnated sinter$_{ref}$ disk with a longitudinally ground steek disk was established as the reference system. The experimental procedure outlined in Section 2.3 also includes other material–surface combinations to evaluate the influence of lubrication methods, material pairing, solid lubricant addition, and surface condition.

2.3. Test Method

The test method involves two steps under line contact conditions at the twin-disk tribometer (see Section 2.2), first starting with measuring friction curves of self-lubricating oil-impregnated sinter material and surface combinations compared to conventional injection lubrication, followed by long-term operation tests to evaluate the operating stability for stationary operating systems.

2.3.1. Friction Curves

For all tests, two new disk pairs with identical specifications are used. Prior to each test under self-lubrication, an initial oil volume of $V_{init} = 0.02$ mL is applied to the dry tissue-cleaned contact surfaces to prevent initial damage during running-in. The measurements with injection lubrication are carried out with an initial oil temperature at $\vartheta_{oil} = 25\ °C$. All measurements are started at disk bulk temperatures $\vartheta_M \approx \vartheta_{ambient}$ with a running-in of $t = 60$ s without an imposed pressure $p_H < 500$ N/mm^2 at moderate speed $v_\Sigma = 1$ m/s and $s = 2\%$. While testing, termination criteria are used to detect possible failures of a disk pair. These criteria are a bulk temperature ϑ_M greater than the critical bulk temperature $\vartheta_{crit} > 140\ °C$ or a CoF µ greater than the critical CoF $\mu_{crit} > 0.12$. Light microscopy images of the specimen surfaces are captured before and after the tests to evaluate the lubricating effect and the corresponding surface alteration. The disk pair of a longitudinally ground sinter$_{ref}$ disk impregnated with the PAO100 with a longitudinally ground steel disk was established as the reference tribosystem for all measurements.

The friction curves are measured at operating points with a constant pressure of $p_H = 600$ N/mm^2 and different sum velocities $v_\Sigma = \{2; 4; 8\}$ m/s. Each measurement at a sum velocity is made at discrete, increasing slip ratio $s = \{0; 2; 5; 10; 20; 30; 40; 50\}\%$. The measurements with each disk pair are repeated once, with the 2nd measurement taken without re-impregnation of the sinter disk with oil. The repeatability of the two measurements is also analyzed to investigate the difference in the tribological behavior of a variant. Therefore, a repetition test is performed with a 2nd new disk pair, and both initial and subsequent measurements are made. Each operating point on the friction curve is maintained until quasi-stationary operating conditions are reached. An operating point is considered quasi-stationary when the bulk temperature difference $\Delta\vartheta_M$ remains within $\Delta\vartheta_M < 1$ K for a period of t = 60 s. If a stationary operating condition is achieved, the average of the measurement data from the last t = 30 s is used. To evaluate the tribological behavior, μ as well as ϑ_M is plotted for each sum velocity v_Σ at each prevailing slip ratio s. The measurements are started with the lowest sum velocity v_Σ, starting with pure rolling at s = 0% at disk bulk temperatures ϑ_M close to $\vartheta_{ambient}$. Then, as the system reaches a stationary operating point, the slip ratio s is gradually increased. The friction curve for the next higher sum velocity v_Σ is run after cooling the system back to $\vartheta_{ambient}$. The procedure is continued until all operating points have been investigated or until a termination criterion is reached. To achieve the objectives of this study, the experiments are performed sequentially with the following structure:

- Demonstration of the functionality of self-lubricating rolling-sliding contacts across a wide range of operating points and comparison with injection lubricated rolling sliding contacts.
- Determination of the tribologically optimal number of oil-impregnated elements in a self-lubricating rolling-sliding contact.
- Investigation of the influence of surface condition and finish on self-lubricating rolling-sliding contacts.

2.3.2. Long-Term Tests

In the long-term operation tests, constant operating points are maintained. Hereafter referred to as the reference operating conditions, the operating point at a pressure of p_H = 1043 N/mm^2, a sum velocity of v_Σ = 4 m/s, and slip ratio of s = 20% is used for evaluation of the test results. The maximum duration of the tests is t = 48 h, which corresponds to a maximum load cycle number $N_2 \approx 1.2$ m on the faster rotating disk, provided that the automatic shutdown mechanism does not terminate the test prematurely due to reaching the termination criteria ϑ_{crit} or μ_{crit}. The repeatability of the long-term test results is indicated by the maximum difference of μ or N_2 of the first and a repetition test performed with a 2nd new disk pair. To further investigate the study's objectives, the long-term operating tests are conducted as follows:

- Demonstration of long-term durability of self-lubricating contacts and comparison with injection-lubricated contacts.
- Investigation of the influence of surface finish on the frictional and stress behavior in self-lubricating contacts.
- Investigation of the improvement of the frictional behavior in self-lubricating contacts through solid lubricant additives in the sinter material.
- Determination of the performance of self-lubricating contacts under higher load and speed.
- Evaluation of the transferability to gear applications with typical tooth flank surfaces.

The measurements of each variant are examined immediately after the running-in. The test run is conducted intermittently in order to enhance the detection of surface alterations of the upper disk after a defined number of load cycles N_2, using the Infinite Focus Sensor R25 from Alicona Imaging GmbH (Raaba, Austria), an optical measuring device based on the principle of focus variation for high-resolution surface imaging. The number of load cycles between two surface measurements is shorter at the beginning of the tests to document the running-in-like process of the first load cycles and increases after each measurement. An overview of the measurement intervals, expressed in terms of load cycle numbers and the corresponding times in hours, is presented in Table 4. During the intermittent measurement intervention, the system is interrupted for t < 5 min to prevent excessive cooling of the bulk temperature ϑ_M. The surface measurement of the upper disk captures individual areas in the shape of a square with an edge length of a = 1 mm and a

20-fold magnification to analyze the local surface condition. To obtain a global qualitative impression of surface alteration across the entire disk width from the composite local images, six overlapping surface images are taken along the direction of the cleaned area from one disk edge to the other. In this study, the surface image in the middle of the upper disk width is evaluated. Moreover, an expert qualitatively evaluated a surface area in the middle of the upper disk width using the optical appearance of pores in 2D images and the matching geometrical hole-like shape of pores in 3D topography measurements.

Table 4. Intermittent measurement intervals during long-term operation tests.

	Measurement Intervals						
Load cycle number N_2	>2.5 k	>10 k	>30 k	<100 k	>250 k	>750 k	>1200 k
Time t in h	0.08	0.33	1	3	8	24	48

Each test run is evaluated by its levels and differences of friction μ and bulk temperature ϑ_M as well as its operating behavior coefficient Ξ according to Ebner [25]. The operating behavior coefficient Ξ is calculated by the division of the load cycles during the longest test sequence, with a maximum difference of the minimum and maximum of μ being smaller than $\Delta\mu < 0.005$, and the load cycles for the whole lifetime of the disk pair. An operating behavior is defined as unstable when the load cycle of the upper disk for the whole lifetime is $N_2 < 0.25$ M, as metastable when $N_2 > 0.25$ M and $\Xi < 0.40$, and stable when $N_2 > 0.25$ M and $\Xi > 0.40$.

3. Results

In the following section, first, the friction curves and, second, the long-term operation tests with applicability promising disk pairs are shown.

3.1. Friction Curve Tests

At first, the functionality of self-lubricating rolling-sliding contacts with oil-impregnated sinter material across a wide range of operating points is shown and compared to an injection-lubricated pure steel contact. Further on, the tribologically optimal number of oil-impregnated sinter disks in a self-lubricating rolling-sliding contact and the influence of surface condition are evaluated.

3.1.1. Demonstration of Self-Lubricating Rolling-Sliding Contacts and Comparison with Injection-Lubricated Rolling-Sliding Contacts

Figure 3a shows the measured friction curves of the reference disk pair with a longitudinally ground, oil-impregnated sinter$_{ref}$ disk and a longitudinally ground steel disk under self-lubrication. Additionally, the maximum values μ_{max} of each curve are shown in a bar diagram. The reference disk pair can be evaluated at all operating points as it did not exceed the termination criteria μ_{crit} or ϑ_{crit} during the tests. The measurements with the 1st and the 2nd disk pair both show the same trend: μ decreases with increasing v_Σ. The friction curves of the 1st disk pair increase steadily within the first three measured slip ratios of $s \leq 5\%$ and maintain a plateau at $v_\Sigma = 2$ m/s over the remaining slip ratios at approximately $\mu \approx 0.064$. At $v_\Sigma = 4$ m/s and 8 m/s, the CoF curves decrease again after reaching their respective local maxima at s = 10%, before increasing again at the highest slip ratio at s = 50% to the global maximum $\mu_{max} = 0.051$ and 0.047, respectively. The friction curves of the tests with the 2nd disk pair are at a continuously lower level compared to the CoFs of the 1st disk pair, with plateau-like curves at $v_\Sigma = 4$ m/s and 8 m/s, reaching $\mu_{max} = 0.042$ and 0.047 at the highest slip ratio, respectively. Figure 3b shows the measured bulk temperature curves and the maximum values $\vartheta_{M,max}$, which correspond to the increasing frictional power depending on the sum velocity, so that the levels of the bulk temperature are at $v_\Sigma = 2$ m/s at the lowest, at 4 m/s at the middle, and at 8 m/s at the highest level. Within the respective sum velocities, the bulk temperature also increases with increasing slip ratio. In the bulk temperature curves, the temperatures of the 1st disk pair are higher than the temperatures of the 2nd disk pair. Specifically, the bulk temperature maxima of the 1st disk pair are $\vartheta_{M,max}$ = 56 °C, 76 °C, and 109 °C, while with the 2nd disk pair they are 53 °C, 64 °C, and 109 °C. Figure 4a shows the measured friction curves of the 1st and 2nd test with the 1st disk pair, which was re-run in the 2nd test after the 1st test without re-impregnation. Additionally, the maximum difference of μ between the 1st and the 2nd test is shown for each sum velocity. As for the 1st test, every operating point of the 2nd test can be evaluated, because no termination criteria were reached. The friction curve of the 2nd test at v_Σ = 2 m/s increases with increasing slip ratio up to s ≤ 10% to a higher level than in the 1st test and, at s = 10%, reaches a local maximum for μ. However, with increasing slip ratio up to s ≤ 20%, μ decreases before increasing again to μ_{max} = 0.083 at s = 40%. At the highest slip ratio s = 50%, μ decreases again. However, the friction curves of the 2nd test at sum velocities of v_Σ = 4 m/s and 8 m/s follow a similar trend as the 1st test, with smaller increases in the curves up to the highest slip ratio at s = 50%. The global maxima of the curves at v_Σ = 4 m/s and 8 m/s are μ_{max} = 0.051 and 0.046, respectively. Additionally, Figure 4b shows the bulk temperature curves ϑ_M and the maximum values $\vartheta_{M,max}$ of the 1st and 2nd test with the 1st disk pair. The curves of both tests are very similar at v_Σ = 8 m/s. At a velocity of 4 m/s, minimal temperature differences between the 1st and 2nd test are observed over most of the range of slip ratio variation, with slightly lower temperatures of the 2nd test at the highest level of slip ratio at s = 50%. In comparison, at v_Σ = 2 m/s, higher bulk temperatures are measured at the 2nd test over the entire range of slip ratio variation. The maximum bulk temperatures of the 2nd test recorded for v_Σ = 2 m/s, 4 m/s, and 8 m/s are ϑ_M = 60 °C, 67 °C, and 104 °C, respectively.

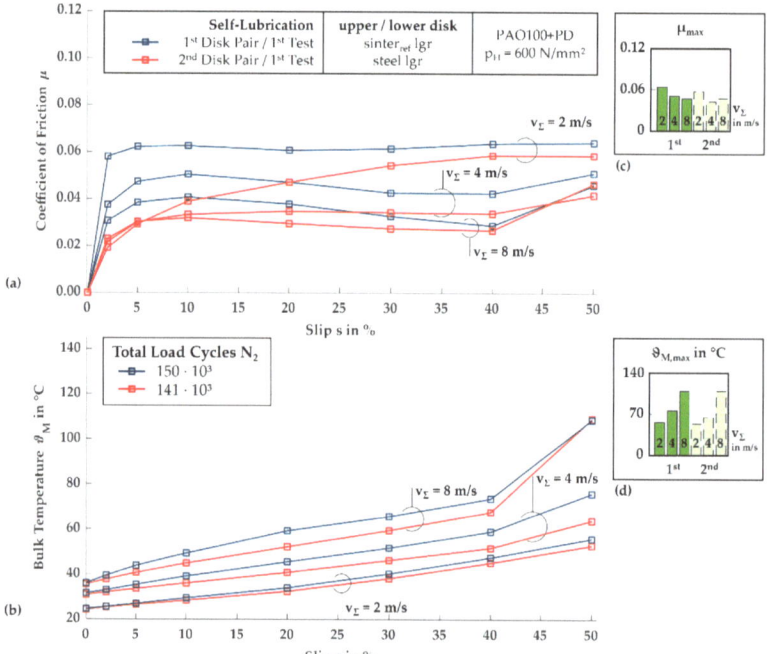

Figure 3. Friction curves (**a**) and bulk temperature curves (**b**) and each of their maximum level (**c**,**d**) of the 1st test of two disk pairs of a longitudinally ground sinter$_{ref}$ disk and a longitudinally ground steel disk under self-lubrication.

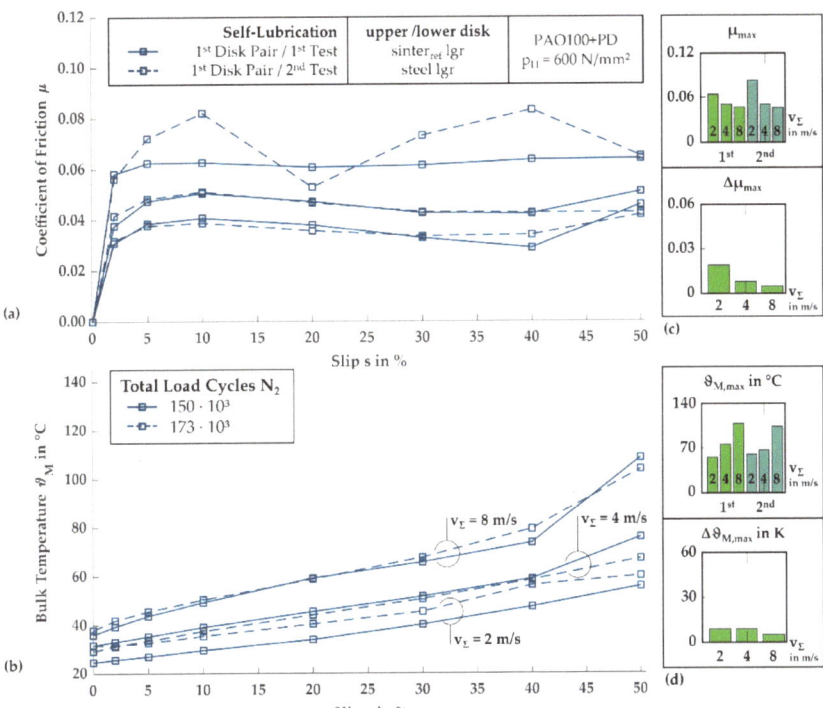

Figure 4. Friction curves (**a**) and bulk temperature curves (**b**) and each of their maximum level as well as maximum difference (**c**) and (**d**) of two tests of the 1st disk pair of a longitudinally ground sinter$_{ref}$ disk and a longitudinally ground steel disk under self-lubrication.

Figure 5a shows the measured friction curves of the 1st and 2nd test with the 1st disk pair with two longitudinally ground steel disks under injection lubrication. The friction curves initially increase with larger gradients over the slip ratios of s = 2–5%, then exhibit a smaller increase. The maximum coefficients of friction are μ_{max} = 0.057, 0.038, and 0.035 in the 1st test run and μ_{max} = 0.058, 0.034, and 0.033 in the 2nd test run. Figure 5b shows the bulk temperature curves of both tests, the 1st and 2nd test with longitudinally ground steel disks under injection lubrication, correlating to the curves shown in Figure 5a. The bulk temperature differences between the sum velocities remain largely constant over the slip ratios, with almost constant temperature differences. The 2nd test at v_Σ = 2 m/s resulted in slightly higher bulk temperatures, approximately $\Delta\vartheta_M$ = +2 K higher than those of the 1st test. The maximum bulk temperatures at v_Σ = 2 m/s, 4 m/s, and 8 m/s are $\vartheta_{M,max}$ = 38 °C, 43 °C, and 52 °C in the 1st test and 40 °C, 43 °C, and 52 °C in the 2nd test.

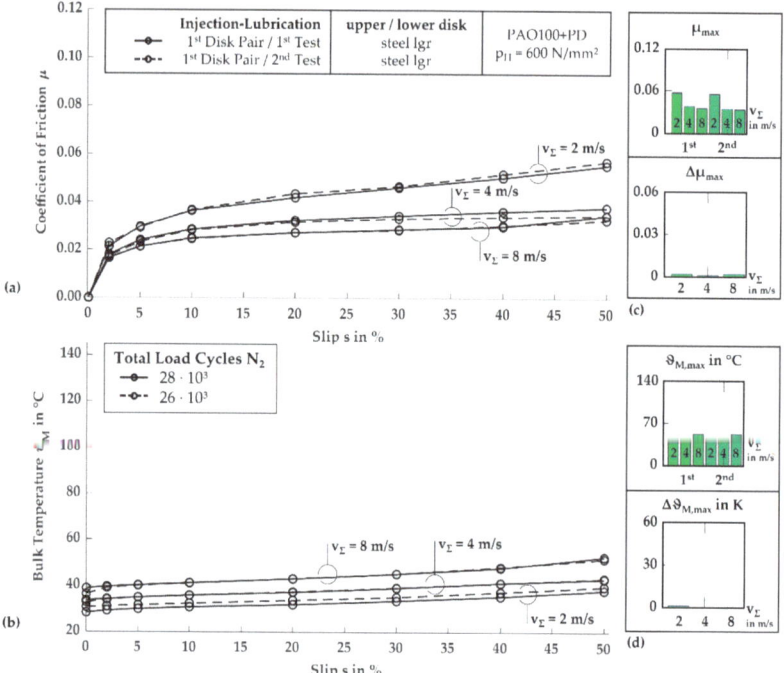

Figure 5. Friction curves (**a**) and bulk temperature curves (**b**) and each of their maximum level as well as maximum difference (**c**) and (**d**) of two tests of the 1st disk pair of two longitudinally ground steel disks under injection lubrication.

3.1.2. Determination of the Tribologically Optimal Number of Oil-Impregnated Sinter Elements in Self-Lubricating Rolling-Sliding Contacts

Figure 6a shows the measured friction curves of the 1st tests of two polished steel disk pairs, each lubricated only once with V_{init}. Because of an expected harsh mixed lubrication regime in once-lubricated EHL tribosystems due to the low amount of prevailing oil, the steel disks were mechanically polished to reduce the influence of the surface roughness on the relative lubricant film thickness and thus improve the expected friction to be better performing compared to once-lubricated EHL tribosystems with longitudinally ground surfaces. Due to the termination criterion of $\mu > 0.12$ reached at $v_\Sigma = 2$ m/s with $s = 20\%$ or 30%, both tests can only be evaluated for a few operating points. Figure 6b shows the bulk temperature curves of both tests correlating to Figure 6a, which both show increasing ϑ_M with increasing slip ratios. Repeating the test with the 1st or the 2nd disk pair is not possible because of surface damage.

Figure 7a shows the measured friction curves and maxima μ_{max} of the 1st tests of two disk pairs, each with two longitudinally ground, oil-impregnated sinter$_{ref}$ disks. The trends observed with the 1st disk pair are clearly related to the sum velocity. The data shows that μ gradually decreases with increasing v_Σ, with all curves starting with a steep gradient of μ within the initial slip ratios, then decreasing after reaching a global maximum μ_{max}. For instance, at $v_\Sigma = 2$ m/s, the global maximum CoF is $\mu_{max} = 0.076$ at $s = 10\%$. Similarly, at $v_\Sigma = 4$ m/s, the global maximum of CoF is $\mu_{max} = 0.060$ at $s = 20\%$, and at $v_\Sigma = 8$ m/s, it is $\mu_{max} = 0.052$ at $s = 5\%$. The bulk temperatures shown together with $\vartheta_{M,max}$ in Figure 7b appear to increase as the sum velocity increases, with relatively constant increases for each sum velocity. At $v_\Sigma = 2$ m/s, the maximum bulk temperature is $\vartheta_{M,max} = 51$ °C, at 4 m/s it is 63 °C, and at 8 m/s it is 79 °C. The 1st test with a 2nd disk pair of the same specification shows trends that follow less clearly the previous patterns of sum velocity and slip ratio variation (see Figure 7a,b). In particular, at $v_\Sigma = 2$ m/s, the coefficient of friction is initially higher at s ≤ 5% and reaches a maximum of μ_{max} = 0.072 at s = 5%, before decreasing at s ≤ 40% and remaining approximately constant up to s = 50%. At v_Σ = 4 m/s, the coefficient of friction initially follows the trend of the 1st disk pair before a peak of μ_{max} = 0.079 at s = 20%, which is higher than the maximum at v_Σ = 2 m/s, and then declining to a level comparable to the 1st test. The friction curve at v_Σ = 8 m/s lies, as typically expected, at the lowest level of the sum velocities within the tests of one disk pair. The curve reaches a local maximum at s = 5%, before decreasing again to s = 20%, then increasing significantly and reaching the global maximum at s = 50% with μ_{max} = 0.044. The correlation between the bulk temperatures and the coefficient of friction curves is evident. The curve at v_Σ = 2 m/s is the lowest with $\vartheta_{M,max}$ = 51 °C, the curve at v_Σ = 4 m/s is higher with $\vartheta_{M,max}$ = 78 °C and shows a significant increase at the peak of the CoF at s = 20%, and the curve at v_Σ = 8 m/s is comparatively the highest, increasing significantly from the slip ratio s = 20% until it reaches a maximum with $\vartheta_{M,max}$ = 122 °C, significantly higher than that of the 1st test.

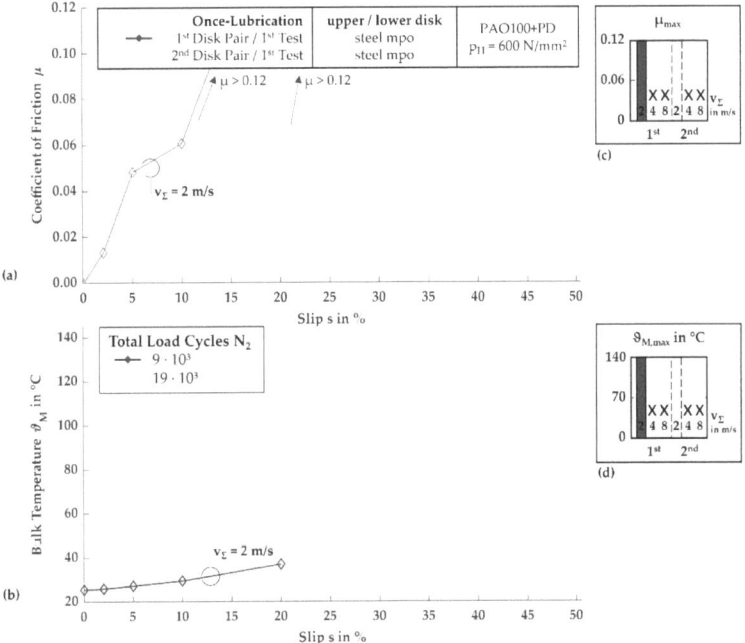

Figure 6. Friction curves (**a**) and bulk temperature curves (**b**) and each of their maximum level (**c,d**) of the 1st test of two disk pairs of two mechanically polished steel disks under once-lubrication.

Figure 8a shows the friction curves and maxima μ_{max} measured with the 1st and 2nd test of the 1st disk pair of two longitudinally ground, oil-impregnated sinter$_{ref}$ disks. The 2nd test of the 1st disk pair can be evaluated at all operating points but shows significantly higher μ soon after initially lower values at all three sum velocities investigated, reaching near the termination criterion at v_Σ = 2 m/s with μ_{max} ≈ 0.12 and increasing to similar values at v_Σ = 4 m/s and 8 m/s, with μ_{max} = 0.063 and 0.062, respectively. Figure 8b shows the bulk temperature curves and maxima $\vartheta_{M,max}$ measured during the 2nd test of the disk pair shown in Figure 8a. The bulk temperatures initially start at slightly higher levels before increasing along with the significant increases in the coefficients of friction, all reaching the bulk temperature maxima at s = 50% at v_Σ = 2 m/s, 4 m/s, and 8 m/s, at $\vartheta_{M,max}$ = 66 °C, 90 °C, and 136 °C, respectively.

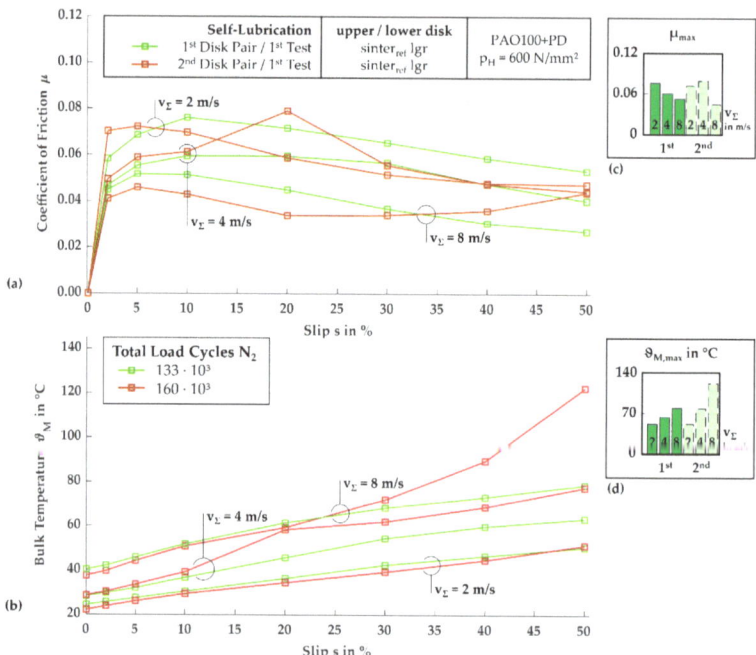

Figure 7. Friction curves (**a**) and bulk temperature curves (**b**) and each of their maximum level (**c**,**d**) of the 1st test of two disk pairs of two longitudinally ground sinter$_{ref}$ disks under self-lubrication.

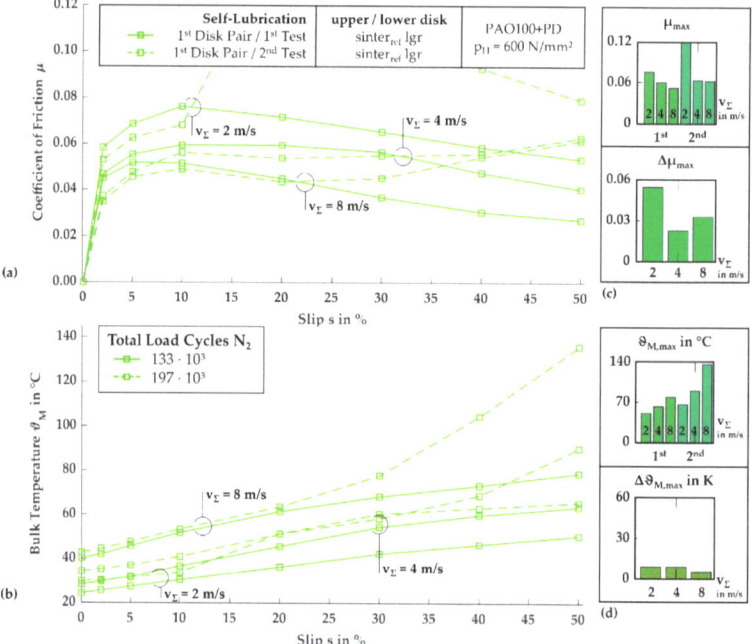

Figure 8. Friction curves (**a**) and bulk temperature curves (**b**) and each of their maximum level as well as maximum difference (**c**) and (**d**) of two tests of the 1st disk pair with two longitudinally ground sinter$_{ref}$ disks under self-lubrication.

3.1.3. Investigation of the Influence of Surface Finish on Self-Lubricating Rolling-Sliding Contacts

The investigation of the surface influence on the tribological behavior of self-lubricating, oil-impregnated sintered rolling-sliding contacts is carried out systematically. Initially, a mechanically polished steel disk surface is paired with a longitudinally ground, oil-impregnated $sinter_{ref}$ disk. Subsequently, a mechanically polished steel disk is paired with different surface variants of the oil-impregnated $sinter_{ref}$ disk. Therefore, the surface of the $sinter_{ref}$ disk is investigated in different conditions for the reference material, including superfinished and mechanically polished surfaces. Furthermore, the influence of a plasma-nitrided $sinter_{pni}$ disk with high hardness is investigated.

Figure 9a,b summarize the evaluated maximum μ_{max} and $\vartheta_{M,max}$ from measured friction and bulk temperature curves of the two test runs with the 1st disk pair (1st and 2nd test), as well as the 1st test with a 2nd disk pair. Additionally, Figure 10a,b show the maximum difference of the friction curves $\Delta\mu_{max}$ and of the bulk temperature curves $\Delta\vartheta_{M,max}$ of the 1st and 2nd test with the 1st disk pair. The results obtained with the reference disk pair with a longitudinally ground steel disk and a longitudinally ground, oil-impregnated $sinter_{ref}$ disk are also included for comparison.

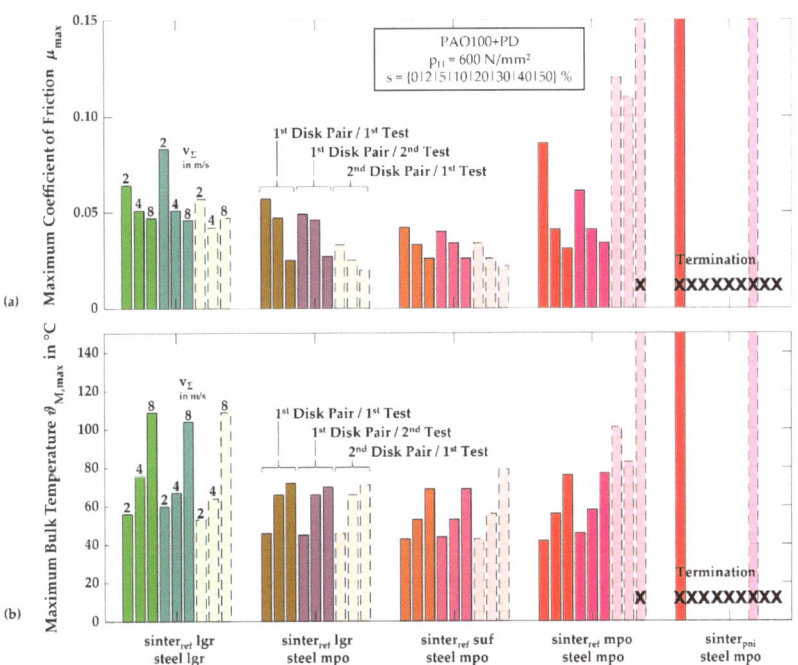

Figure 9. Friction curve maximum μ_{max} (**a**) and bulk temperature maximum $\vartheta_{M,max}$ (**b**) of the friction curve tests with sinter material and surface variants under self-lubrication.

The surface variation of the steel disk from a longitudinally ground to a polished condition shows improved tribological behavior compared to the reference measurements with both disk surfaces in a longitudinally ground condition, indicated by low levels of μ_{max} and $\vartheta_{M,max}$ in the 1st test runs of the 1st and 2nd disk pair and especially with the 2nd disk pair, with $\mu_{max} = \{0.033, 0.025, 0.020\}$ and $\vartheta_{M,max} = \{46, 66, 71\}$ °C. There was good repeatability of the 1st test with the 1st disk pair through the 1st test with the 2nd disk pair, with very similar $\vartheta_{M,max}$ and smaller μ_{max} at all v_Σ (see Figure 9a,b), and a low maximum difference of $\Delta\mu_{max} = 0.003 \ldots 0.010$ and $\Delta\vartheta_{M,max} \leq 2$ K between the 1st and 2nd test with the 1st disk pair (see Figure 10a,b).

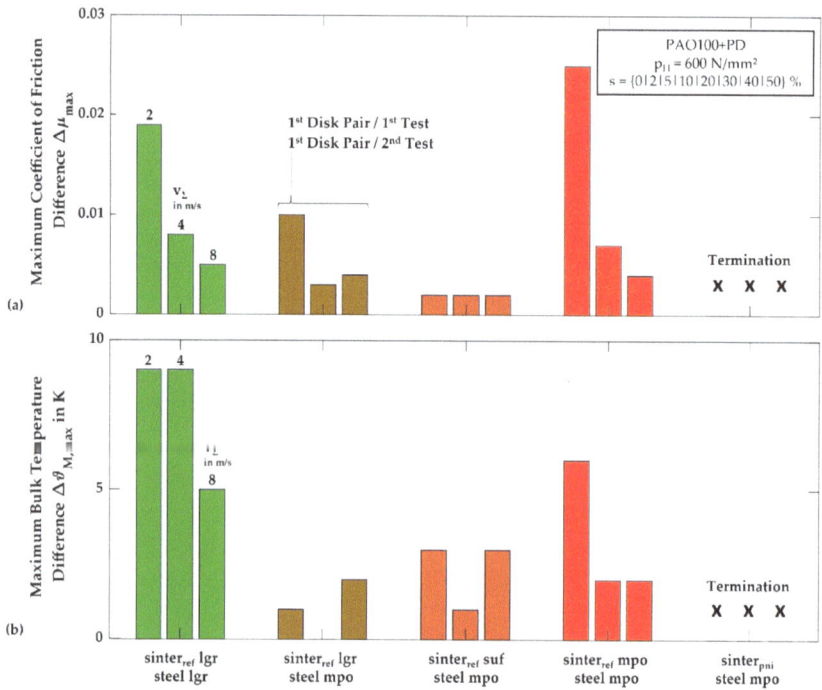

Figure 10. Maximum difference of the friction curves $\Delta\mu_{max}$ (**a**) and the bulk temperature curves $\Delta\vartheta_{M,max}$ (**b**) of the friction curve tests with sinter material and surface variants under self-lubrication.

Figures 11a and 12a show the measured friction curves with disk pairs with mechanically polished steel and superfinished sinter$_{ref}$ disk surfaces; the curves with a superfinished sinter$_{ref}$ disk surface show, on one hand, a similar tribological beneficial trend to the curves with the mechanically polished steel disk surface, and on the other hand, the best tribological behavior compared to the other surface variants. Measurements taken with superfinished contact surfaces show $\mu_{max} \leq 0.042$ for all tests with both disk pairs down to $\mu_{max} \leq 0.022$ at $v_\Sigma = 8$ m/s with the 2nd disk pair as well as a very good repeatability between the 1st tests with the 1st and 2nd disk pair, with slightly lower levels of μ_{max} with the 2nd disk pair (see Figure 11a) and the least maximum difference $\Delta\mu_{max} = 0.002$ between the 1st and 2nd test with the same 1st disk pair (see Figure 12a). Figures 11b and 12b additionally show the measured bulk temperature curves of the disk pairs with mechanically polished steel and superfinished sinter$_{ref}$ disk surfaces. Overall, the levels of the maximum bulk temperature $\vartheta_{M,max}$ are the lowest compared to the tests with the other surface variants, with the lowest maximum bulk temperatures at the 1st test with the 2nd disk pair with $\vartheta_{M,max} = \{43, 53, 69\}$ °C. Comparing the 1st and 2nd test with the 1st disk pair, a low maximum difference of $\Delta\vartheta_{M,max} \leq 3$ K is shown.

In contrast, the tests with disk pairs with mechanically polished steel and sinter$_{ref}$ disk surfaces show, at the 1st tests of the 1st and 2nd disk pair, initially at a sum velocity of $v_\Sigma = 2$ m/s, high to very high CoF, with $\mu_{max} = 0.086$ and $= 0.12$ (see Figure 9a). During both test runs (1st and 2nd test) with the 1st disk pair, the friction curve levels decrease with increasing sum velocity, while the 2nd test shows a lower level of μ_{max} at $v_\Sigma = 2$ m/s and comparable levels of μ_{max} at $v_\Sigma = 4$ m/s and $= 8$ m/s. During the 1st test with the 2nd disk pair, the friction curve level first slightly decreases with an increasing sum velocity at $v_\Sigma = 4$ m/s before leading to termination at the highest sum velocity at $v_\Sigma = 8$ m/s. The bulk temperatures of the two tests with the 1st disk pair shown in Figure 9b are comparable to the bulk temperature levels with the disk pairs of a polished steel disk and ground sinter disk and show moderate to low maximum difference with $\Delta\vartheta_{M,max} \leq 6$ K. The bulk temperatures of the 1st test with the 2nd disk pair follow the trend of µ and, compared to the tests with the 1st disk pair, also start at a high level at $v_\Sigma = 2$ m/s with $\vartheta_{M,max} = 101$ °C, and decrease slightly to $\vartheta_{M,max} = 83$ °C at $v_\Sigma = 4$ m/s before thermally increasing at $v_\Sigma = 8$ m/s. With the comparison of the tests of the two disk pairs with identical specifications, which can show both a good frictional behavior with a low to moderate difference of the maximum $\Delta\mu_{max} \leq 0.025$ and $\Delta\vartheta_{M,max} \leq 6$ K between the 1st and 2nd test with one disk pair (see Figure 10a,b), and a non-functional behavior with another disk pair, a significantly worse repeatability of the tribological behavior of disk pairs with polished sinter disk surfaces is shown.

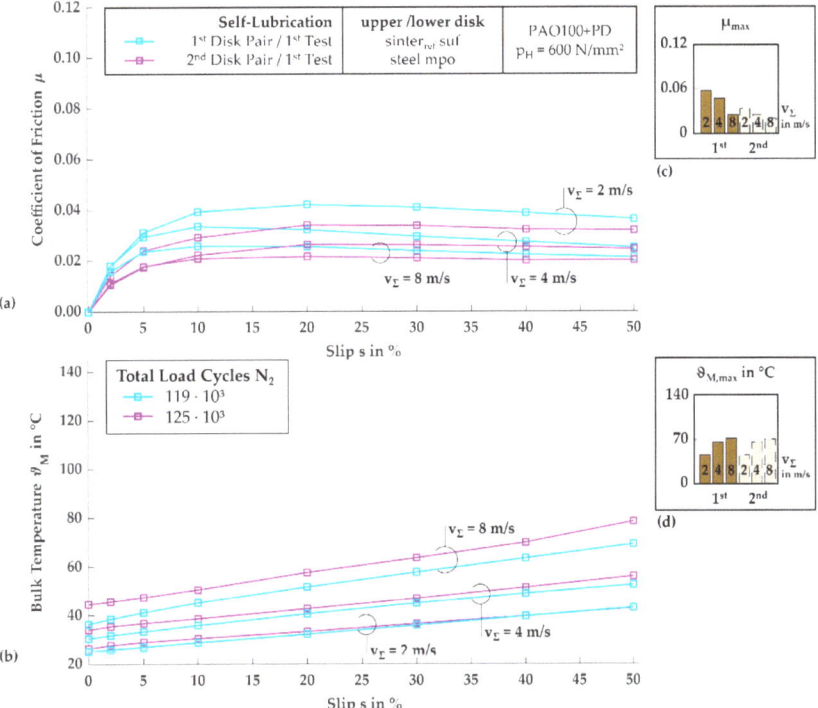

Figure 11. Friction curves (**a**) and bulk temperature curves (**b**) and each of their maximum level (**c**,**d**) of the 1st test of two disk pairs of a polished steel disk and a superfinished sinter$_{ref}$ disk under self-lubrication.

The tests with the disk pairs with a polished steel disk and a plasma-nitrided sinter$_{pni}$ disk with high hardness cannot be evaluated due to a failure within the initial sum velocity of $v_\Sigma = 2$ m/s in the 1st tests of the 1st and 2nd disk pair. The failure is indicated by an acceleration detector, which causes the tribometer to stop immediately. An analysis of the sinter$_{pni}$ disk surfaces revealed delamination of the compound layer from the underlying diffusion layer.

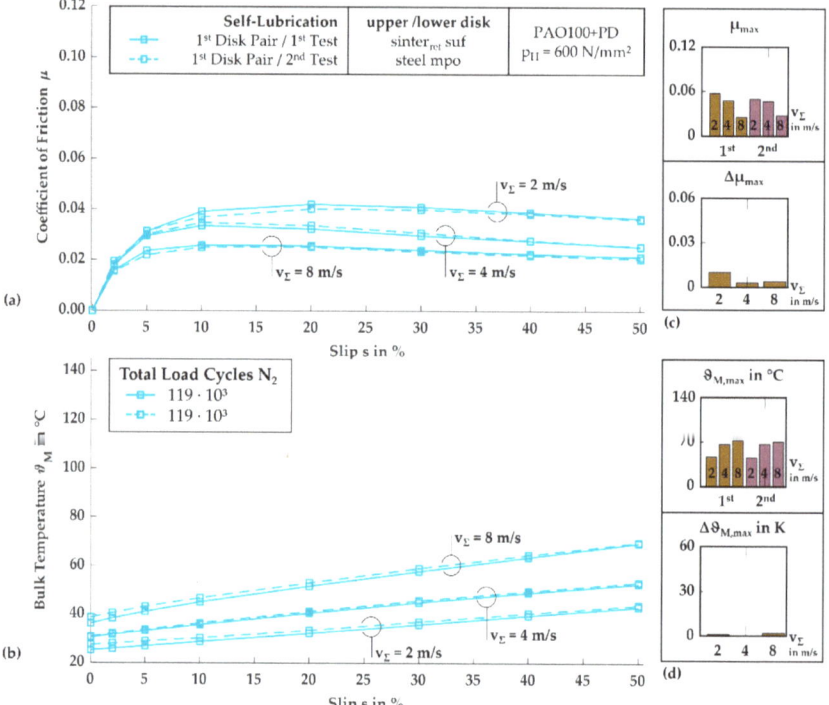

Figure 12. Friction curves (**a**) and bulk temperature curves (**b**) and each of their maximum level as well as maximum difference (**c,d**) of two tests of the 1st disk pair with a polished steel disk and a superfinished sinter$_{ref}$ disk under self-lubrication.

3.1.4. Summary of the Friction Curve Tests

Table 5 summarizes the evaluated data of the friction curve tests (see Sections 3.1.1–3.1.3), including the total test duration N_2 of each friction curve test of the two disk pairs.

The results of the friction curve tests were used to reduce the number of variants to the most promising disk pairs for long-term testing. As a result, the disk pair variants terminating after one test under self-lubrication with a mechanically polished sinter$_{ref}$ disk as well as with a sinter$_{pni}$ disk are not considered for the long-term tests. The disk pair of two sinter$_{ref}$ disks with its increased material preparation effort for oil impregnation by simultaneously increased friction and bulk temperature levels at all tests and increased differences of friction and bulk temperature between the 1st and 2nd test with the 1st disk pair are also not considered for the long-term tests.

Table 5. Maximum friction μ_{max}, maximum bulk temperature $\vartheta_{M,max}$, and total load cycles N_2 of the upper disk of all friction curve tests with both the 1st and the 2nd disk pair as well as maximum difference of the friction $\Delta\mu_{max}$ and maximum difference of the bulk temperature $\Delta\vartheta_{M,max}$ between the 1st and the 2nd test of the 1st disk pair.

Lubrication Method Upper/Lower Disk $p_H = 600$ N/mm² $s = 0\ldots50\%$		μ_{max} at v_Σ			$\vartheta_{M,max}$ in °C at v_Σ			N_2 in M
		2 m/s	4 m/s	8 m/s	2 m/s	4 m/s	8 m/s	Total
once-lubrication steel mpo/steel mpo	1.1 [4]	>0.120 [1]	x [2]	x [2]	>140 [1]	x [2]	x [2]	x [2]
	1.2 [5]	x [2]	x [2]	x [2]	x	x [2]	x [2]	x [2]
	$\Delta\mu_{max}$	x [2]	x [2]	x [2]	x	x [2]	x [2]	
	2.1 [6]	>0.120 [1]	x [2]	x [2]	>140 [1]	x [2]	x [2]	x [2]
injection lubrication steel lgr/steel lgr	1.1 [4]	0.057	0.038	0.035	38	43	52	0.03
	1.2 [5]	0.055	0.034	0.033	40	43	52	0.03
	$\Delta\mu_{max}$	0.002	0.001	0.002	1	0	0	
	2.1 [6]	_ [3]	_ [3]	_ [3]	_ [3]	_ [3]	_ [3]	_ [3]
self-lubrication sinter$_{ref}$ lgr/steel lgr	1.1 [4]	0.064	0.051	0.047	56	76	109	0.15
	1.2 [5]	0.083	0.051	0.046	60	67	104	0.17
	$\Delta\mu_{max}$	0.019	0.008	0.005	9	9	5	
	2.1 [6]	0.057	0.042	0.047	53	64	109	0.14
self-lubrication sinter$_{ref}$ lgr/sinter$_{ref}$ lgr	1.1 [4]	0.076	0.060	0.052	51	63	79	0.13
	1.2 [5]	0.120	0.063	0.062	66	90	90	0.20
	$\Delta\mu_{max}$	0.055	0.023	0.033	17	17	15	
	2.1 [6]	0.072	0.079	0.044	51	88	122	0.16
self-lubrication sinter$_{ref}$ lgr/steel mpo	1.1 [4]	0.057	0.047	0.025	46	66	72	0.13
	1.2 [5]	0.049	0.046	0.027	45	66	70	0.13
	$\Delta\mu_{max}$	0.010	0.003	0.004	1	0	2	
	2.1 [6]	0.033	0.025	0.020	46	66	71	0.13
self-lubrication sinter$_{ref}$ suf/steel mpo	1.1 [4]	0.042	0.033	0.026	43	53	69	0.12
	1.2 [5]	0.040	0.034	0.026	44	53	69	0.12
	$\Delta\mu_{max}$	0.002	0.002	0.002	3	1	3	
	2.1 [6]	0.034	0.026	0.022	43	56	79	0.13
self-lubrication sinter$_{ref}$ mpo/steel mpo	1.1 [4]	0.086	0.041	0.031	42	56	76	0.14
	1.2 [5]	0.061	0.041	0.034	46	58	77	0.13
	$\Delta\mu_{max}$	0.025	0.007	0.004	6	2	2	
	2.1 [6]	0.120	0.110	>0.120 [1]	101	83	>140 [1]	0.08 [1]
self-lubrication sinter$_{pni}$/steel mpo	1.1 [4]	>0.120 [1]	x [2]	x [2]	>140 [1]	x [2]	x [2]	x [2]
	1.2 [5]	x [2]	x [2]	x [2]	x [2]	x [2]	x [2]	x [2]
	$\Delta\mu_{max}$	x [2]	x [2]	x [2]	x [2]	x [2]	x [2]	
	2.1 [6]	>0.120 [1]	x [2]	x [2]	>140 [1]	x [2]	x [2]	x [2]

[1]: termination at operating point; [2]: no value because of termination; [3]: no test conducted; [4]: 1st disk pair, 1st test; [5]: 1st disk pair, 2nd test; [6]: 2nd disk pair, 1st test.

3.2. Long-Term Tests

This section presents the results of the long-term tests with the selected material pairing and surface condition variants from Section 3.1. The operating behavior of the reference disk pair with a longitudinally ground steel and a longitudinally ground sinter$_{ref}$ disk is compared to that of an injection-lubricated disk pair with two longitudinally ground steel disks. Furthermore, the influence of surface finish on the operating behavior during the long-term tests is examined, followed by an investigation of the influence of the solid lubricant for friction reduction in both once-lubricated and self-lubricating disk pairs. Also, the performance at higher loads is evaluated, and the functionality of gear-application-like surfaces on the disks is examined.

3.2.1. Demonstration of Long-Term Durability of Self-Lubricating Contacts and Comparison to Injection Lubrication

Figure 13 shows the measured friction curves and the calculated stability coefficients of the two long-term tests with a self-lubricating EHL tribosystem with a polished steel disk and an oil-impregnated ground sinter$_{ref}$ disk as well as an EHL tribosystem with longitudinally ground steel disk pairs under injection lubrication without temperature regulation. Both the 1st and 2nd test of the injection-lubricated disk pair show for the whole test run of 48 h (i.e., N$_{2|1st}$ and N$_{2|2nd}$ > 1.20 M) very good stability according to Ebner [25], with Ξ = 0.95 and = 0.76. Initially, each of the curves starts with a coefficient of friction level of μ = 0.045 and exhibits the level for the main time of the test duration. In comparison, both tests with the self-lubricating EHL tribosystem start with a lower CoF of μ = 0.037. Then, the test with the 1st disk pair without intermediate interruptions for surface measurements increases steadily with a small gradient, passing the level of CoF of the injection-lubricated tribosystem after t ≈ 25 h and reaching μ$_{max}$ = 0.052. The determined stability coefficient is Ξ = 0.51 and indicates stable operating behavior. However, the test with the 2nd disk pair slightly decreases at the first section of the tests and is maintained at a comparably low level of μ = 0.035 for at least t ≈ 24 h, when the level slowly starts increasing to μ$_{max}$ = 0.045. The operating behavior of the 2nd disk pair can be classified as stable, as the determined stability coefficient is Ξ = 0.71. In addition to the comparable lower μ and the stable operating behavior, the EHL tribosystem under self-lubrication shows less repeatability because of the lower level of μ of the 2nd test, with Δμ$_{max}$ = 0.020, whereas both tests with the injection-lubricated EHL tribosystem show a difference of Δμ$_{max}$ = 0.005 from the beginning of the tests and a neglectable difference for the rest of the measurement time.

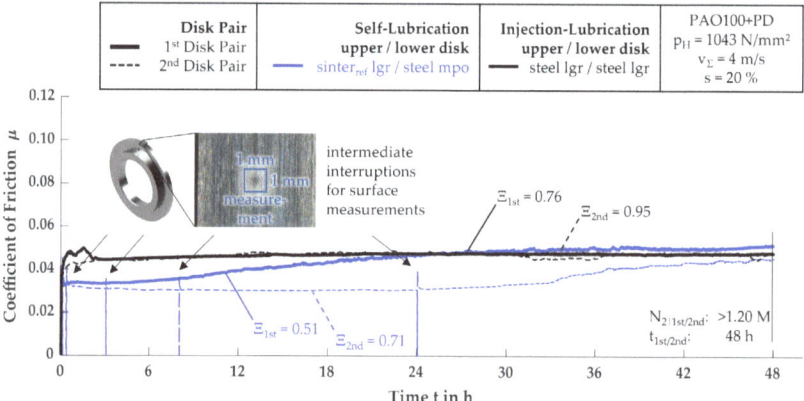

Figure 13. Friction curves of the long-term tests of two longitudinally ground steel disk pairs under injection lubrication and two disk pairs with a polished steel disk and a longitudinally ground sinter$_{ref}$ disk under self-lubrication.

As there are intermediate interruptions for surface measurements, the following EHL tribosystem behavior after continuing the test has to be evaluated to verify the friction and bulk temperature trends deriving from the disk pair specifications and not from cool-down periods. At the intermittently interrupted test with the self-lubricating disk pair shown in Figure 13, one can see the very short peaks of CoF directly after re-starting the test and, after a few load cycles, the return of the friction curves to the previous level the moment before the interruption of the twin-disk tribometer. With the curves returning to the previous state before the interruption, the influence of the test interruptions on the trend of the tribosystem operating behavior can be assumed as negligible.

3.2.2. Investigation of the Influence of Surface Finish on the Frictional and Stress Behavior in Self-Lubricating Contacts

Figure 14a shows the measured friction curves and the calculated stability coefficients of the long-term tests with self-lubricating disk pairs with the reference EHL tribosystem of a longitudinally ground steel disk and a longitudinally ground oil-impregnated sinter$_{ref}$ disk and surface variants of the disk pair with disk pairs of a mechanically polished steel disk and an oil-impregnated longitudinally ground as well as a mechanically polished sinter$_{ref}$ disk. When varying the surface finish of the pairing disks in the self-lubricating EHL tribosystem, the trends of the measured friction during the long-term tests are comparable to the trends obtained at the friction curve tests, with no termination until the end of the 48 h test runs at $N_{2|1st}$ and $N_{2|2nd} > 1.20$ M (see Figure 9 in Section 3.1.3). In nearly all tests, the disk pairs with at least one smoothed surface show lower levels of μ compared to the reference system with longitudinally ground surfaces, except in the 1st test with the disk pair of a mechanically polished steel disk and a longitudinally ground sinter$_{ref}$ disk. When taking the pair of the polished steel disk and the superfinished sinter$_{ref}$ disk evaluated as the best during the friction curve tests into account, the comparably very low level of $\mu \leq 0.040$ and the very stable operating behavior of both disk pairs with $\Xi = 1.00$ and $= 0.69$ as well as the good repeatability with $\Delta\mu_{max} < 0.008$ verify the trend of the best-performing self-lubricating EHL tribosystem. Further, the bulk temperatures ϑ_M measured during the long-term tests are shown in Figure 14b, in which every curve of ϑ_M correlates to the trend of μ evaluated, leading to the lowest ϑ_M measured with the disk pair of a mechanically polished steel disk and a superfinished sinter$_{ref}$ disk. The short intermediate interruptions for surface measurements lead at all tests to a difference of the bulk temperature of $\Delta\vartheta_{M,int} \leq 37.5$ K while cooling down. In every test run, the offset $\Delta\vartheta_{M,int}$ is compensated at a few load cycles and has little significant influence on the further levels as well as trends of the bulk temperature curves and friction curves measured before the interruption.

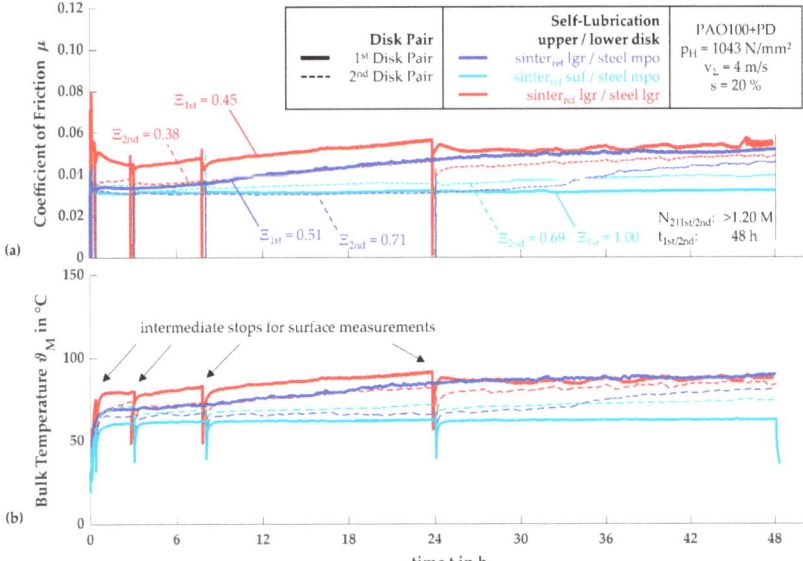

Figure 14. Friction curves (**a**) and bulk temperature curves (**b**) of the long-term tests with disk pairs of a longitudinally ground or mechanically polished steel disk and a longitudinally ground or superfinished sinter$_{ref}$ disk under self-lubrication.

3.2.3. Investigation of the Improvement of the Frictional Behavior in Self-Lubricating Contacts through Solid Lubricant Additives in the Sinter Material

Additionally to the measurements made with the sinter$_{ref}$ material, tests with an oil-impregnated and a non-impregnated solid lubricant alloyed sinter$_{sl}$ material are made. Figure 15 shows the measured friction curves and the stability coefficients during long-term tests with disk pairs of a mechanically polished steel disk and variants of the longitudinally ground sinter disk in terms of a non-oil-impregnated sinter$_{ref}$ or sinter$_{sl}$ disk for investigations under once-lubrication as well as an oil-impregnated sinter$_{ref}$ or sinter$_{sl}$ disk for investigations under self-lubrication.

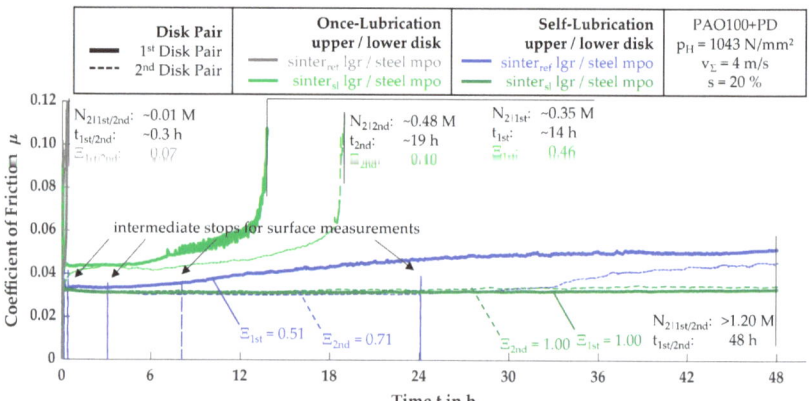

Figure 15. Friction curves of the long-term tests with disk pairs of a mechanically polished steel disk and a longitudinally ground sinter$_{ref}$ or sinter$_{sl}$ disk under once-lubrication and self-lubrication.

Both tests with the once-lubricated steel-sinter$_{ref}$ disk pairs show unstable operating behavior with $\Xi = 0.07$, leading to termination after just a few load cycles of the upper disk. The operating behavior obtained with the once-lubricated EHL tribosystems with a non-oil-impregnated sinter$_{sl}$ disk is characterized by an initial friction curve level of $\mu \approx 0.044$, which is slightly higher than the friction level of the EHL tribosystem with a disk pair of a mechanically polished steel disk and a longitudinally ground, oil-impregnated sinter$_{ref}$ disk under self-lubrication. During the initial sequence of <7 h (i.e., $N_{2|1st}$ and $N_{2|2nd}$ < 0.15 M) stable operating behavior is obtained. After that, μ increases in the test with the 1st disk pair with a steep gradient and demonstrates unstable operating behavior, promptly reaching the termination criterion after t $\approx$ 14 h at $N_{2|1st} \approx 0.35$ M. The test with the 2nd disk pair reaches unstable operating behavior, and thus its termination limit is after a longer period of increasing μ, after t $\approx$ 19 h at $N_{2|2nd}$ < 0.50 M. When additionally impregnating the sinter$_{sl}$ disk with PAO100+PD, the solid lubricant additive consistently results in low levels of μ over the whole measurement time. Additionally, it exhibits repeatably larger load cycle ranges without any increase in μ, thus demonstrating the most stable operating behavior with $\Xi = 1.00$. The difference between the tests with the two disk pairs of $\Delta\mu_{max}$ < 0.002 is the lowest when compared to all other tests.

3.2.4. Determination of the Performance of Self-Lubricating Contacts under Higher Load and Speed

Based on the long-term tests under reference operating conditions, maximum load tests are carried out at both higher v_Σ and higher p_H. The variation in kinematics is tested with the promising self-lubricating EHL tribosystems with a polished steel disk and, as the best-performing sinter material variant evaluated, a longitudinally ground sinter$_{sl}$ disk. Figure 16 shows the friction curves of the long-term tests with disk pairs of a mechanically polished steel disk and a longitudinally ground sinter$_{sl}$ disk under self-lubrication at v_Σ

and p_H, each at a higher level compared to the reference operating conditions shown in Sections 3.2.1–3.2.3.

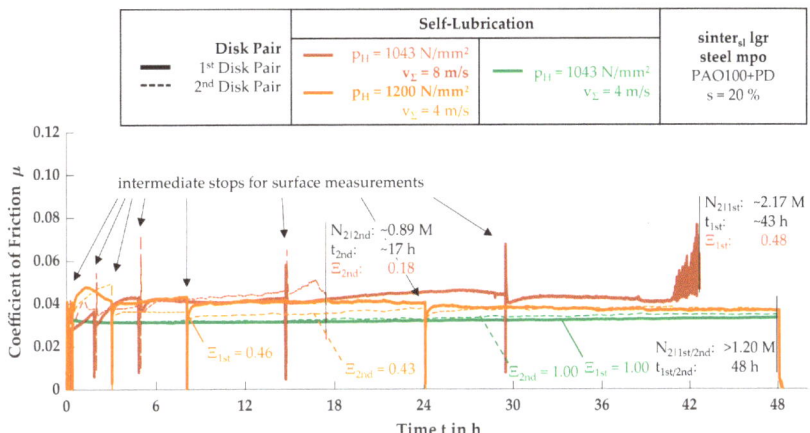

Figure 16. Friction curves of the long-term tests at different v_Σ or p_H levels with disk pairs of a mechanically polished steel disk and a longitudinally ground sinter$_{sl}$ disk under self-lubrication.

When the total speed is increased to $v_\Sigma = 8$ m/s (equivalent to a maximum of $N_{2|1st}$ and $N_{2|2nd} > 2.4$ M for a full 48 h test run), the tests with the 1st and 2nd steel–sinter$_{sl}$ disk pair reach their termination limits at different times, after $t \approx 43$ h at $N_{2|1st} \approx 2.17$ M and after $t \approx 17$ h at $N_{2|2nd} \approx 0.89$ M due to exceeding $\vartheta_{M,max} = 140$ °C. Due to the higher v_Σ and, therefore, the sliding velocity, the measured CoF of $\mu \approx 0.043$ is comparably higher than at the lower sum velocity of $v_\Sigma = 4$ m/s. Whereas the 1st disk pair shows stable operating behavior with $\Xi = 1.00$, the operating behavior of the earlier terminating 2nd disk pair is evaluated as metastable with $\Xi = 0.18$. The difference between the tests with the two disk pairs is $\Delta\mu_{max} < 0.04$. Nevertheless, the test with the 1st disk pair shows a 244% higher lifetime than the test with the 2nd disk pair.

At tests at higher pressure $p_H = 1200$ N/mm^2, the two tests with a steel–sinter$_{sl}$ disk pair initially differ from each other more ($\Delta\mu_{max} = 0.011$) than after further test progression, which leads to a measured μ in both tests at the same level ($\Delta\mu_{max} \approx 0$) of $\mu = 0.038$. The overall trend shows initially metastable operating behavior with $\mu = 0.040$, which changes to a more stable operating behavior with a decreasing μ after increasing load cycles, N. Although the frictional power is higher, both tests reach the measurement time limitation at $N_{2|1st}$ and $N_{2|2nd} > 1.2$ M without termination.

3.2.5. Evaluation of the Functionality of Self-Lubrication with Tooth Flank Similar Surfaces

Figure 17 shows the friction curves measured during the long-term tests under self-lubrication with disk pairs of an axially ground steel disk and a sinter$_{ref}$ as well as a sinter$_{sl}$ material variant of an axially ground, oil-impregnated sinter disk. The axially ground surface structure orientation represents the typical surface structure of cylindrical gear tooth flank surfaces.

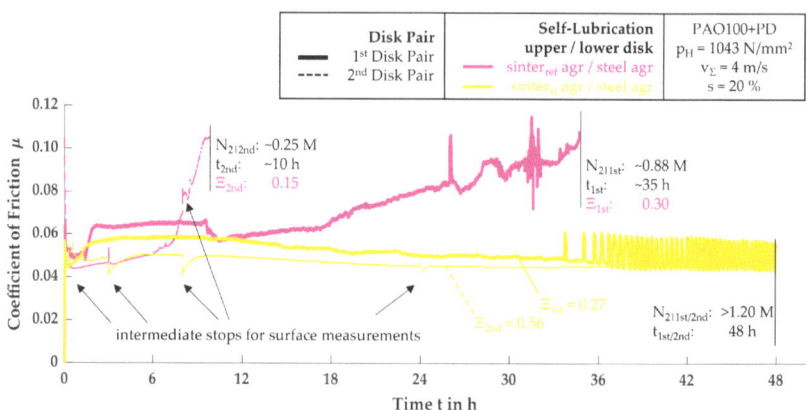

Figure 17. Friction curves of the long-term tests with axially ground disk pairs of a steel disk and a sinter$_{ref}$ or sinter$_{sl}$ disk under self-lubrication.

In both variants of the EHL tribosystem with axially ground sinter$_{ref}$ and sinter$_{sl}$ material, the measured friction curves of all tests are consistently higher than the measured friction curves of the tests with the reference EHL tribosystem with longitudinally ground surface structures. The two tests with the sinter$_{ref}$ disk both terminate due to exceeding the upper limits of $\vartheta_{M,max}$ after different times with sequences of short-term stable, metastable, and unstable operating behavior, approximately after $t \approx 35$ h at $N_{2|1st} \approx 0.88$ M and after $t \approx 10$ h at $N_{2|2nd} \approx 0.25$ M of the upper disk. The test with the 1st disk pair shows stabilization of operating behavior after an initial increase in μ. This is demonstrated by the formation of plateaus in the measured friction curves. However, there is a brief decrease in μ after $t > 9$ h at $N_{2|1st} \approx 0.20$ M, followed by an increase of μ with stronger metastable operating behavior from $t \geq 10$ h at $N_{2|1st} \approx 0.22$ M until the test run terminates, which leads to an overall stability coefficient of $\Xi = 0.30$, characterizing a metastable operating behavior, although the system is terminated. The test with the 2nd disk pair shows a continuous increase in μ and, with this, an overall stability coefficient of $\Xi = 0.15$, characterizing an unstable operating behavior. In both tests with the axially ground disk pair with a sinter$_{sl}$ disk, the solid lubricant additive of sinter$_{sl}$ proves advantageous. In both test runs, the measured levels of μ are lower compared to the tests with the axially ground disk pairs with a sinter$_{ref}$ disk, resulting in both the 1st and the 2nd test lasting until the end of the designated test time of $t = 48$ h at $N_{2|1st}$ and $N_{2|2nd} > 1.2$ M. Generally, a temporarily decreasing level of μ can be observed in both tests, which leads to the stability coefficients $\Xi = 0.27$ and $= 0.56$. During the 1st test only, the operating behavior changed from a stable to a metastable status, with rapidly increasing and then decreasing trends of μ starting from $t \approx 34$ h at $N_{2|1st} \approx 0.83$ M, with the decreasing trend continuing and the experiment not having to be terminated.

3.2.6. Summary of the Long-Term Tests

Table 6 shows the evaluated data of the long-term tests conducted at the reference operating conditions (see Sections 3.2.1–3.2.3 and 3.2.5) as well as at operating conditions at increased load and increased sum velocity (see Section 3.2.4). The lifetime reached by each disk pair is shown in percentage of the reached load cycles $N_{2|1st}$ and $N_{2|2nd}$ compared to the aimed load cycle $N_{2|48h}$ after $t = 48$ h. For the tests with $v_\Sigma = 4$ m/s, the load cycles of a test without termination are $N_{2|48h} > 1.2$ M, and for the tests with $v_\Sigma = 8$ m/s, $N_{2|48h} > 2.4$ M.

Table 6. Mean friction $\bar{\mu}_{1st}$ and $\bar{\mu}_{2nd}$, mean bulk temperature $\bar{\vartheta}_{M|1st}$ and $\bar{\vartheta}_{M|2nd}$, relative load cycles of the upper disk $N_{2|1st}$ and $N_{2|1st}$, and stability coefficients Ξ_{1st} and Ξ_{2nd} of all long-term tests and maximum difference of friction $\Delta\mu_{max}$ between the long-term tests with the 1st and 2nd disk pairs.

Lubrication Method Upper/Lower Disk $p_H = 1043$ N/mm² $v_\Sigma = 4$ m/s $s = 20\%$	$\bar{\mu}_{1st}$ $\bar{\mu}_{2nd}$	$\Delta\mu_{max}$	$\bar{\vartheta}_{M|1st}$ $\bar{\vartheta}_{M|2nd}$ in °C	$N_{2|1st}/N_{2|48h}$ $N_{2|2nd}/N_{2|48h}$ in %	Ξ_{1st} Ξ_{2nd}
once-lubrication sinter_ref lgr/steel mpo	0.094 0.052	0.059	92 52	<1 <1	0.07 0.07
once-lubrication sinter_sl lgr/steel mpo	0.051 0.047	0.057	93 86	≈29 ≈40	0.46 0.48
injection lubrication steel lgr/steel lgr	0.047 0.047	0.005	56 56	100 100	0.76 0.95
self-lubrication sinter_ref lgr/steel lgr	0.051 0.043	0.013	85 78	100 100	0.45 0.38
self-lubrication sinter_ref lgr/steel mpo	0.045 0.035	0.020	82 70	100 100	0.51 0.71
self-lubrication sinter_ref suf/steel mpo	0.031 0.036	0.008	62 70	100 100	1.00 0.69
self-lubrication sinter_sl lgr/steel mpo	0.032 0.033	0.002	65 68	100 100	1.00 1.00
self-lubrication sinter_ref agr/steel agr	0.073 0.057	0.041	110 92	≈73 ≈21	0.30 0.15
self-lubrication sinter_sl agr/steel agr	0.053 0.047	0.010	91 87	100 100	0.27 0.56
self-lubrication [1] sinter_sl lgr/steel mpo	0.045 0.039	0.011	86 87	100 100	0.46 0.43
self-lubrication [2] sinter_sl lgr/steel mpo	0.042 0.042	0.016	117 116	≈90 ≈37	0.48 0.18

[1]: operating condition variation at $p_H = 1200$ N/mm²; [2]: operating condition variation at $v_\Sigma = 8$ m/s.

None of the disk pairs of the two disk pair variants tested under once-lubrication reach the aimed lifetime, whereas the disk pair with a sinter_ref disk shows a negligible lifetime with unstable operating conditions, and the disk pair with a sinter_sl disk shows a lifetime of minimum 29% to maximum 40% of the aimed lifetime with a stability coefficient of Ξ_{1st} and $\Xi_{2nd} > 0.40$, leading to a stable operating condition according to Ebner [25]. Further, seven of the eleven disk pair variants reach the aimed lifetime of $N_{2|48h} > 1.2$ M and >2.4 M after t = 48 h, even the disk pair tested at an increased applied load. All the disk pairs reaching the aimed lifetime operate in stable operating conditions, except the metastable operating 2nd disk pair with the both longitudinally ground sinter_ref and steel disks and the 1st disk pair with the both axially ground sinter_sl and steel disks. The lowest friction results from the disk pairs with a mechanically polished steel disk and a superfinished sinter_ref disk or a longitudinally ground sinter_sl disk, with $\bar{\mu}_{1st} = 0.031$ and $\bar{\mu}_{2nd} = 0.036$, and $\bar{\mu}_{1st} = 0.032$ and $\bar{\mu}_{2nd} = 0.033$, respectively.

4. Discussion

In the following section, the influences of the material, the surface finish, and operating conditions on the self-lubricating EHL tribosystem are discussed.

4.1. Influence of the Material on the Operating Behavior

Tests were conducted on different material pairings, i.e., steel–steel, steel–sinter, and sinter–sinter material disk pairs. To evaluate the operating behavior of the EHL tribosystem according to the considered material combination, the measured friction and bulk temperature curves of the friction curve tests (see Section 3.1) and the long-term tests (see Section 3.2) as well as surface measurements are analyzed. For this, Figure 18a–d shows the surface alteration of the upper disk of the 1st disk pair after the 2nd friction curve test run compared to the new surfaces. Additionally, the absolute and relative difference of the arithmetic mean surface roughness ΔRa before and after the test is determined.

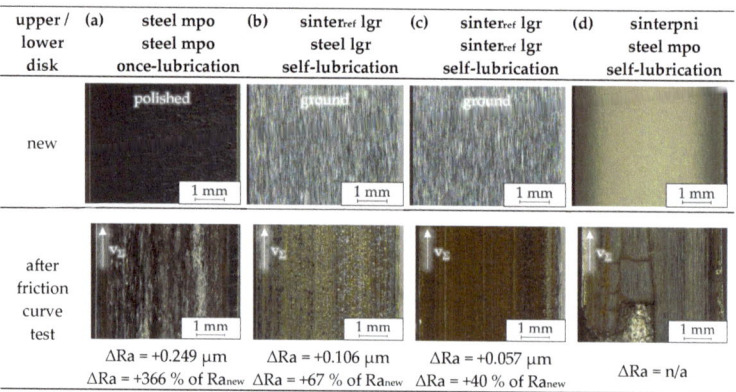

Figure 18. Surface images and absolute and relative difference of arithmetic mean roughness ΔRa of the upper disk surface of the 1st disk pair before the friction curve tests and after the 2nd test run.

4.1.1. EHL Tribosystem without Oil-Impregnated Sinter Disks under Once-Lubrication and with One Oil-Impregnated sinter$_{ref}$ Disk under Self-Lubrication

The tests with once-lubricated disk pairs with two steel disks or with one steel and one sinter$_{ref}$ disk without oil impregnation but with the initially added oil quantity V_{init} could not be fully completed in both friction curve (see Figure 6a,b) and long-term tests (see Figure 15), due to continuously increasing μ and ϑ_M reaching the termination limits after a comparatively short number of load cycles (see Tables 5 and 6). With the failure of the once-lubricated tribosystem without an oil-impregnated partner, the functionality of self-lubrication due to extruding oil of the pores supplementing the initial lubricant quantity V_{init} could be verified as expected and could be separated from a possible presumption that the measured frictional behavior with at least one oil-impregnated specimen is achieved just by the PAO100 oil with its PD additives. Further, the surfaces of the EHL tribosystems under self-lubrication via oil-impregnated sinter disks show less severe surface alteration compared to the once-lubricated EHL tribosystems (see Figure 18a,b), which show heavy alteration characterized by abrasively and/or adhesively altered regions of the initially very smooth polished steel disk surface after the comparatively short test runs. The heavy alteration is also verified by a significant relative increase of surface roughness of $\Delta Ra = 366\%$ compared to the moderate increase of surface roughness of $\Delta Ra = 67\%$ of the self-lubricating disk pair with one longitudinally ground, oil-impregnated sinter$_{ref}$ disk. With this, an improvement of the heavy mixed lubrication regime by thermal expansion-, elastic deformation-, and centrifugal force-driven extruding oil out of the material structure into the tribocontact (see [25]) can be assumed, which can increase the lubricant film thickness to a sufficient film thickness (see [27]) to further separate the mating surfaces, which consequently results in less surface alteration.

To further evaluate the operating behavior due to surface alteration analysis during the long-term tests, surface measurements were taken at intermittent intervals (see Section 2.3). Figure 19 shows the surface images sequentially taken during the long-term test with the 1st disk pair with the longitudinally ground steel and sinter$_{ref}$ disk pair under self-lubrication. Initially, nearly no pores are optically visible in the 2D images of the 1 mm² square area in the center of the disk surface. However, immediately after the first load cycles, there is an optically recognizable increasing change in the surface appearance, with increasing visible pores in terms of number and size. By the end of the test, the surface appearance is characterized by regions smoothed due to wear but also longitudinally oriented marks scattered with numerous pores as qualitatively determined. Areas around the pores appear less stressed, indicating local oil extrusion and possible locally improved lubrication regimes, which was also seen in the lubricant film investigation with an optical ball-on-disk tribometer by Ebner [47]. To verify the subjective identification of pores based on the 2D representation of the surface images using simultaneously measured 3D data, the depth information from the 3D topography measurement of the surface can be used. Round shaped areas deeper than the reference level in the form of holes in the material surface represent potential oil extruding pores connected to the inner oil-storing pore structure. Figure 20 shows the topography measurements of the sinter$_{ref}$ disk of the 1st disk pair with a longitudinally ground steel and a longitudinally ground sinter$_{ref}$ disk before the test, after $N_2 = 2.5$ k, and at the end of the test (here: $N_2 > 1.2$ M). The sequential measurements show that initially, even with the grinding marks, pores can be detected through round holes that are deeper than the marks. Thus, initially there are a few randomly distributed pores on the unworn surface. After the first load cycles, the number of identifiable pores increases slightly. In addition, an influenced area of oil transported to the surface can be seen around the pores, where the initial grinding structure is still visible, indicating locally relatively low solid contact ratios and ongoing abrasion. By the end of the test, the number of pores has increased slightly again, with the diameter of the pores appearing larger. The surface seems to be smoothed due to alteration, but the initially uniform grinding direction remains over the whole test.

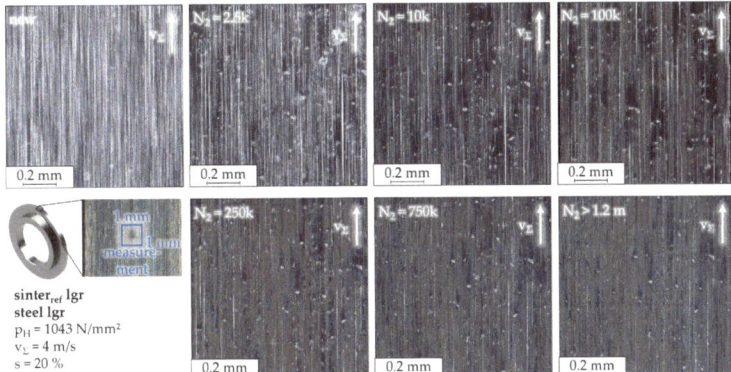

Figure 19. 2D surface images of the upper disk surface of the 1st disk pair with a longitudinally ground steel and sinter$_{ref}$ disk sequentially made during the long-term tests under self-lubrication.

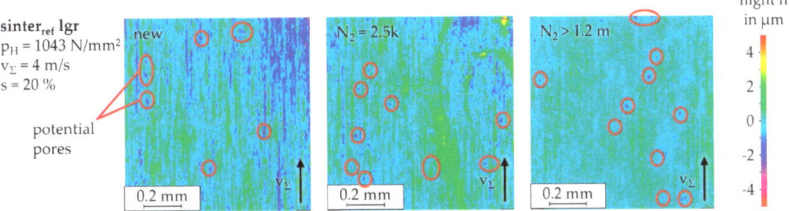

Figure 20. 3D surface measurement of the upper disk surface of the 1st disk pair with a longitudinally ground steel and sinter$_{ref}$ disk sequentially made during the long-term tests under self-lubrication.

4.1.2. EHL Tribosystem with One Oil-Impregnated sinter$_{pni}$ Disk under Self-Lubrication

The friction curve tests with sinter$_{pni}$ were automatically and abruptly terminated by the test rig acceleration detector after a few load cycles (see Section 3.1.3). Further, Figure 18d shows the surface images of the sinter$_{pni}$ disk before and after the friction curve test. Before the test, the images of the sinter$_{pni}$ disk surfaces of the new disks indicate a low to no presence of surface pores that enable the oil to flow across the surface, thus enabling the self-lubrication principle, leading to predicted failure of the test. The failure presumably resulted from the surface pores covered by the compound layer. Figure 21a,b shows the sectional images of a new longitudinally ground sinter$_{ref}$ and a sinter$_{pni}$ disk and verifies the presumption of the covered near-surface pores by the compound layer, which prevents oil flow to the surface, resulting in insufficient support for the lubricant film and reduced friction-reducing effects as with, e.g., sinter$_{ref}$ disk pairs. As a consequence, the compound layer delaminated during the friction curve tests (see image of sinter$_{pni}$ disk after friction curve test in Figure 18d), which led to heavy surface damage, causing significant vibration.

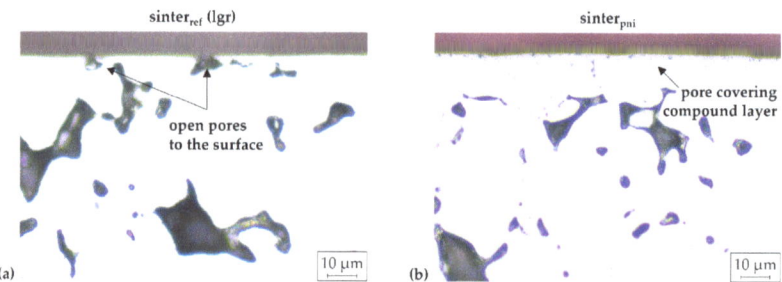

Figure 21. Sectional image of the near-surface porous structure of a longitudinally ground sinter$_{ref}$ disk (**a**) and a sinter$_{pni}$ disk without surface finish (**b**).

4.1.3. EHL Tribosystem with Two Oil-Impregnated sinter$_{ref}$ Disks under Self-Lubrication

The test with self-lubricating disk pairs of two longitudinally ground, oil-impregnated sinter$_{ref}$ material resulted in no improvement of tribological performance due to mainly higher levels of μ_{max} and $\vartheta_{M,max}$ in all tests with both disk pairs as well as higher differences of $\Delta\mu_{max}$ and $\Delta\vartheta_{M,max}$ between the 1st and the 2nd tests with the 1st disk pair compared to the reference EHL tribosystem with just one oil-impregnated sinter$_{ref}$ disk paired with a steel disk (see Figure 7a,b and Figure 8a,b, Tables 5 and 6). Surface alteration of both disk pair variants, the sinter$_{ref}$–sinter$_{ref}$ disk pair (see Figure 18c) and the reference steel–sinter$_{ref}$ disk pair (see Figure 18b), can be seen in a subjectively determined local increase of pore numbers, pore size, and regions with tribofilms, as well as an objectively measured relative increase of $\Delta Ra = 40\%$ and $\Delta Ra = 67\%$, respectively. A disk pair comprising two oil-impregnated sinter disks exhibits less benefits in tribological behavior, i.e., primarily repeatable low levels of μ and ϑ_M, than a disk pair with one steel disk and one sinter disk. Consequently, the latter disk pair is considered the most effective. This can be due to a one-sided flow of the oil extruding from the surface for EHL lubricant film formation as well as intrusion back from the tribocontact under high hydrodynamic pressure into the porous material. Presumably because of the back-intrusion in permeable bodies, a smaller lubricant film thickness-reducing effect dominates the tribocontact compared to the tribocontact within an EHL tribosystem with two porous structures and thus a two-sided oil intrusion with a higher total permeability of the tribocontact, which led in Ebner's [25] calculations of self-lubricating EHL tribosystem configurations with increasing permeability to a decreasing lubricant film thickness.

4.1.4. EHL Tribosystem with One Non-Impregnated sinter$_{sl}$ Disk under Once-Lubrication or One Oil-Impregnated sinter$_{sl}$ Disk under Self-Lubrication

Sinter$_{sl}$ material with MoS$_2$ and WS$_2$ solid lubricant additives showed in once-lubricated long-term tests at the reference operating conditions compared to the unstable running once-lubricated steel–steel tribosystem a longer lifetime up to $N_2 = 0.35$ M at moderate operating conditions with at least half of the lifetime running with stable operating behavior before failing, with decent increasing friction and bulk temperatures (see Figure 15). With this, the expected solid friction reducing influence of the solid lubricant in the sinter material is proven. When impregnating the sinter$_{sl}$ material additionally with oil, the effect of solid friction reduction via a solid lubricant as well as liquid lubrication via extruded oil out of the pores are combined, which repeatably leads to the lowest levels of μ and ϑ_M measured, with the lowest difference between the tests with two identical disk pair specifications, and continuously stable, quasi-stationary operating behavior (see Table 6).

4.2. Influence of the Surface on the Operating Behavior

Test were also conducted on different surface finish pairings, i.e., longitudinally and axially ground, superfinished, and mechanically polished. To evaluate the influence of surface finishes on the frictional behavior, stability of the operating behavior, and lifetime behavior of self-lubricating rolling-sliding contacts with oil-impregnated sintered material, again the measured friction and bulk temperature curves of the friction curve tests (see Section 3.1) and the long-term tests (see Section 3.2) as well as surface measurements are analyzed. For conventionally lubricated EHL tribosystems, the direct link between surface roughness and frictional behavior is commonly known. For self-lubricating EHL tribosystems with oil-impregnated sintered materials, the study suggests that smoother surfaces may result in lower friction, but surface finishing can affect the possible deviations of surface porosity and thus the prevailing oil flow within the tribocontact. This, in turn, affects the global and local lubrication and friction regime. For this, Figure 22a–c shows the surface alteration of the upper disk of the 1st disk pair after the 2nd friction curve test run compared to the new surfaces, and the absolute and relative difference of the arithmetic mean surface roughness ΔRa before and after the test is determined.

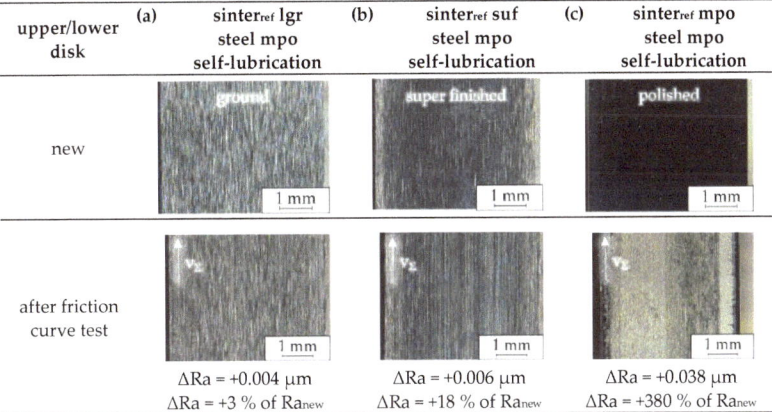

Figure 22. Surface images and absolute and relative difference of arithmetic mean roughness ΔRa of the upper disk surface of the 1st disk pair before the friction curve tests and after the 2nd test run.

4.2.1. EHL Tribosystem with Ground Steel Disk and Sinter Disk Surfaces under Self-Lubrication

Tests with ground EHL tribosystem partners were conducted with longitudinally and axially ground disk surfaces.

The tests with the reference tribosystem with longitudinally ground steel and sinter$_{ref}$ disks show in the friction curve tests a typical tribological behavior of injection-lubricated rolling-sliding contacts, with a transition from a linear increase of the coefficient of friction at a low slip ratio to a thermal regime with decreasing coefficient of friction at a higher slip ratio, as well as lower values of μ and higher ϑ_M with increasing v_Σ (see Figure 3a,b and Figure 4a,b). All operating points can be tested in both test runs with the 1st disk pair as well as with 2nd disk pair. In long-term tests under reference conditions, the reference tribosystem (see Figure 14) shows slightly higher μ and ϑ_M compared to injection-lubricated disk pairs (see Figure 13), which can be caused by less cooling via convection by the surrounding external oil as well as lower oil amounts for sufficient lubricant film formation. However, the reference tribosystem reaches the aimed end of the test at $N_2 > 1.2$ M without termination. The tests of the self-lubricating EHL tribosystem with an axially ground steel and sinter$_{sl}$ disk with both surfaces similar to gear tooth flank surfaces could be completed without termination (see Figure 17) as well as with a proper tribological performance with decreasing levels of friction with test progress and a mostly stable to metastable operating behavior. The EHL tribosystem with a steel disk and a sinter$_{ref}$ disk shows severe tribological performance with a metastable to unstable operating behavior, leading to friction and bulk temperature increase with test progress and, thus, termination at N2 < 0.88 M and < 0.25 M.

4.2.2. EHL Tribosystem with a Smoothed Steel Disk Surface Paired with a Longitudinally Ground Sinter Disk Surface under Self-Lubrication

The variation in the steel disk surface, from ground to polished finish, expectedly results in a significant improvement in the tribological behavior of a self-lubricating contact with a ground oil-impregnated sinter material. This improvement can be seen in the better repeatability of the 1st friction curve tests of the two disk pairs, as well as the lower difference between the 1st and 2nd test with the 1st disk pair and the overall slightly lower levels of μ and ϑ_M compared to the reference tribosystem (see Figure 9a,b and Figure 10a,b and Table 5). Compared to the operating behavior of the injection-lubricated EHL tribosystem in the long-term tests (see Figure 13), the levels of μ and ϑ_M with the self-lubricating disk pair with a polished steel disk and a longitudinally ground sinter$_{ref}$ disk are lower for part of the test duration up to the whole test duration. This might be mainly caused by the improved lubrication regime, due to lower arithmetic mean surface roughness of the contact partners. Because there is no heat convection by an external oil, the friction-induced temperature of the tribosystem increases during operation, which decreases the oil viscosity and thereby increases the influence of the surface roughness of both mating partners on the lubricant film thickness. This consequently leads to an increase of solid friction and thus an increase of total friction measured. This can also be seen in the documentation in Figure 22a of the surface alteration, in which the surface of the ground sinter$_{ref}$ disk shows nearly no alteration in terms of tribofilm formation, pore number, or size increase as well as ΔRa (ΔRa = 3%).

4.2.3. EHL Tribosystem with a Smoothed Steel Disk Surface Paired with a Smoothed sinter$_{ref}$ Disk Surface under Self-Lubrication

The friction curve test results with an additional polished steel smoothed surface of the sinter disk also show the trend of an improved tribological behavior of a reduction in surface roughness by superfinishing or polishing, resulting in lower measured levels of μ and ϑ_M compared to tests with conventionally ground surfaces (see Figure 9a,b, Figure 10a,b, Figure 11a,b and Figure 12a,b and Table 5). In the following sections, the influence of the finish variants of mechanically polishing and superfinishing of the sinter$_{ref}$ disk surface on the operating behavior is evaluated in detail.

Superfinishing of the sinter$_{ref}$ Disk Surface

In addition to the results that the tests with disk pairs with a superfinished sinter$_{ref}$ disk surface show better repeatability of the 1st tests of two disk pairs and lower difference of the measured μ and ϑ_M between the test of the 1st and the 2nd disk pair compared to the reference system as well as the lowest levels of μ and the most stable operating behavior during the long-term tests compared to the other surface variants in Figure 14a,b (see also Table 6), Figure 22b shows the low surface roughness increases with ΔRa = 18%, with an absolute increase of ΔRa = 0.006 μm, which is negligible. Further, Figure 23 shows the 2D image documentation of the surface alteration during the long-term tests at reference conditions with the mechanically polished steel and superfinished sinter$_{ref}$ disk pair. Similar to the tribosystem with two longitudinally ground disks, at the initial state, only a few surface pores are visibly recognizable. Additionally, there is a less uniform direction of the surface structure obtained. After a few load cycles, the surface pores are numerous and increase in size, whereas the initial surface structure shows for a longer time less global alteration. Furthermore, the surface area surrounding the pores shows significant deviations locally in the pore-free surface regions. The initial state of the altered porosity and surface modification after the first load cycles persists, with a similar appearance throughout the whole test. Figure 24 shows, additionally, the topography measurements of the sinter$_{ref}$ disk of the 1st disk pair with a mechanically polished steel disk and a longitudinally ground sinter$_{ref}$ disk before the test, after N_2 = 2.5 k, and at the end of the test. The measurements confirm the observation, as evidenced by the identifiable pores, verified by depth information, which initially increase slightly and then remain constant over time. The less severe alteration of the disk surfaces can be interpreted as another result of the comparably best tribological behavior evaluated within the surface variants, which again verifies Ebner's findings [25].

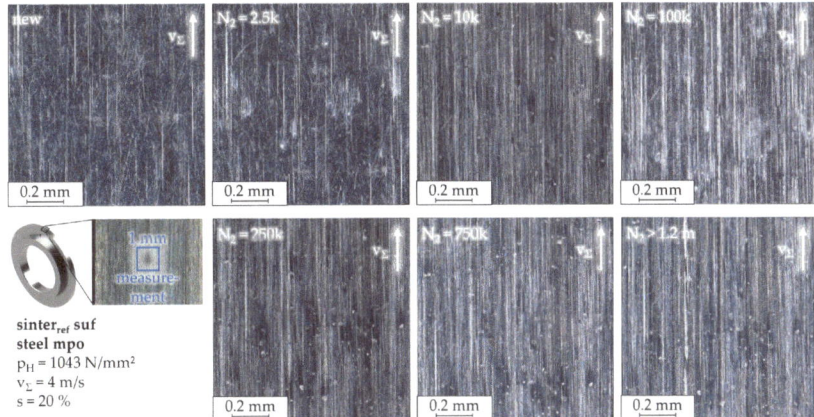

Figure 23. 2D surface images of the upper disk surface of the 1st disk pair with a mechanically polished steel and a superfinished sinter$_{ref}$ disk sequentially made during the long-term tests under self-lubrication.

With the surface images and topography measurements of the best-performing disk pair surface combinations, it can be stated that in tests that show good tribological behavior and do not lead to premature termination due to metastable or unstable operating behavior, the porosity becomes more uncovered initially and, thus, is more exposed, which can be interpreted as a running-in process of oil-impregnated sinter materials with surface finish. Subsequently, once there is sufficient oil exchange between the inner material structure and the tribocontact through the opened surface pores, there are no significant changes in surface porosity.

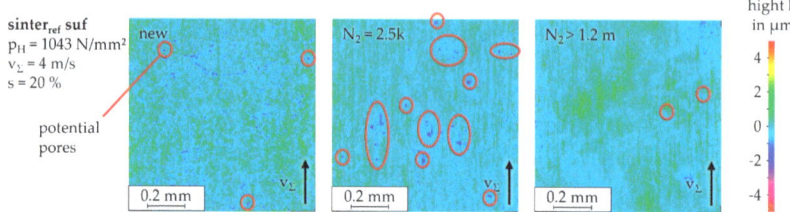

Figure 24. 3D surface measurement of the upper disk surface of the 1st disk pair with a mechanically polished steel and a superfinished sinter$_{ref}$ disk sequentially made during the long-term tests under self-lubrication.

Mechanical Polishing of the sinter$_{ref}$ Disk Surface

In contrast, the friction curve test with the comparably smoothest sinter$_{ref}$ disk, a mechanically polished surface, shows greater difference between the 1st and 2nd tests with the 1st disk pair and a worse repeatability of the tests with two disk pairs compared to an EHL tribosystem with a mechanically polished steel disk and a longitudinally ground or superfinished sinter$_{ref}$ disk. It is postulated that the manual polishing process results in inconsistencies in abrasive material removal and pore closure. While this facilitates the requisite smoothing of the surfaces and thus a comparatively high increase of the minimum lubricant film thickness, it also gives rise to unexpected and divergent measurement results due to local surface porosity emphasis. Further, the additional effort of mechanically smoothing the sinter disk surface is less useful when evaluating the surface alteration shown in Figure 22c, which shows a significant optical change of the surface with a measured increase of $\Delta Ra = 380\%$. With this, the surface finish of mechanically polishing the sinter disks was excluded for the subsequently run long-term tests.

4.3. Influence of the Operating Conditions on the Operating Behavior

Tests with disk pairs of a mechanically polished steel disk and a longitudinally ground sinter$_{ref}$ disk were also conducted with a comparably higher load or sum velocity. To evaluate the operating behavior of the EHL tribosystem according to the prevailing operating condition, the measured friction and bulk temperature curves of the friction curve tests (see Section 3.1) and the long-term tests (see Section 3.2) as well as surface measurements are analyzed.

4.3.1. EHL Tribosystem under Comparably Higher Load

The increase in transmitted power during the long-term tests at higher loads initially exhibits areas with metastable operating behavior, which transitions into stable operating behavior after further load cycles in both tests of the two disk pairs (see Figure 16). Initially, there is a measurable difference of µ between the tests of the two disk pairs, which gradually converges to nearly identical stable operating behavior at the same level. Thus, the influence of load increase in long-term tests results in a slightly higher level of both µ and ϑ_M compared to the tests at the reference load (see Table 6), but a functional tribosystem is still maintained throughout the entire intended test. The increased measured friction is primarily attributed to the higher specific friction in the tribocontact at higher pressure.

4.3.2. EHL Tribosystem under Comparably Higher Sum Velocity

In contrast to the increased load, the impact of an increased speed of $v_\Sigma = 8$ m/s exerts a more pronounced influence on the operating behavior of the self-lubricating EHL tribosystem. During the tests shown in Figure 16, the severe operating condition results in varying lifetimes, and in both tests with the two disk pairs, the test runs fail due to the emergence of unstable operating behavior at disparate times. The primary cause of this phenomenon is not only the higher specific friction power but also the increased dissipated energy generated over time, which results in increased system heating and a concomitant decrease of oil viscosity. This decrease of viscosity also reduces the lubricant film thickness, potentially increasing solid contacts. Further, Figures 25 and 26 show, through 2D images and 3D topography measurements, the surface alteration of the sinter$_{sl}$ disk of the 1st long-term tests at the increased sum velocity of v_Σ = 8 m/s with the disk pairs of a polished steel and a longitudinally ground sinter$_{sl}$ disk pair that terminated before the designated test time. Visually, there is only a slight alteration in the appearance of the ground surface within the first load cycles up to about $N_2 \approx$ 100k. However, at N_2 = 250k, corresponding to an abrupt decrease in μ, the porosity increases significantly in both number and size. At all subsequent time intervals when the surface condition was recorded, the increasing porosity correlates with the increasing measured friction until the tribosystem fails, resulting in highly pronounced surface damage characterized by numerous visible and measurable large surface pores as well as, presumably, severe wear.

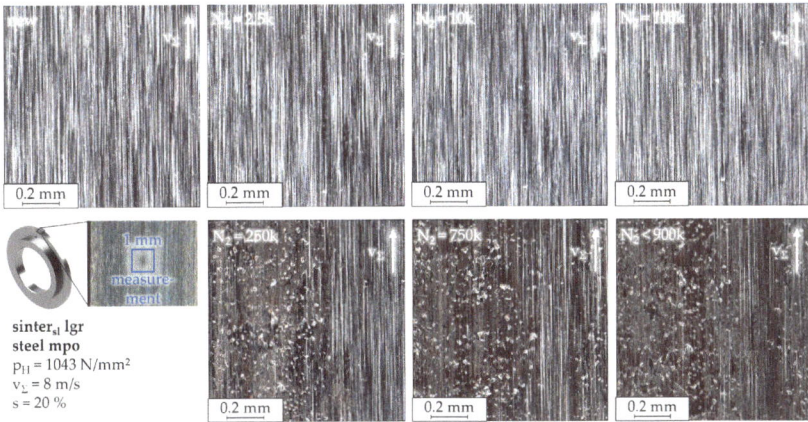

Figure 25. 2D surface images of the upper disk surface of the 1st disk pair with a mechanically polished steel and a longitudinally ground sinter$_{ref}$ disk sequentially made during the long-term tests at higher v_Σ under self-lubrication.

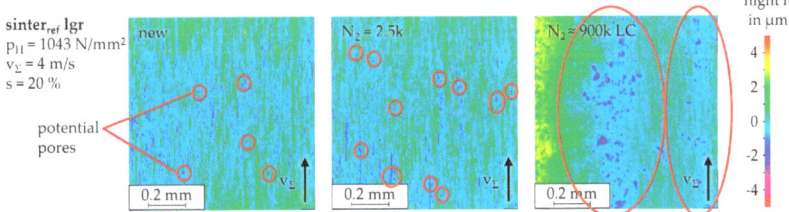

Figure 26. 3D surface measurement of the upper disk surface of the 1st disk pair with a mechanically polished steel and a longitudinally ground sinter$_{ref}$ disk sequentially made during the long-term tests at higher v_Σ under self-lubrication.

With the surface images and topography measurements of the disk pairs that terminated during the long-term tests, it can be stated that in tribosystems that experience more severe tribological stress, due to factors such as the initial limited presence of surface pores or operating conditions that exceed the performance limit of the self-lubricating EHL tribosystem, porosity increases significantly. This, in turn, enables more oil to intrude from the tribocontact back into the inner material structure, reducing the lubricant film thickness. As a result, solid contacts increase, leading to increased friction and wear as well as temperature increase, which reduces oil viscosity, further promoting the formation of thinner lubricant film thickness and enhancing oil intrusion. Under these conditions, the

EHL tribosystem will continue to degrade, leading also to thermal and chemical alteration of the oil with further load cycles, until the EHL tribosystem reaches its limits and fails.

The consistently good measurements of the tribological behavior of the self-lubricating EHL tribosystem at moderate to high reference operating conditions, together with the stable or metastable operating behavior at higher pressures and/or speeds, suggests the suitability of self-lubricated tribosystems for use in various operating modes, including, e.g., intermittent high-load operation followed by less loaded or unloaded cool-down periods.

5. Conclusions

This study investigated the functionality and influence of specific material and surface specifications on the operating behavior of oil-impregnated, self-lubricating sintered rolling-sliding contacts under moderate to high operating conditions. A twin-disk tribometer was used to evaluate the tribological behavior of various material disk pairs and surface finishes. Friction curve tests were performed to measure the coefficient of friction and bulk temperature, while long-term tests were performed to evaluate the lifetime behavior of promising material disk pairs and surface finishes, including friction and bulk temperature measurements. The main findings are as follows:

- A once-lubricated EHL tribosystem without external oil lubrication performs better when a solid lubricant is added to the sintering powder prior to sintering.
- A self-lubricating EHL tribosystem with longitudinally ground surfaces and one oil-impregnated sinter disk exhibits adequate friction behavior and, thus, mostly stable operating behavior during the tests. Two oil-impregnated sinter disks show worse tribological performance.
- Smoothing the surface of the steel disk in the self-lubricating EHL tribosystem improves tribological performance, while finishing both contact partner surfaces by mechanical polishing of the steel disk and superfinishing of the sinter disk provides the best repeatable tribological performance.
- For axially ground surfaces similar to typical gear tooth flanks, the best tribological performance of a self-lubricating EHL tribosystem can be achieved with a solid lubricant addition to the sintered material prior to sintering.
- The surface alteration during the tests with EHL tribosystems with good tribological performance is characterized by a consistent running-in-like increase in pore size and quantity within the initial load cycles and an only slight additional change during the further lifetime.
- Nevertheless, great differences in operating behavior and surface alteration are observed for disk pairs manufactured with identical requirements for mechanical properties and surface conditions.

Systematic studies of material and surface porosity, the resulting permeability, and the influence of manufacturing methods such as heat treatment, surface finishing, etc., on the surface porosity as well as on the sinter material density in near-surface regions are further needed to understand these differences in operating behavior and surface alteration.

The tests of this study were conducted under stationary operating conditions. The results demonstrated that, for a sufficient number of load cycles, self-lubricating EHL tribosystems can operate within specific load ranges, with a highly stable operating behavior and low levels of μ and ϑ_M. A stationary operating application with medium-high loads, such as geared motors for logistic conveyor belts, could be a viable option. Further investigations at adapted model contact tests are required to ascertain the suitability of the self-lubrication technology for promising applications under intermittent operating conditions with peaks of higher load or speeds and operating sequences under low load to cool down, e.g., in machine tools such as hand screwdrivers. In addition, the so-far limited characterization of surface porosity reveals the need for a quantitative method to determine pore size distribution and quantity, allowing for a more objective characterization of surface porosity, as the surface is considered one of the main components of future mathematical models for the design or prediction calculations of self-lubricating EHL tribosystems with oil-impregnated sintered materials.

Author Contributions: Conceptualization, N.S., T.L.; methodology, N.S., T.L.; experiments, N.S.; validation, N.S.; formal analysis, T.L.; writing—original draft preparation, N.S.; writing—review and editing, T.L., K.S.; supervision, T.L., K.S.; project administration, T.L., K.S.; funding acquisition, K.S. All authors have read and agreed to the published version of the manuscript.

Funding: The presented results are based on the research project 03LB3001A, supported by the Federal Ministry for Economic Affairs and Climate Action (BMWK) and supervised by the Project Management Jülich (PtJ). The authors are grateful for the sponsorship and support received from BMWi and Project Management Agency PtJ.

Data Availability Statement: The raw data supporting the conclusions of this article will be made available by the authors on request.

Acknowledgments: The authors would like to express their thanks to the project partners Miba Sinter Group AG for their support with the sinter material, OSK Kiefer GmbH for surface finishing support, as well as Getriebebau NORD GmbH and Hilti AG for their continued support during the project.

Conflicts of Interest: The authors declare no conflicts of interest.

References

1. Ebner, M.; Lohner, T.; Michaelis, K.; Höhn, B.-R.; Stahl, K. Self-Lubricating Gears with Oil-Impregnated Sintered Materials. *Forsch. Im Ingenieurwesen* **2017**, *2*, 13–28.
2. Niemann, G.; Winter, H. *Maschinenelemente 2: Getriebe Allgemein, Zahnradgetriebe—Grundlagen, Stirnradgetriebe*, 2nd ed.; revised; Springer: Berlin/Heidelberg, Germany, 2003; ISBN 978-3-662-11874-0.
3. DIN 51509-1; Selection of Lubricants for Gears; Gear Lubricating Oils. Beuth GmbH: Berlin, Germany, 1976.
4. Bartel, D. *Simulation von Tribosystemen: Grundlagen und Anwendungen*; Postdoctoral Thesis; 1st ed.; Vieweg + Teubner; University of Magdeburg: Wiesbaden, Germany, 2010; ISBN 978-3-8348-1241-4.
5. Conrades, V. Proteins of the Matrix in Tissue Engineered Meniscus. Ph.D. Thesis, Technical University of Munich, München, Germany, 2007; pp. 12–20.
6. Schatt, W. *Pulvermetallurgie: Technologien und Werkstoffe*, 2nd ed.; revised and extended; Springer: Berlin/Heidelberg, Germany, 2007; ISBN 3-540-23652-X.
7. DIN 1850-3; Plain Bearings: Part 3: Sintermetal Bushes. Beuth GmbH: Berlin, Germany, 1998.
8. Roberts, M. *A Study of Oil Circulation in the R4 Spin-Axis Bearing with Sintered Nylon Ball Retainer: Proc. Gyro. Spin-Axis Hydrodynamic*; Massachusetts Institute of Technology: Cambridge, UK, 1966.
9. Fote, A.A.; Slade, B.A.; Feuerstein, S. The behavior of thin oil films in the presence of porous lubricant reservoirs. *Wear* **1978**, *46*, 377–385. [CrossRef]
10. Bertrand, P.A.; Carré, D.J. Oil Exchange between Ball Bearings and Porous Polyimide Ball Bearing Retainers. *Tribol. Trans.* **1997**, *40*, 294–302. [CrossRef]
11. Marchetti, M.; Meurisse, M.-H.; Vergne, P.; Sicre, J.; Durand, M. Lubricant Supply by Porous Reservoirs in Space Mechanisms. In Proceedings of the 26th Leeds-Lyon Symposium on Tribology, Leeds, UK, 14-17 September 1999; Volume 38, pp. 777–785. [CrossRef]
12. Marchetti, M.; Meurisse, M.-H.; Vergne, P.; Sicre, J.; Durand, M. Analysis of oil supply phenomena by sintered porous reservoirs. *Tribol. Lett.* **2001**, *10*, 163–170. [CrossRef]
13. Bertrand, P.A.; Carré, D.J.; Bauer, R. Oil Exchange Between Ball Bearings and Cotton-Phenolic Bail-Bearing Retainers. *Tribol. Trans.* **1995**, *38*, 342–352. [CrossRef]
14. Scheichl, B.; Neacsu, I.A.; Kluwick, A. A novel view on lubricant flow undergoing cavitation in sintered journal bearings. *Tribol. Int.* **2015**, *88*, 189–209. [CrossRef]
15. Morgan, V.T.; Cameron, A. Mechanism of lubrication in porous metal bearings. In Proceedings of the Conference on Lubrication and Wear, London, UK, 1–3 October 1957; Volume 89, pp. 151–157.
16. Dizdar, S. Pitting Resistance of Sintered Small-Module Gears: Proceedings of the institution of Mechanical Engineers. *Part J J. Eng. Tribol.* **2013**, *227*, 1225–1240. [CrossRef]
17. Lipp, K. *Rolling Contact Fatigue of Sintered Steels under Constant and Variable Hertzian Pressure and Sliding*; LBF: Darmstadt, Germany, 1997.
18. Dlapka, M.; Danninger, H.; Gierl, C.; Lindqvist, B. Defining the pores in PM components. *Met. Powder Rep.* **2010**, *65*, 30–33. [CrossRef]

19. Li, X.; Olofsson, U. A study on friction and wear reduction due to porosity in powder metallurgic gear materials. *Tribol. Int.* **2017**, *110*, 86–95. [CrossRef]
20. Balasoiu, A.M.; Braun, M.J.; Moldovan, S.I. A parametric study of a porous self-circulating hydrodynamic bearing. *Tribol. Int.* **2013**, *61*, 176–193. [CrossRef]
21. Manoylov, A.V.; Borodich, F.M.; Evans, H.P. Modelling of Elastic Properties of Sintered Materials. *Proc. R. Soc. A* **2013**, *469*. [CrossRef]
22. Zapf, G. *Handbook of Manufacturing Technology Vol. 1: Handbuch der Fertigungstechnik*; Carl Hanser, München Wien: München, Germany, 1981.
23. Ebner, M.; Lohner, T.; Michaelis, K.; Stemplinger, J.-P.; Höhn, B.-R.; Stahl, K. Self-Lubricated Elastohydrodynamic (EHL) Contacts with Oil-Impregnated Sintered Materials. In *TAE 2016*; Technische Akademie Esslingen: Ostfildern, Germany, 2016.
24. Ebner, M.; Lohner, T.; Weigl, A.; Michaelis, K.; Stemplinger, J.-P.; Höhn, B.-R.; Stahl, K. Hochbelastete und schmierstoffgetränkte Wälzpaarungen aus Sintermaterial ohne externe Schmierstoffzuführung. *Tribol. Schmier.* **2016**, *63*, 22–30.
25. Ebner, M. Self-Lubrication of Highly-Loaded Gear Contacts with Oil-Impregnated Porous Ferrous Metals. Ph.D. Thesis, Technical University of Munich, Munich, Germany, 2021.
26. Ebner, M.; Lohner, T.; Stahl, K. *Konstruktionselemente mit Hertz'scher Punkt- oder Linienlast und Schlupf in der Kontaktfläche ohne äußere Schmierung: DFG Koselleck Final Report*; German Research Foundation: München, Germany, 2019.
27. Ebner, M.; Schwarz, A. *Einfluss Poröser Oberflächen auf die Selbstschmierung: 2. Tribologie-Kolloquium des GfT-Arbeitskreises München*; Gesellschaft für Tribologie e.V.: Jülich, Germany, 2019.
28. Omasta, M.; Ebner, M.; Sperka, P.; Lohner, T.; Krupka, I.; Hartl, M.; Höhn, B.-R.; Stahl, K. Film formation in EHL contacts with oil-impregnated sintered materials. *Ind. Lubr. Tribol.* **2018**, *70*, 612–619. [CrossRef]
29. Ebner, M.; Schwarz, A. *Selbstschmierende hochbelastete Wälzpaarungen: 2. Tribologie-Kolloquium des GfT-Arbeitskreises München*; Gesellschaft für Tribologie e.V.: Jülich, Germany, 2019.
30. Zhang, S.; Li, Y.; Hu, L.; Feng, D.; Wang, H. AntiWear Effect of Mo and W Nanoparticles as Additives for Multialkylated Cyclopentanes Oil in Vacuum. *J. Tribol.* **2017**, *139*, 021607. [CrossRef]
31. Rabaso, P.; Ville, F.; Dassenoy, F.; Diaby, M.; Afanasiev, P.; Cavoret, J.; Vacher, B.; Le Mogne, T. Boundary lubrication: Influence of the size and structure of inorganic fullerene-like MoS2 nanoparticles on friction and wear reduction. *Wear* **2014**, *320*, 161–178. [CrossRef]
32. Srinivas, V.; Rao, C.K.R.; Abyudaya, M.; Jyothi, E.S. Extreme Pressure Properties of 600 N Base Oil Dispersed with Molybdenum Disulphide Nano Particles. *Univers. J. Mech. Eng.* **2014**, *2*, 220–225. [CrossRef]
33. Bakunin, V.N.; Suslov, A.Y.; Kuzmina, G.N.; Parenago, O.P.; Topchiev, A.V. Synthesis and Application of Inorganic Nanoparticles as Lubricant Components ÔÇô a Review. *J. Nanoparticle Res.* **2004**, *6*, 273–284. [CrossRef]
34. Zhu, S.; Cheng, J.; Qiao, Z.; Yang, J. High temperature solid-lubricating materials: A review. *Tribol. Int.* **2019**, *133*, 206–223. [CrossRef]
35. Vazirisereshk, M.R.; Martini, A.; Strubbe, D.A.; Baykara, M.Z. Solid Lubrication with MoS2: A Review. *Lubricants* **2019**, *7*, 57. [CrossRef]
36. Kovalchenko, A.M.; Fushchich, O.I.; Danyluk, S. The tribological properties and mechanism of wear of Cu-based sintered powder materials containing molybdenum disulfide and molybdenum diselenite under unlubricated sliding against copper. *Wear* **2012**, *290*, 106–123. [CrossRef]
37. Dhanasekaran, S.; Gnanamoorthy, R. Dry sliding friction and wear characteristics of Fe–C–Cu alloy containing molybdenum di sulphide. *Mater. Des.* **2007**, *28*, 1135–1141. [CrossRef]
38. *DIN 30910-6*; Sintered Metal Materials: Sint-Material Specifications: Hot-Forged Sintered Steels for Structural Parts. Beuth GmbH: Berlin, Germany, 1990.
39. *DIN EN ISO 5755*; Sintered Metal Material: Specifications. Beuth GmbH: Berlin, Germany, 2022.
40. *DIN EN ISO 13565-1*; Geometrical Product Specifications (GPS)—Surface Texture: Profile Method—Surfaces Having Stratified Functional Properties: Part 1: Filtering and General Measurement Conditions. Beuth GmbH: Berlin, Germany, 1998.
41. Cai, J.; Jin, T.; Kou, J.; Zou, S.; Xiao, J.; Meng, Q. Lucas-Washburn Equation-Based Modeling of Capillary-Driven Flow in Porous Systems. *Langmuir* **2021**, *37*, 1623–1636. [CrossRef] [PubMed]
42. Hamraoui, A.; Nylander, T. Analytical approach for the Lucas-Washburn equation. *J. Colloid Interface Sci.* **2002**, *250*, 415–421. [CrossRef] [PubMed]
43. Lados, D.; Apelian, D.; Semel, F.J. *Open and Closed Porosity in P/M Materials: Measurement and Variation with Density Levels and Sintering Conditions*; Euro PM 2005 Tools for Improving PM; European Powder Metallurgy Association: Prague, Czech Republic, 2005.
44. Lohner, T.; Merz, R.; Mayer, J.; Michaelis, K.; Kopnarski, M.; Stahl, K. On the Effect of Plastic Deformation (PD) Additives in Lubricants. *Tribol. Und Schmier.* **2015**, *62*, 13–24.
45. UK Castrol Ltd. *Castrol Optigear Synthetic PD. . .ES: Product Data Sheet*; BP Europa SE: Hamburg, Germany, 2019.

46. Vojacek, H. Das Reibungsverhalten von Fluiden unter Elastohydrodynamischen Bedingungen. Einfluss der Chemischen Struktur des Fluides, der Werkstoffe und der Makro- und Mikrogeometrie der Gleit/Wälzkörper. Ph.D. Thesis, Technical University of Munich, Munich, Germany, 1984.
47. Ebner, M.; Omasta, M.; Lohner, T.; Sperka, P.; Krupka, I.; Hartl, M.; Michaelis, K.; Höhn, B.-R.; Stahl, K. Local Effects in EHL Contacts with Oil-Impregnated Sintered Materials. *Lubricants* **2019**, *7*, 1. [CrossRef]

Disclaimer/Publisher's Note: The statements, opinions and data contained in all publications are solely those of the individual author(s) and contributor(s) and not of MDPI and/or the editor(s). MDPI and/or the editor(s) disclaim responsibility for any injury to people or property resulting from any ideas, methods, instructions or products referred to in the content.

Article

Numerical Simulations and Experimental Validation of Squeeze Film Dampers for Aircraft Jet Engines

Markus Golek [1], Jakob Gleichner [1], Ioannis Chatzisavvas [2,*], Lukas Kohlmann [2], Marcus Schmidt [3], Peter Reinke [3] and Adrian Rienäcker [1]

[1] Institute for Powertrain and Vehicle Technology, University of Kassel, Moenchebergstr. 7, 34125 Kassel, Germany; markus.golek@uni-kassel.de (M.G.); jakob.gleichner@uni-kassel.de (J.G.); adrian.rienaecker@uni-kassel.de (A.R.)
[2] MTU Aero Engines AG, Dachauer Str. 665, 80995 Munich, Germany; lukas.kohlmann@mtu.de
[3] Faculty of Engineering and Health, University of Applied Sciences and Arts Hildesheim/Holzminden/Goettingen, Von-Ossietzky-Straße 99, 37085 Goettingen, Germany; marcus.schmidt@hawk.de (M.S.); peter.reinke@hawk.de (P.R.)
* Correspondence: ioannis.chatzisavvas@mtu.de

Abstract: Squeeze film dampers are used to reduce vibration in aircraft jet engines supported by rolling element bearings. The underlying physics of the squeeze film dampers has been studied extensively over the past 50 years. However, the research on the SFDs is still ongoing due to the complexity of modeling of several effects such as fluid inertia and the modeling of the piston rings, which are often used to seal SFDs. In this work, a special experimental setup has been designed to validate the numerical models of SFDs. This experimental setup can be used with various SFD geometries (including piston ring seals) and simulate almost all conditions that may occur in an aircraft jet engine. This work also focuses on the inertia forces of the fluid. The hydrodynamic pressure distribution of a detailed 3D-CFD model is compared with the solution of the Reynolds equation including inertia effects. Finally, the simulation results are compared with experimental data and good agreement is observed.

Keywords: squeeze film damper; Reynolds equation; experimental validation; inertia effects; vibrations; rotor dynamics

1. Introduction

Squeeze film dampers (SFDs) are an essential component in today's aircraft engines, serving to attenuate rotor vibrations excited mainly by mass unbalance or geometric imperfections. High-pressure rotors of modern aircraft engines, for instance, are typically supported with low stiffness in the bearing positions. This places modes with high rotor strain energy well above the operational speed range. In this design concept, the rotor must pass through two rigid body modes at low speeds, which requires damping of the vibration amplitudes at the bearing positions as provided by SFDs. In Figure 1, a typical aircraft jet engine is shown.

The oil–pressure distribution of squeeze film dampers, like standard journal bearing, is calculated using the Reynolds equation [1]. This linear partial differential equation can be approximated with closed-form analytical solutions [2–6]. Due to the limitations of these closed-form solutions, numerical approximations of the full Reynolds equation are usually employed using the Finite Element, the Finite Difference or the Finite Volume methods. Approaches using global ansatz functions such as the global Galerkin approach [7–9] have also been used to find approximate solutions of the Reynolds equation. However, their implementation in complex geometries is more complicated than the local Finite Element approaches.

Figure 1. Geared turbofan engine PW1100G-JM [10].

Cavitation in hydrodynamic lubrication plays a significant role in the pressure distribution derived from the Reynolds equation. Several approaches have been used in the literature (see, for example, [11]); however, the mass-conserving algorithms such as this described by Kumar and Booker are well established [12].

The thermal effects in the oil are known to influence the pressure distribution in the oil film and thus the hydrodynamic bearing forces. In the work of Dowson [13], the generalized Reynolds equation was developed. A set of integro-differential equations is solved, including the generalized Reynolds and the energy equation of the oil, to obtain the hydrodynamic pressure distribution as well as the 3D temperature field. A detailed discussion on this topic can also be found in [14]. In [15], the thermal effects in the oil including inertia effects have been developed.

The effect of fluid inertia, although typically neglected for journal bearings, may be significant for SFDs. In [16,17], the temporal inertia terms were included in the Reynolds equation. In [18,19], an approach was developed to include both the temporal and the convective inertia terms in the Reynolds equation. This approach increases the computational effort for the calculation of the pressure distribution in the oil, but may significantly improve the physical accuracy of the calculated pressure.

Dedicated experimental results of SFDs are rather scarce in the literature, despite their widespread use. The pioneering work of San Andres focused on the experimental validation of squeeze film damper models (see, for example, [20–22]). The aim of this work is to develop an experimental device similar to the work of [20], which will be able to reproduce all the kinematic conditions (eccentricities, velocities and accelerations) and all the oil-supply conditions (oil-supply pressures and oil-supply temperatures). With minor modifications, this test rig can be used for any SFD geometry, which may include a circumferential groove and/or oil-supply holes or even more complicated geometries. In addition, the influence of the typical sealing mechanisms (piston rings, o-rings, etc.) can be easily quantified. Therefore, this test rig and the results obtained can be used to validate SFD models. In this paper, efficient thermo-hydrodynamic SFD models developed for rotor dynamic simulations and detailed CFD models have been validated.

The main contributions of this work are summarized below:

- A special experimental setup is presented that can be used to validate the SFD bearing forces for all operating conditions that may occur in an aircraft jet engine.
- The SFD bearing forces obtained from the experimental results are compared with those obtained from the numerical simulations and a good agreement is found.
- The 2D solution of the Reynolds equation is compared with a 3D-CFD solution for thin-film lubrication conditions. The influence of inertia effects is shown to be significant for the specific parameter used.

In Section 2, the theory of thermo-hydrodynamic lubrication is presented, including inertia effects and a mass-conserving cavitation algorithm. In Section 3, the experimental setup and its capabilities are discussed. In Section 4, the results are presented, and finally, in Section 5, the main conclusions of this work are summarized.

2. Hydrodynamic Lubrication in Squeeze Film Dampers

A typical SFD combined with a rolling element bearing is shown in Figure 2.

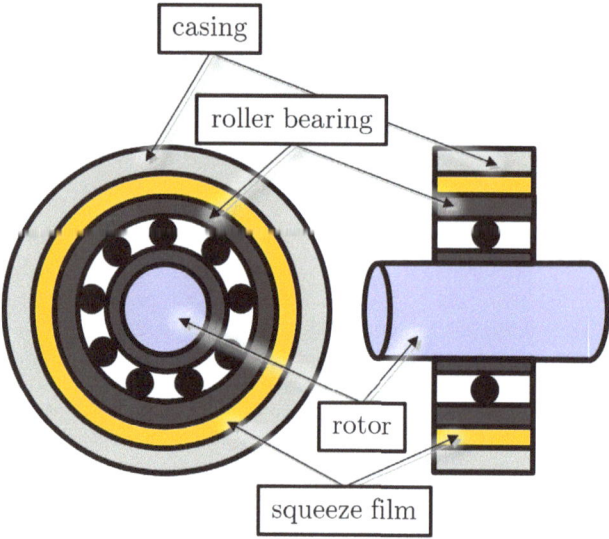

Figure 2. Sketch of a rotor supported by a roller bearing with a non-centralized SFD.

In aircraft jet engine applications, SFDs can be centralized or non-centralized (floating rolling element bearing). Centralized SFDs are usually equipped with a squirrel cage (see, for example [23]). The current experimental setup can capture the effects of both SFD designs. In the experimental setup, and therefore in all simulations, oil from a typical oil company for aircraft jet engines was used, namely Mobil Jet Oil II. The dynamic viscosity and density of the oil were used for the hydrodynamics. The oil density, specific heat capacity and thermal conductivity were used in the thermal model.

2.1. Thermo-Hydrodynamic Modeling

The thermo-hydrodynamic modeling described in this work is based on the generalized Reynolds equation and the Energy equation developed in [13] combined with a mass-conserving cavitation algorithm from [12], including the inertia effects from [18].

2.1.1. Generalized Reynolds Equation with a Mass-Conserving Cavitation Algorithm

The generalized Reynolds equation is solved using the Finite Element Method on a 2D mesh. As the oil cannot withstand high negative pressures, a cavitation model is used in combination with the Reynolds equation. The simplest approach is to use the Gümbel cavitation approach, which simply sets all negative pressures to zero [14,24], violating the conservation of mass. To ensure mass conservation, Kumar and Booker [12] provide an algorithm for tracking the density $\bar{\rho}$ of a mixture of oil and gas in the spatial/temporal diverging gap. The viscosity of the mixture is assumed to behave in the same way: $\bar{\rho}/\bar{\rho}_{oil} = \bar{\eta}/\bar{\eta}_{oil}$. This prevents non-physical oil flow in the cavitation region as observed with non mass-conserving procedures. The algorithm divides the fluid film into different regions that must be identified:

- Region 1a ($\bar{\rho} = \bar{\rho}_{oil}$, $\partial\bar{\rho}/\partial t = 0$);

- Region 1b ($\bar{\rho} = \bar{\rho}_{oil}$, $\partial\bar{\rho}/\partial t < 0$);
- Region 2 ($\bar{\rho} < \bar{\rho}_{oil}$).

The transition between the full-film regions 1a and 1b is a complementary problem that is solved iteratively. After using an Euler-explicit time integration scheme, nodes are moved between regions 1a and 2. Equation (1) shows the generalized Reynolds equation used to calculate the hydrodynamic pressure in the oil:

$$\frac{\partial}{\partial x}\left(\bar{\rho} F_1 \frac{\partial p_0}{\partial x}\right) + \frac{\partial}{\partial y}\left(\bar{\rho} F_1 \frac{\partial p_0}{\partial y}\right) = \frac{\partial}{\partial x}(\bar{\rho} u_J F_2) + h\frac{\partial \bar{\rho}}{\partial t} + \bar{\rho}\frac{\partial h}{\partial t}, \tag{1}$$

where h is the gap function and p_0 is the pressure distribution. The x-coordinate represents the circumferential direction and the y-coordinate represents the axial direction of the SFD. The Reynolds equation balances the fluid flows caused by pressure gradients on the left-hand side of the equation and by shearing, density changes and squeezing on the right-hand side. Figure 3a shows the pressure distribution in the oil of an SFD, where the cavitation area is determined by the Kumar–Booker algorithm. Figure 3b compares the pressure distribution over the circumferential coordinate at the axial centerline $b = 0$. The Kumar–Booker algorithm provides the spatial and temporal evolution of the cavitation area. The size of the cavitation area can have a large effect on the pressure distribution and therefore on the hydrodynamic bearing forces. Due to mass conservation, the Kumar–Booker model is required to determine the oil flow out of the SFD. Therefore, all experimental results are compared with the simulation results using the Kumar–Booker cavitation model.

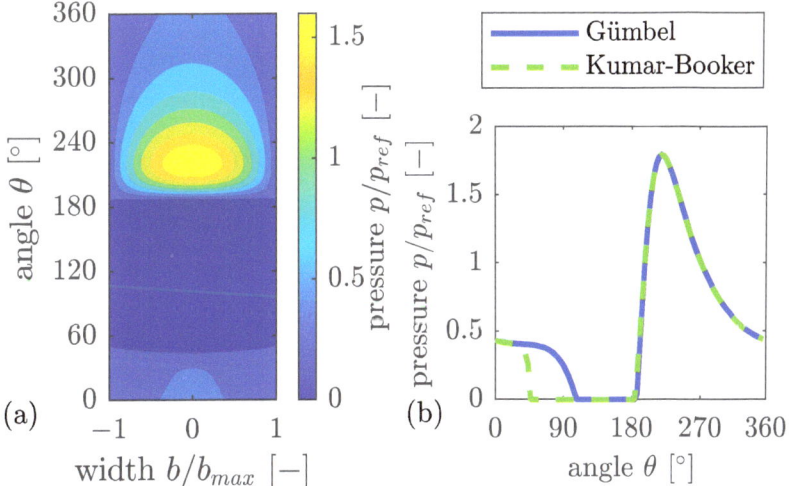

Figure 3. (**a**) Oil–pressure distribution in the SFD with Kumar–Booker cavitation model. (**b**) Comparison of oil–pressure with different cavitation models.

Figure 4 shows the evolution of the density in the SFD over the circumferential coordinate at the axial centerline $b = 0$. At the beginning of the orbit at time t_0, the gap is completely filled with oil. In this case, the density and the cavitation area evolve within a third of an orbit (t_1 to t_4) and from then on follow the high-pressure field, only changing their position according to the orbit (compare t_4 and t_5).

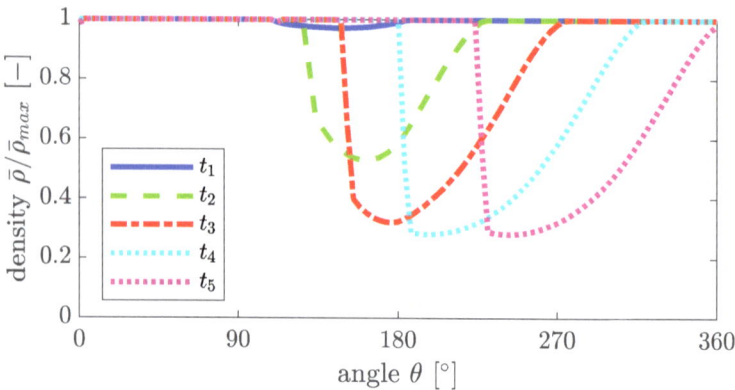

Figure 4. Density evolution with Kumar-Booker cavitation model.

The temperature distribution in the oil film is not constant. Isothermal models may offer simplicity in the implementation and in the solution procedure but they neglect the variable temperature field in the oil film and cannot account for effects such as temperature differences of the oil inlet, journal and casing. To account for these effects, integrals of viscosity across over the gap height are introduced at each node of the 2D mesh [13], as shown in Equation (2):

$$F_1 = J_{2h} - \frac{J_{1h}^2}{J_{0h}}, \quad F_2 = \frac{J_{1h}}{J_{0h}}, \quad J_{0h} = \int_0^h \frac{1}{\eta} dz, \quad J_{1h} = \int_0^h \frac{z}{\eta} dz, \quad J_{2h} = \int_0^h \frac{z^2}{\eta} dz. \quad (2)$$

Isothermal modeling with constant viscosity over gap height leads to the more commonly known equation derived in [13,25] with $F_1 = h^3/12\eta$ and $F_2 = h/2$. Here, the oil viscosity is not constant across the gap height (z-direction). J_{0h}, J_{1h} and J_{2h} are necessary to integrate viscosity considering various exponents of the gap height h. By introducing a parabolic 'global' ansatz function for the temperature across the gap, the need to include a 3D mesh for the energy equation is avoided, heavily reducing the computational cost. In order to determine the temperature distribution along and across the oil film, the energy equation for fluids has to be solved simultaneously.

2.1.2. Temporal and Convective Inertia in the Reynolds Equation

Hamzehlouia in [18] developed a model that can approximate the pressure distribution in fluid film bearings including inertia effects. Based on a perturbation calculation and assuming that inertia has no effect on the fluid velocities, Hamzehlouia derived an extended Reynolds Equation (3) from the Navier–Stokes equations that includes inertia terms.

$$\frac{\partial}{\partial x}\left(\bar{\rho}F_1 \frac{\partial p_1}{\partial x}\right) + \frac{\partial}{\partial y}\left(\bar{\rho}F_1 \frac{\partial p_1}{\partial y}\right) = \frac{\partial}{\partial x}(\bar{\rho}u_J F_2) + h\frac{\partial \bar{\rho}}{\partial t} + \bar{\rho}\frac{\partial h}{\partial t} + G_1(x,y) + G_2(x,y). \quad (3)$$

where G_1 and G_2 are the extended parts and depend on the pressure distribution without inertia effects. G_1 in Equation (4) describes the temporal (fluid flow caused by change in fluid velocity over time) inertia and G_2 in Equation (5) takes convective (fluid flow caused by a change in the fluid velocity due to its change in position) inertia into account:

$$G_1(x,y) = \frac{\bar{\rho}^2}{12\bar{\eta}}\left(h^2\frac{\partial^2 h}{\partial t^2} + \frac{h}{6\bar{\eta}}\frac{\partial h}{\partial x}\left(h^3\frac{\partial^2 p_0}{\partial t \partial x} + 3h^2\frac{\partial h}{\partial t}\frac{\partial p_0}{\partial x}\right)\right). \quad (4)$$

$$\begin{aligned}
G_2(x,y) = \frac{\bar{\rho}^2}{12\bar{\eta}^3}\bigg(&-\frac{5h^5}{20}\left(\frac{\partial h}{\partial x}\right)^2\left(\frac{\partial p_0}{\partial x}\right)^2 - \frac{5h^6}{144}\frac{\partial^2 h}{\partial x^2}\left(\frac{\partial p_0}{\partial x}\right)^2 - \frac{5h^6}{72}\frac{\partial h}{\partial x}\frac{\partial p_0}{\partial x}\frac{\partial^2 p_0}{\partial x^2} \\
&-\frac{7h^6}{72}\frac{\partial h}{\partial x}\frac{\partial p_0}{\partial x}\frac{\partial^2 p_0}{\partial x^2} - \frac{h^7}{72}\left(\frac{\partial^2 p_0}{\partial x^2}\right)^2 - \frac{h^7}{72}\frac{\partial p_0}{\partial x}\frac{\partial^3 p_0}{\partial x^3} \\
&-\frac{7h^6}{144}\frac{\partial h}{\partial x}\frac{\partial^2 p_0}{\partial x\partial y}\frac{\partial p_0}{\partial y} - \frac{h^7}{144}\frac{\partial^3 p_0}{\partial x^2\partial y}\frac{\partial p_0}{\partial y} - \frac{h^7}{144}\left(\frac{\partial^2 p_0}{\partial x\partial y}\right)^2 \\
&-\frac{7h^6}{144}\frac{\partial h}{\partial x}\frac{\partial p_0}{\partial x}\frac{\partial^2 p_0}{\partial y^2} - \frac{h^7}{144}\frac{\partial^2 p_0}{\partial x^2}\frac{\partial^2 p_0}{\partial y^2} - \frac{h^7}{144}\frac{\partial p_0}{\partial x}\frac{\partial^3 p_0}{\partial x\partial y^2} \\
&-\frac{5h^6}{144}\frac{\partial h}{\partial x}\frac{\partial^2 p_0}{\partial x\partial y}\frac{\partial p_0}{\partial y} - \frac{5h^6}{144}\frac{\partial h}{\partial x}\frac{\partial p_0}{\partial x}\frac{\partial^2 p_0}{\partial y^2} - \frac{h^7}{144}\frac{\partial^3 p_0}{\partial x\partial y^2}\frac{\partial p_0}{\partial x} \\
&-\frac{h^7}{144}\left(\frac{\partial^2 p_0}{\partial x\partial y}\right)^2 - \frac{h^7}{144}\frac{\partial^2 p_0}{\partial y^2}\frac{\partial^2 p_0}{\partial x^2} - \frac{h^7}{144}\frac{\partial p_0}{\partial y}\frac{\partial^3 p_0}{\partial x^2\partial y} - \frac{h^7}{72}\left(\frac{\partial^2 p_0}{\partial y^2}\right)^2 \\
&-\frac{h^7}{72}\frac{\partial p_0}{\partial y}\frac{\partial^3 p_0}{\partial y^3}\bigg).
\end{aligned} \quad (5)$$

The extension term G_1 consists primarily of the gap function and the time derivatives of the gap function. G_2 depends mainly on the pressure gradients, the gap function and the gradient of the gap function in circumferential direction. The gap is assumed to be constant in axial direction [18]. Solving Equation (1) gives the pressure distribution p_0 without inertia effects. It is used in Equation (3) to return the pressure distribution p_1, including the complete inertia. San Andrés provides the implementation of temporal inertia effects with linearization without a perturbation method and by neglecting the pressure gradients, as shown in Equation (6) [26]:

$$G_1(x,y) = \frac{\bar{\rho}^2 h^2}{12\bar{\eta}}\frac{\partial^2 h}{\partial t^2}; \quad G_2(x,y) = 0. \quad (6)$$

This model is more time efficient, and compared to Hamzehloiuia's temporal inertia effects, the results are very similar. Figure 5 shows the forces F_x and F_y for a circular orbit around the centerline.

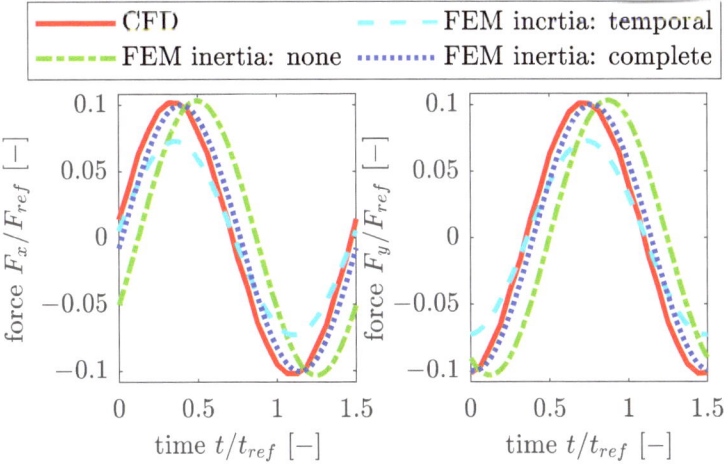

Figure 5. Comparison of inertia models for circular orbits around centerline.

A 3D-CFD simulation is assumed to provide the reference solution for the hydrodynamic bearing forces. The 3D-CFD simulations of the two-phase fluid flow inside the SFD were performed using the code OpenFOAM, which is based on the finite volume method and uses time-dependent, three-dimensional, incompressible Navier–Stokes equations. The computational domain of the simulation is enclosed by three geometrical boundaries: the inner moving solid wall, the outer fixed solid wall and the axial open ends of the fluid film. In the numerical setup, individual boundary conditions must be specified for the flow variables. Zero gradient conditions apply for pressure and volume fraction at the fixed solid walls, as well as no-slip conditions for the velocity. At the open ends, the ambient pressure is specified and the volume fraction is set to one. The dynamic mesh motion is defined by the displacement of the boundary points. For this purpose, the temporal and spatial displacement of the inner moving wall is implemented and a velocity distribution of each individual mesh point is obtained. The moving mesh is characterized by stretching and squeezing of the volume cells within the fluid film, which results in a change in the local film thickness. In the previous work of Schmidt et al. [27] and Reinke et al. [28], it was determined that a minimum number of cells must be applied to the squeeze film in the radial direction. The sensitivity test of a squeezed lubrication film carried out showed that six cells applied across the film achieves the acceptable radial resolution for the expected flow conditions. The overall mesh size contains 2.8 million cells with respect to the cell aspect value due to cell deformation, which is below the maximum value of 10 proposed by Kistner [29].

FEM solutions without any inertia effects, with temporal inertia effects only and with complete (temporal + convective) inertia effects are generated. The FEM solution without inertia effects has almost the same amplitude as the FEM solution with complete inertia, but there is a phase shift between these two solutions. The complete inertia model shows a better agreement with the 3D-CFD results. The temporal inertia model shows the largest deviation from the 3D-CFD results in this case. In Figure 6, the same comparison is performed as in Figure 5 but with an off-centered orbit.

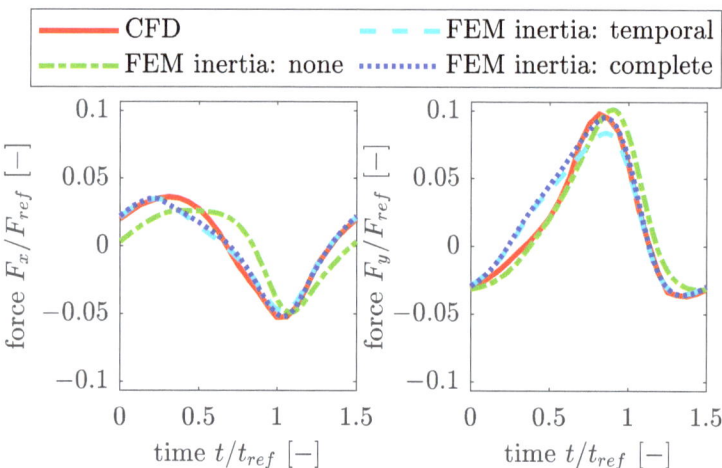

Figure 6. Comparison of inertia models for circular orbits around off-center position.

The complete inertia model once again shows a better agreement with the reference 3D-CFD solution.

2.1.3. Energy Equation in the Oil Film

The energy equation of the oil shown in Equation (7) is simplified by considering similar order-of-magnitude assumptions as in the Reynolds equation [30]:

$$h\bar{\rho}\bar{c}_p\left(\bar{u}\frac{\partial \overline{T}}{\partial x} + \bar{v}\frac{\partial \overline{T}}{\partial y}\right) - \frac{\partial}{\partial x}\left(h\bar{\lambda}\frac{\partial \overline{T}}{\partial x}\right) - \frac{\partial}{\partial y}\left(h\bar{\lambda}\frac{\partial \overline{T}}{\partial y}\right) - \bar{\lambda}\left[\frac{\partial T}{\partial z}\right]_0^h = \bar{\phi}_{hyd}. \tag{7}$$

On the left-hand side are the convection and conduction terms, and on the right-hand side is the hydrodynamic dissipation as a source of energy. The equation is integrated over the gap height to solve it on the same 2D mesh as the pressure distribution. The fluid velocities (Equation (8)) and the hydrodynamic dissipation (Equation (9)) are integrated over the gap height using the Dowson integrals from Equation (2):

$$\begin{aligned}\bar{u} &= -\frac{F_1}{h}\frac{\partial p_1}{\partial x} + \frac{u_J}{h}F_2, \\ \bar{v} &= -\frac{F_1}{h}\frac{\partial p_1}{\partial y},\end{aligned} \tag{8}$$

$$\bar{\phi}_{hyd} = \frac{u_J^2}{J_{0h}} + F_1\left(\frac{\partial p_1}{\partial x}\right)^2 + F_1\left(\frac{\partial p_1}{\partial y}\right)^2. \tag{9}$$

To calculate these terms, the pressure distribution p_1 and the velocity of the journal u_J are required [31].

2.2. Finite Element Formulation of the Thermo-Hydrodynamic Equations

The Reynolds equation is solved using the Finite Element Method on a 2D mesh. The FE formulation is given in Equation (10) [25,31]:

$$\int_{\Omega}\bar{\rho}F_1\left(\frac{\partial \psi_i}{\partial x}\frac{\partial \psi_j}{\partial x} + \frac{\partial \psi_i}{\partial y}\frac{\partial \psi_j}{\partial y}\right)d\Omega \cdot p_{1j} = $$
$$\int_{\Omega}\left(\bar{\rho}_u u_J F_2\frac{\partial \psi_i}{\partial x} + \frac{\partial \bar{\rho}}{\partial t}h\psi_i + \bar{\rho}\frac{\partial h}{\partial t}\psi_i + G_1(x,y)\psi_i + G_2(x,y)\psi_i + \dot{m}_{\Omega}\psi_i\right)d\Omega + \int_{\Gamma}\dot{m}_{\Gamma}\psi_i d\Gamma. \tag{10}$$

The Reynolds equation is now a hyperbolic differential equation with a space-and-time-dependent density. If the problem has dominant convection, the numerical solution will show non-physical oscillations. Therefore, upwinding techniques are necessary to remove these oscillations [32]. Following this path, $\bar{\rho}_u$ is the upwind density. To integrate the density $\bar{\rho}_u$, a term is added to the shape function ψ_i to move the integration point streamline upwards. The added term mostly depends on the stream direction $u_J/|u_J|$. This is called the streamline upwind Petrov–Galerkin (SUPG) method [33]. Using this method, oscillations in the axial direction may still occur if the density gradient in the axial direction is high due to pressure/density boundaries at the SFD ends and lower cavitation pressure. Discontinuity Capturing solves this problem by adding another term to ψ_i, which is controlled by the sign of the density gradient in axial direction $(\partial\bar{\rho}/\partial y)/|\partial\bar{\rho}/\partial y|$ [34]. Equation (11) shows the FE formulation of the energy equation for fluids.

$$\int_{\Omega}\left\{W_i\left(h\bar{\rho}\bar{c}_p\left(\bar{u}\frac{\partial \psi_j}{\partial x} + \bar{v}\frac{\partial \psi_j}{\partial y}\right) + \frac{\bar{\lambda}}{h}12\psi_j\right) + h\bar{\lambda}\left(\frac{\partial \psi_i}{\partial x}\frac{\partial \psi_j}{\partial x} + \frac{\partial \psi_i}{\partial y}\frac{\partial \psi_j}{\partial y}\right)\right\}d\Omega \cdot \overline{T}_j = $$
$$\int_{\Omega}\left\{W_i\left(\bar{\phi}_{hyd} + 6\frac{\bar{\lambda}}{h}(T_J + T_C)\right)\right\}d\Omega. \tag{11}$$

It is solved on the same 2D mesh as the hydrodynamic pressure equation. This diffusion-convection problem is also typically convection-dominated, making upwinding necessary to obtain useful results [31]. SUPG is also used here to remove oscillations. The shape function ψ_i is extended by a perturbation term P_i to move the integration points streamline upwards like $W_i = \psi_i + P_i$. Stream direction can be identified by the fluid velocities $\tilde{u}$ and $\tilde{v}$ to account for circumferential and axial flow. Unlike the density in the Reynolds equation, all shape functions of the energy equation are modified by the upwinding term [31,32].

3. Experimental Setup

Despite the significant efforts to understand and mathematically describe the mechanics of squeeze film damping, the SFD system is complicated and—in the presence of piston ring sealing and deep circumferential grooves—exhibits areas which are not yet well understood and analyzed. An experimental validation is therefore imperative to demonstrate the benefits and shortcomings of the analytical techniques employed. In Figure 7, a novel test rig, operated at the University of Kassel (UKS), is shown in order to perform full-scale tests of squeeze film dampers (SFD). It consists of a shaft that is enabled by the design of the suspension to perform an orbital motion.

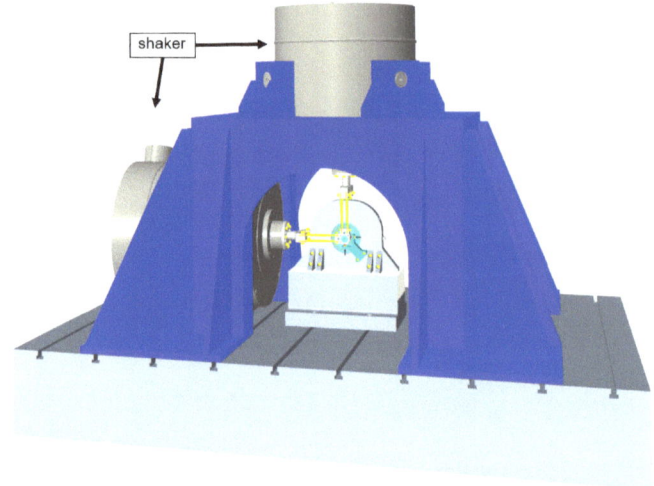

Figure 7. CAD view.

Two electromagnetic shakers, one for vertical and one horizontal direction, cause the shaft to move on a pre-described orbit. During operation, shaft and housing are separated by a thin pressurized oil film, which simulates restoring forces comparable to an unbalanced loading of a rotating shaft. Due to its modular design, the test rig can be operated with different bearing geometries regarding clearance, groove or oil-supply holes. Similar to the test rig of San Andrés, the circular motion is realized by control of the input of the two shakers [20–22].

In comparison to the test rig of San Andrés, the shaft of the UKS test rig executes the circular movement and is directly mounted to the shakers without additional support. The shaft has similar dimensions to San Andrés, and under operation, the relative eccentricities can reach up to 0.9. Additionally, the dimensions of the test rig has been optimized to shift the natural frequencies of the moving parts above the operating range. In this way, tests at high operating frequencies can be performed, which cover the typical range of rotational speeds of aero engines.

The main mechanical components of the system under test are shaft, housing, force measuring flanges and spring sheets connecting the shaft to the shakers (Figure 8). The orbital motion of the shaft is enabled by use of spring sheets with a high axial stiffness to transmit the forces provided by the shakers. Laterally, the plates are flexible to minimize the reaction forces on the perpendicular axes. The scaling of the smaller shaft, in contrast to the larger shaft, allows the test stand to reach all operating points. In the case of the the larger shaft, occurring accelerations and forces render reaching operating points impossible. The oil is pre-heated within a heat exchanger and supplied to the pressurized film by an oil pump. Prior to test runs, the test rig is heated by the oil flow for several hours until constant temperature of shaft and housing is reached. In addition to temperature, different inlet oil pressures, operating frequencies and eccentricities can be set. Additionally, the SFD test rig can be supplemented with piston rings and a circumferential oil-supply groove in the housing. Figure 9 shows the test rig setup at the University of Kassel.

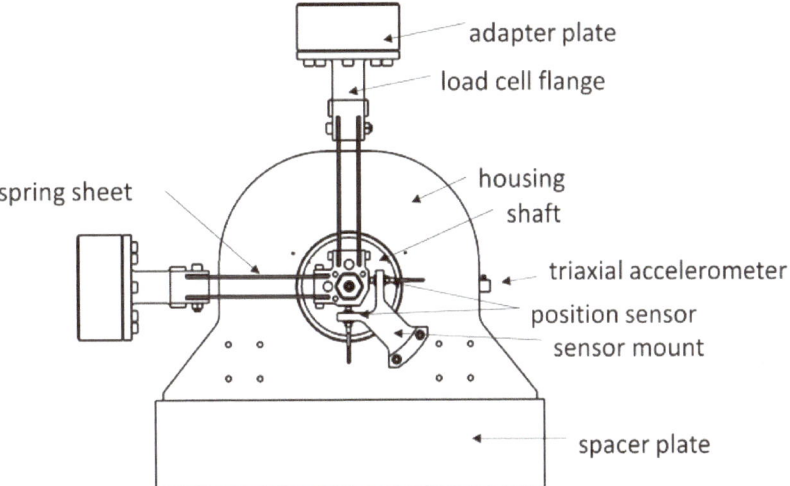

Figure 8. Schematic view.

Figure 9. Squeeze film damper test rig at the University of Kassel.

The central parameters on this test stand are measured and recorded at various points. This includes 17 temperature measuring points located inside and outside of the shaft housing, the shaft and oil supply. Additionally, sensors for volume flow and oil pressure (3×) are mounted within the oil system. Measurement setup is completed by five accelerometers located on housing, each shaker table and frame. The most important measurement variables include force and displacement. Figure 10 shows a detailed view of the test rig, highlighting the position of the measured variables.

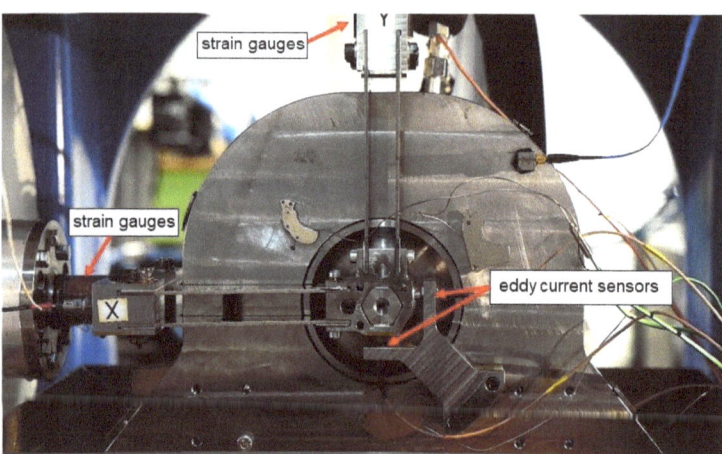

Figure 10. Detailed view of the housing with shaft.

Both shakers are connected to the spring plates by a force measuring flange. These flanges are equipped with strain gauges and have been force-calibrated prior to the measurement campaign in a universal testing machine. The orbit of the shaft is measured on both sides of the housing by two eddy current sensors and used to control the shakers. For each test, an inlet oil pressure and temperature is set and both stationary and transient operating points of frequency and eccentricity can be conducted by the dynamic control scheme. In addition, a second control loop can be applied to move the center line of the shaft statically in the housing and maintaining the desired position throughout the test. Positioning of the shaft is a crucial step carried out after every modification to the experimental setup. The position of the shaft in the housing is of central importance for the results of the test. Even small changes will result in a large effect on the resulting forces and the position of the orbits. However, it is not possible to determine the absolute position of the shaft in relation to the housing as the eddy current sensors can only measure the relative position of the shaft. The absolute position of the shaft relative to the housing is determined by the alignment curve. The alignment curve, which is also important for determining the static and dynamic eccentricity of each test point, is traced through the static movement of the shaft in the housing over the full circumference under a defined force.

In order to compare hydrodynamic forces of the test rig results with simulation, the inertia forces of the moving components have to be subtracted from the measured forces. The forces caused by bending of the spring sheets are very small compared to the fluid forces and can be neglected.

4. Results and Discussion

The comparison of measurement and simulation covers the range of one orbit around a slightly off-centered position (see Figure 11). The amplitudes of the measured eccentricities in both directions, x and y, are almost equal, which means the orbit is more circular than elliptical. The phase shift between e_x and e_y is 90°.

Figure 12 shows the meshed SFD model which is used to solve the Reynolds and energy equation. The experimental setup used a typical aircraft engine SFD length (L)-to-diameter (D) ratio of $L/D = 0.2$. At the side ends of the SFD, pressure is set to atmospheric pressure. Oil flows into the SFD through three feed holes where the measured flow is set as the natural boundary condition. The algorithm of Kumar and Booker exhibits a large cavitation area (marked with dark blue). At inlet holes, the cavitation area decreases over time as the oil flows in.

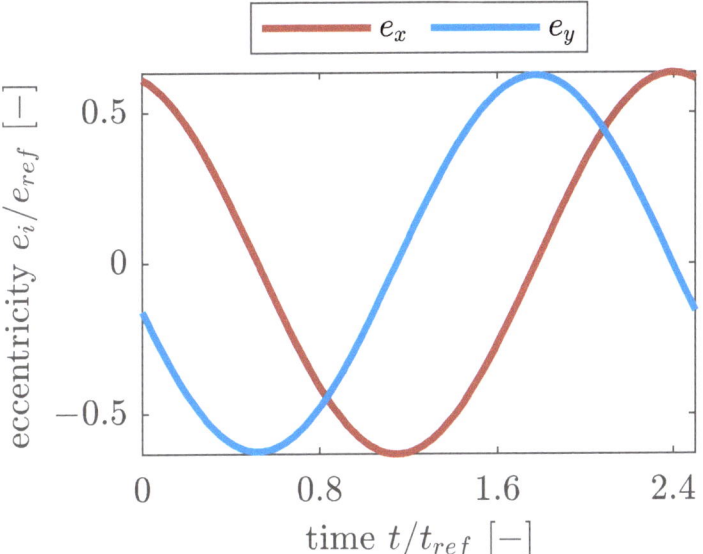

Figure 11. Eccentricity of measurement.

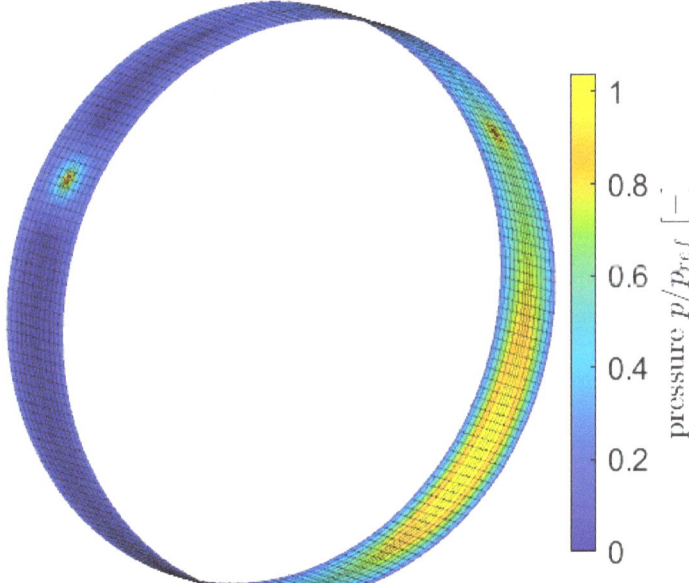

Figure 12. SFD mesh with exemplary pressure distribution.

For the given motion as given in Figure 11, both the resulting forces within the experiment and the calculated forces from the simulation can be compared. Figure 13 shows a good correlation of measurement and simulation for F_x and F_y. Except for the deviation around $t/t_{ref} = 1$, amplitude matches well and the phase is almost identical.

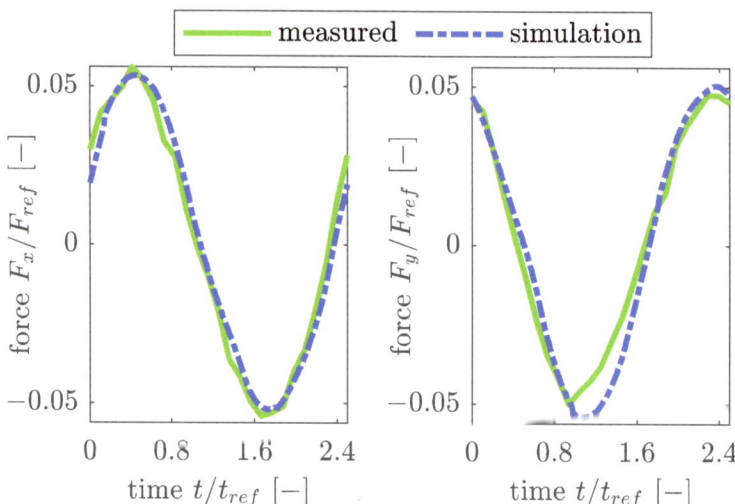

Figure 13. Forces F_x and F_y of measurement.

Oil flow out of the SFD's ends is shown in Table 1. It is averaged over one orbit. The deviation of measurement and simulation is shown below.

Table 1. Comparison of oil flow out of the SFD.

Measured q/q_{ref} [−]	Simulation q/q_{ref} [−]	Rel. Deviation
0.450	0.422	6.22%

5. Conclusions

In this work, an experimental setup for the validation of models of squeeze film dampers has been presented. Detailed thermo-hydrodynamic SFD bearing models were developed by focusing on inertia, cavitation and thermal effects. The 2D FEM models based on the Reynolds equation were compared with 3D-CFD models. It was shown that the inertia effects should be taken into account when high physical accuracy is required. Furthermore, the numerical simulation results were compared with experimental data and showed a very good agreement. This work is currently being extended to squeeze film dampers with a circumferential groove and piston rings. Different oil-supply pressures, oil-supply temperatures, different bearing eccentricities and different orbiting speeds are considered. In order to compare between different SFDs, the inertia and damping coefficients will be extracted from the experimental data. A comparison will also be made with the coefficients obtained from the numerical simulations. Finally, back-to-back experiments will be performed with and without oil in the SFD to identify the influence of the SFD forces.

Author Contributions: Conceptualization, I.C. and A.R.; methodology, M.G., I.C. and A.R.; software, M.G., M.S. and A.R.; validation, J.G., M.G., L.K. and A.R.; writing—original draft preparation, M.G., J.G. and I.C.; writing—review and editing, I.C., L.K., M.S. and A.R.; visualization, M.G. and J.G.; supervision, I.C., L.K., A.R. and P.R.; project administration, I.C. All authors have read and agreed to the published version of the manuscript.

Funding: This research was partially funded by WTD 61/Bundeswehr.

Data Availability Statement: Data are not made available due to confidentiality reasons.

Acknowledgments: The authors would like to thank MTU Aero Engines for the permission to publish this work.

Conflicts of Interest: Authors Ioannis Chatzisavvas and Lukas Kohlmann were employed by the company MTU Aero Engines AG. The remaining authors declare that the research was conducted in the absence of any commercial or financial relationships that could be construed as a potential conflict of interest.

References

1. Reynolds, O. IV. On the theory of lubrication and its application to Mr. Beauchamp tower's experiments, including an experimental determination of the viscosity of olive oil. *Philos. Trans. R. Soc. Lond.* **1886**, *177*, 157–234.
2. Dubois, G.B.; Ocvirk, F.W. The short bearing approximation for plain journal bearings. *Trans. Am. Soc. Mech. Eng.* **1955**, *77*, 1173–1178. [CrossRef]
3. Yamamoto, T.; Ishida, Y. *Rotordynamics: A Modern Treatment with Applications*; Wiley: Hoboken, NJ, USA, 2001.
4. Della Pietra, L.; Adiletta, G. The Squeeze Film Dampers over Four Decades of Investigations. Part I: Characteristics and Operating Features. *Shock Vib. Dig.* **2002**, *34*, 3–26.
5. Hori, Y. *Hydrodynamic Lubrication*, 1st ed.; Springer: Tokyo, Japan, 2006. [CrossRef]
6. Szeri, A.Z. *Fluid Film Lubrication*; Cambridge University Press: Cambridge, UK, 2010.
7. Someya, T. Stabilität einer in zylindrischen Gleitlagern laufenden, unwuchtfreien Welle: Beitrag zur Theorie des instationär belasteten Gleitlagers. *Ingenieur-Archiv* **1963**, *33*, 85–108. [CrossRef]
8. van Buuren, S. *Modeling and Simulation of Porous Journal Bearings in Multibody Systems*; KIT Scientific Publishing: Karlsruhe, Germany, 2014; Volume 21.
9. Chatzisavvas, I. Efficient Thermohydrodynamic Radial and Thrust Bearing Modeling for Transient Rotor Simulations. Ph.D. Thesis, Technische Universität Darmstadt, Darmstadt, Germany, 2018.
10. PW1100G-JM: Geared Turbofan Engine. Available online: https://www.mtu.de/engines/commercial-aircraft-engines/narrowbody-and-regional-jets/gtf-engine-family/ (accessed on 13 May 2024).
11. Dowson, D.; Taylor, C. Cavitation in bearings. *Annu. Rev. Fluid Mech.* **1979**, *11*, 35–65. [CrossRef]
12. Kumar, A.; Booker, J. A Finite Element Cavitation Algorithm: Application/Validation. *J. Tribol.* **1991**, *113*, 255–260. [CrossRef]
13. Dowson, D. A generalized Reynolds equation for fluid-film lubrication. *Int. J. Mech. Sci.* **1962**, *4*, 159–170. [CrossRef]
14. Pinkus, O.; Sternlicht, B.; Saibel, E. Theory of hydrodynamic lubrication. *J. Appl. Mech.* **1962**, *29*, 221–222. [CrossRef]
15. Hamzehlouia, S.; Behdinan, K. Thermohydrodynamic Modeling of Squeeze Film Dampers in High-Speed Turbomachinery. *SAE Int. J. Fuels Lubr.* **2018**, *11*, 129–146. [CrossRef]
16. San Andrés, L.; Vance, J.M. Effects of Fluid Inertia and Turbulence on the Force Coefficients for Squeeze Film Dampers. *J. Eng. Gas Turbines Power* **1986**, *108*, 332–339. [CrossRef]
17. San Andrés, L.; Vance, J.M. Effects of Fluid Inertia on Finite-Length Squeeze-Film Dampers. *ASLE Trans.* **1987**, *30*, 384–393. [CrossRef]
18. Hamzehlouia, S. Squeeze Film Dampers in High-Speed Turbomachinery: Fluid Inertia Effects, Rotordynamics, and Thermohydrodynamics. Ph.D. Thesis, Mechanical and Industrial Engineering, University of Toronto, Toronto, ON, Canada, 2017.
19. Hamzehlouia, S.; Behdinan, K. A study of lubricant inertia effects for squeeze film dampers incorporated into high-speed turbomachinery. *Lubricants* **2017**, *5*, 43. [CrossRef]
20. San Andrés, L. Force coefficients for a large clearance open ends squeeze film damper with a central feed groove: Experiments and predictions. *Tribol. Int.* **2014**, *71*, 17–25. [CrossRef]
21. San Andrés, L.; Jeung, S.H.; Den, S.; Savela, G. Squeeze Film Dampers: An Experimental Appraisal of Their Dynamic Performance. In Proceedings of the First Asia Turbomachinery and Pump Symposium, Singapore, 22–25 February 2016.
22. San Andrés, L.; Koo, B.; Jeung, S.H. Experimental Force Coefficients for Two Sealed Ends Squeeze Film Dampers (Piston Rings and O-Rings): An Assessment of Their Similarities and Differences. *J. Eng. Gas Turbines Power* **2018**, *141*, 021024. [CrossRef]
23. Chatzisavvas, I.; Arsenyev, I.; Grahnert, R. Design and optimization of squirrel cage geometries in aircraft engines toward robust whole engine dynamics. *Appl. Comput. Mech.* **2023**, *17*, 93–104. [CrossRef]
24. Lang, O.R.; Steinhilper, W. *Gleitlager: Berechnung und Konstruktion von Gleitlagern Mit Konstanter und Zeitlich Veränderlicher Belastung*; Springer: Berlin/Heidelberg, Germany, 1978.
25. Rienäcker, A. Instationäre Elastohydrodynamik von Gleitlagern Mit Rauhen Oberflächen und Inverse Bestimmung der Warmkonturen. Ph.D. Thesis, RWTH Aachen, Aachen, Germany, 1995.
26. San Andrés, L.; Delgado, A. A Novel Bulk-Flow Model for Improved Predictions of Force Coefficients in Grooved Oil Seals Operating Eccentrically. *J. Eng. Gas Turbines Power* **2012**, *134*, 052509. [CrossRef]
27. Schmidt, M.; Reinke, P.; Rabanizada, A.; Umbach, S.; Rienäcker, A.; Branciforti, D.; Philipp, U.; Preuß, A.C.; Pryymak, K.; Matz, G. Numerical Study of the Three-Dimensional Oil Flow Inside a Wrist Pin Journal. *Tribol. Trans.* **2020**, *63*, 415–424. [CrossRef]
28. Reinke, P.; Schmidt, M. *Lokale, Hochauflösende 3D-CFD-Simulation der Schmierspaltströmung in Einem Instationär Belasteten Radialgleitlager*; Final Report, No. 1154, FVV e.V.; FVV: Frankfurt/M, Germany, 2014; Volume 1073.
29. Kistner, B. Modellierung und Numerische Simulation der Nachlaufstruktur von Turbomaschinen am Beispiel einer Axialturbinenstufe. Ph.D. Thesis, Technische Universität Darmstadt, Darmstadt, Germany, 1999.
30. Khonsari, M.M.; Booser, E.R. *Applied Tribology: Bearing Design and Lubrication*; John Wiley & Sons, Ltd.: Hoboken, NJ, USA, 2017.

31. Jaitner, D. Effiziente Finite-Elemente-Lösung der Energiegleichung zur Thermischen Berechnung Tribologischer Kontakte. Ph.D. Thesis, University of Kassel, Kassel, Germany, 2017.
32. Hughes, T.J.; Tezduyar, T. Finite element methods for first-order hyperbolic systems with particular emphasis on the compressible Euler equations. *Comput. Methods Appl. Mech. Eng.* **1984**, *45*, 217–284. [CrossRef]
33. Brooks, A.N. A Petrov-Galerkin Finite Element Formulation for Convection Dominated Flows. Ph.D. Thesis, California Institute of Technology, Pasadena, CA, USA, 1981.
34. Codina, R. A discontinuity-capturing crosswind-dissipation for the finite element solution of the convection-diffusion equation. *Comput. Methods Appl. Mech. Eng.* **1993**, *110*, 325–342. [CrossRef]

Disclaimer/Publisher's Note: The statements, opinions and data contained in all publications are solely those of the individual author(s) and contributor(s) and not of MDPI and/or the editor(s). MDPI and/or the editor(s) disclaim responsibility for any injury to people or property resulting from any ideas, methods, instructions or products referred to in the content.

Article

Modelling of Static and Dynamic Elastomer Friction in Dry Conditions

Fabian Kaiser *, Daniele Savio and Ravindrakumar Bactavatchalou

Freudenberg Technology Innovation SE & Co. KG, Hoehnerweg 2-4, 69469 Weinheim, Germany; daniele.savio@freudenberg.com (D.S.)
* Correspondence: fabian.kaiser@freudenberg.com

Abstract: Understanding the tribological behavior of elastomers in dry conditions is essential for sealing applications, as dry contact may occur even in lubricated conditions due to local dewetting. In recent decades, Persson and co-authors have developed a comprehensive theory for rubber contact mechanics and dry friction. In this work, their model is implemented and extended, particularly by including static friction based on the bond population model by Juvekar and coworkers. Validation experiments are performed using a tribometer over a wide range of materials, temperatures and speeds. It is shown that the friction model presented in this work can predict the static and dynamic dry friction of various commercial rubber materials with different base polymers (FKM, EPDM and NBR) with an average accuracy of 10%. The model is then used to study the relevance of different elastomer friction contributions under various operating conditions and for different roughness of the counter surface. The present model will help in the development of novel optimized sealing solutions and provide a foundation for future modeling of lubricated elastomer friction.

Keywords: rubber; elastomer; friction; simulation; seals; viscoelasticity; roughness; contact mechanics

1. Introduction

Many academic institutions and companies have a long history in the research of sealing components. Nevertheless, there are still some significant aspects of rubber seals that are not fully understood. One example is the physical interactions happening in the frictional contacts of dynamical seals. From the perspective of Freudenberg as a sealing and lubricant manufacturing company, improved understanding of such tribological phenomena in sealing contacts is crucial.

When designing new sealing elements and novel elastomers for sealing applications, engineers must strike a balance between several, sometimes conflicting, requirements. On the one hand, tightness, i.e., the primary sealing function, must be ensured. On the other hand, efficient designs with high lifetime, i.e., with controlled friction and wear, are also desired. Ultimately, developers must rely on a good understanding of tribological and sealing behavior under a wide range of operating conditions to ensure design efficiency and reliability. This need applies to several sealing components. The classical example is the radial shaft seal, which heavily relies on friction at the sealing edge to ensure proper operation during shaft rotation [1]. However, even static seals like O-rings can undergo motion once systems are pressurized or depressurized [2], typically after long standstill times.

Predictive friction models for elastomers in dry and lubricated conditions will help speed up the development process in terms of seal design and material selection, improve the transferability between tests and actual product operation and give valuable guidelines for the development of new materials. All of this will reduce costs and/or increase sales.

Such models must consider several aspects of a tribological contact. First, typical engineering surfaces are rough, and, when pressed together, do not come into contact

Citation: Kaiser, F.; Savio, D.; Bactavatchalou, R. Modelling of Static and Dynamic Elastomer Friction in Dry Conditions. *Lubricants* **2024**, *12*, 250.
https://doi.org/10.3390/lubricants12070250

Received: 29 May 2024
Revised: 21 June 2024
Accepted: 24 June 2024
Published: 9 July 2024

Copyright: © 2024 by the authors. Licensee MDPI, Basel, Switzerland. This article is an open access article distributed under the terms and conditions of the Creative Commons Attribution (CC BY) license (https://creativecommons.org/licenses/by/4.0/).

everywhere, which has been well known since the ground-breaking work by Greenwood and Williamson [3]. Adhesive or abrasive friction processes take place at these contact spots [4]; close to the contacts, additional losses occur through viscoelastic damping in elastomers when they slide on a hard, rough counterface. Extensive theoretical developments on the calculation of the contact area between rough surfaces were performed by Persson [5]. In the past 20 years, his theory has been extensively tested [6] and extended, e.g., for adhesion [7], layered materials [8] or plasticity [5,9]. It has also been applied successfully to calculate the static leakage of polymeric [10–12] and metallic seals [13], the latter including elastoplastic material behavior, as a function of surface roughness. Other numerical methods have also been used to predict sealing performance of systems under operation, e.g., Boundary Element Methods for metallic face seals [14] or Finite Element Analysis for rubber sealing cylinders [15].

In terms of solid body friction, Persson's seminal paper from 2001 already considered viscoelastic dissipation in rubbery materials [5]. More recent developments also consider other sliding friction contributions acting in the contact spots [16,17]. It should be noted that most of the papers on friction focus on tire compounds, seeing limited application to sealing elastomers [18] up to now.

An additional aspect to consider is the presence of fluids in the contact—be it specialized lubricants or simply the sealed media—which can reduce solid-body contact and friction through different mechanisms. At high speeds, hydrodynamic lift separates the surfaces, resulting in the well-known Stribeck curve [19]. This is modeled through the well-established Reynolds equation [19]. The effect of surface roughness on flow is either modeled deterministically or accounted for by flow factors from homogenization approaches [20] and from contact mechanics results [21]. At low sliding speeds, the boundary lubrication regime is deeply affected by wettability effects (i.e., the spreading coefficient) between the elastomer, counterface and lubricant [22] (Figure 1). First modeling approaches are presented in the literature [23,24], but many aspects, particularly regarding the interaction with roughness, are still unknown.

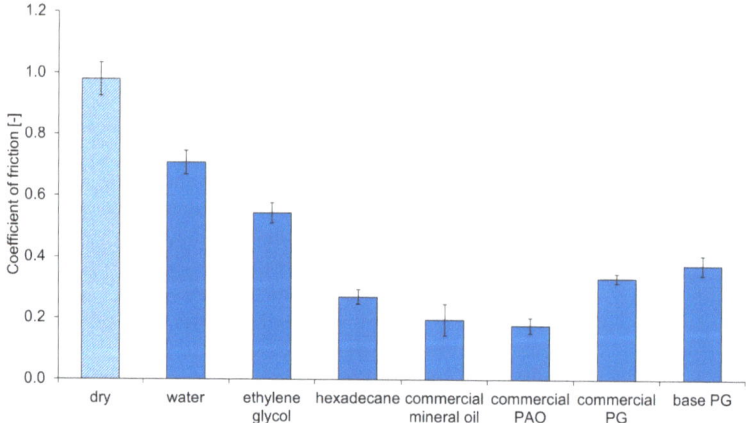

Figure 1. Boundary friction coefficient for FKM 1 with simple fluids and commercial oils (PAO = polyalphaolefine, PG = polyglycol).

Overall, the state of the art lacks a theory coupling realistic roughness contact, material properties and surface–fluid wetting interactions which determine friction under boundary lubrication. To fill this gap, a necessary step is to improve our modeling of dry elastomer friction, since this constitutes the main reference to compare the lubricating action of different fluids (see Figure 1). Moreover, unlubricated contact spots will likely persist even in presence of lubricants [23], and their contact and friction behavior must be correctly quantified before envisioning an extension to surface–fluid wetting.

In this work, we tackle the following aspects of dry elastomer friction. First, we test the applicability of state-of-art friction models [16] to typical sealing elastomers in a wide range of sliding speeds and temperatures and compare their results to tribological experiments. Second, we analyze which friction contributions are the most relevant in typical sealing systems. Third, we propose an extension of the current theory to quantify breakloose friction as a function of standstill time.

2. Materials and Methods
2.1. Experimental Methods
2.1.1. Elastomer Materials and Their Characterization

We consider four commercial fully formulated elastomer compounds with various base polymers (FKM, NBR and EPDM). An overview is provided in Table 1: all materials except FKM 2 were provided by Freudenberg Sealing Technologies. FKM 2 was acquired from a different supplier. All materials come in 2 mm thick slabs, from which different samples for material characterization or tribological testing are punched out.

Table 1. Overview of the elastomer materials. The reported glass transition temperature T_g was obtained from Dynamic Mechanical Analysis performed at 0.1 Hz.

Material	Typical Application	Color	Cross-Linker	Fillers	Shore Hardness	T_g (°C)
FKM 1	Radial shaft seals	Red–brown	Bisphenolic	Mineral	75	$-8\,°C$
FKM 2	O-rings	Dark brown	Bisphenolic	Mineral	90	$-6\,°C$
NBR	Radial shaft seals	Blue	Sulphuric	Mineral	75	$-20\,°C$
EPDM	O-rings	Black	Peroxidic	Carbon black	75	$-44\,°C$

Dynamic Mechanical Analysis (in shear mode, with a 1% strain amplitude) is performed for each elastomer using a dynamical–mechanical material tester EPLEXOR 2000N (GABO Qualimeter GmbH, Germany). The machine performs temperature and frequency sweeps, which are automatically combined into master curves through a WLF shift procedure [25]. A typical example is provided in Figure 2, showing the frequency evolution of the storage and loss moduli $G'(\omega)$ and $G''(\omega)$ and the loss tangent $\tan(\delta)$ for FKM 1. We consider rubber incompressible (i.e., Poisson ratio of $\approx$0.5) in all conditions.

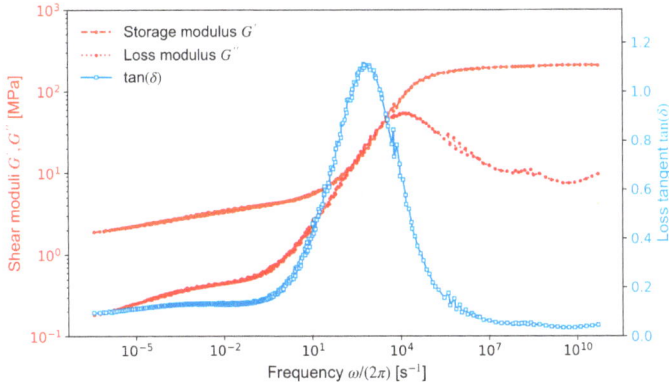

Figure 2. Master curves of the shear moduli G', G'' and the loss tangent $\tan(\delta)$ as a function of excitation frequency for FKM 1 at 20 °C.

2.1.2. Surfaces and Their Characterization

Surface topography data for the elastomers and the steel counterface are measured using White Light Interferometry (WLI) on a Bruker NPFLEX device. The measured domains were approximately 1.3×1.7 mm^2 large, with a lateral resolution of 1.27 µm in both epitaxial directions. On each material, measurements were performed 2–3 times to ensure that statistically representative surface patches were selected.

Figure 3a shows the ground 1.2842 (90MnCrV8) steel countersurface used in the tribological experiments: one can note the anisotropic features typical of its manufacturing process.

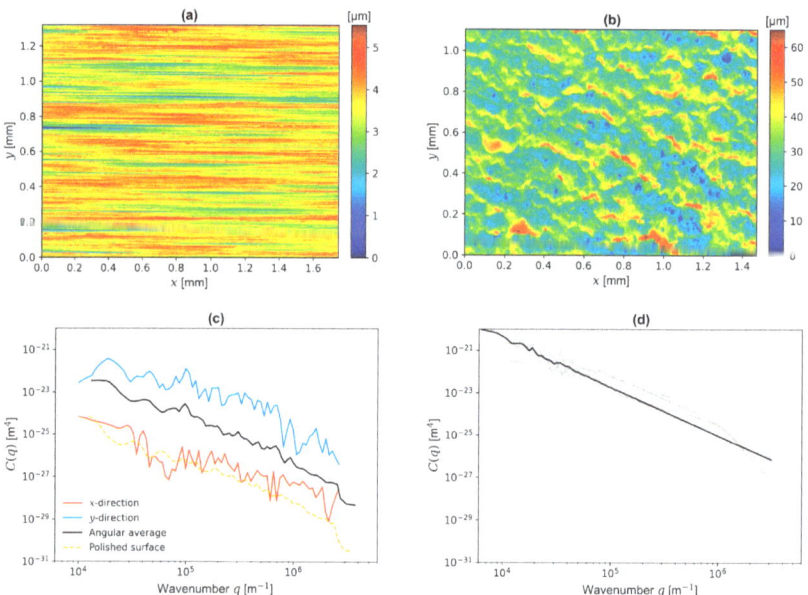

Figure 3. Roughness topography of (**a**) the ground counterface and (**b**) worn rubber. (**c**) Directional and angular average PSDs of the ground steel counterface. A polished surface is also shown for comparison purposes. (**d**) PSDs of worn rubber surfaces (gray) and their average (black).

For the elastomers, topography measurements were performed mainly on worn surfaces. In fact, our tribological tests show that the skin from the rubber molding process is usually removed after very few sliding cycles, so a worn state is representative of most of the experiment. Figure 3b presents a typical example showcasing the isotropic nature of the worn elastomer roughness.

In our simulations based on Perssons' theory, height data needs to be transformed into their power spectral density [9]—also called PSD or $C(\boldsymbol{q})$, where $\boldsymbol{q} = (q_x, q_y)$ is the wavevector. After removing tilt and curvature from measured data, followed by Hann windowing [26], the PSD can be calculated through an FFT algorithm [27]. Radially averaged power spectra for selected worn elastomers are shown in Figure 3d. Due to their similar magnitude and slope, all curves were averaged into one (the solid black line), which we used in our calculations as a representative worn rubber surface. The radially averaged PSD for ground steel is also depicted in Figure 3c, together with the PSD curves for the x and y directions. In the calculations, we employed the full $C(q_x, q_y)$ to account for anisotropy. Note that the PSD of the roughness parallel to the surface "grooves" (x-direction) is comparable to the angular average of a polished surface.

2.1.3. Tribological Testing

A custom-built reciprocating test rig was used to measure static and dynamic friction of the rubber materials. In this setup (see Figure 4), the elastomer samples are normally loaded using a pressurized pneumatic bellow. The reciprocating counterface motion relative to the rubber is driven by a linear motor with speeds ranging from 0.5 mm/s to 1 m/s. The setup is installed in a climate chamber (Binder MKF 720) that allows testing of friction and wear between $-40\ °C$ and $150\ °C$. The experimental parameters used for the rubber materials are listed in Table 2.

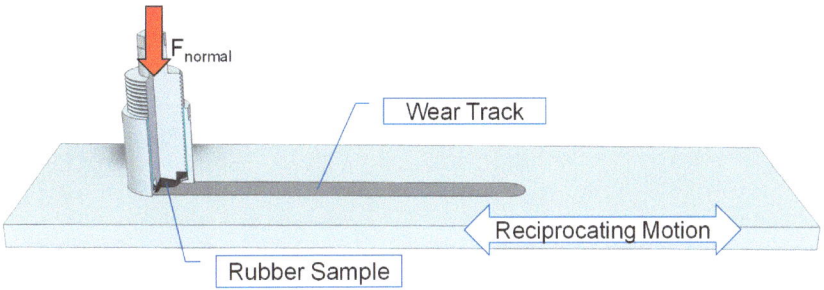

Figure 4. Schematic illustration of test setup.

Table 2. Test conditions.

Parameters	Conditions
Stroke	50 mm
Sliding Speeds	1–300 mm/s
Waiting Times Between Strokes	1–3600 s
Temperatures	-40–$100\ °C$ ($80\ °C$ for NBR)
Normal Load	40 N (approx. 1.3 MPa nominal contact pressure)

The normal and friction force are measured using force transducers (Burster 8524, Burster Präzisionsmesstechnik, Gernsbach, Germany and ME KD80s, ME-Meßsysteme, Henningsdorf, Germany, respectively). The static friction value is defined as the maximum friction force when relative movement starts after waiting, and the dynamic friction value is averaged over the central 30 mm of the 50 mm stroke when conditions are steady.

Punched elastomer discs (2 mm thickness and 20 mm diameter) are installed in a specially designed pin. After mounting, the rubber samples assume a spherical shape with a radius of curvature of approx. 22 mm to the rubber (see Figure 4). This way, reliable friction testing can be achieved using readily available rubber sheets, while avoiding edge effects typical of flat pins with sharp borders. The geometry and chosen loading conditions result in an average contact pressure of approx. 1.3 MPa, which is in line with typical sealing applications. The setup allows for reproducible measurements without noticeable friction instabilities, e.g., stick-slip, over the whole speed and temperature range.

2.2. Simulation Methods

2.2.1. Contact Area Calculation

When the rubber is pushed against the counterface with a given external load F_N the nominal contact area A_0 and pressure $\sigma_0 = F_N / A_0$ can easily be calculated using analytical (e.g., Hertzian) formulations or finite element methods.

Friction processes are, however, strongly dependent on the true area of contact A_{true}, generally a small fraction of A_0 in the case of rough surfaces. Both A_{true} and the relative contact area $A_{rel} = A_{true}/A_0$ between the elastomer surface and steel counterface can be predicted using Persson's theory. The reader can refer to several papers for its derivation [5,28] and comparisons to other calculation methods [6]. At its core, this theory involves solving a diffusion equation for the probability distribution of contact stresses $P(\sigma, q)$ with increasing high-frequency surface roughness content:

$$\frac{\partial P}{\partial q} = f(q) \frac{\partial^2 P}{\partial \sigma^2}. \tag{1}$$

The diffusion coefficient $f(q)$ depends on the nominal contact pressure $\sigma_0 = F_N/A_0$, the combined roughness power spectral density $C(q) = C(q)_{\text{elastomer}} + C(q)_{\text{steel}}$ and the equivalent contact modulus E^*. As the rubber accommodates most of the deformation, $E^* = \frac{1+2\nu}{1+\nu^2} G^*(T, \omega)$, where $G^*(T, \omega)$ is the temperature and frequency dependent shear modulus of the rubber from DMA and $\nu = 0.5$ is its Poisson's ratio.

Our numerical implementation of Equation (1) includes quartic correction factors for the elastic energy and stress broadening from [29]. Once solved for the probability distribution of contact stresses $P(\sigma, q)$, the relative contact area is obtained:

$$A_{\text{rel}}(q) = \int P(\sigma, q) dq. \tag{2}$$

2.2.2. Dynamic Elastomer Friction and Its Contributions

The elastomer friction model used in this work bases on recent works by Persson and coworkers [16]. Here, three main contributions to friction are considered:

$$\mu_{\text{total}} = \mu_{\text{contact}} + \mu_{\text{viscoelastic}} + \mu_{\text{crack}}. \tag{3}$$

Their physical meaning, where they occur and details on their modeling are discussed below.

μ_{contact}: This contribution acts in the real contact area between rough surfaces, and can be expressed as follows [16]:

$$\mu_{\text{contact}} = \frac{\tau_f A_{\text{rel}}}{\sigma_0} + \mu_{\text{transfer}}. \tag{4}$$

The two terms correspond to separate friction mechanisms. In the first, polymer chains attach to the countersurface, are stretched and finally detach, dissipating the stored energy as friction losses. According to [16], the resulting interfacial shear stress τ_f is expressed as a semi-empirical master curve:

$$\tau_f(v) = \tau_0 \exp\left(-c \left[\log_{10}\left(\frac{v}{v_0}\right)\right]^2\right), \tag{5}$$

where v is the current sliding speed and $v_0 = 1$ mm/s is a fixed reference velocity. Both the dimensionless constant $c \approx 0.25$ and the maximum shear stress τ_0 at $v = v_0$ are elastomer-dependent parameters.

The master curve in Equation (5) relies on an Arrhenius-type shifting to calculate the frictional shear stress $\tau_f(v, T) = \tau_f(a_T'(T) \cdot v)$ at different temperatures. The shift factor a_T' is given by [16]:

$$\ln(a_T') = \frac{\epsilon}{k_B} \left(\frac{1}{T} - \frac{1}{T_0} - \frac{1}{(T_g - 20K)} + \frac{1}{T_{g,0}} \right). \tag{6}$$

Here, the reference temperatures $T_0 = 20$ °C for the master curve Equation (5) and $T_{g0} = -38$ °C for the glass transition are fixed [30]. T_g is the glass transition temperature of the bulk elastomer, with a correction of 20 K accounting for the lower T_g of thin polymer films compared to the bulk material [16]. Finally, the activation energy ϵ is an elastomer-dependent parameter of the order of 1 eV.

For the commercial rubber materials for sealing applications tested in this work, we found that τ_0 in Equation (5) can be expressed as:

$$\tau_0(T) = \frac{E^*(T)}{k}, \tag{7}$$

where the factor k is a material-dependent constant of order unity.

Overall, the resulting bonding friction coefficient $\tau_f A_{\text{rel}}/\sigma_0$ shows a bell-like dependence with the sliding speed. Although the model from Equations (5)–(7) appears of empirical nature, it can match more complex theories accounting for the molecular mechanisms of the bonding process in the contact zones. In [31], rate constants for bond formation and breakage were related to the bond activation energy and the strain energy acting on stretched chains during sliding, ultimately resulting in a dimensionless expression for the shear stress:

$$\hat{\tau} = \frac{\tau(T,v)}{\tau_0 \tau_*} = \frac{\left(\frac{1}{\hat{V}}\right)\exp\left(\frac{u}{\hat{V}}\right)}{1+\left(\frac{1}{\hat{V}}\right)\exp\left(\frac{u}{\hat{V}}\right)E_1\left(\frac{u}{\hat{V}}\right)}\left[G_1\left(\frac{u}{\hat{V}}\right) - \ln\left(\frac{u}{\hat{V}}\right)E_1\left(\frac{u}{\hat{V}}\right)\right], \tag{8}$$

where E_1 and G_1 are exponential integral functions defined as:

$$E_1(x) = \int_x^\infty \frac{e^{-y}}{y}dy, \quad G_1(x) = \int_x^\infty \frac{e^{-y}}{y}\ln(y)dy. \tag{9}$$

Other quantities in Equation (8) are the dimensionless sliding velocity $\hat{V}(T) = a'_T(T) \cdot v/v_0$, the bond activation energy parameter u and the shear stress scaling factor τ_*.

As the models proposed in [16,31] both describe the bonding friction contribution as a function of the sliding velocity, they should provide similar results. One can thus seek suitable relationships between the parameters of Equations (5)–(7) and u, τ_* in Equations (8) and (9). Based on a numerical parametric study, the following heuristic expressions were obtained for the calculation of u and τ_*:

$$u = \exp\left[-(0.578c + 0.0325)^{-1}\right], \quad \tau_* = (1.8c + 0.0384)^{-1}. \tag{10}$$

Table 3 summarizes all friction model parameters for the four considered elastomers.

Table 3. Friction model parameters of Equations (5)–(7) and Equations (8) and (9) for the four elastomer materials. Here, k, c and ϵ are determined from the friction experiments in Section 3.1, while u and τ_* are calculated from Equation (10).

Material	k [-]	c [-]	ϵ [eV]	u [-]	τ_* [-]
FKM 1	2.2	0.16	0.89	3.35×10^{-4}	3.06
FKM 2	2.0	0.19	0.89	9.12×10^{-4}	2.62
NBR	2.8	0.29	1.1	6.73×10^{-3}	1.78
EPDM	1.7	0.19	0.77	9.12×10^{-4}	2.62

The second friction contribution in Equation (4), i.e., μ_{transfer}, accounts for the formation and shearing of an elastomer transfer film in the contact spots. This process occurs in almost all tribological applications involving dry sliding of rubber, where the molding skin of the elastomer is quickly worn and deteriorated polymer chains adhere to the counter surface, forming a thin rubbery layer. In our experiments, such transfer films form and contribute to the friction coefficient at all speeds and temperatures—i.e., regardless of whether the elastomer displays a rubbery or a glassy response. The exact physical origin of this contribution is still unclear: it has been attributed to hard fillers of the elastomer scratching the counter surface [17] or the thermoplastic-like behavior of the rubber material in its glassy state [16]. We speculate that it may be due to inter-chain friction between the chains still attached to the polymer and those transferred to the counterface. This would explain why it is present at all temperatures and speeds, as well as with elastomers that have no hard fillers capable of scratching the counterface. In any case, friction related to the shearing of the transfer film can be best quantified at temperatures below glass transition, where other contributions to the friction coefficient vanish. For the simulations presented here, we use $\mu_{transfer} = 0.4$, in good agreement with [16].

$\mu_{viscoelastic}$: Viscoelastic friction occurs in a bulk elastomer in contact with a rough hard countersurface, whose asperities lead to a pulsating excitation and energy losses in the rubber as sliding occurs. The viscoelastic dissipation is highest at excitation frequencies corresponding to the maximum of the loss modulus of the elastomer. Persson and coworkers have developed a comprehensive model to calculate this contribution in prior decades [5], including a model to calculate the local temperature increase due to the sliding [32,33] which is also incorporated in the present model.

$$\mu_{viscoelastic} = \frac{1}{2}\int_q q^3 C(q) S(q) A_{rel}(q) \left[\int_0^{2\pi} \cos\phi \, Im\left[\frac{E^*(qv\cos\phi, T_q)}{\sigma_0} \right] d\phi \right] dq, \qquad (11)$$

where the correction factor $S(q)$ has been adapted from [29]:

$$S(q) = 1 - \frac{2}{9}\left(1 - A_{rel}(q)^2\right) - \frac{2}{3}\left(1 - A_{rel}(q)^4\right). \qquad (12)$$

In Equation (11), $Im(E^*)$ is the loss modulus of the elastomer at the contact temperature T_q and frequencies related to the sliding speed v and roughness wavevector q. The factor $\cos\phi$ and the related integral account for the angle between the sliding direction and roughness at different wavelengths, so that the impact of anisotropic roughness on viscoelastic friction is modeled correctly. One can also recognize further contributions of the surface roughness in $C(q)$ and of the relative contact area $A_{rel}(q)$.

μ_{crack}: the crack opening contribution to friction occurs at the edges of the real contact area as two surfaces are separated under pull-off, rolling or sliding. The opening of the contact edge is similar to a Mode-I crack propagating in the rubber material, causing viscoelastic losses depending on the crack radius, propagation speed and the dimension of the contact spots [34,35]. This friction contribution is relevant for rolling contacts, but under pure sliding it is expected to only play a minor role on total friction [35].

2.2.3. Static Friction Model and Its Contributions

We now propose an extension of the dry dynamic friction model to account for breakloose friction as a function of different standstill times $t_{standstill}$. Coherent with the experiments, the static friction coefficient $\mu_{static} = \tau_{static}/\sigma_0$ is calculated from the maximum shear stress τ_{static} occurring at the onset of sliding.

One should note that, although established in tribology, the wording "static friction" is a misnomer, as it implies the absence of motion in the system. In reality, microscopic movements occur at the onset of sliding, be it through a deformation of the bulk materials [36] and localized siding [37,38], or creep [39] at the sliding interface. Ultimately, it is reasonable to assume that the same friction phenomena occurring in the contact zones under dynamic conditions, i.e., under macroscopic sliding, also apply to the microscopic motion of the breakloose process.

Thus, our static friction model is based on the contributions discussed in the previous section, with some changes:

$$\mu_{static} = \frac{\tau_{f,static}}{\sigma_0} A_{rel,static} + \mu_{transfer} + \mu_{viscoelastic}(A_{rel,static}) + \mu_{crack}(A_{rel,static}). \qquad (13)$$

The main difference is a modified contact area $A_{rel,static}$ accounting for relaxation and creep of the elastomer during standstill under a given normal load. Its calculation employs the softer contact modulus $E^*(t_{standstill})$, which leads to an increased contact area compared to sliding conditions. Apart from the constant μ_{transfer}, all other friction contributions scale up with the ratio $A_{\text{rel,static}}/A_{\text{rel,sliding}}$, which can become large when the elastomer has a glassy response under sliding and a rubbery state during standstill.

The static friction model also employs a modified definition of the bonding shear stress $\tau_{f,\text{static}}$, based on an adapted bond population model from [39]. Here, the time evolution of interfacial shear stress is expressed in terms of bond formation, bond breaking due to strain and the resulting interfacial creep velocity through the following set of dimensionless differential equations (adapted from [39]):

$$\frac{d\hat{N}_w}{d\hat{t}} = [1 - \hat{N}_w - \hat{N}_s] - \hat{N}_w\, u \exp\left(\frac{\hat{\tau}_w}{\hat{N}_w}\right), \quad \frac{d\hat{N}_s}{d\hat{t}} = -\hat{N}_s\, u \exp\left(\frac{\hat{\tau}_s}{\hat{N}_s}\right),$$
$$\frac{d\hat{\tau}_w}{d\hat{t}} = \frac{d\hat{N}_w}{d\hat{t}}\frac{\hat{\tau}_w}{\hat{N}_w} + \hat{N}_w\, \hat{V}_c, \quad \frac{d\hat{\tau}_s}{d\hat{t}} = \frac{d\hat{N}_s}{d\hat{t}}\frac{\hat{\tau}_s}{\hat{N}_s} + \hat{N}_s\, \hat{V}_c, \quad (14)$$
$$\frac{d(\hat{\tau}_w + \hat{\tau}_s)}{d\hat{t}} = \hat{V} - \hat{V}_c.$$

Here, $\hat{t}$ is the dimensionless time, $\hat{N}$ is the number of interfacial bonds, u the bond activation energy parameter, $\hat{V}$ the applied speed, $\hat{V}_c$ the interfacial creep velocity and $\hat{\tau}$ the shear stress. Indexes s and w refer to two types of interfacial bonds: so-called strong bonds, created in the standstill phase, and weak bonds, created in the breakloose phase, i.e., the time required to achieve macroscopic motion.

Initial conditions for Equation (14) are $\hat{N}_w = \hat{\tau}_w = \hat{\tau}_s = 0$ and $\hat{N}_s(\hat{t}=0) = \hat{N}_{\text{init}}$, i.e., the number of strong bonds created during standstill. A best fit of experimental data gives $\hat{N}_{\text{init}} \approx 0.025 \log 10(t_{\text{standstill}}) < 0.1$ for typical experimental standstill times ranging from seconds up to a few hours. The other parameters and scaling for the shear stress in Equation (14) are the same as in the dynamic friction model and can be found in Table 3.

The differential equations are solved for $\hat{t} < \left(\frac{1}{\hat{V}}\right)\exp\left(\frac{u}{\hat{V}}\right)E_1\left(\frac{u}{\hat{V}}\right)$, i.e., the life expectancy of bonds from [31] to account for spontaneous bond desorption during the breakloose phase and subsequent sliding at vanishing velocities. Finally, the breakloose stress peak corresponds to the maximum stress $\hat{\tau}_{f,\text{static}} = \max(\hat{\tau}_w(\hat{t}) + \hat{\tau}_s(\hat{t}))$. Ultimately, the evolution of $\tau_{f,\text{static}}$ with the applied sliding velocity v shows a similar bell curve as the equivalent friction contribution τ_f from the dynamic model at low speeds, while at high speeds $\tau_{f,\text{static}} \propto \ln(v/u)$ in accordance with [39].

In engineering, static friction is usually modeled as a constant independent on the operating conditions. It may seem surprising that in the proposed model μ_{static} actually depends on the applied sliding velocity, yet this behavior is coherent with the literature and our experiments (see Section 3.1). This is because both bond formation and breaking are rate processes, which introduce a time and speed dependence of the shear stress at the sliding interface. Ultimately, this is the same velocity-dependent bonding contribution found in dynamic friction, but considered in a different sliding phase: while dynamic friction focuses on the steady state, static friction models the breakloose process prior to macroscopic motion.

This can be understood as the shearing equivalent to pull-off experiments for adhesion testing: Polymeric solids are known to exhibit a strong dependence of the pull-off force versus the speed [40], linked to interfacial bond breaking and crack opening [41]. Literally everyone knows that one way to minimize pain when tearing off a bandage from the skin is to pull very slowly to allow the glue to un-stick.

3. Results & Discussion

3.1. Model Parametrization and Comparison with Experiments

Figure 5 shows typical dynamic friction curves for the four selected elastomers in dry conditions. Here, the friction coefficient is plotted as a function of the sliding speed, which varies in the experiments by 2.5 decades. Furthermore, the wide temperature range (from $-20\,°C$ up to 80–$100\,°C$) covers the rubbery regime for all elastomers and extends to below the glass transition for the FKM and NBR materials. In Figure 5, the points correspond to experimental data and the lines to the simulation model.

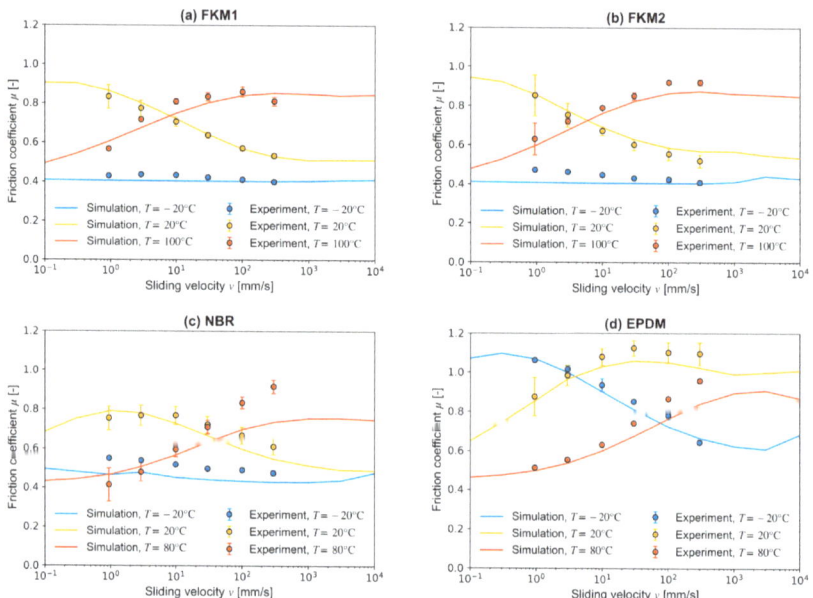

Figure 5. Comparison of the dry dynamic elastomer friction model with experimental data: (**a**) FKM 1, (**b**) FKM 2, (**c**) NBR and (**d**) EPDM.

In the latter, most parameters are fixed, but the constants k, c and ϵ of Table 3 are elastomer-dependent and need to be adjusted to experimental data. First, the transfer film, viscoelastic and crack propagation contributions are determined from the friction model in Section 2.2.2. They are then subtracted from the total experimental friction coefficient to obtain the bell-shaped curve for the bonding friction contribution. The height, width and location of the maximum of this bell-shaped curve relate to k, c and ϵ, respectively, allowing to calculate the three friction model parameters through a numerical fit.

In this work, experimental data at all considered temperatures were used as a base for the fitting procedure. However, we found that the friction model can be well parameterized from data at a single temperature—generally around ambient—if the chosen (T, v) range covers adequately the bonding friction contribution and its maximum.

The quantitative agreement of experiments and simulations is generally very good, with an average deviation between experiment and simulation below 10%, although in specific conditions (e.g., NBR at 80 °C) larger differences are seen. Overall, the model appears capable of predicting dry elastomer friction for virtually any speed and temperature, with a relatively small parametrization effort requiring only few tribological experiments.

Furthermore, once the elastomer-dependent parameters k, c and ϵ have been determined for sliding conditions, the model can also be applied to quantify breakloose friction. Figure 6 shows how the static friction coefficient depends on the applied macroscopic sliding velocity (a–c) and standstill time (d). Additionally, in this case the simulations capture the main features of the experimental data, with a comparable accuracy to the dynamic friction model.

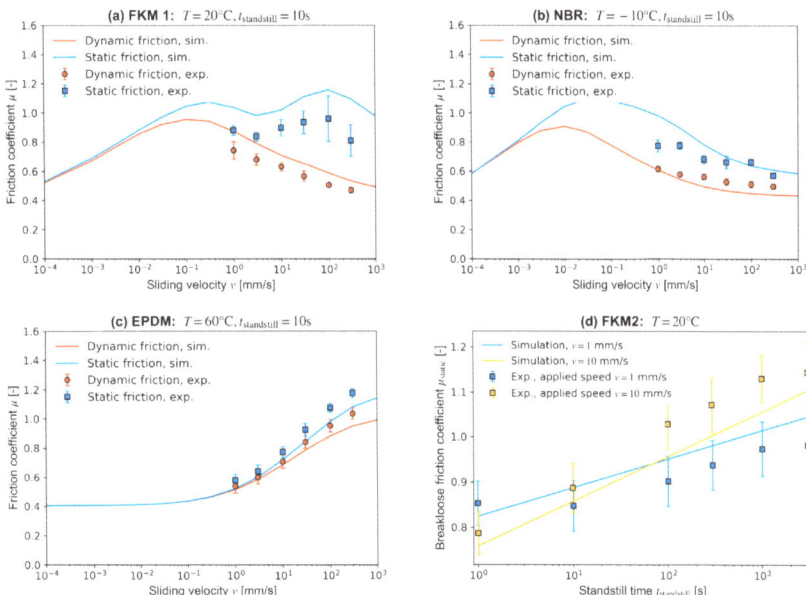

Figure 6. Comparison of the elastomer friction model with experimental data: (**a**–**c**) Dependence of dynamic and breakloose friction on the applied sliding velocity in the experiment. (**d**) Dependence of static friction on standstill time, for two given sliding velocities.

3.2. Relevance of Friction Contributions

In the development framework for new sealing product and rubbers, the validated dry friction model can be used to gain new insights on the tribological behavior of elastomers. First, one can study the system under conditions which are difficult to access experimentally, e.g., to quantify friction at very low or high sliding velocities, or for long standstill times of the order of days or weeks.

Furthermore, the model can be used to analyze which friction contributions are relevant under given tribological conditions, ultimately providing guidance on how to optimize a certain system, for instance, by tuning the elastomer properties or the surface finishing, to obtain a desired friction behavior. We discuss this aspect in the following, focusing on the operating conditions in temperature and sliding speed first.

Figure 7a shows the simulated friction for FKM 1 at −20 °C, well below its glass transition temperature. As expected, the transfer film μ_transfer dominates the friction process, with the other terms vanishing at sliding speeds above 1 mm/s. At 20 °C (see Figure 7b), contact friction due to interfacial bonds and the transfer film is the most important contribution at sliding speeds of up to 10 mm/s, while at higher velocities the viscoelastic term also becomes significant. The contribution of crack opening at the asperity scale is small throughout the analyzed range of sliding speeds. Increasing the temperature to 100 °C (see Figure 7c) shifts the peak of all contributions to higher sliding velocities, in accordance with the shift factors a'_T for the interfacial bonds and a_T for the viscoelastic friction. This results in μ_contact being the only relevant contribution in the considered velocity range. At even higher sliding speeds beyond 1 m/s, $\mu_\text{viscoelastic}$ may also become significant, but the elastomer would heat up and fail rapidly in dry conditions, which makes this practically irrelevant for technical systems.

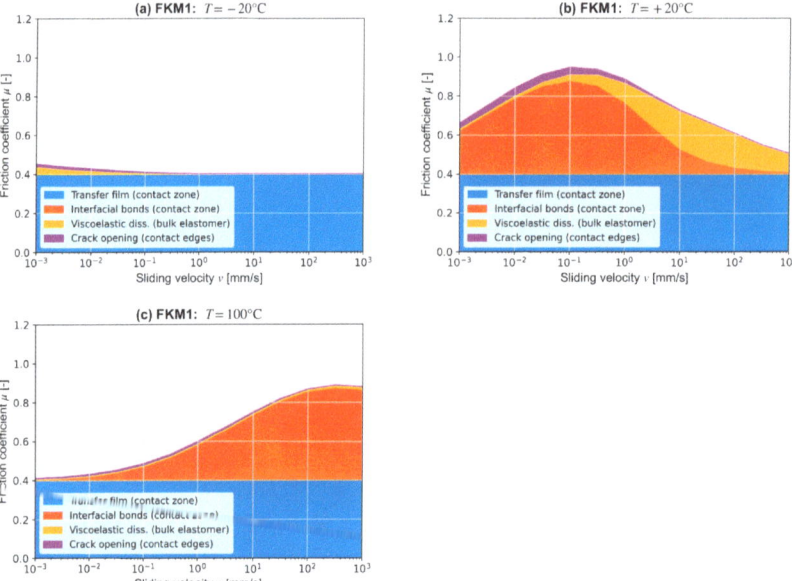

Figure 7. Dynamic friction contributions for FKM1 at different temperatures: (**a**) −20 °C, (**b**) 20 °C and (**c**) 100 °C.

In Figure 8, dynamic friction for FKM 2 is compared to the contributions for breakloose friction at standstill times of a few seconds and a few hours. According to the proposed model, μ_{transfer} is unchanged, μ_{crack} remains negligible over the whole velocity range and the bonding and viscoelastic terms change significantly at applied speeds exceeding 1 mm/s. Both μ_{bonding} and $\mu_{\text{viscoleastic}}$ scale up with the contact area ratio $A_{\text{rel,static}}/A_{\text{rel,sliding}}$. As shown in Figure 8d for the two considered standstill times, this quantity becomes significantly larger than unity for $v > 10$ mm/s, where FKM 2 at $T = 20$ °C showcases a glassy response under sliding and a rubbery behavior during standstill.

In this velocity range, the bonding friction contribution μ_{bonding} decays rapidly under dynamic conditions, as the rate for bond formation is slow compared to the characteristic time of sliding. However, during standstill strong bonds can form with the counterface (see Section 2.2.3), which must be sheared during the breakloose process, causing increased friction. This additional contribution to the bonding term is visualized through the dotted line in Figure 8b,c. At standstill times of a few seconds or minutes, its impact on the total friction coefficient remains minor. However, it becomes significant at longer $t_{\text{standstill}}$ periods, which allow more strong bonds to form at the interface over an increased contact area $A_{\text{rel,static}}$.

Finally, the viscoelastic contribution also scales up with the ratio $A_{\text{rel,static}}/A_{\text{rel,sliding}}$ compared to dynamic friction, leading to a term of the same order of magnitude of the bonding friction contribution in the $v > 10$ mm/s velocity range. When summing all static friction contributions, a second peak can appear at high applied velocities in addition to the maximum of the bonding friction bell curve. This is, for instance, the case for FKM 1 at $T = 20$ °C seen in Figure 6a and for FKM 2 in Figure 8c.

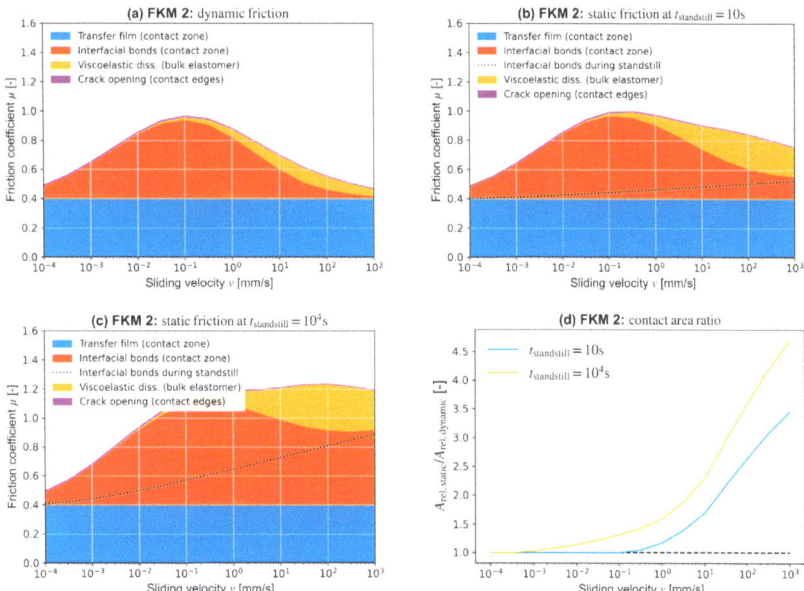

Figure 8. (a) Dynamic friction contributions for FKM 2 at 20 °C. (b,c) Static friction contributions under the same conditions for two different standstill times. (d) Ratio of contact areas $A_{\text{rel,static}} / A_{\text{rel,sliding}}$ as a function of the applied velocity v.

We now investigate how the dynamic friction contributions vary as a function of surface roughness. Up to now, the reference surface was the ground steel sample mentioned in Section 2.1.2, with the sliding direction set as perpendicular to the grinding pattern. The results are recalled in Figure 9a for reference. In Figure 9b, a similar ground surface is used, but this time the rubber slides along the surface "grooves". While the overall friction coefficient stays approximately constant, two significant changes occur in the friction contributions. First, the viscoelastic term is strongly reduced as the pulsating excitations due to roughness and related energy losses in the rubber are suppressed. Second, the bonding friction contribution rises in the velocity range between 1 mm/s and 1 m/s. Here, in absence of a high frequency excitation, the elastomer showcases a rubbery response with a softer contact modulus, leading to an increased contact area and higher $\mu_{bonding}$. Similar trends can be observed when considering a polished counter surface: the viscoelastic term vanishes completely, leaving contact friction as the only contribution of the friction coefficient (Figure 9c).

Table 4. Metrics for different countersurfaces: The root mean square roughness height h_{rms} and slope h'_{rms} are defined in [12]. U_{el}/A_0 is the elastic energy per unit area to form full contact, calculated according to [9] with a representative contact modulus $E^* = 13.3$ MPa.

Surface	h_{rms} [μm]	U_{el}/A_0 [J/m^2]	h'_{rms} [-]
Ground steel (perpendicular case)	0.47	0.25	0.28
Ground steel (parallel case)	0.48	0.19	0.18
Polished steel	0.059	0.0071	0.049
Concrete road	141	247	1.22

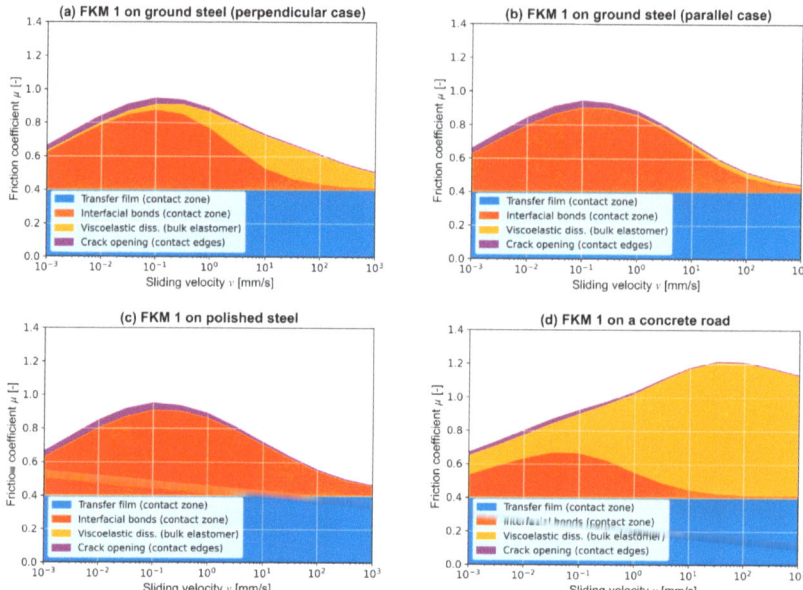

Figure 9. Dynamic friction contributions for FKM 1 sliding over different countersurfaces at 20 °C, with their main surface metrics listed in Table 4. (**a**) Standard configuration with a ground steel surface. Here, the sliding direction is perpendicular to the roughness pattern of the metal counterface. (**b**) A similar ground steel surface, with sliding occurring parallel to the roughness pattern. (**c**) Polished steel surface. (**d**) Concrete road surface.

Based on these results, it may seem surprising that Persson developed a model for viscoelastic rubber friction first [5], with the contact model following more than a decade later [30]. Other groups have also focused on the viscoelastic contribution [42]. It should be noted that most literature on dry rubber friction has mainly treated tire–road contacts. To understand what difference this makes in terms of friction contributions, we have performed simulations using data for a concrete road surface from [16], as seen in Figure 9d. Its root mean square roughness $h_{\mathrm{rms}} \approx 141$ μm is about 300 times larger than the ground steel surfaces where $h_{\mathrm{rms}} \approx 0.5$ μm. A comparison of several important surface metrics in Table 4 highlights the differences.

Between Figure 9a with Figure 9d, one can see important differences. First, the viscoelastic contribution is much bigger at all sliding speeds for the concrete road surface, leading to a significantly increased friction peak at approx. 50 mm/s where the viscoelastic losses in the elastomer are maximal. This is due to the larger dynamic excitation of the rubber bulk material during sliding through the increased roughness of the road countersurface. The latter also leads to a reduction in contact area A_{true} under the same given external load compared to the reference case of Figure 9a, so that the interfacial bond contribution becomes secondary for the total friction coefficient. Finally, in tire–road contacts, the rolling motion also leads to an increased crack opening contribution. This aspect is not considered in our model, which has been fine-tuned for the sliding motion typical of sealing applications.

Ultimately, it is crucial to ensure that any simulation model provides an adequate description for the system under study. Specifically for elastomer friction, when looking into tire contacts, $\mu_{\mathrm{viscoelastic}}$ is indeed a key contribution due to the large roughness of road surfaces. However, typical engineering surfaces used in sealing applications are orders of magnitude smoother, such that the contact friction contribution becomes dominant due to the increased contact area and the reduction of viscoelastic losses.

4. Summary and Conclusions

In this work, we present a physically based model for the static and dynamic friction of elastomers in dry conditions. For steady state sliding, we base our approach on the works of Persson and coworkers. The total friction coefficient is divided into several contributions, each representative of a separate friction process. Both interfacial bonding and shearing of a transfer film occur in the true contact area between the elastomer and counterface. Viscoelastic dissipation occurs in the elastomer bulk, while a crack opening process takes place at contact edges during sliding.

We then propose an extension of the model to treat static friction. Here, the friction contributions scale up with the increased contact area due to elastomer relaxation during standstill. Additionally, the interfacial bonding contribution is improved to account for creep motion in the breakloose phase, based on bond population models by Juvekar and coworkers.

After parameterization of only three elastomer-dependent constants, the model can predict friction for a given rubber–counterface contact at several applied speeds, temperatures or standstill times. The model is compared to tribological experiments, showing good accuracy over a wide range of conditions for four commercial rubber compounds with different base polymers.

For typical engineering surfaces used in sealing applications, the most important friction contribution comes from contact area between the rubber and counterface, i.e., from interfacial bonding and the transfer film. However, the relative importance of individual friction phenomena can change drastically depending on the roughness of the hard counterbody. When considering sliding over a road surface, viscoelastic friction increases significantly and becomes the dominant contribution.

From our perspective as a sealing and lubricant manufacturing company, this information is crucial to develop novel products and optimized elastomer formulations for customer applications, as the required changes will depend on the most relevant friction contributions in each specific case.

Accessing this knowledge is very difficult with traditional approaches based purely on experiments. The viscoelastic and bonding contributions, both bell-shaped curves with the sliding velocity, can differ significantly in terms of amplitude and width. These differences are large enough to change the frictional behavior, depending on the operating conditions. However, the two contributions overlap over a wide range of speeds and feature similar temperature shift factors, so it is impossible to accurately discern the two terms by looking at the total, i.e., experimental friction coefficient.

Conversely, the friction contributions in each tribological system can be quickly predicted in a quantitative way from the current simulation model for dry elastomer friction. This will constitute a powerful tool to speed up material and product development, allowing for better transferability between simple tribological tests and actual product operation, and providing crucial knowledge for elastomer optimization.

Finally, this model can also be used as a solid foundation to model lubricated conditions where dry contact can occur locally depending on the elastomer–fluid wetting properties [23].

Author Contributions: Conceptualization, F.K. and D.S.; methodology, F.K. and D.S.; software, F.K. and D.S.; validation, F.K.; writing—original draft preparation, D.S. and F.K.; writing—review and editing, F.K., D.S. and R.B.; visualization, F.K. and D.S.; supervision, R.B.; project administration, F.K. and R.B.; funding acquisition, F.K. and R.B. All authors have read and agreed to the published version of the manuscript.

Funding: This research received no external funding.

Data Availability Statement: Some data presented in this article are not readily available for confidentiality reasons. Requests to access the datasets should be directed to F.K.

Acknowledgments: We would like to thank Michele Scaraggi for the fruitful discussions regarding theory development and implementation. The authors would also like to thank Freudenberg for permission to publish this work for the 175th anniversary of the company.

Conflicts of Interest: The test specimens examined in this study are the property of Freudenberg Technology Innovation SE & Co. KG. The authors are employees of Freudenberg Technology Innovation SE & Co. KG. The results were obtained in the course of their work.

References

1. Kammüller, M. Zum Abdichtverhalten von Radial-Wellendichtringen. Ph.D. Thesis, University of Stuttgart, Stuttgart, Germany, 1986.
2. Müller, H.K.; Nau, B. *Fluid Sealing Technology*; Marcel Dekker: New York, NY, USA, 1998.
3. Greenwood, J.A.; Williamson, J.B.P. Contact of Nominally Flat Surfaces. *Proc. R. Soc. A Math. Phys. Eng. Sci.* **1966**, *295*, 300–319.
4. Persson, B.N.J. Theory of powdery rubber wear. *J. Phys. Condens. Matter* **2009**, *21*, 485001. [CrossRef] [PubMed]
5. Persson, B.N.J. Theory of rubber friction and contact mechanics. *J. Chem. Phys.* **2001**, *115*, 3840–3861. [CrossRef]
6. Müser, M.H.; Dapp, W.B.; Bugnicourt, R.; Sainsot, P.; Lesaffre, N.; Lubrecht, T.A.; Persson, B.N.J.; Harris, K.; Bennett, A.; Schulze, K.; et al. Meeting the Contact-Mechanics Challenge. *Tribol. Lett.* **2017**, *65*, 118. [CrossRef]
7. Persson, B.N.J.; Scaraggi, M. Theory of adhesion: Role of surface roughness. *J. Chem. Phys.* **2014**, *141*, 124701. [CrossRef] [PubMed]
8. Persson, B.N.J. Contact mechanics for layered materials with randomly rough surfaces. *J. Phys. Condens. Matter* **2012**, *24*, 095008. [CrossRef] [PubMed]
9. Persson, B.N.J. Contact mechanics for randomly rough surfaces. *Surf. Sci. Rep.* **2006**, *61*, 201–227. [CrossRef]
10. Persson, B.N.J.; Yang, C. Theory of the leak-rate of seals. *J. Phys. Condens. Matter* **2008**, *20*, 315011. [CrossRef]
11. Tiwari, A.; Persson, B.N.J. Physics of suction cups. *Soft Matter* **2019**, *15*, 9482–9499. [CrossRef]
12. Persson, B.N.J. Fluid Leakage in Static Rubber Seals. *Tribol. Lett.* **2022**, *70*, 31. [CrossRef]
13. Fischer, F.J.; Schmitz, K.; Tiwari, A.; Persson, B.N.J. Fluid Leakage in Metallic Seals. *Tribol. Lett.* **2020**, *68*, 125. [CrossRef]
14. Pérez-Ràfols, F.; Larsson, R.; Almqvist, A. Modelling of leakage on metal-to-metal seals. *Tribol. Int.* **2016**, *94*, 421–427. [CrossRef]
15. Jin, L.; Cheng, Y.; Zhang, K.; Xue, Z.; Liu, J. Axisymmetric model of the sealing cylinder in service: Analytical solutions. *J. Mech.* **2021**, *37*, 404–414. [CrossRef]
16. Tiwari, A.; Miyashita, N.; Espallargas, N.; Persson, B.N.J. Rubber friction: The contribution from the area of real contact. *J. Chem. Phys.* **2018**, *148*, 224701. [CrossRef]
17. Tolpekina, T.V.; Persson, B.N.J. Adhesion and Friction for Three Tire Tread Compounds. *Lubricants* **2019**, *7*, 20. [CrossRef]
18. Wohlers, A.; Heipl, O.; Persson, B.N.J.; Scaraggi, M.; Murrenhoff, H. Numerical and Experimental Investigation on O-Ring-Seals in Dynamic Applications. *Int. J. Fluid Power* **2009**, *10*, 51–59. [CrossRef]
19. Dowson, D. *History of Tribology*; Wiley: Hoboken, NJ, USA, 1998.
20. Almqvist, A. On the Effects of Surface Roughness in Lubrication. Ph.D. Thesis, Luleå University of Technology, Luleå, Sweden, 2006.
21. Persson, B.N.J.; Scaraggi, M. Lubricated sliding dynamics: Flow factors and Stribeck curve. *Eur. Phys. J. E* **2011**, *34*, 113. [CrossRef] [PubMed]
22. Bongaerts, J.H.H.; Foutouni, K.; Stokes, J.R. Soft-tribology: Lubrication in a compliant PDMS–PDMS contact. *Tribol. Int.* **2007**, *40*, 1531–1542. [CrossRef]
23. Persson, B.N.J.; Mugele, F. Squeeze-out and wear: Fundamental principles and applications. *J. Phys. Condens. Matter* **2004**, *16*, 295–355. [CrossRef]
24. Martin, A.; Clain, J.; Burguin, A.; Brochard-Wyart, F. Wetting transitions at soft, sliding interfaces. *Phys. Rev. E* **2002**, *65*, 031605. [CrossRef]
25. Williams, M.L.; Landel, R.F.; Ferry, J.D. The Temperature Dependence of Relaxation Mechanisms in Amorphous Polymers and Other Glass-forming Liquids. *J. Am. Chem. Soc.* **1955**, *77*, 3701–3707. [CrossRef]
26. Jacobs, T.B.; Junge, T.; Pastewka, L. Quantitative characterization of surface topography using spectral analysis. *Surf. Topogr. Metrol. Prop.* **2017**, *5*, 13001. [CrossRef]
27. Persson, B.N.J.; Albohr, O.; Tartaglino, U.; Volokitin, A.I.; Tosatti, E. On the nature of surface roughness with application to contact mechanics, sealing, rubber friction and adhesion. *J. Phys. Condens. Matter* **2005**, *17*, R1. [CrossRef]
28. Yang, C.; Persson, B.N.J. Contact mechanics: Contact area and interfacial separation from small contact to full contact. *J. Phys. Condens. Matter* **2008**, *20*, 215214. [CrossRef]
29. Wang, A.; Müser, M.H. Gauging Persson Theory on Adhesion. *Tribol. Lett.* **2017**, *65*, 103. [CrossRef]
30. Lorenz, B.; Oh, Y.R.; Nam, S.K.; Jeon, S.H.; Persson, B.N.J. Rubber friction on road surfaces: Experiment and theory for low sliding speeds. *J. Chem. Phys.* **2015**, *142*, 194701. [CrossRef]
31. Singh, A.K.; Juvekar, V.A. Steady dynamic friction at elastomer–hard solid interface: A model based on population balance of bonds. *Soft Matter* **2011**, *7*, 10601–10611. [CrossRef]
32. Fortunato, G.; Ciaravola, V.; Furno, A.; Lorenz, B.; Persson, B.N.J. General Theory of frictional heating with application to rubber friction. *J. Phys. Condens. Matter* **2015**, *27*, 175008. [CrossRef]

33. Persson, B.N.J. Rubber friction: Role of the flash temperature. *J. Phys. Condens. Matter* **2006**, *18*, 7789–7823. [CrossRef]
34. Persson, B.N.J. Crack propagation in finite-sized viscoelastic solids with application to adhesion. *Europhys. Lett.* **2017**, *119*, 18002. [CrossRef]
35. Tiwari, A.; Dorogin, L.; Tahir, M.; Stöckelhuber, K.W.; Heinrich, G.; Espallargas, N.; Persson, B.N.J. Rubber contact mechanics: Adhesion, friction and leakage of seals. *Soft Matter* **2017**, *13*, 9103–9121. [CrossRef] [PubMed]
36. Lengiewicz, J.; de Souza, M.; Lahmar, M.A.; Courbon, C.; Dalmas, D.; Stupkiewicz, S.; Scheibert, J. Finite deformations govern the anisotropic shear-induced area reduction of soft elastic contacts. *J. Mech. Phys. Solids* **2020**, *143*, 104056. [CrossRef]
37. Lorenz, B.; Persson, B.N.J. On the origin of why static or breakloose friction is larger than kinetic friction, and how to reduce it: The role of aging, elasticity and sequential interfacial slip. *J. Phys. Condens. Matter* **2012**, *24*, 225008. [CrossRef] [PubMed]
38. Maegawa, S.; Itoigawa, F.; Nakamura, T. New insight into the mechanism of static friction: A theoretical prediction of the effect of loading history on static friction force based on the static friction model proposed by Lorenz and Persson. *Tribol. Int.* **2016**, *102*, 532–539. [CrossRef]
39. Soni, P.; Singh, A.; Katiyar, J. Experiments and Prediction of Hold Time-Dependent Static Friction of a Wet Granular Layer. *Tribol. Lett.* **2023**, *71*, 75. [CrossRef]
40. Creton, C.; Ciccotti, M. Fracture and adhesion of soft materials: A review. *Rep. Prog. Phys.* **2016**, *79*, 046601. [CrossRef]
41. Müser, M.H.; Persson, B.N.J. Crack and pull-off dynamics of adhesive, viscoelastic solids. *Europhys. Lett.* **2022**, *137*, 36004. [CrossRef]
42. Le Gal, A.; Klüppel, M. Investigation and modelling of rubber stationary friction on rough surfaces. *J. Phys. Condens. Matter* **2008**, *20*, 015007. [CrossRef]

Disclaimer/Publisher's Note: The statements, opinions and data contained in all publications are solely those of the individual author(s) and contributor(s) and not of MDPI and/or the editor(s). MDPI and/or the editor(s) disclaim responsibility for any injury to people or property resulting from any ideas, methods, instructions or products referred to in the content.

Article

Effect of Hydrogen Pressure on the Fretting Behavior of Rubber Materials

Géraldine Theiler [1,*], Natalia Cano Murillo [1] and Andreas Hausberger [2]

1 Bundesanstalt für Materialforschung und -prüfung (BAM), 12203 Berlin, Germany; natalia.murillo@bam.de
2 Polymer Competence Centre Leoben GmbH (PCCL), 8700 Leoben, Austria
* Correspondence: geraldine.theiler@bam.de

Abstract: Safety and reliability are the major challenges to face for the development and acceptance of hydrogen technology. It is therefore crucial to deeply study material compatibility, in particular for tribological components that are directly in contact with hydrogen. Some of the most critical parts are sealing materials that need increased safety requirements. In this study, the fretting behavior of several elastomer materials were evaluated against 316L stainless steel in an air and hydrogen environment up to 10 MPa. Several grades of cross-linked hydrogenated acrylonitrile butadiene (HNBR), acrylonitrile butadiene (NBR) and ethylene propylene diene monomer rubbers (EPDM) were investigated. Furthermore, aging experiments were conducted for 7 days under static conditions in 100 MPa of hydrogen followed by rapid gas decompression. Fretting tests revealed that the wear of these compounds is significantly affected by the hydrogen environment compared to air, especially with NBR grades. After the aging experiment, the friction response of the HNBR grades is characterized by increased adhesion due to elastic deformation, leading to partial slip.

Keywords: fretting wear; rubbers; hydrogen; high-pressure

Citation: Theiler, G.; Cano Murillo, N.; Hausberger, A. Effect of Hydrogen Pressure on the Fretting Behavior of Rubber Materials. *Lubricants* **2024**, *12*, 233. https://doi.org/10.3390/lubricants12070233

Received: 30 April 2024
Revised: 31 May 2024
Accepted: 14 June 2024
Published: 23 June 2024

Copyright: © 2024 by the authors. Licensee MDPI, Basel, Switzerland. This article is an open access article distributed under the terms and conditions of the Creative Commons Attribution (CC BY) license (https://creativecommons.org/licenses/by/4.0/).

1. Introduction

With the development of hydrogen infrastructure, a growing interest is devoted to research on material compatibility with hydrogen due to high safety requirements in distribution and dispensing infrastructure. Some of the most critical parts are sealing materials that need increased reliability specifications to avoid any leakage, which would lead to an imminent risk of serious damage but also mitigate the acceptance of this technology [1,2].

For static and dynamic seals, polymeric materials are used as sealing components in a wide range of conditions, such as O-rings and piston rings in high-pressure and/or cryogenic hydrogen. Rubber O-ring seals have been commonly used in high-pressure hydrogen storage systems for preventing the leakage of hydrogen gas. However, failure of a sealing component can occur due to swelling induced by dissolved hydrogen [3,4]. It has been reported that most damage to the sealing materials occurs during rapid gas decompression (RGD). Therefore, more and more studies are dedicated to the influence of high-pressure hydrogen and RGD on the properties of rubber materials [5–19]. Most of the works reported the effect of the fillers and additives on the physical properties of the elastomer compounds.

Volume change upon RGD inevitably induces motion of the sealing material, particularly during compression cycles. Further low amplitude oscillating motion can also occur in the reciprocating O-ring seal [15]. It is therefore crucial to characterize the friction and wear of the rubber materials in hydrogen as well as after RGD where more effects occur. Up to now, most literature refers to the effect of the hydrogen environment on the sliding friction and wear of polymers [20–26] or rubbers [27,28].

Works have been published on EPDM and NBR grades under 28 MPa hydrogen in a linear reciprocating custom-built apparatus [27,28]. Both rubber materials showed a higher coefficient of friction (COF) and drastically reduced wear behavior in all hydrogen exposed tests compared with the test results in ambient air conditions. It was suggested that the variations in the friction and wear behavior of EPDM and NBR are due to a combined effect of high-pressure hydrogen and the plasticizer and fillers [28].

Choi et al. studied the friction and wear behavior of NBR rubber under continuous sliding in air after high-pressure exposure in hydrogen [29]. They reported that the hydrogen-induced damage was related to the tribological properties of filled NBR.

Reports on fretting experiments in hydrogen are rare [30]. Zhou et al. developed a numerical model to investigate the fretting characteristic of O-ring seals under the action of swelling behavior induced by dissolved hydrogen [30]. They concluded that the amplitude of reciprocating motion affects the fretting state of rubber seals, and that a high-pressure hydrogen atmosphere promotes a sticking regime because of swelling of the rubber.

The scope of this study is to characterize the fretting behavior of selected rubber materials against 316L steel under hydrogen atmosphere. The focus lies here on the friction behavior and the characterization of the wear scar of the rubber materials. The core of the study deals with the influence of the hydrogen pressure on the friction performance compared to air. Tests were performed in gaseous hydrogen at 0.1 MPa and 10 MPa. In addition, aging experiments were conducted under static conditions in 100 MPa hydrogen, followed by further fretting tests under hydrogen immediately and 24 h after decompression. This work gives an inside view of the relevant mechanisms in seal contact under a hydrogen atmosphere and thus make an important contribution to material and seal development.

2. Materials and Methods

2.1. Materials

This study includes two different types of cross-linked hydrogenated acrylonitrile butadiene (HNBR). Both are filled with CB in different compositions and the acrylonitrile (ACN) content is relatively low, at 21%, to keep a low glass transition temperature and to have good low temperature properties. Furthermore, four grades of nitrile butadiene rubber specially developed for this project were investigated. The effect of vulcanization, filler and plasticizer have been considered in this study as indicated in Table 1. In addition, two sulfur-cured EPDM rubber grades with different carbon black were tested. All materials were provided as a 2 mm sheet by Arlanxeo Deutschland GmbH, Dormagen, Germany.

Table 1. Overview of the materials with the curing system, filler composition and resulting shore A hardness.

Materials	Curing	Fillers	Hardness (ShA)
HNBR-CB	Peroxide	carbon black (75 phr)	79
HNBR-CB-PA	Peroxide	carbon black (67 phr) + PA (10 phr)	82
NBR-Sil	Sulfur	silica (60 phr) + SCA 2 phr	70
NBR-CB-perox	Peroxide	carbon black (70 phr) (MgO)	81
NBR-CB	Sulfur	carbon black (75 phr)	76
NBR-CB-plast	Sulfur	carbon black (95 phr) + plast. (10 phr)	78
EPDM1	Sulfur	carbon black (100 phr)	77
EPDM2	Sulfur	carbon black (120 phr)	81

The samples were cut to 15 mm × 10 mm, cleaned with isopropanol and exposed to dry heat at 60 °C for 48 h. This step was undertaken to remove any moisture and outgassing of the samples that might affect the friction properties.

2.2. Test Method

The test method is described in Figure 1. "As-received sample" means here the cleaned and dried sample. As-received samples were tested at ambient temperature in air and in hydrogen at 0.1 MPa and 10 MPa. Some samples were aged in high-pressure autoclaves according to the methodology described in [13]. After aging experiments, further fretting tests were performed immediately after decompression and 24 h later.

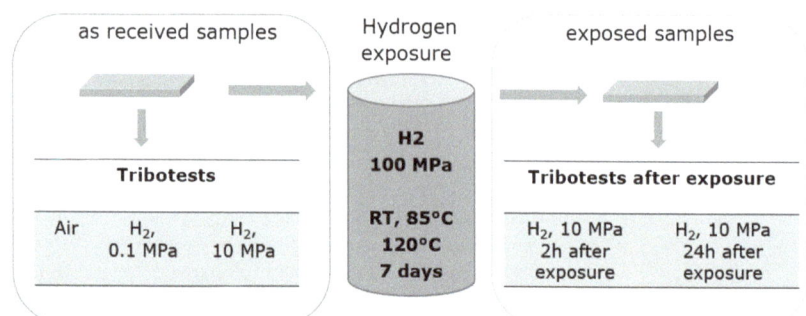

Figure 1. Exposure and test methodology.

Friction tests were performed in a fretting tribometer developed at BAM for hydrogen environments up to 10 MPa (Figure 2). The tribometer is designed as an insert for a stainless-steel autoclave with a maximum gas pressure of 10 MPa. The normal and frictional forces are measured using strain gauges that are attached to deformation bodies in a bridge circuit. The normal force is generated by a spring using an electric motor stretched via a spindle drive. The tribometer is driven by a voice coil drive, which generates the reciprocating motion. Both the normal force and the tangential displacement are set via PID control loops.

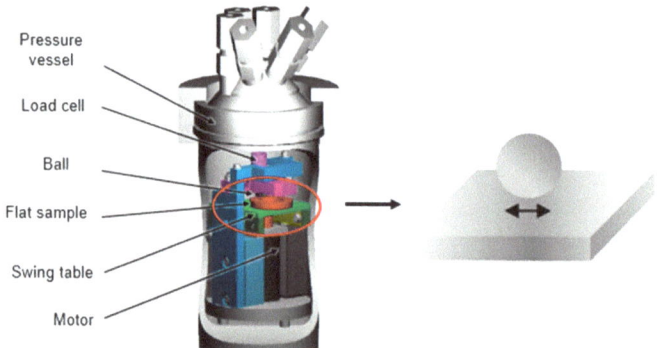

Figure 2. PT1 tribometer and ball-on-disc configuration.

The upper specimen is a 316L stainless steel ball with a diameter of 6.0 mm (G300). The lower specimen is the rubber sample fixed on a steel flat.

The friction measurement of rubber materials is complex and often raises some issues with regards to the setup. Since most rubber samples are available as a plate, friction tests are mostly conducted with a flat rubber material sliding against a steel ball or cylinder in a fretting test [27,31–34] or against a flat counterpart in a linear motion [35]. Other configurations found in the literature are a hemispherical rubber pin or a rubber cylinder against a steel disc [36,37]. In this study, flat rubber samples were used.

The normal load was set to 5 N, the frequency to 10 Hz and the stroke to 2 mm, with a total of 20,000 cycles. Tests were performed at room temperature in air (moisture range 40–60%) and in gaseous hydrogen (H_2O < 5 ppm) at 0.1 MPa and 10 MPa. Three tests were performed for each condition.

2.3. Friction Measurement

During the test, friction force, Ff, values were recorded along with time and displacement. The hysteresis data contains 1000 values per second, leading to 100 points for each cycle. Therefore, after 10 min, the values were recorded every 20 s.

The average friction force (Ff mean) was calculated as follows:

$$\text{Ff mean} = \frac{\sum_{i=1}^{n}|\text{Ff}\, i|}{n}, \qquad (1)$$

The average friction coefficient (COF mean) was calculated as follows:

$$\text{COF mean} = \frac{\text{Ff mean}}{\text{FN}} \qquad (2)$$

where FN is the normal load.

2.4. Surface Analyses and Wear Measurement

Worn surfaces of the rubbers and transfer film formed on the discs were inspected by means of an optical microscope (VHX-500, Keyence Deutschland GmbH, Neu-Isenburg, Germany), a confocal 3D profilometer (μsurf expert, NanoFocus AG, Oberhausen, Germany) and a scanning electron microscope (SEM, Supra™40, Carl Zeiss AG, Oberkochen, Germany) equipped with an energy dispersive X-ray spectrometer. Chemical analyses were performed by micro-ATR-IR attenuated total reflection (Hyperion 3000, Bruker Optics, Ettlingen, Germany) with a germanium crystal. The wavelength range was set from 4000 cm^{-1} to 500 cm^{-1}. An atmospheric compensation, an extended ATR correction and a baseline correction were carried out for all spectra. After the test, the wear scar on the rubber surface was characterized by a 3D profilometer and the wear volume was measured based on the 3D profile, considering the area below the reference line (Figure 3).

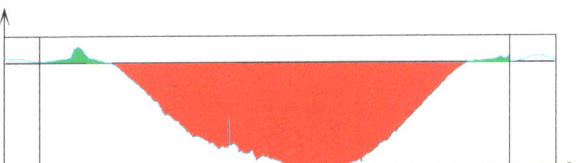

Figure 3. Profile of the wear scar; wear volume corresponds to the red area.

3. Results and Discussion

3.1. Friction Tests on as Received Samples: Effect of Hydrogen Environment

3.1.1. Friction Force Curves

Figure 4 depicts the variation of the friction force (Ff) versus relative displacement (D) as a function of time. The friction loops characterize friction and deformation behavior at the interface. The area of the loops corresponds to the energy dissipated during the cycles, but this was not calculated in this study. As described in [37–39], the fretting mechanism can be deduced from the shape of the fretting hysteresis, which can be linear, elliptical and parallelogram shapes. This corresponds to the three fretting regimes, namely partial slip, mixed fretting and gross slip regime [39].

As shown in Figure 4, the loops of the rubber materials described an ellipse at the beginning of the tests and move toward a quasi-parallelogram, whereby the slope corresponds to the static friction and the horizontal segment to the sliding regime. The fretting regime changes from partial to gross slip. The transitional phase takes place after several cycles for most materials.

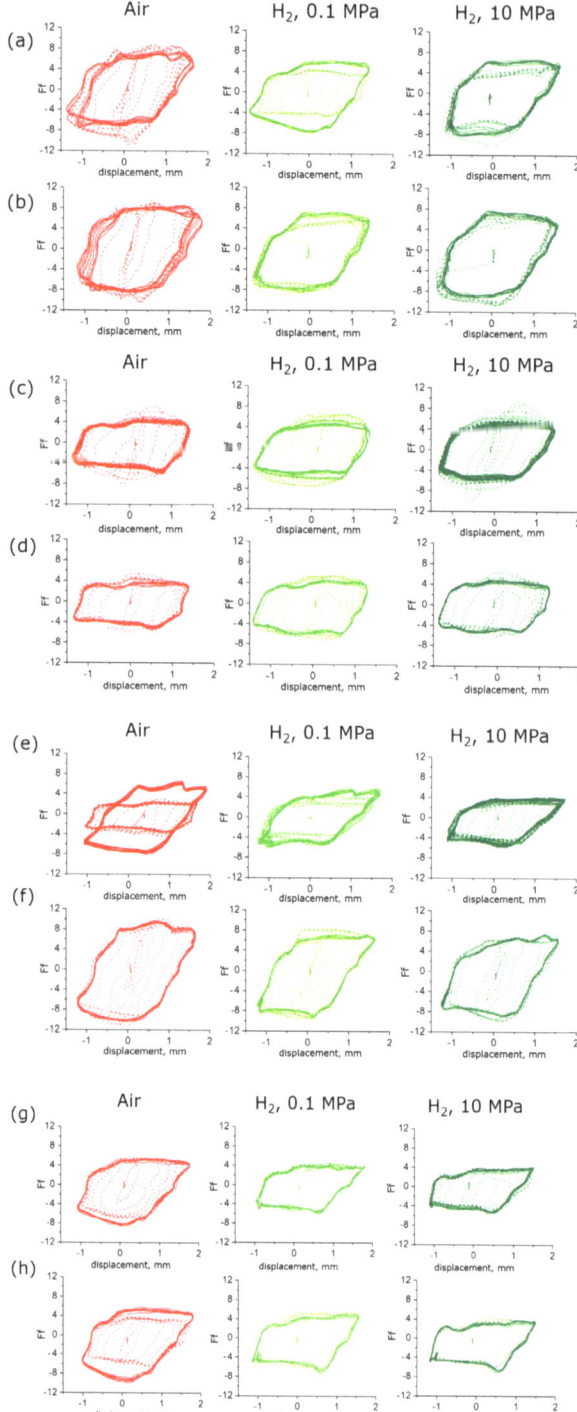

Figure 4. Friction force–D curves of HNBR-CB (**a**), HNBR-CB-PA (**b**), EPDM1 (**c**), EPDM2 (**d**), NBR-Sil (**e**), NBR-CB-perox (**f**), NBR-CB (**g**) and NBR-CB-plast (**h**) in air, and in hydrogen at 0.1 MPa and 10 MPa (dashed line: begin; solid line: end of the test).

For the HNBR and NBR grades, the loop is rather uneven. The inclined slope reveals a large sticking phase on one hand, and the sliding regime is relatively unstable on the other hand.

For NBR-Sil, the frictional force is characterized by a first loop at the beginning of the test, which significantly increases with time. After that, the fretting regime reaches a stable state.

In comparison, EDPM materials have a relatively stable sliding friction, although the maximum friction force at the beginning of the test is higher than that at the end. This may be due to the increase of hardness within the first cycle, leading to the decrease of adhesion and hysteresis resistance between the rubber and the steel metal ball.

For each rubber grade, the shape of the loop is rather similar in air and hydrogen. The main differences are related to the maximal friction force and the calculated friction coefficient.

3.1.2. Average Friction Coefficient

Figure 5 shows the exemplary evolution of the friction coefficient as a function of time of several grades tested in ambient air and in hydrogen at 0.1 MPa and 10 MPa. As indicated in Figure 5a,b, the friction coefficient of HNBR-CB and EPDM2 are higher in a hydrogen environment compared to the values in air. For both rubber materials, the COF increases rapidly at first before decreasing to a stable value for EPDM2. The COF of HNBR-CB, however, gradually increases in hydrogen after this descent stage.

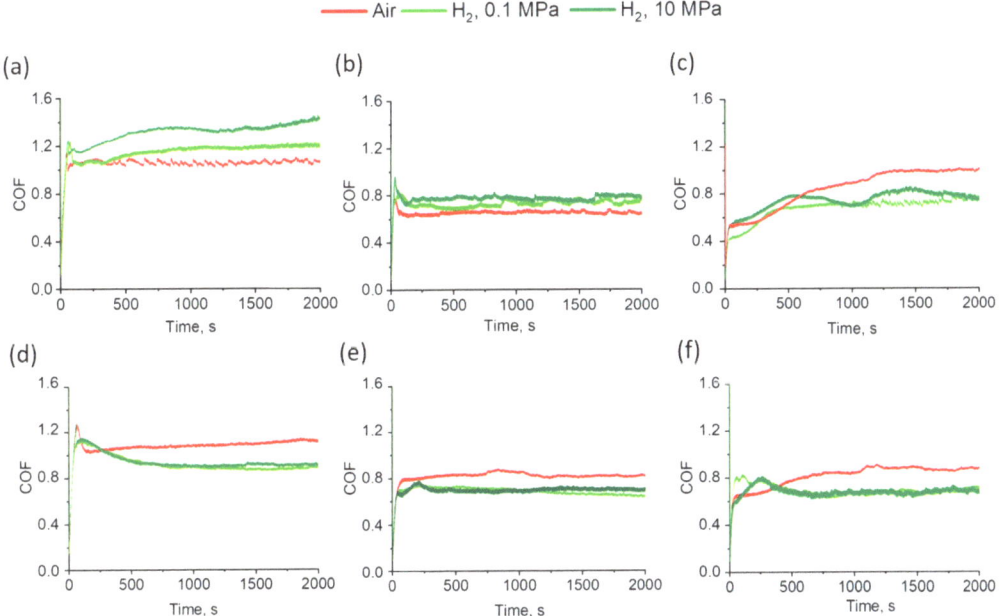

Figure 5. Coefficient of friction curves as a function of time of HNBR-CB (**a**), EPDM2 (**b**), NBR-Sil (**c**), NBR-CB-perox (**d**), NBR-CB (**e**) and NBR-CB-plast (**f**) in ambient air and in hydrogen at 0.1 MPa and 10 MPa.

With regard to the NBR grades, the COF in air is higher than in the hydrogen environment (Figure 5c–f), with similar friction curves independent of the hydrogen pressure. It is noticeable that the COF of NBR-Sil increases gradually and slowly in all conditions, reaching a steady state friction only after 1000 s (10,000 cycles). This may be related to the lower hardness of NBR-Sil compared to the other grades, leading to a larger deformation and an increased contact area, as observed in [40]. As a comparison, the friction curves of CB-filled NBR grades reach a stable friction state after a short running in phase.

An overview of the mean friction coefficient of all rubbers measured at the end of the tests is presented in Figure 6a. The COF of HNBR and EPDM increase slightly with hydrogen pressure, while lower values were obtained in hydrogen with all NBR grades. Comparing the HNBR grades, the addition of a PA filler leads to a slight increase of the friction coefficient, independent of the environment, while increasing the CB content in the EPDM grades results in a minor decrease in the COF. Among the sulfur-cured rubber, both CB-filled grades, NBR-CB and NBR-CB-plast, have a lower COF compared to the SiO_2-filled NBR-Sil. This is consistent with another previous study performed in air [41]. Further, it is noticeable that the peroxide-cured NBR-CB-perox grade has a higher COF than the sulfur-cured grades under both air and hydrogen conditions. As mentioned in [42], the peroxide-cured CB-filled NBR grade shows the stiffest material behavior with the highest crosslink density among these NBR grades. Due to the restriction of its polymer chain movements, a higher shear strength is expected, which could lead to increased adhesion.

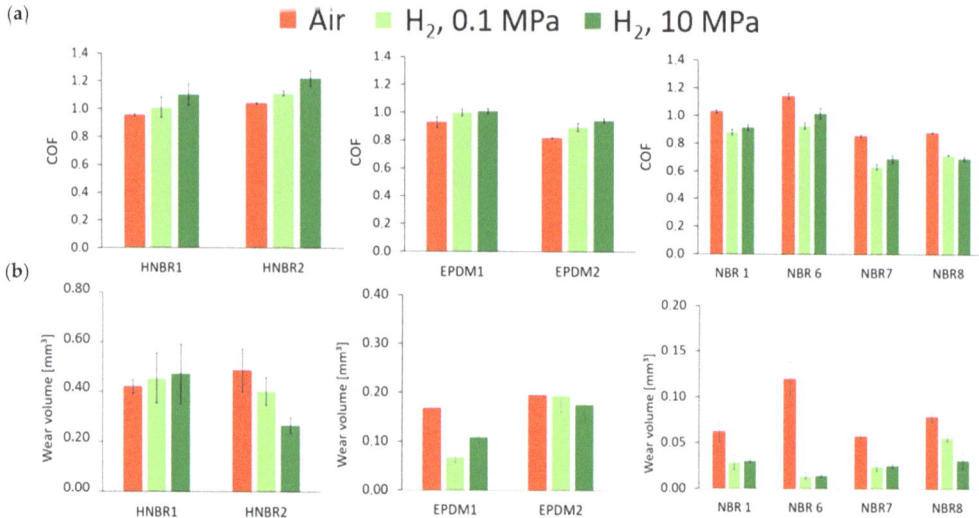

Figure 6. Average friction coefficient at the end of the tests (**a**) and wear volume of rubber materials (**b**) after tests in ambient air and in hydrogen at 0.1 MPa and 10 MPa (with different scales).

The influence of hydrogen on the friction coefficient of EPDM and HNBR grades are in accordance with the work done by Kuang et al. [28], who found that the COF of EPDM against steel is larger under high-pressure hydrogen than that measured in ambient air under reciprocating sliding. Concerning NBR materials, however, discrepancies in the results were found compared to [28]. This is possibly due to different formulations and testing conditions, namely sliding friction and hydrogen pressure. Wang [43] observed different trends between reciprocating and fretting behavior of several polymers in air, due to more energy generated at the surface during fretting. On the other hand, at very high-pressure hydrogen the sticking regime of rubber materials should be promoted because of the swelling effect [30]. In any case, as mentioned in [44], the fretting characteristic of polymers is strongly related to the frictional heat produced during fretting and the heat resistance of the materials. It is therefore reasonable to suggest that the cooling effect of the hydrogen environment, which has a higher thermal conductivity than air, affects the deformation and therefore the fretting behavior of these rubber materials. However, this should not be the only influencing parameter since different results were observed depending on the materials.

3.1.3. Effect of Hydrogen on the Wear Volume

Figure 6b gives an overview of the wear volume obtained by 3D profilometry. For most rubber materials, the wear values are similar or reduced in a hydrogen environment, which is in good agreement with [28]. The wear value of the rubbers is rather stable for HNBR-CB and EPDM2 but decreases for HNBR-CB-PA and EPDM1, which have respectively a lower amount of carbon black compared to the former grades.

Similarly, by comparing the results of the CB-filled sulfur-cured NBR grades, it is noticeable that increasing the CB content from 75 ppm to 95 ppm leads also to higher wear in hydrogen. This is possibly related to the restriction of the polymer chain with increased CB loading, although the addition of a plasticizer to the NBR-CB-plast compensates the stiffness of higher filler loading as reported in [42].

On the other hand, although the silica-filled rubber (NBR-Sil) has a higher friction coefficient compared to the CB-filled NBR, the wear values of both rubber materials were similar in air, as well as in hydrogen.

Therefore, other factors should be taken into consideration. HNBR-CB-PA contains polyamide fillers and as reported in [23], the tribological properties of polyamides are influenced by the environment. NBR-CB-plast contains plasticizers which may also have an effect on the wear behavior in air and in hydrogen.

Further, most significant effects were obtained with NBR-CB-perox, which is a peroxide-cured rubber with additives. It has the highest wear values among the NBR materials in air and the lowest in hydrogen. It is therefore appropriate to suggest that either the curing process and/or the additive have a major influence on the wear properties of these rubber materials in hydrogen. This will be discussed in the following section.

3D profile images of the wear scar were used to evaluate the wear damage of the rubbers. Figure 7 illustrates examples of some of them. Deeper wear scars were obtained after the experiment in air compared to those performed in hydrogen.

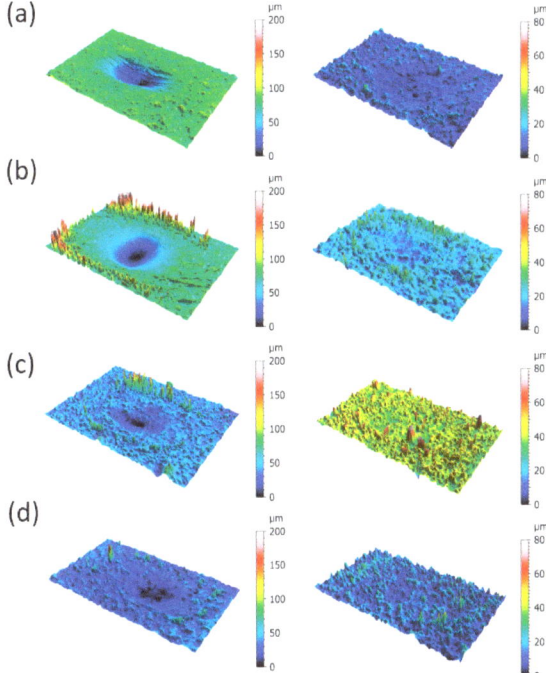

Figure 7. 3D profilometer images of the wear scar of (**a**) NBR-Sil, (**b**) NBR-CB-perox, (**c**) NBR-CB and (**d**) NBR-CB-plast after tests in air (**left**) and in 10 MPa H_2 (**right**).

3.1.4. Surface Morphology and Wear Mechanism

Optical microscopy of the wear scar of selected rubber materials and associated counterfaces after testing in air and hydrogen at 10 MPa are compared in Figure 8.

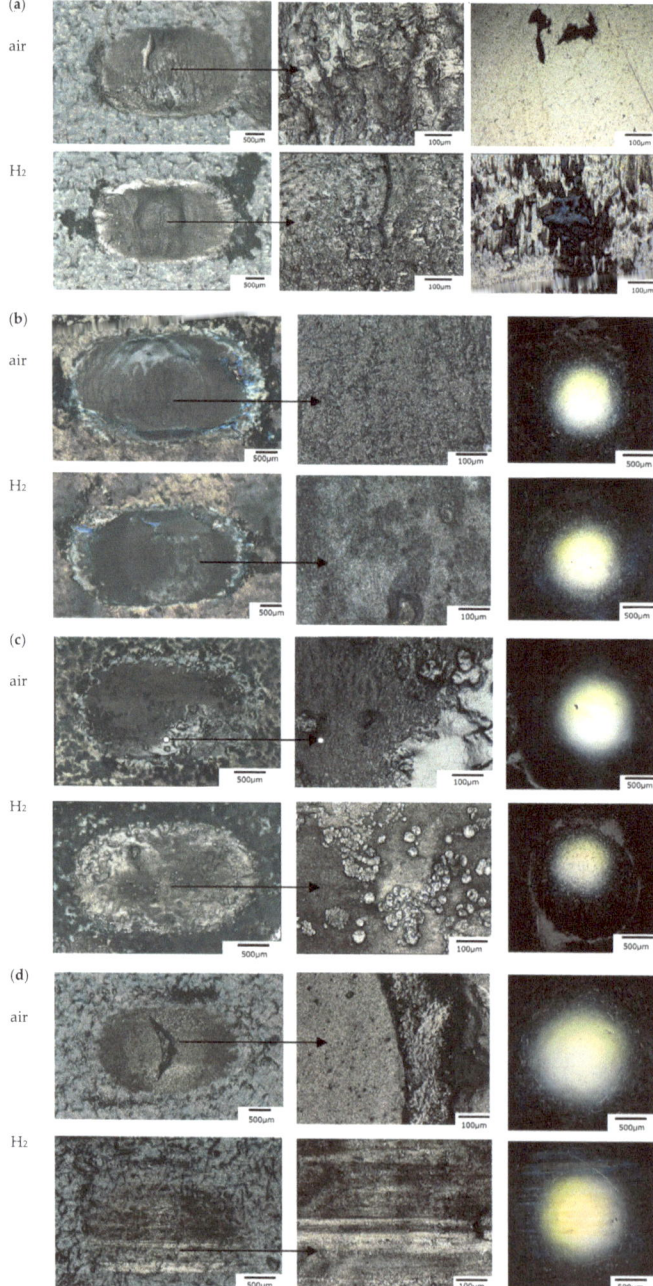

Figure 8. Optical microscope images of the wear scar (**left** and **middle**) and of the ball (**right**) after tests in air and in 10 MPa H_2: (**a**) HNBR-CB, (**b**) EPDM2, (**c**) NBR-CB and (**d**) NBR-CB-perox.

As shown in Figure 8a, the wear scar of HNBR-CB is perpendicular to the fretting direction and forms a wavy structure, which is associated with adhesive wear [33]. Cracks are observed on the surface and propagated perpendicularly to the sliding direction after tests in air and hydrogen. Wear debris is present outside the contact in both testing conditions. The main difference is observed on the counterface. A large amount of rubber stuck to the steel ball after the hydrogen experiment. The friction mechanism is therefore mainly adhesive in hydrogen. This may explain the higher coefficient of friction for HNBR grades in hydrogen.

In comparison, the wear damage of EPDM2 is visibly milder than for the HNBR grade. No visible cracks were detected at the surface of the rubber but a typical wave pattern perpendicular to the fretting direction was observed (Figure 8b). Wear particles are seen on the ball in both conditions.

Figure 8c displays the worn surface of the sulfur-cured NBR filled with CB (NBR-CB). While a wave pattern is detected after testing in air, the wear scar in hydrogen is covered with some wear debris. These fine particles are also present on the counterface and more abundant in hydrogen conditions. These particles may act as a solid lubricant in the contact area, reducing the friction and wear of the rubber [36].

Figure 8d presents the surface images of the CB-filled peroxide-cured NBR grade (NBR-CB-perox). Significant differences are observed compared to the previous sulfur-cured NBR. After testing in air, a torn tongue perpendicular to the sliding direction is observed, similar to HNBR. It is therefore likely that the curing process affects the material properties as mentioned in [41], and therefore the friction behavior. After the hydrogen test, however, ploughing marks parallel to the sliding direction are present, with no visible stick regions between the two contacting surfaces. This suggests that abrasive wear was more likely to occur. The formation of a thin transfer film on the counterface leads to reduced friction and wear in hydrogen.

To characterize further the fretting mechanisms and the influence of additives and fillers in the NBR grades, further EDX was performed on the wear scar of the rubber after testing in air and hydrogen at 10 MPa. Figure 9 collects the EDX maps of NBR-Sil, NBR-CB-perox and NBR-CB-plast.

From Figure 9a, a homogenous distribution of the silica fillers in the wear scar after testing in both conditions can be deduced. ZnO, however, appears more pronounced at the surface of the rubber after the hydrogen experiment. Similarly, more additives (ZnO and MgO) were detected on the surface of the wear scar of NBR-CB-perox after hydrogen exposure (Figure 9b). The EDX analyses of the NBR-CB-plast indicate a significant agglomeration of additives and plasticizers (identified by Zn and S, respectively) after the hydrogen experiment (Figure 9c). It seems that hydrogen promotes the migration and agglomeration of the additives towards the surface of the material.

Additional ATR-IR spectroscopy of selected rubber materials were performed to detect possible chemical reactions during the friction process.

Figure 10 depicts the ATR-IR spectra of the HNBR, EPDM and NBR grades. No significant chemical reactions could be observed but some increase in signal intensity was detected for several rubbers.

Wear scar analysis of HNBR-CB through FT-IR reveals a general broadening of the IR signal after testing in hydrogen in the region of 1521 and 860 cm^{-1}, possibly due to more exposed CB and higher absorption of the IR signal. As reported in [13], the exposure to hydrogen can cause a tendency of the CB fillers to come to the surface, absorbing in this range of IR. ATR-IR of the wear scar of EPDM1 indicates marginal changes; however, a small peak in 1735 cm^{-1} corresponding to carbonyl group was detected on the sample tested in air, which could possibly be related to some oxidation on the worn surface.

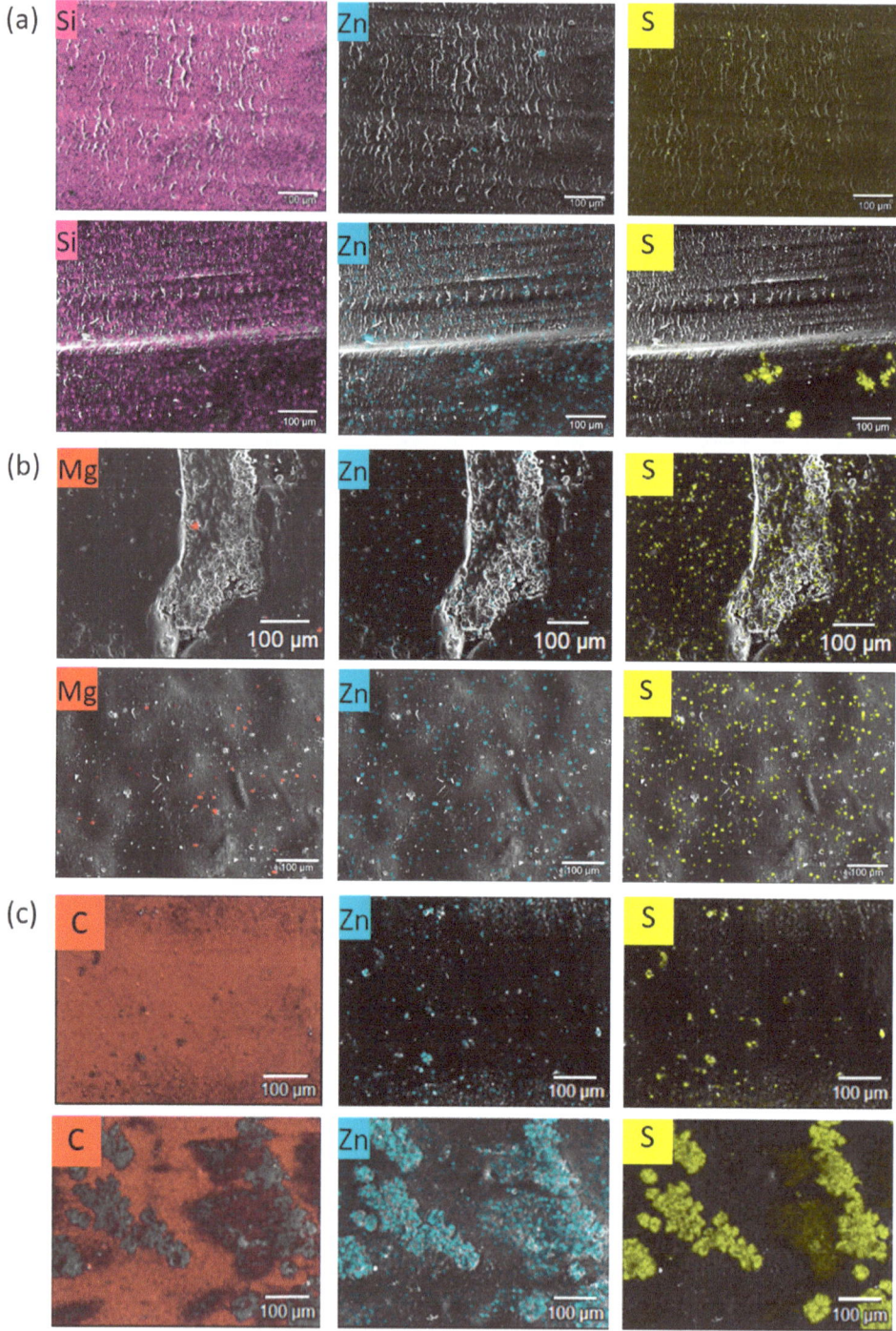

Figure 9. EDX maps of the wear scar after tests in air (**top**) and in 10 MPa H_2 (**bottom**): (**a**) NBR-Sil, (**b**) NBR-CB-perox and (**c**) NBR-CB-plast.

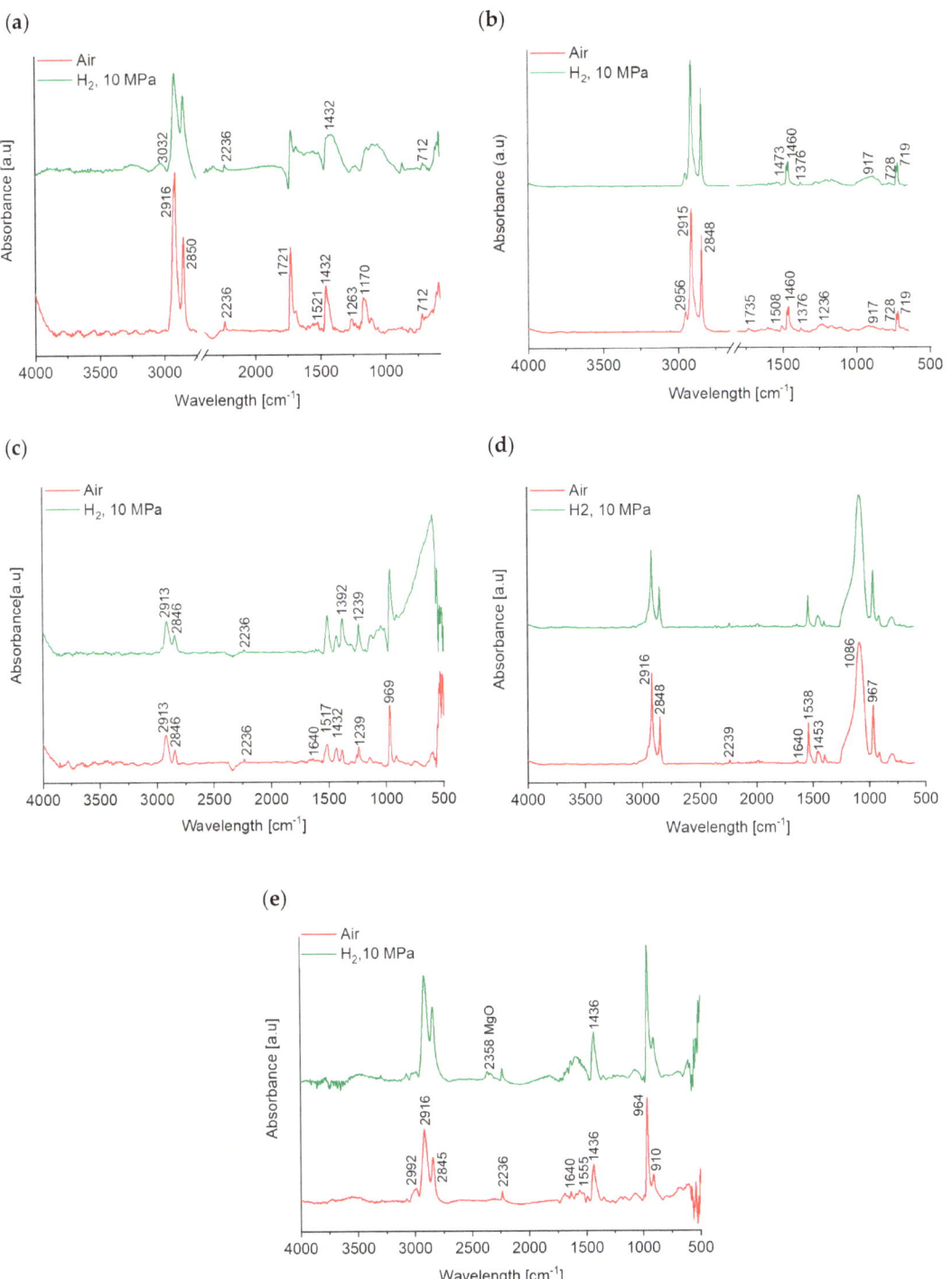

Figure 10. ATR-IR of the wear scar after tests in air and in 10 MPa H_2: (**a**) HNBR-CB, (**b**) EPDM1, (**c**) NBR-CB, (**d**) NBR-Sil and (**e**) NBR-CB-perox.

Regarding the NBR-CB, testing in hydrogen brings small changes in the IR spectra; the 1517 cm^{-1} band might be related to additives, such as antioxidants. The 969 cm^{-1} band corresponds to the C-H out-of-plane bending band of the butadiene double bond. The cyano group in air and after testing in hydrogen, however, shows no significant change.

While no chemical changes could be detected for NBR-Sil (the same for the broad peak of the silica filler at 1086 cm^{-1} that is present in both conditions), some small changes are detected for the NBR-CB-perox materials; the peaks between 1690 and 1000 cm^{-1} in NBR-CB-perox mostly related to CB, increasing slightly after testing in hydrogen, while 2358 cm^{-1} is related to MgO. The 964 cm^{-1} band corresponds to the C-H out-of-plane bending band of the butadiene double bond. The 910 cm^{-1} band is related to the peroxide crosslinking additive. ATR-IR therefore confirms the EDX results, suggesting that the presence of a curing activator and an increase of CB at the surface. The dissolved H_2 may act as plasticizer in the rubber materials, increasing the motion of the polymer and fillers. Upon friction, these particles move towards the surface, acting favorably at the friction contact.

3.2. Friction Tests on Exposed Samples: Effect of High-Pressure Exposure Followed by Rapid Gas Decompression

Following the test method described in Figure 1, further samples were aged in high-pressure hydrogen at 100 MPa for 7 days at 120 °C for HNBR and at 85 °C for the EPDM grades, respectively. A day before the end of the exposure period, the temperature was switched off to cool down the specimens before taking them out and the pressure was adjusted to replenish the pressure loss due to the cooling effect. This step is necessary to allow the characterization of the cooled sample immediately after decompression, without a temperature effect.

After exposure at 100 MPa, friction tests were performed at 10 MPa immediately after decompression and one day after. Figure 11a shows the hysteresis of HNBR-CB-PA immediately after exposure. The loop has an elliptical shape corresponding to partial slip. The displacement occurs only on the edge part and is mainly due to elastic deformation of the elastomer. The effect of the high-pressure hydrogen on the physical and mechanical properties were studied and published in [13]. Accordingly, the swelling of the rubber immediately after the RGD significantly decreases the hardness of the rubber. This friction process is mainly ascribed to the deformation of the rubber. By repeating the friction tests 24 h later, the hysteresis almost retrieves a parallelogram shape as the volume of the rubber recovers (Figure 11b). The elastic deformation, however, is still predominant.

As a comparison, further tests were performed with the EPDM2 grades. The friction response in 10 MPa hydrogen was similar to before (Figure 4) and after aging in high-pressure hydrogen (Figure 11c,d). This can be related to a relatively stable behavior of the physical and mechanical properties of EPDM2, as reported in [45]. Other works revealed similar trends regarding the aging of elastomers in hydrogen [46], where EPDM material showed the least degradation compared to the HNBR grade.

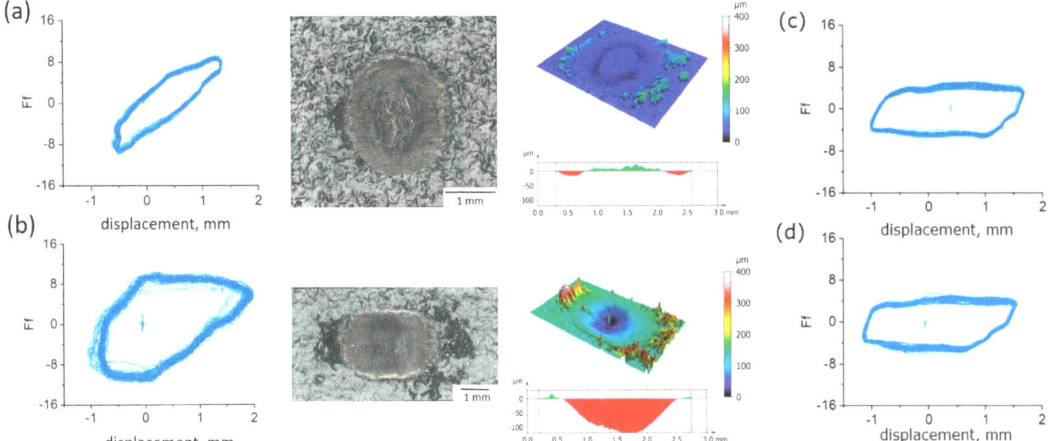

Figure 11. Friction force versus displacement for HNBR-CB-PA immediately (**a**) and 24 h (**b**) after finishing the high-pressure exposure with corresponding optical microscopy images and 3D profile measurements of the wear scar; Friction force versus displacement for (**c**) EPDM2 immediately and 24 h (**d**) after finishing the high-pressure exposure.

4. Conclusions

In this study, the fretting behavior of several elastomer materials against a 316L ball were evaluated in an air and hydrogen environment up to 10 MPa. Several grades of cross-linked hydrogenated acrylonitrile butadiene (HNBR), acrylonitrile butadiene (NBR) and ethylene propylene diene monomer rubbers (EPDM) were investigated.

The influence of the hydrogen environment was studied at first on as-received samples. Based on the friction results and wear characterization, the following conclusions can be drawn:

- Under the testing conditions (a normal load of 5 N, frequency of 10 Hz, a 2 mm stroke), the fretting behavior was in the gross slip regime for all rubber materials. The shape of the loop was similar in hydrogen compared to air, and the influence of hydrogen pressure was relatively small, although some effects were seen on the HNBR and EPDM grades. The average friction coefficient and the wear volume, however, were affected differently by the hydrogen conditions depending on the materials. This suggests that the cooling effect of the hydrogen environment is not the only influencing factor.
- The friction of the HNBR and EPDM grades increased with hydrogen pressure. Adhesive wear is predominant in hydrogen in CB filled HNBR, while the addition of PA fillers reduced wear.
- Concerning NBR grades, reduced friction and wear were measured for all grades in hydrogen. It was found that the curing process and the additive have a major influence on the wear properties of these rubber materials in hydrogen. Most significant effects were obtained with the peroxide-cured rubber, having the highest wear value among the NBR materials in air and the lowest one in hydrogen. Among the sulfur-cured NBR grades, lower friction was achieved with CB compared to SiO_2 fillers.
- No significant chemical reactions were detected by means of ATR-IR, apart from a possible oxidation on the worn EPDM surface tested in air. However, both EDX and ATR-IR analyses revealed migration and agglomeration of the additives at the friction contact of the NBR grades in hydrogen, acting favorably on the friction and wear resistance of the rubbers.

The influence of high-pressure exposure followed by rapid gas decompression was studied on aged samples. Significant effects were observed immediately after decompression for the HNBR material, which is related to the softening of the rubber after RGD. EPDM rubber, however, was less affected by the aging experiment and therefore might be more suitable for high-pressure applications than HNBR grades.

Author Contributions: Conceptualization, methodology, writing, supervision, G.T.; conducting of surface analyses and interpretation, N.C.M.; funding acquisition, A.H. All authors have read and agreed to the published version of the manuscript.

Funding: The research work was performed within the COMET-project "Polymers4Hydrogen" (project-no.: 21647053) at the Bundesanstalt für Materialforschung und -prüfung (BAM) within the framework of the COMET-program of the Federal Ministry for Climate, Action, Environment, Energy, Mobility, Innovation and Technology and the Federal Ministry for Digital and Economic Affairs with contributions by Polymer Competence Center Leoben GmbH (PCCL, Austria), Montanuniversität Leoben (Department Polymer Engineering and Science, Chair of Chemistry of Polymeric Materials, Chair of Materials Science and Testing of Polymers), Technical University of Munich, Tampere University of Technology, Politecnico di Milano and Arlanxeo Deutschland GmbH, ContiTech Rubber Industrial Kft., Peak Technology GmbH, SKF Sealing Solutions Austria GmbH, and Faurecia.

Data Availability Statement: The datasets generated for this study are available upon request from the corresponding authors.

Acknowledgments: The authors would like to thank Kaiser from Arlanxeo for providing the materials, Heidrich, Slachciak and Optiz for performing the tests and microscopy characterization as well as Hidde and Weimann for the FT-IR and EDX analyses.

Conflicts of Interest: Author Andreas Hausberger was employed by the company Polymer Competence Centre Leoben GmbH (PCCL). The remaining authors declare that the research was conducted in the absence of any commercial or financial relationships that could be construed as a potential conflict of interest.

References

1. Balasooriya, W.; Clute, C.; Schrittesser, B.; Pinter, G. A review on applicability, limitations, and improvements of polymeric materials in high-pressure hydrogen gas atmospheres. *Polym. Rev.* **2021**, 175–209. [CrossRef]
2. Zheng, Y.; Tan, Y.; Zhou, C.; Chen, G.; Li, J.; Liu, Y.; Liao, B.; Zhang, G. A review on effect of hydrogen on rubber seals used in the high-pressure hydrogen infrastructure. *Int. J. Hydrogen Energy* **2020**, *45*, 23721–23738. [CrossRef]
3. Nishimura, S. Rubbers and elastomers for high-pressure hydrogen seal. *Soc. Polym. Sci.* **2015**, *64*, 356–357.
4. Yamabe, J.; Nishimura, S. Influence of fillers on hydrogen penetration properties and blister fracture of rubber composites for O-ring exposed to high-pressure hydrogen gas. *Int. J. Hydrogen Energy* **2009**, *34*, 1977–1989. [CrossRef]
5. Yamabe, J.; Nishimura, S.; Koga, A. A study on sealing behavior of rubber O-ring in high pressure hydrogen gas. *SAE Int. J. Mater. Manuf.* **2009**, *2*, 452–460. [CrossRef]
6. Yamabe, J.; Nishimura, S.; Nakao, M.; Fujiwara, H. Blister fracture of rubbers for O-ring exposed to high pressure hydrogen gas. In Proceedings of the 2008 International Hydrogen Conference—Effects of Hydrogen on Materials, Grand Teton, WY, USA, 7–10 September 2009; pp. 389–396.
7. Yamabe, J.; Nishimura, S. Nanoscale fracture analysis by atomic force microscopy of EPDM rubber due to high pressure hydrogen decompression. *J. Mater. Sci.* **2011**, *46*, 2300–2307. [CrossRef]
8. Yamabe, J.; Koga, A.; Nishimura, S. Failure behavior of rubber Oring under cyclic exposure to high-pressure hydrogen gas. *Eng. Fail. Anal.* **2013**, *35*, 193–205. [CrossRef]
9. Castagnet, S.; Mellier, D.; Nait-Ali, A.; Benoit, G. In-situ X-ray computed tomography of decompression failure in a rubber exposed to high-pressure gas. *Polym. Test.* **2018**, *70*, 255–262. [CrossRef]
10. Castagnet, S.; Ono, H.; Benoit, G.; Fujiwara, H.; Nishimura, S. Swelling measurement during sorption and decompression in a NBR exposed to high-pressure hydrogen. *Int. J. Hydrogen Energy* **2017**, *42*, 19359–19366. [CrossRef]
11. Simmons, K.L.; Kuang, W.; Burton, S.D.; Arey, B.W.; Shin, Y.; Menon, N.C.; Smith, D.B. H-Mat hydrogen compatibility of polymers and elastomers. *Int. J. Hydrogen Energy* **2021**, *46*, 12300–12310. [CrossRef]
12. Kulkani, S.; Choi, K.; Kuang, W.; Menon, N.; Mills, B.; Soulami, A.; Simmons, K. Damage evolution in polymer due to exposure to high-pressure hydrogen gas. *Int. J. Hydrogen Energy* **2021**, *46*, 19001–19022. [CrossRef]
13. Theiler, G.; Cano Murillo, N.; Halder, K.; Balasooriya, W.; Hausberger, A.; Kaiser, A. Effect of high-pressure hydrogen environment on the physical and mechanical properties of elastomers. *Int. J. Hydrogen Energy* **2024**, *58*, 389–399. [CrossRef]
14. Menon, N.; Alvine, K.; Kruizenga, A.; Nissen, A.; San Marchi, C.; Brooks, K. Behaviour of polymers in high pressure environments as applicable to the hydrogen infrastructure. In Proceedings of the Pressure Vessels and Piping Conference—American Society of Mechanical Engineers, Vancouver, BC, Canada, 17–21 July 2016.

15. Zhou, C.; Zheng, J.; Gu, C.; Zhao, Y.; Liu, P. Sealing performance analysis of rubber O-ring in high-pressure gaseous hydrogen based on finite element method. *Int. J. Hydrogen Energy* **2017**, *42*, 11996–12004. [CrossRef]
16. Jeon, S.K.; Kwon, O.H.; Tak, N.H.; Chung, N.K.; Baek, U.B.; Nahm, S.H. Relationships between properties and rapid gas decompression (RGD) resistance of various filled nitrile butadiene rubber vulcanizates under high-pressure hydrogen. *Mater. Today Commun.* **2022**, *30*, 103038. [CrossRef]
17. Ono, H.; Nait-Ali, A.; Kane Diallo, O.; Benoit, G.; Castagnet, S. Influence of pressure cycling on damage evolution in an unfilled EPDM exposed to high-pressure hydrogen. *Int. J. Fract.* **2018**, *210*, 137–152. [CrossRef]
18. Fujiwara, H.; Ono, H.; Nishimura, S. Effects of fillers on the hydrogen uptake and volume expansion of acrylonitrile butadiene rubber composites exposed to high pressure hydrogen: Property of polymeric materials for high pressure hydrogen devices (3). *Int. J. Hydrogen Energy* **2022**, *47*, 4725–4740. [CrossRef]
19. Kuang, W.; Nickerson, E.K.; Li, D.; Clelland, D.T.; Seffens, R.J.; Ramos, J.L.; Simmons, K.L. An in-situ view cell system for investigating swelling behavior of elastomers upon high-pressure hydrogen exposure. *Int. J. Hydrogen Energy* **2024**, *71*, 1317–1325. [CrossRef]
20. Chen, Q.; Morita, T.; Sawae, Y.; Fukuda, K.; Sugimura, J. Effects of trace moisture content of tribofilm formation, friction and wear of CF-filled PTFE in hydrogen. *Tribol. Int.* **2023**, *188*, 108905. [CrossRef]
21. Theiler, G.; Gradt, T. Tribological characteristics of polyimide composites in hydrogen environment. *Tribol. Int.* **2015**, *92*, 162–171. [CrossRef]
22. Theiler, G.; Gradt, T. Environmental effects on the sliding behaviour of PEEK composites. *Wear* **2016**, *368–369*, 278–286. [CrossRef]
23. Theiler, G.; Gradt, T. Comparison of the Sliding Behavior of Several Polymers in Gaseous and Liquid Hydrogen. *Tribol. Online* **2023**, *18*, 217–231. [CrossRef]
24. Sawae, Y.; Morita, T.; Takeda, K.; Onitsuka, S.; Kaneuti, J.; Yamaguchi, T.; Sugimura, J. Friction and wear of PTFE composites with different filler in high purity hydrogen gas. *Tribol. Int.* **2021**, *157*, 106884. [CrossRef]
25. Sawae, Y.; Fukuda, K.; Miyakoshi, E.; Doi, S.; Watanabe, H.; Nakashima, K.; Sugimura, J. Tribological Characterization of Polymeric Sealing Materials in High-Pressure Hydrogen Gas. In Proceedings of the STLE/ASME 2010 International Joint Tribology Conference, San Francisco, CA, USA, 17–20 October 2010.
26. Sawae, Y.; Nakashima, K.; Doi, S.; Murakami, T.; Sugimura, J. Effects of High-Pressure Hydrogen on Wear of PTFE and PTFE Composite. In Proceedings of the ASME/STLE 2009 International Joint Tribology Conference, Memphis, Tennessee, USA, 19–21 October 2009; pp. 233–235.
27. Duranty, E.R.; Roosendaal, T.J.; Pitman, S.G.; Tucker, J.C.; Owsley, S.L., Jr.; Suter, J.D. In situ high pressure hydrogen tribological testing of common polymer materials used in the hydrogen delivery infrastructure. *JoVE* **2018**, *133*, e56884. [CrossRef]
28. Kuang, W.; Bennett, W.D.; Roosendaal, T.J.; Arey, B.W.; Dohnalkova, A. In situ friction and wear behavior of rubber materials incorporating various fillers and/or a plasticizer in high-pressure hydrogen. *Tribol. Int.* **2021**, *153*, 106627. [CrossRef]
29. Choi, B.L.; Jung, J.K.; Baek, U.B.; Choi, B.-H. Effect of Functional Fillers on Tribological Characteristics of Acrylonitrile Butadiene Rubber after High-Pressure Hydrogen Exposures. *Polymers* **2022**, *14*, 861. [CrossRef] [PubMed]
30. Zhou, C.; Chen, G.; Xiao, S.; Hua, Z.; Gu, C. Study on fretting behavior of rubber O-ring seal in high-pressure gaseous hydrogen. *Int. J. Hydrogen Energy* **2019**, *44*, 22569–22575. [CrossRef]
31. CSA/ANSI CHMC 2:19; Test Methods for Evaluating Material Compatibility in Compressed Hydrogen Applications—Polymers. CSA Group: Toronto, ON, Canada, 2019.
32. Zhang, T.; Su, J.; Shu, Y.; Shen, F.; Ke, L. Fretting Wear Behavior of Three Kinds of Rubbers under Sphere-On-Flat Contact. *Materials* **2021**, *14*, 2153. [CrossRef] [PubMed]
33. Wang, C.; Hausberger, A.; Berer, M.; Pinter, G.; Grün, F.; Schwarz, T. An investigation of fretting behavior of thermoplastic polyurethane for mechanical seal application. *Polymer Testing* **2018**, *72*, 271–284. [CrossRef]
34. Karger-Kocsis, J.; Mousa, A.; Major, Z.; Bekesi, N. Dry friction and sliding wear of EPDM rubbers against steel as a function of carbon black content. *Wear* **2008**, *264*, 359–367. [CrossRef]
35. ISO 15113:2015; Rubber—Determination of Frictional Properties. ISO: Geneva, Switzerland, 2015.
36. Karger-Kocsis, J.; Felhös, D.; Xu, D.; Schlarb, A.K. Unlubricated sliding and rolling wear of thermoplastic dynamic vulcanizates (Santoprene R) against steel. *Wear* **2008**, *265*, 292–300. [CrossRef]
37. Shen, M.; Peng, X.; Meng, X.; Zheng, J.; Zhu, M. Fretting wear behavior of acrylonitrile–butadiene rubber (NBR) for mechanical seal applications. *Tribol. Int.* **2016**, *93*, 419–428. [CrossRef]
38. Zhou, Z.R.; Nakazawa, K.; Zhu, M.-H.; Maruyama, N.; Kapsa, P.; Vincent, L. Progress in fretting maps. *Tribol. Int.* **2006**, *39*, 1068–1073. [CrossRef]
39. Vingsbo, O.; Soderberg, S. On fretting maps. *Wear* **1988**, *126*, 131–147. [CrossRef]
40. Adam, A.; Paulkowski, D.; Mayer, B. Friction and Deformation Behavior of Elastomers. *Mater. Sci. Appl.* **2019**, *10*, 527–542. [CrossRef]
41. Mokhtari, M.; Schipper, D.J.; Tolpekina, T.V. On the Friction of Carbon Black- and Silica-Reinforced BR and S-SBR Elastomers. *Tribol. Lett.* **2014**, *54*, 297–308. [CrossRef]
42. Clute, C.; Balasooriya, W.; Cano Murillo, N.; Theiler, G.; Kaiser, A.; Fasching, M.; Schwarz, T.; Hausberger, A.; Pinter, G.; Schlögl, S. Morphological investigations on silica and carbon-black filled acrylonitrile butadiene rubber for sealings used in high-pressure H_2 applications. *Int. J. Hydrogen Energy* **2024**, *67*, 540–552. [CrossRef]

43. Wang, Q.; Wang, Y.; Wang, H.; Fan, N.; Yan, F. Experimental investigation on tribological behavior of several polymer materials under reciprocating sliding and fretting wear conditions. *Tribol. Int.* **2016**, *104*, 73–82. [CrossRef]
44. Guo, Q.; Luo, W. Mechanisms of fretting wear resistance in terms of material structures for unfilled engineering polymers. *Wear* **2002**, *249*, 924–931. [CrossRef]
45. Cano Murillo, N.; Kaiser, A.; Balasooriya, W.; Meinel, D.; Theiler, G. Effect of hydrogen environments on the physical and mechanical properties of elastomers. Bundesanstalt für Materialforschung und -prüfung (BAM), Berlin, Germany. 2024, *to be published*.
46. Zaghdoudi, M.; Kommling, A.; Bohning, M.; Jaunich, M. Ageing of elastomers in air and in hydrogen environment: A comparative study. *Int. J. Hydrogen Energy* **2024**, *63*, 207–216. [CrossRef]

Disclaimer/Publisher's Note: The statements, opinions and data contained in all publications are solely those of the individual author(s) and contributor(s) and not of MDPI and/or the editor(s). MDPI and/or the editor(s) disclaim responsibility for any injury to people or property resulting from any ideas, methods, instructions or products referred to in the content.

Article

Effect of Harmful Bearing Currents on the Service Life of Rolling Bearings: From Experimental Investigations to a Predictive Model

Volker Schneider [1,*], Marius Krewer [1], Gerhard Poll [1] and Max Marian [1,2]

1. Institute of Machine Design and Tribology, Leibniz University Hannover, An der Universität 1, 30823 Garbsen, Germany; krewer@imkt.uni-hannover.de (M.K.); poll@imkt.uni-hannover.de (G.P.); marian@imkt.uni-hannover.de (M.M.)
2. Department of Mechanical and Metallurgical Engineering, School of Engineering, Pontificia Universidad Católica de Chile, Vicuña Mackenna 4860, Macul, Región Metropolitana 6904411, Chile
* Correspondence: schneider@imkt.uni-hannover.de; Tel.: +49-511-762-2245

Abstract: This study investigates the effects of harmful bearing currents on the service life of rolling bearings and introduces a model to predict service life as a function of surface roughness. Harmful bearing currents, resulting from electrical discharges, can cause significant surface damage, reducing the operational lifespan of bearings. This study involves comprehensive experiments to quantify the extent of electrical stress caused by these currents. For this purpose, four series of tests with different electrical stress levels were carried out and the results of their service lives were compared with each other. Additionally, a novel model to correlate the service life of rolling bearings with varying degrees of surface roughness caused by electrical discharges was developed. The basis is the internationally recognized method of DIN ISO 281, which was extended in the context of this study. The findings show that the surface roughness continues to increase as the electrical load increases. In theory, this in turn leads to a deterioration in lubrication conditions and a reduction in service life.

Keywords: electrical bearing current; electrical bearing damage; electrical discharge; bearing surface wear; service life calculation

1. Introduction

Parasitic currents and electric discharges have been subject to research for more than a century. Since the early days of electric motor development, the phenomena of shaft voltages and associated bearing damage has been known [1–4]. Due to the increased interest in electric drives, this has once again become a focus of research in recent years [5]. Modern frequency converters with high switching frequencies induce different types of parasitic currents to bearings situated in the motors and gearboxes of electric drivetrains. These currents occur in the form of electric discharge machining (EDM) currents and circular bearing currents [6,7].

In rolling bearings, the concentrated contact between the rolling element and raceway is usually highly stressed. Together with the the usage of lubricants, a so-called elastohydrodynamically lubricated (EHL) contact is formed, whereby the solid bodies are seperated by a lubricating intermediate medium. A schematic representation of the EHL contact with a typical pressure distribution is shown in Figure 1a. The electrical characteristics of this contact can be modeled using an equivalent circuit comprising the contact's capacitance and resistance, as shown in Figure 1b. The total capacitance includes the Hertz'ian capacitance C_{Hertz} from the Hertz'ian contact area and the outside capacitance $C_{Outside}$ from the inlet and outlet zones. Various theoretical models on the capacitance and impedance of rolling bearings have been developed [8–12].

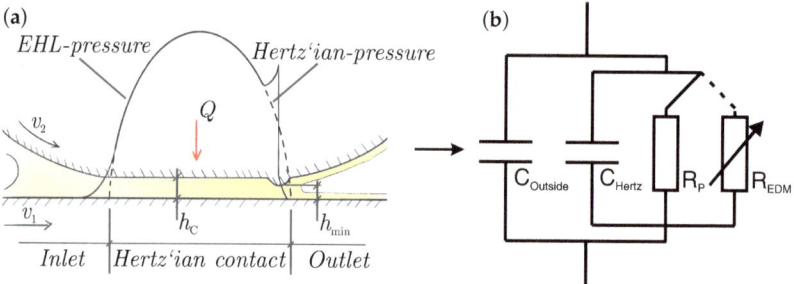

Figure 1. (**a**) An EHL contact with its (**b**) electrical equivalent circuit.

In the presence of a separating lubricating film, the parallel resistance R_P is very high. Once the applied voltage exceeds a specific threshold, a breakdown occurs, leading to a discharge through the substantially lower resistance R_{EDM}, which varies over time. TISCHMACHER [13] simulated and measured these discharges. A typical discharge scenario is displayed in Figure 2, whereby the voltage continues to rise until it reaches a critical value; in this case, at just under 30 volts. Until this point, the current is relatively constant around zero. Subsequently, as the discharge occurs, an abrupt decline in voltage ensues, together with a rise of the current to a peak of almost 2.5 A. Subsequently, both signals oscillate until they stabilize again.

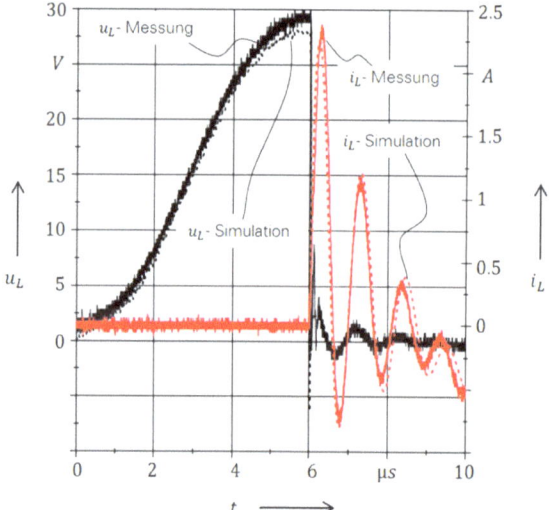

Figure 2. Measured and simulated discharge voltage and current from [13].

Depending on the circumstances, these discharges can lead to diverse forms of damage to rolling bearings [14]. In the case of full-film lubrication, EDM can occur and the temperatures in an arc can peak at up to 3000 K, which not only melts but also vaporizes parts of the material near the surface and has the potential to significantly alter the surface roughness [15,16]. As described by Furtmann [8,14,17], the short-term melting with consecutive oil-quenching alternates with repeated over-rollings, creating a "grey frosted" surface. The term refers to the visual appearance of the surface, which, on a macroscopic scale, has similarity to mechanically induced micro pitting. However, there are microscopic differences as the grey frosting shows smoothly flowing edges of the melted areas instead of rough fractures as seen in Figure 3.

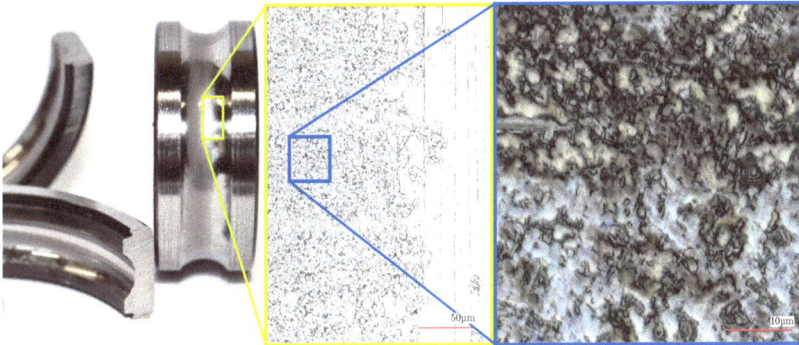

Figure 3. Macroscopic and microscopic view of electrically grey frosted surface.

More severe damage, such as fluting, can occur as a result of repeated discharges, leading to increased vibration, accelerated wear, and potential bearing failure. Fluting manifests as alternating lighter and darker areas corresponding to valleys and peaks on the bearing surface [13]. In addition to surface damage, the lubricant itself can be adversely affected, resulting in degraded performance, such as a reduction in dielectric strength [18].

While the damage patterns described are well established and have been subject of research for a while, the influence of electrical pre-stress on the service life of rolling bearings has not yet been part of investigations. Therefore, the questions that need to be investigated are, firstly, to what extent does harmful current passage affect the service life of rolling bearings? And secondly, how can this influence be quantified and made predictable? This prompted the initiation of novel experimental investigations, the results of which are presented in this study. The findings led to the development of a new extension to an existing model for calculating the service life of rolling bearings, which now incorporates the impact of electrical currents on service life.

2. Materials and Methods

The primary objective of the experimental investigations was to apply controlled electrical pre-stress to cylindrical rolling bearings and subsequently test their service life without further electrical stress. At first, the influence on the effect of a single steel rolling element in an otherwise hybrid bearing was investigated. After these preliminary studies, the electrical pre-stressing of the bearings took place. Finally, this was followed by the service-life tests themselves. Two test rigs were used for the experimental investigations: the modified universal test rig and four-bearing radial fatigue life test rig, both of which are described below. All bearings underwent a run-in procedure to make sure that the machining marks were flattened [19] and consistent experimental conditions were maintained during all experimental investigations.

2.1. Test Equipment

The influence investigations and the induction of controlled electrical pre-stress were conducted using a modified universal bearing test rig. This rig accommodates various bearing types, including cylindrical roller bearings and deep groove ball bearings, under different loading conditions and controlled-temperature environments. Different lubrication methods, such as oil immersion and grease lubrication, can also be used. As shown in Figure 4, the test head was modified with insulating bearing seats as well as hybrid support bearings with ceramic rolling elements made of zirconium oxide. These measures create a controlled current path through the test bearing, which was equipped with both ceramic elements and a varying number of steel rolling elements. The universal bearing test rig enabled the measurement of frictional torques, shaft speed, temperatures, and mechanical loads of the bearing. It was also possible to measure the capacitance and film thickness as described in [12,20].

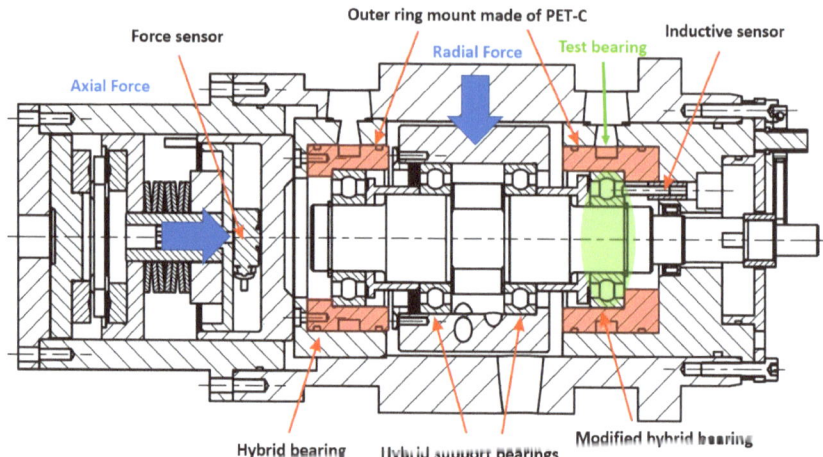

Figure 4. Universal test rig modified to investigate electrical effects in bearings.

The voltage application is facilitated by an Aegis shaft grounding ring connected to one of the outputs of the voltage source, thereby applying a potential to the shaft. Contact with the outer ring is established through a spring-loaded aluminum electrode, completing the circuit by connecting back to the voltage source. This configuration ensures a short current path with minimal unwanted losses. To further minimize losses, all cables were kept as short as possible (refer to Figure 5).

To measure and record the bearing voltage, a passive probe (Tektronix TPP0250, Beaverton, OR, USA, 250 MHz) from a four-channel oscilloscope (Tektronix MDO 3024, Beaverton, OR, USA, 200 MHz, 2.5 GS/s) was used, which came into contact with the shaft via an additional Aegis shaft grounding ring. This setup ensures that any losses in the cables from the voltage source to the shaft are excluded from the measurement. A second channel monitors the bearing current using a current sensor (Pearson 6595, Pearson Electronics Inc., Palo Alto, CA, USA) placed in the return path.

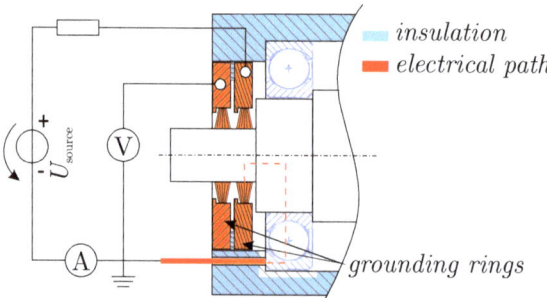

Figure 5. Electrical setup at the universal test rig.

For the service life tests, a four-bearing radial fatigue life test rig, as seen in Figure 6, was employed to determine the service life of radial bearings under electrical pre-stress conditions. Bearing failures were detected through vibration-based condition monitoring [21]. This rig could apply radial loads up to 25 kN per bearing, corresponding to a C/P-value of 2.5 for NU206 cylindrical roller bearings, and accommodated various lubrication types and controlled temperature conditions.

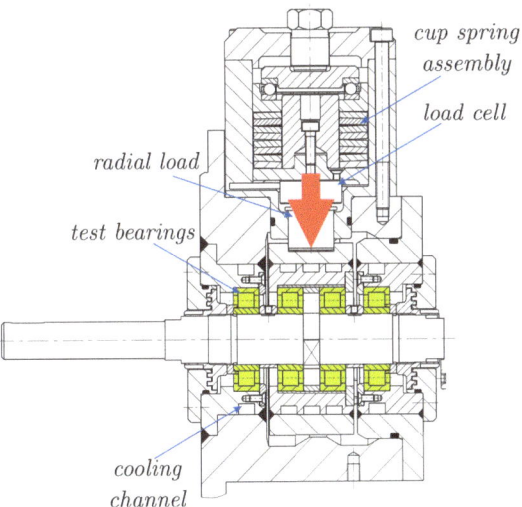

Figure 6. Four bearing radial fatigue life test rig.

Before and after each step of the investigations the bearings were thoroughly cleaned in an ultrasonic bath using both polar (isopropanol) and non-polar (benzine) solvents, followed by drying in an oven. After the service life tests the bearings were examined under a Keyence laser scanning microscope (Keyence VK-X200, Keyence, Osaka, Japan) and a reflective light microsocope (Keyence VHX 600, Keyence, Osaka, Japan). In particular, it was used to investigate the area surface roughness of the tested bearings.

2.2. Influence of Loaded Zone on Bearing Current

To prepare for the service life tests, the influence of radial load on the electrical pre-stress of the bearings was investigated. The mechanical load affects the size of the loaded zone and the number of rolling elements forming a high capacitive Hertz'ian contact. To understand how load distribution influences electric discharge phenomena, both theoretical and experimental studies were conducted.

The correlation between the azimuth angle of the radially loaded bearing and its theoretical local capacitances, discharge voltages, and discharge energies is graphically illustrated in Figure 7. Within the loaded zone, the capacitance C is increased due to the reduced thickness of the contact's lubricating film h_c. Essentially, the capacitance is a function of the electrical permittivity of the lubricant ϵ and the ratio of the Hertz'ian contact area A_{Hertz} and film thickness $C_{contact} = \epsilon \times \frac{A_{Hertz}}{h_c}$. Extended methods to calculate this capacitance can be found in a number of works, e.g., [12,20,22]. As is apparent from Equation (1), the lower minimum film thickness h_{min} inside the loaded zone compared to the non-loaded zone also results in a lower discharge voltage $U_{discharge}$, since the insulating property of the lubricant, the dielectric strength, is exceeded more easily by the applied field strength E_{crit} than in regions with higher film thicknesses. The energy W stored in a capacitor then occurs as a result of its capacitance and the applied voltage. Since the energy at the time of discharge $W_{discharge}$ is of interest, the critical discharge voltage must be inserted for the voltage term, and therefore Equation (2) is obtained. It should be noted that if the film thickness decreases, the capacitance will rise and the critical breakdown voltage will decrease. However, it should also be kept in mind that the breakdown voltage in Equation (2) is squared and therefore has a more significant influence on the total energy than the capacitance. This results in the red curve shown in Figure 7.

$$U_{discharge} = E_{crit} \times h_{min} \tag{1}$$

$$W_{\text{discharge}} = \frac{1}{2} C_{\text{bearing}} \times U_{\text{discharge}}^2 \qquad (2)$$

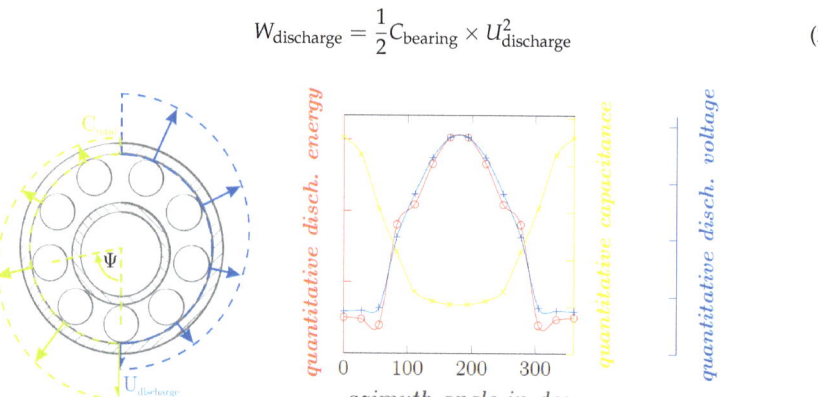

Figure 7. Theoretical distribution of capacitance (green), discharge voltage (blue), and discharge energy (red).

In summary, the likelihood of electric discharges is higher in the loaded zone, but each discharge releases a relatively low amount of energy compared to those outside the loaded zone, leading to different surface mutation characteristics. To test this hypothesis, an experiment was conducted with a hybrid NU206 cylindrical roller bearing equipped with a single steel rolling element under a radial load. After running in and applying electrical stress, the non-rotating outer ring was examined under a microscope. As expected, the loaded zone showed a "grey frosted" surface with numerous small craters, while the non-loaded zone exhibited fewer but larger craters, indicating a higher amount of discharge energy.

In summary, the likelihood of electric discharges is higher in the loaded zone, but each discharge releases a relatively low amount of energy compared to those outside the loaded zone, leading to different surface mutation characteristics. To test this hypothesis, an experiment was conducted with a hybrid NU206 cylindrical roller bearing equipped with a single steel rolling element under radial load. After running in and applying electrical stress, the non-rotating outer ring was examined under a microscope. The result is shown in Figure 8. As the theoretical preliminary considerations suggested, inside the loaded zone, a grey frosted surface with numerous craters and regions affected by over-rolling can be observed. Opposite the loaded zone are fewer but larger craters, while relative large areas of the surface remain unaltered, which indicates a higher discharge energy than that inside the loaded zone. The rotating inner ring and the steel rolling element are subject to the full bandwidth of the electrical load, which is distributed over the entire outer ring, and exhibit a damage pattern regardless of the azimuth angle.

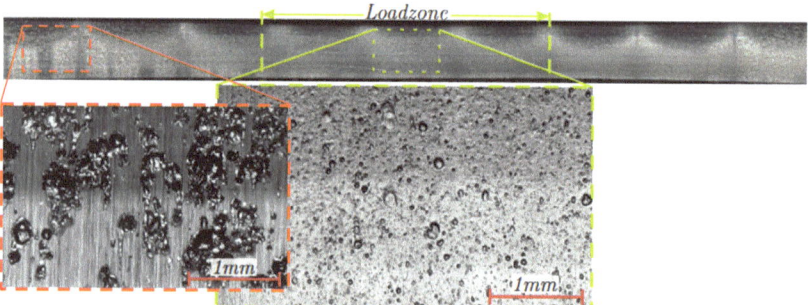

Figure 8. Dependence of the degree of surface mutation in the loaded zone of a non-rotating outer ring of a cylindrical roller bearing.

2.3. Experimental Service Life Investigations

To explore a range of electrical pre-stresses, the number of steel rolling elements in the modified hybrid bearings, as well as the value and duration of the discharge current from the current/voltage source, were adjusted to achieve the discharge energies listed in Table 1. These adjustments were monitored using the described measurement equipment. In the first test series, efforts were made to exceed the limits defined by TISCHMACHER and MÜTZE [13,17]. The initial two experimental series used a laboratory voltage source, while subsequent series 3 and 4 employed an EDM source, leading to the ranges of electrical load shown in Table 1.

Table 1. Discharge energies, apparent powers, and current densities of different experiment series.

Parameter	Energy	Apparent Power	Current Density
Series 1	0.8–3 µJ	20–210 VA	0.5–1.5 A/mm^2
Series 2	16–264 nJ	0.83–19.4 VA	0.05–0.5 A/mm^2
Series 3	197 nJ	12.9–13.8 VA	0.38 A/mm^2
Series 4	1 µJ	64–70.4 VA	1.93 A/mm^2

During pre-stressing, the test bearings were equipped with at least two steel rolling elements, while the rest of the bearing was filled with ceramic rolling elements to accelerate the electrical pre-stress process. This setup prevented electrical discharges in the load-free zone, as excessive electrical surface mutation would result in damage that no longer corresponded to grey frosting. After the pre-stressing, the modified hybrid bearing was disassembled, cleaned, and equipped with non-stressed steel rolling elements except for one pre-stressed steel rolling element, ensuring we also took the influence of current-damaged rolling elements into consideration with regard to the service life investigations. The pre-stressed bearings of the different experiment series were then tested until failure in the four-bearing radial fatigue test rig. Failure detection was carried out via vibration sensors, and any failed bearings were replaced with new, non-pre-stressed bearings to continue the test cycle for the remaining pre-stressed bearings. The consistency of the test parameters was maintained across all experiment series and the parameters themselves are listed in Table 2. The only difference between the test series is the electrical pre-stress induced on the universal test rig.

Table 2. Operating parameters for the service life tests.

Parameter	Value
Bearing	NU206
Load C/P	2.5
Rotational speed in 1/min	2500
Outer ring temperature in °C	50
Viscosity ratio κ	3
Maximum contact pressure in GPa	3.0
Amount of steel rolling elements	2
Lubricant	Renolin CLP 68
Lubrication type	Injection lubrication
Shutdown criterion	Vibration

The listed κ-value of 3 used in the tests was calculated for normal bearings that are not electrically stressed. It was observed that the pre-stressed bearings showed higher vibration and noise levels compared to non-stressed reference bearings. Despite anticipated full lubrication conditions for the used value of $\kappa = 3$, consistent lubrication was not always achieved, especially in the test series involving high discharge energies. This was verified through capacitive lubricating film measurements during operation [10,20].

3. Results of Experimental Service Life Investigations

While analyzing the failed bearing surfaces, it was apparent that the failure patterns showed a similar appearance throughout the majority of the tests. An example of such damage is shown in Figure 9. It was recorded using a reflective light microscope and a laser-scanning microscope. The damage consistently featured a shallow spalling region with a depth of approx. 40 µm and a wave-like structure, which seemed to propagate opposite to the rolling direction. As shown in the rolling direction, some of these wave-like structures also had a deep spalling behind them.

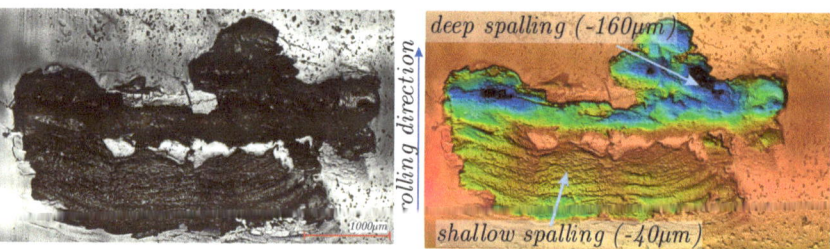

Figure 9. Surface spalling after 3.6 million revolutions from test series 1.

One potential explanation for this wave-like structure lies in the melting and resolidification processes induced by electrical discharges. Tensile residual stress could have been introduced into the material to a depth of 0–40 µm [23]. In addition, the material hardness below the hard and brittle surface was potentially reduced through the heat input of the discharges [24]. This could facilitate subsurface crack formation, which is more likely to occur due to the softened steel beneath the surface, resulting in a core crushing mechanism [25].

The service life test results are illustrated in Weibull diagrams, with each test series described briefly below.

3.1. Service Life of Test Series 1

Test series 1 had the highest applied electrical stress among all test series. It is worth noting that no full-film lubrication could be detected capacitively during operation in this series. This can be attributed to the heavily roughened surface, which led to continuous contact between rough surfaces, preventing the development of a measurable capacitive charging curve. All observed surface spallings were found on the inner ring, similar to the one shown in Figure 9. The Weibull distribution of all counted experiments from test series 1 is shown in Figure 10. A total of 24 pre-stressed bearings were tested, with 20 categorized as failed items and the remaining four considered to be survivors. The experimental life of $B_{10} = 2.74 \times 10^6$ revolutions and Weibull slope of $\beta = 2.47$ were obtained.

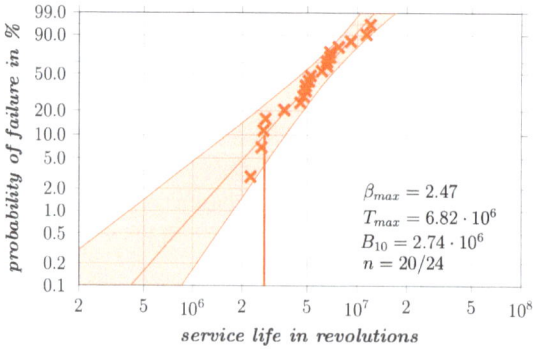

Figure 10. Weibull distribution of test series 1.

3.2. Service Life of Test Series 2

In test series 2, the electrical stress was significantly reduced. A total of eight bearings were tested, of which six failed, and two were rated as survivors. The experimental life of $B_{10} = 10.5 \times 10^6$ revolutions and Weibull slope of $\beta = 1.83$ were obtained and shown in Figure 11.

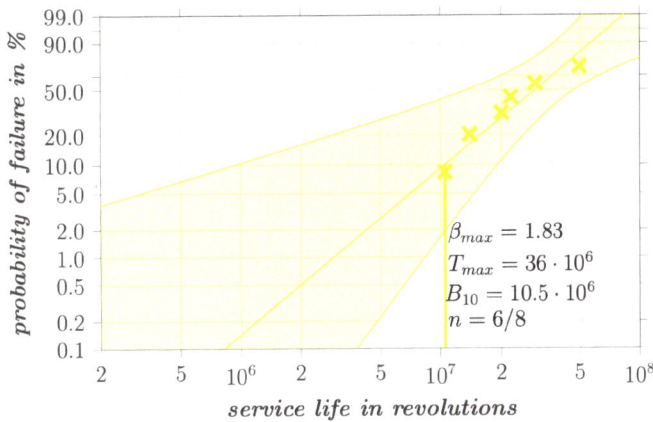

Figure 11. Weibull distribution of test series 2.

3.3. Service Life of Test Series 3

The electrical stress in test series 3 was even further reduced compared to series 1 and 2. An experimental life of $B_{10} = 14.45 \times 10^6$ revolutions and Weibull slope of $\beta = 2.73$ for eight tested bearings were obtained and shown in Figure 12.

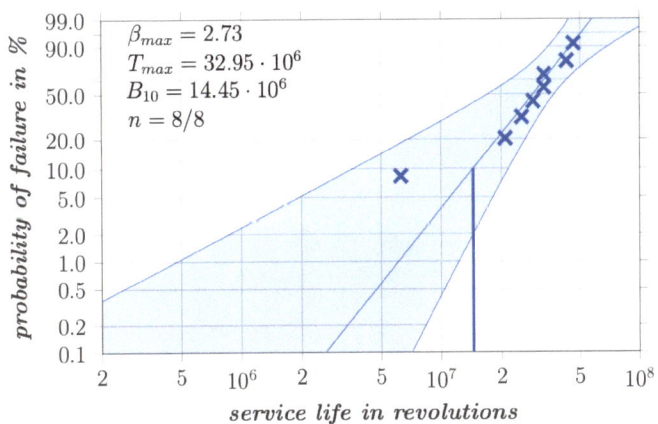

Figure 12. Weibull distribution of test series 3.

3.4. Service Life of Test Series 4

In the last test series the electrical stress was higher than in series 2 and 3 but lower than in series 1 (Table 1). A total of four bearings were tested in this series, resulting in a large standard deviation. The obtained experimental service life is $B_{10} = 7.36 \times 10^6$ revolutions with a Weibull slope of $\beta = 2.16$ and shown in Figure 13.

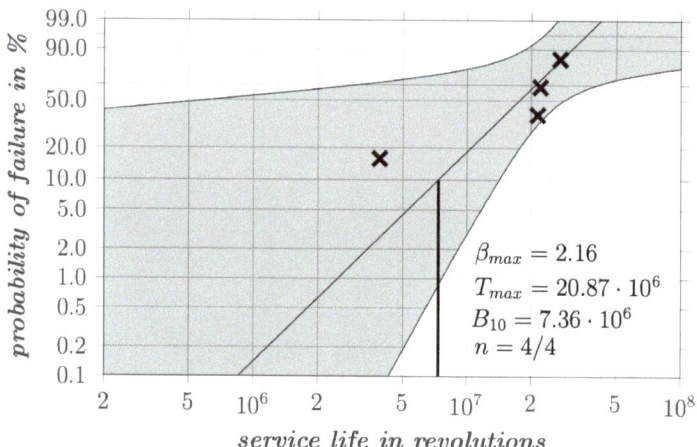

Figure 13. Weibull distribution of test series 4.

3.5. Discussion of All Test Series

The experiments comprise the service life investigation of a total of 44 electrically pre-stressed cylindrical roller bearings. The overall results are summarized in a Weibull Graph in Figure 14. The test series with higher electrical loads and therefore more severe pre-stress in form of grey frosting showed a shorter service life. This indicates that the failure of electrically stressed rolling bearings can be attributed to mechanisms other than fatigue and that current passage can indeed exert an influence on the service life that scale with the severity of the discharge energy. In comparison, reference tests from KEHL [26] obtained a service life of $B_{10} = 21.66 \times 10^6$ revolutions and Weibull slope of $\beta = 1.86$. He used the same test rig with the same mechanical loads, but with a viscosity ratio of $\kappa = 2$ instead of 3. However, it should be pointed out again that this κ value refers to reference (non-electrically stressed) bearings. The importance of this is discussed in the following section.

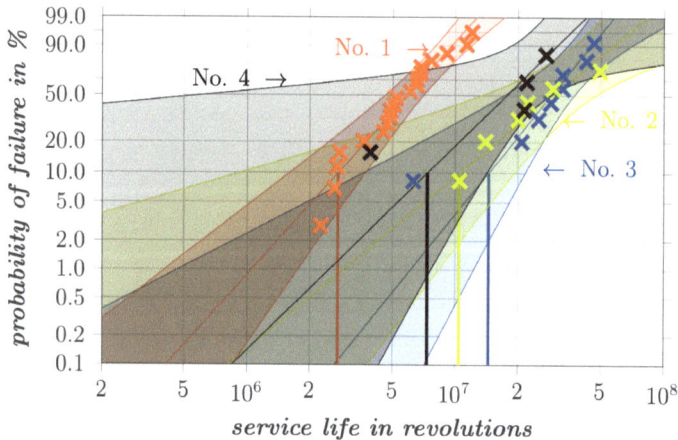

Figure 14. Weibull distribution of all test series.

4. Extended Service Life Model

As shown in the previous section, it is apparent that electrical stress has an impact on the service life of rolling bearings. Consequently, the question arises: How can this influence be quantified and subsequently integrated into the calculation of rolling bearing service life? The conducted test series lead to the assumption that in the context of electrically stressed bearings, the surface roughness is an essential factor in the reduction of service life.

An internationally accepted method to calculate the service life is described in DIN ISO 281 [27]. The equation for calculating the fatigue life of cylindrical roller bearings is based on the slice method, where the load distribution along one rolling element for n_s slices is used to calculate the modified rating life L_{10mr}.

$$L_{10mr} = \left\{ \sum_{k=1}^{n_s} \left\{ \left[a_{\text{ISO}}\left(\frac{e_C C_U}{P_{ks}}, \kappa \right) \right]^{-9/8} \left[\left(\frac{q_{kci}}{q_{kei}} \right)^{-9/2} + \left(\frac{q_{kce}}{q_{kee}} \right)^{-9/2} \right] \right\} \right\}^{-8/9} \quad (3)$$

The service life calculation is based on various parameters, such as the viscosity ratio κ, the contamination coefficient e_C, the fatigue load limit C_U, dynamic equivalent load P_{ks}, and the dynamic load rating q_{kc}, as well as the dynamic load rating q_{ke} of each k-th slice. Unfortunately, the influencing factors in the equation do not include the influence of the surface roughness. To include surface roughness, the viscosity ratio κ (Equation (4)) must be modified. This ratio characterizes the relationship between operational ν and nominal kinematic viscosity ν_1, thereby reflecting the lubrication condition in the fatigue life equation. Under normal circumstances, it is assumed that the surface roughness corresponds to that of non-electrically stressed bearings.

$$\kappa = \frac{\nu}{\nu_1} \quad (4)$$

To link the viscosity ratio κ with surface roughness, the parameter Λ^* is introduced [28,29]. It includes the elastic deformation of micro-roughness peaks, minimum h_m and central film thickness h_c and the area-related roughness value of the reduced peak height Spk. Bearings that were electrically stressed and run-in not-stressed reference bearings were measured with a laser scanning microscope to initially calculate the Λ^* value. The measured values from each test series and the reference to be able to apply the method from HANSEN to calculate Λ^* are listed in Table 3.

Table 3. Calculation results for the application example.

Series Number	Spk in μm	r/R	Λ^*
1	0.347	0.0275	0.42
2	0.19	0.055	0.88
3	0.174	0.091	1.01
4	0.32	0.05	0.52
reference	0.15	0.098	1.23

With the result of the reference bearing, it is possible to establish a relationship between Λ^* and κ, since the κ value would always refer to a normal surface finish. As shown in Figure 15a, an approximation line can be effectively fitted. Based on the results, it is assumed that the correlation between both values is linear and can be described by the following equation:

$$\kappa(\Lambda^*) = 3.125 \times (\Lambda^* - 0.29) \quad (5)$$

Using the measured roughness values from the electrically mutated surfaces, it is possible to determine the Λ^*-value and to use Equation (5) to determine the corresponding $\kappa(\Lambda^*)$ value, as shown in Figure 15b. This can then be used as an input for Equation (3).

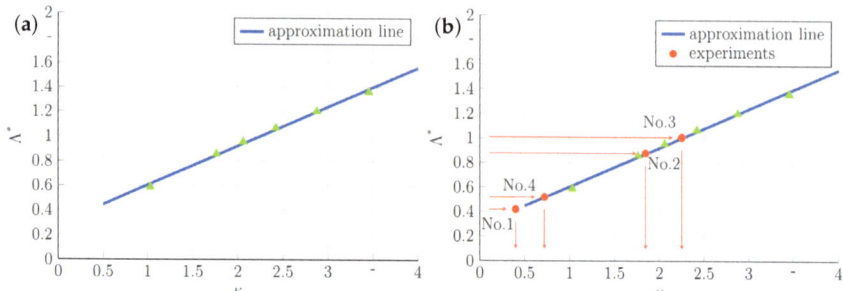

Figure 15. (a) Relation between Λ^* and κ and (b) classification of electrically mutated surfaces.

To summarize the outlined procedure, the four steps that must be performed sequentially are briefly listed as follows:

1. Measurement of surface roughness.
2. Calculation of Λ^*.
3. Calculation of $\kappa(\Lambda^*)$.
4. Determine theoretical service life with method from DIN ISO 281.

Using the new viscosity ratios, it is now possible to correlate the experimentally determined service life of the electrically stressed and reference bearings from [26] with the calculated service life from Equation (3). The values with the associated experimental service life are listed in Table 4.

Table 4. Calculation results for the application example.

Series Number	$\kappa(\Lambda^*)$	B_{10} in 10^6 rev.
1	0.4	2.74
2	1.844	10.5
3	2.25	14.45
4	0.719	6.9
reference mixed lub.	0.5	7.14
reference full lub.	2	21.66

The service life plotted against the viscosity ratio is shown in Figure 16. It can be seen that the κ-values of the electrically pre-stressed bearings differed substantially between the test series, even though a viscosity ratio for non-stressed bearings of $\kappa = 3$ was set. In particular, for experiment series 1 and 4, which were characterized by higher electrical loads, the κ-value was significantly lower than those for series 2 and 3, indicating mixed lubrication. This was also confirmed via capacitive measurements during the experiments. It is also apparent that the level of electrical stress directly translated to the degree of surface roughness. The higher the electrical stress, the higher the surface roughness. The newly categorized service lives of the pre-stressed bearings matched the calculated service life quite well. However, a considerable reduction in service life was observed in the area of full lubrication compared to the reference tests. Even test series 3, which had the lowest electrical load, achieved a comparably shorter service life with $B_{10} = 14.45 \times 10^6$ rev. than the reference test with similar lubrication conditions of $B_{10} = 21.66 \times 10^6$ rev. In the regime of mixed lubrication, the lubrication regime exerts a predominant influence on the bearing's performance. Under these conditions, the differences in service life between electrically pre-stressed bearings from test series 4 and reference bearings are minimal.

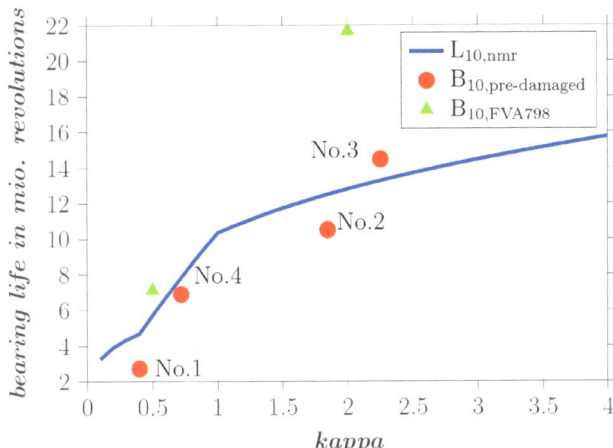

Figure 16. Calculated (blue line) and experimentally determined service lives of electrically pre-stressed bearings (red circles) and reference experiments without harmful bearing currents (green triangles) from KEHL [26] plotted against lubrication condition κ.

5. Conclusions and Outlook

The experimental investigations presented in this study provide insights into the impact of electrical pre-stressing on the service life of rolling bearings. The results indicate that electrical discharges, which manifest in the form of EDM currents and circular bearing currents, can lead to notable surface mutations and reduce the service life of the bearings. The induced surface damage, characterized by grey frosting, craters, and in severe cases, fluting, results in increased vibration and noise levels, ultimately affecting the operational integrity and longevity of the bearings.

The experimental data derived from controlled electrical pre-stressing and subsequent service life tests reveal a clear correlation between the severity of electrical pre-stress and the reduction in bearing life. The Weibull analyses of the different test series demonstrate that higher discharge energies correlate with shorter bearing service lives. For instance, test series 1, which had the highest electrical load, exhibited a significantly reduced service life ($B_{10} = 2.74 \times 10^6$ revolutions) compared to the less electrically stressed bearings in the test series 3 ($B_{10} = 14.45 \times 10^6$ revolutions).

The surface damages observed in the bearings after the service life test was stopped included shallow spalling with wave-like structures, which is indicative of the melting and resolidification processes induced by electric discharges. These structural changes may introduce tensile residual stress and reduce material hardness, contributing to the reduced fatigue life. The study also highlights the importance of considering the influence of surface roughness and lubrication conditions when evaluating the service life of electrically stressed bearings, as it seems to be the main reason for a shortened service life.

An extended service life model was proposed in this study, which incorporates the effect of electrical surface mutations by modifying the viscosity ratio κ. This modification accounts for the altered lubrication conditions due to increased surface roughness, providing a more accurate prediction of the bearing's service life under electrical stress. Using this test methodology, it can be concluded that the service life under mixed friction conditions exhibits minimal deviation from the reference tests. However, significant differences are observed under full lubrication, suggesting that the failure mechanisms in electrically pre-stressed bearings diverge from traditional fatigue mechanisms.

In conclusion, the findings underscore the necessity for further research into the mechanisms of electric discharge-induced damage in rolling bearings and the development of advanced models for service life prediction.

Author Contributions: Conceptualization, V.S. and M.K.; methodology, V.S. and M.K.; software, V.S. and M.K.; validation, V.S. and M.K.; formal analysis, V.S. and M.K.; investigation, V.S. and M.K.; resources, G.P. and M.M.; data curation, V.S. and M.K.; writing—original draft preparation, V.S. and M.K.; writing—review and editing, V.S., M.K., G.P. and M.M.; visualization, V.S.; supervision, G.P. and M.M.; project administration, G.P.; funding acquisition, G.P. All authors have read and agreed to the published version of the manuscript.

Funding: This research was funded by the German joint industrial research (Industrielle Gemeinschaftsforschung IGF) with the Grant No. 22079N.

Data Availability Statement: The original contributions presented in the study are included in the article, further inquiries can be directed to the corresponding author.

Conflicts of Interest: The authors declare no conflicts of interest.

References

1. Pöhlmann, A. *Wälzlager in Elektrischen Lokomotiven und Diesellokomotiven*; SKF Kugellagerfabriken Schweinfurt: Schweinfurt, Germany, 1959.
2. Punga, F.; Hess, W. Eine Erscheinung an Wechsel- und Drehstromgeneratoren. *Elektrotechnik Maschinenbau* **1907**, *25*, 615 618.
3. Fleischmann, L. Ströme in Lagern und Wellen. *Elektr. Kraftbetriebe Bahnen* **1909**, *7*, 352–353.
4. Alger, P.L.; Samson, H.W. Shaft Currents in Electric Machines. *Trans. Am. Inst. Electr. Eng.* **1924**, *XLIII*, 235–245. . [CrossRef]
5. Habibullah, M.; Lu, D.D.; Xiao, D.; Rahman, M.F. Finite-State Predictive Torque Control of Induction Motor Supplied From a Three-Level NPC Voltage Source Inverter. *IEEE Trans. Power Electron.* **2017**, *32*, 479–489. [CrossRef]
6. Hausberg, V. Elektrische Lagerbeanspruchung Umrichtergespeister Induktionsmaschinen. Ph.D. Thesis, Leibniz Universität Hannover, Hanover, Germany, 2001.
7. Stockbrügger, J.O. Analytische Bestimmung parasitärer Kapazitäten in elektrischen Maschinen. Ph.D. Thesis, Institutionelles Repositorium der Leibniz Universität Hannover, Hanover, Germany, 2021. [CrossRef]
8. Furtmann, A. Elektrische Belastung von Maschinenelementen Im Antriebsstrang. Ph.D. Thesis, Leibniz Universität Hannover, Hanover, Germany, 2017.
9. Gemeinder, Y. Lagerimpedanz und Lagerschädigung bei Stromdurchgang in umrichtergespeisten elektrischen Maschinen. Ph.D. Thesis, Technische Universität Darmstadt, Darmstadt, Germany, 2016.
10. Wittek, E. Charakterisierung Des Schmierungszustandes Im Rillenkugellager Mit Dem Kapazitiven Messverfahren. Ph.D. Thesis, Leibniz Universität Hannover, Hanover, Germany, 2016.
11. Radnai, B. Wirkmechanismen Bei Spannungsbeaufschlagten Wälzlagern. Ph.D. Thesis, Technische Universität Kaiserslautern, Kaiserslautern, Germany, 2016.
12. Schneider, V.; Liu, H.C.; Bader, N.; Furtmann, A.; Poll, G. Empirical Formulae for the Influence of Real Film Thickness Distribution on the Capacitance of an EHL Point Contact and Application to Rolling Bearings. *Tribol. Int.* **2021**, *154*, 106714. [CrossRef]
13. Tischmacher, H. Systemanalysen zur elektrischen Belastung von Wälzlagern bei umrichtergespeisten Elektromotoren. Ph.D. Thesis, Leibniz Universität Hannover, Hanover, Germany, 2017.
14. Preisinger, G. Cause and Effect of Bearing Currents in Frequency Converter Driven Electrical Motors: Investigations of Electrical Properties of Rolling Bearings. Ph.D. Thesis, TU Wien, Vienna, Austria, 2002.
15. Harder, A.; Zaiat, A.; Becker-Dombrowsky, F.M.; Puchtler, S.; Kirchner, E. Investigation of the Voltage-Induced Damage Progression on the Raceway Surfaces of Thrust Ball Bearings. *Machines* **2022**, *10*, 832. [CrossRef]
16. Zuo, X.; Xie, W.; Zhou, Y. Influence of Electric Current on the Wear Topography of Electrical Contact Surfaces. *J. Tribol.* **2022**, *144*, 071702. [CrossRef]
17. Mütze, A. Bearing Currents in Inverter-Fed AC-Motors. Ph.D. Thesis, Technische Universität Darmstadt, Darmstadt, Germany, 2004.
18. Romanenko, A.; Muetze, A.; Ahola, J. Effects of Electrostatic Discharges on Bearing Grease Dielectric Strength and Composition. *IEEE Trans. Ind. Appl.* **2016**, *52*, 4835–4842. [CrossRef]
19. Leenders, P.; Houpert, L. Study of the Lubricant Film in Rolling Bearings; Effects of Roughness. In *Tribology Series*; Dowson, D., Taylor, C.M., Godet, M., Berthe, D., Eds.; Fluid Film Lubrication—Osborne Reynolds Centenary; Elsevier: Amsterdam, The Netherlands, 1987; Volume 11, pp. 629–638. [CrossRef]
20. Schneider, V.; Bader, N.; Liu, H.; Poll, G. Method for in Situ Film Thickness Measurement of Ball Bearings under Combined Loading Using Capacitance Measurements. *Tribol. Int.* **2022**, *171*, 107524. [CrossRef]
21. Hacke, B. Früherkennung von Wälzlagerschäden in Drehzahlvariablen Windgetrieben. Ph.D. Thesis, Leibniz Universität Hannover, Hanover, Germany, 2011.
22. Gonda, A. Determination of Rolling Bearing Electrical Capacitances with Experimental and Numerical Investigation Methods. Ph.D. Thesis, TU Kaiserslautern, Kaiserslautern, Germany, 2023.
23. Ekmekci, B. Residual Stresses and White Layer in Electric Discharge Machining (EDM). *Appl. Surf. Sci.* **2007**, *253*, 9234–9240. [CrossRef]

24. Karastojkovic, Z.; Janjusevic, Z. Hardness and Structure Changes at Surface in Electrical Discharge Machined Steel C3840. In Proceedings of the 3rd BMC, Ohrid, North Macedonia, 24–27 September 2003.
25. Hwang, J.; Coors, T.; Pape, F.; Poll, G. Simulation of a Steel-Aluminum Composite Material Subjected to Rolling Contact Fatigue. *Lubricants* **2019**, *7*, 109. [CrossRef]
26. Kehl, J. FVA 798: Ermüdungslebensdauer Bei Oberflächenbeschädigungen. *FVA-Forschungsheft* **2022**, *1478*, 1–193.
27. *DIN ISO 281*; Wälzlager—Dynamische Tragzahlen und Nominelle Lebensdauer—Verfahren zur Berechnung der Modifizierten Referenz Lebensdauer für Allgemein Belastete Wälzlager. Beuth Verlag: Berlin, Germany, 2009.
28. Hansen, J.; Björling, M.; Larsson, R. A New Film Parameter for Rough Surface EHL Contacts with Anisotropic and Isotropic Structures. *Tribol. Lett.* **2021**, *69*, 37. [CrossRef]
29. Hansen, J. Elasto-Hydrodynamic Film Formation in Heavily Loaded Rolling-Sliding Contacts: Influence of Surface Topography on the Transition between Lubrication Regimes. Ph.D. Thesis, Lulea University of Technology, Lulea, Sweden, 2021.

Disclaimer/Publisher's Note: The statements, opinions and data contained in all publications are solely those of the individual author(s) and contributor(s) and not of MDPI and/or the editor(s). MDPI and/or the editor(s) disclaim responsibility for any injury to people or property resulting from any ideas, methods, instructions or products referred to in the content.

Article

Ice-versus-Steel Friction: An Advanced Numerical Approach for Competitive Winter Sports Applications

Birthe Grzemba [1,*] and Roman Pohrt [2]

1 Institute for Research and Development of Sports Equipment (FES), 12459 Berlin, Germany
2 Independent Researcher, 10119 Berlin, Germany
* Correspondence: bgrzemba@fes-sport.de

Abstract: Understanding and predicting the friction between a steel runner and an ice surface is paramount for many winter sports disciplines such as luge, bobsleigh, skeleton, and speed skating. A widely used numerical model for the analysis of the tribological system steel on ice is the Friction Algorithm using Skate Thermohydrodynamics (F.A.S.T.), which was originally introduced in 2007 and later extended. It aims to predict the resulting coefficient of friction (COF) from the two contributions of ice plowing and viscous drag. We explore the limitations of the existing F.A.S.T. model and extend the model to improve its applicability to winter sports disciplines. This includes generalizing the geometry of the runner as well as the curvature of the ice surface. The free rotational mechanical mounting of the runner to the moving sports equipment is introduced and implemented. We apply the new model to real-world geometries and kinematics of speed skating blades and bobsleigh runners to determine the resulting COF for a range of parameters, including geometry, temperature, load, and speed. The findings are compared to rule-of-thumb testimonies from athletes, previous numerical approaches, and published experimental results where applicable. While the general trends are reproduced, some discrepancy is found, which we ascribe to the specific assumptions around the formation of the liquid water layer derived from melted ice.

Keywords: ice friction; winter sports; thermohydrodynamics; frictional melting; numerical model

Citation: Grzemba, B.; Pohrt, R. Ice-versus-Steel Friction: An Advanced Numerical Approach for Competitive Winter Sports Applications. *Lubricants* **2024**, *12*, 203. https://doi.org/10.3390/lubricants12060203

Received: 26 April 2024
Revised: 28 May 2024
Accepted: 29 May 2024
Published: 4 June 2024

Copyright: © 2024 by the authors. Licensee MDPI, Basel, Switzerland. This article is an open access article distributed under the terms and conditions of the Creative Commons Attribution (CC BY) license (https://creativecommons.org/licenses/by/4.0/).

1. Introduction

The interplay between ice and runners in winter sports presents a captivating nexus of physics, engineering, and athletic achievement. At the heart of competitive winter sports lies the relentless pursuit of performance optimization. Athletes and equipment manufacturers alike strive to gain a competitive edge. Understanding the complex dynamics governing frictional interactions is paramount for enhancing equipment design and, ultimately, achieving peak athletic performance.

Conducting experiments to study ice–steel friction in winter sports presents a unique set of challenges. The frictional forces involved are low, speeds are high, and results can vary significantly with respect to ice conditions, rendering experimental analysis costly and time-intensive.

Numerical simulations have emerged as indispensable tools in unraveling the intricacies of ice–runner friction. By leveraging computational models, researchers can explore a vast array of parameters and scenarios regarding the runners' geometries, materials, climates, types of ice, kinematics, etc. Numerical simulations can thus offer unparalleled insights into the underlying physics of winter sports.

In this paper, we build on an existing popular model called Friction Algorithm using Skate Thermohydrodynamics (F.A.S.T.) to extend the scenarios to which it is applicable. We start by introducing the previous generations of the F.A.S.T. model and laying out the basic principle of its numerical scheme. The main limitations and challenges resulting from this scheme are highlighted. We then introduce our additions to the model, extending its capabilities but also introducing new steps with significant computational costs. The accompanying challenges are discussed.

In Section 3, we then employ the new model to investigate cases of two major winter sports disciplines: speed skating and bobsleigh. Typical blade to ice configurations in these disciplines can be seen in Figure 1.

(a) (b)

Figure 1. Winter sports racing: (**a**) speed skater at the World Championships 2024 in Inzell and (**b**) bobsleigh pilot at the 2024 Monobob World Championship in Winterberg (©R. Hartnick).

2. Methods

In the following sections, the origin and fundamentals of the F.A.S.T. model are briefly summarized and the extensions made are presented.

2.1. F.A.S.T. Model

We start by looking at the history and current state of the F.A.S.T. model, as it was used as a starting point for our work.

2.1.1. Genesis of the F.A.S.T. Model

The F.A.S.T. model was originally developed in 2007 by Penny et al. for the sport of speed skating [1]. It was implemented for an upright speed skating blade and contained terms for plowing of the ice and shearing of the water layer, which is built through melting, heat conduction, and squeezing. In 2011, Lozowski and Szilder published the results of a corrected and extended version of the code and provided a sensitivity analysis [2].

Poirier adopted the model for the sport of bobsleigh in 2011 [3], translated it from FORTRAN to C++, and made several additions to the code, some discipline-related and some not. Most notably, he made the water layer thickness variable in a lateral direction, which is crucial when considering laterally variable geometries. He also added new estimations for the ice hardness dependent on ice temperature and included the heat transfer between the blade and the water layer. In 2013, Lozowski et al. published their results for an inclined speed skating blade [4] but did not include the additions Poirier had made to the code, which led to deviations in water layer thickness. An application of the model to skeleton runners was published in 2014 by Lozowski et al. [5]; however, except for implementing the skeleton runner geometry, no changes were made to the code. In 2017, Stell delivered an adaptation of the code for the sport of luge for which he used Poirier's C++ code and translated it to Matlab Code [6]. Stell's Matlab Code served as the starting point for this study.

A more recent application of F.A.S.T. for speed skating was developed by Du et al. [7]. Therein, the assumption used in the code for the runner temperature, which is a very sensitive parameter and which was recommended by Poirier [3] for further analysis, was analyzed. Du et al. concluded that Poirier's assumption that the runner has the same temperature as the melt water is valid. Therefore, we adopted this assumption for our studies.

2.1.2. Principles

The F.A.S.T. model accepts a set of parameters for the ice surface, the runner, the ambient conditions, and the loading as well as the sliding speed. It assumes a quasi-static momentary sliding state and predicts the resulting area of real contact, the thickness of the melt layer, and frictional forces. The model accounts for the effects of plowing through ice and the melt water film in two distinct steps, which we will briefly discuss here. For further details and implementation, the reader is referred to Poirier [3].

The first step concerns the plowing of the blade through ice and is based purely on ice hardness, load, and geometry. Elastic deformation of the ice is rightfully neglected; the problem is dominated by plastic deformation.

Using an iterative process, the runner's depth of indentation into the ice is found to satisfy the condition for a normal load F_N.

$$F_N = H_{Ice} \cdot A_c \qquad (1)$$

The underlying assumption is that the ice hardness H_{Ice} is the maximum compressive stress that will occur everywhere inside the contact zone. The stress distribution carrying the vertical load is, therefore, uniform.

The plowing is assumed to remove the ice volume and, therefore, any rear part of the runner cannot come into contact unless it is located lower than any previous geometry features. From Figure 2, it can be seen that the contact zone will thus generally arise in the front part of the runner. For a spherical runner geometry on undisturbed ice, the resulting contact zone will be semi-elliptical when observed from above. In this stage, the plowing force F_P is determined as

$$F_P = H_{Ice} \cdot A_P. \qquad (2)$$

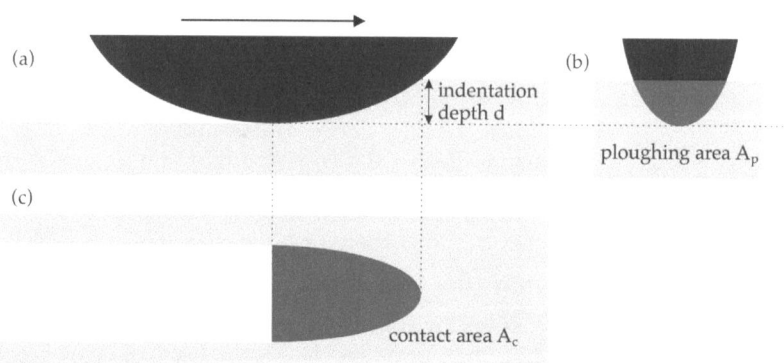

Figure 2. Schematic side (**a**), front (**b**), and top (**c**) view of the plowing action of a simple loaded runner on an undisturbed ice surface inside the F.A.S.T. model. The indentation depth d leads to geometrical overlap where ice material is plowed away in the contact zone. Both vertical load capacity and frictional drag from plowing are determined by the ice's hardness and affected areas.

The plowing stage of the simulation results in the determination of the plowing force and the real contact area, i.e., the footprint of the runner on the ice.

The second stage calculates the thickness of the melt water layer under the consideration of heat transfer and Couette flow.

Consider a particular spot on the ice surface. It is assumed that a preexisting microscopical melt layer is always present on the ice surface (see, e.g., [8]). When the runner arrives, the available heat for the further melting of ice is calculated based on the following aspects.

- Generation of heat by shearing of the melt layer through its viscous properties
- "Slow" heat conduction from the ice surface into the bulk of the ice

- "Fast" heat conduction from the liquid melt layer into the ice surface
- Heat conduction from the runner into the liquid layer

The sum of these contributions produces the available heat. For any point on the ice, it is integrated over the sliding history from the first instance of contact and divided by the density and latent heat of fusion to determine the volume of melted water, i.e., the film thickness.

The longer a point on the ice is in contact, the greater its film thickness becomes.

From the film thickness h, the viscous resistance of the runner F_S can be determined via integration over the contact zone:

$$F_s = v\,\eta \int_{A_c} \frac{1}{h} dA_c \qquad (3)$$

Here, v denotes the sliding velocity and η the dynamic viscosity of water.

An interesting property of the model is that this second stage considers the geometry of the runner and ice only through the two-dimensional shape of the contact zone obtained in the plowing stage.

It should be noted that the F.A.S.T. model described by Poirier and Stell includes an additional procedure to calculate the effect of the squeeze flow out of the contact zone. This flow is directed sideways and reduces the film thickness. However, we found the effect of the additional calculation to be negligible in all relevant cases and have decided to disable it due to its high computation cost.

Furthermore, we would like to point out that the MATLAB implementation given by Stell contains flaws, which lead to inaccurate results. If you intend to work with this implementation, please consider our supplemental material in Appendix A.

2.2. Extensions to F.A.S.T.

Some of the restrictions of the original model make it unsuitable for realistically investigating in sports disciplines. We have extended the model to tackle the most relevant issues. In the following sections, we introduce the new requirements, discuss the implications that follow, and how we implemented them.

2.2.1. Generalizing Geometry

In the Poirier version of F.A.S.T., the geometries are always similar: the ice surface is perfectly flat and the runner is spherical with two radii of curvature. This configuration always leads to a semi-elliptical shape for the contact area, which is calculated analytically inside the program as a function of the indentation depth.

For arbitrary runner geometries, the area of contact is not generally accessible analytically and must instead be found by testing the surface pointwise: A value of the indentation depth is assumed and the entire runner geometry is lowered by that amount.

Then, each line of ice surface is traced from the front of the runner to the back, following the ice geometry. If the runner geometry intersects the surface, the latter is ploughed and the current position is marked as being in contact.

This must be performed for all rows and for each assumed indentation depth.

In order to satisfy Equation (1), the correct indentation depth must again be found iteratively. This process is straightforward but comes with considerable computational cost. In principle, any geometrical particularities of a given sport discipline can be accounted for.

For speed skating, the longitudinal blade radius and inclination angle of the blade are varied; for bobsleigh, the longitudinal geometry consists of a multitude of radii and the track is not flat. For a more detailed discussion of winter sport runner geometries, see the case studies in Sections 3.1 and 3.2. Custom runner geometries can now be used and the track can be curved in both directions.

With arbitrary geometries, it can no longer be assumed that the resulting contact area is contiguous. Figure 3 shows an example.

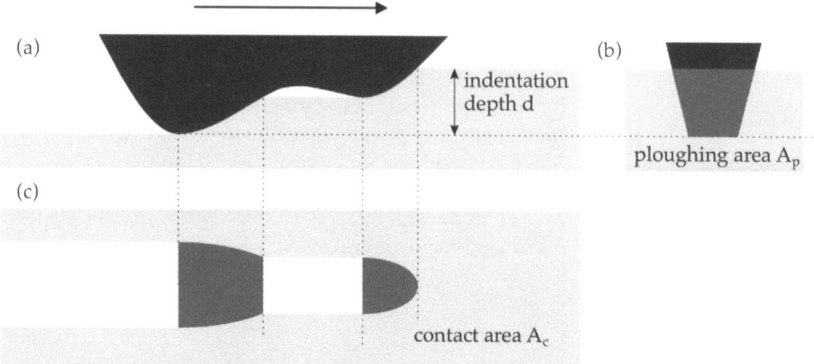

Figure 3. Schematic side (**a**), front (**b**), and top (**c**) view of the indentation configuration of an arbitrarily shaped loaded runner inside the improved F.A.S.T. model. The shape of the runner and the curvature of the ice can lead to multiple non-contiguous contact zones.

2.2.2. Refreezing of Liquid Layer

The non-contiguous contact areas have implications for the second stage of the simulation (film generation) as well. A point on the ice surface can enter into contact, leave contact, and come into contact again towards the rear of the runner.

The question arises of how the film thickness should evolve after the contact has been lost. Looking at the terms contributing to heat in the contact, it is clear that the heat from viscous shearing and from transfer related to the runner is absent when the ice is not in contact. The remaining terms are:

- "Slow" heat conduction from the ice surface into the bulk of the ice;
- "Fast" heat conduction from the liquid melt layer into the ice surface.

These correspond to some outflow of heat and, thus, the refreezing of the film after a loss of contact behind a runner contact section, see Section 3.2 for an example under real conditions.

2.2.3. Lateral Forces

For various disciplines, athletes rely on some lateral load to be carried by the runner, e.g., when passing through curves. We estimated the maximum load capacity in the y-direction by calculating the effective plowing area in the respective direction and multiplying it with the ice hardness. These values can differ between the left and right-hand sides when the runner/ice are tilted with respect to each other (for instance, in a turn inside a curved track).

2.2.4. Rotational Free Mounting of Runners

In the first stage of the F.A.S.T. model, the runner is lowered into the ice until the force balance is satisfied. The process is geometrically governed: an indentation depth is selected for which the resulting normal force is found to match the external load. However, the indentation depth is not the only geometrical parameter, even when 2D. The angle of attack also governs the footprint dramatically. For most applications, the angle of attack is not determined by the design of the sports equipment but rather a consequence of the load distribution front/rear on the runner or, more specifically, the absence of a rotational moment with respect to the y-axis. For speed skating, the point of reference can be understood as the athlete's ankle joint. In bobsleigh, each runner has an individual fixed bearing to allow for a free adjustment of the angle of attack, see Figure 4.

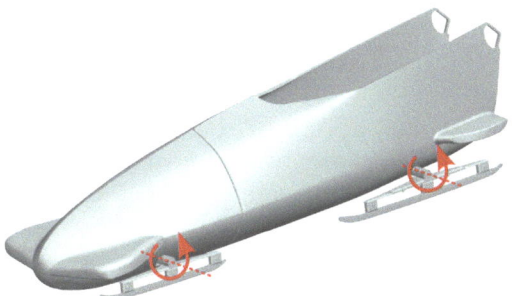

Figure 4. Typical mounting of bobsleigh runners. The runners are mounted in carriers, which have a rotational degree of freedom around the front and rear axle.

In order to account for this, the attack angle as an additional degree of freedom is introduced as well as the additional condition; therefore, the torsional moment, with respect to a given point x_r, vanishes:

$$M_y = 0 = H_{Ice} \int_{A_c} x - x_r dA_c \qquad (4)$$

Numerically, a combination of the indentation depth d and the attack angle α_F must be found iteratively to satisfy both Equations (1) and (4). This is numerically challenging for two major reasons.

Firstly, the resulting normal force and even more so the resulting rotational moment are extremely sensitive to even the smallest adjustments of the selected values for the angle of attack and, to a lesser degree, to the indentation depth, see Figure 5.

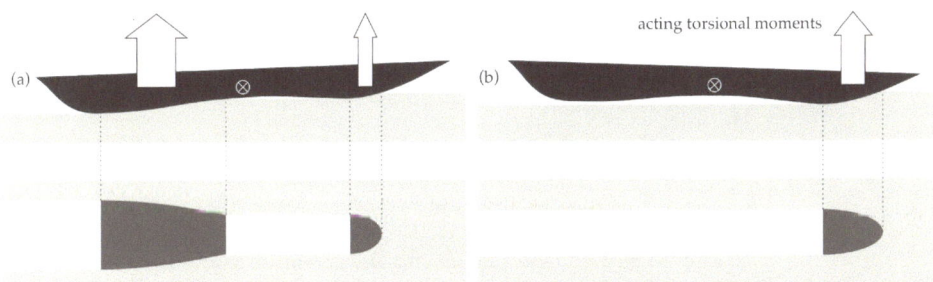

Figure 5. In elongated runners, a slight change in the angle of attack will strongly affect the resulting footprint (compare (**a**) and (**b**)). Its influence on the position and value of the normal forces (arrows) and the resulting rotational moment around the lateral axes (circle with interior cross) is, therefore, extremely pronounced.

Secondly, since the computational domain consists of discrete elements, the contact area (multiplied by the ice hardness) cannot be tuned to match the normal force exactly. Instead, only discrete values can be obtained that correspond to a certain number of surface points identified to be in contact.

Again, for the determination of the resulting rotational torque, the discreteness of the contact area applies, preventing Equation (4) from being satisfied exactly. Both the indentation depth and the angle of attack must thus be chosen to represent a contact configuration (e.g., a subset of surface points to be in contact) which is a viable compromise in both equations.

As a consequence of the discrete nature of the problem, the dependencies are non-smooth and, therefore, cannot be tackled with standard optimization techniques.

In practice, we found this task to be the most challenging, with convergence towards a good compromise being very hard to achieve.

3. Results

The extensions made to the code hold many new possibilities and enable more realistic investigations of the friction behavior of winter sports equipment. As the first case study, we looked at several issues from the sports of speed skating and bobsleigh.

3.1. Speed Skating

As a first case study, we applied the extended F.A.S.T. implementation to the sport of speed skating, which the code was originally developed for.

Our advanced implementation of the F.A.S.T. code for speed skating allows not only for the realistic representation of loads and inclination angles but also for the advanced representation of realistic blade geometries. For example, modern speed skating blades do not only have a longitudinal radius (see R_L in Figure 6) but also a pre-formed lateral radius, where the blade is bent around the vertical z-axis, to allow for better performance in the curves, additionally, the longitudinal geometry does not have to consist of a single radius, but can now be defined by a set of coordinates, which is essential and common for the discipline of short track, which uses variable radii over the length of the blade. Exploration of these variations in geometry will be conducted in future studies.

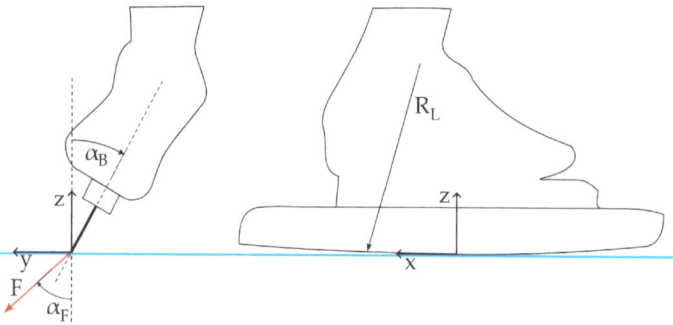

Figure 6. Sketch of a speed skating skate, consisting of a shoe, mounting system, and blade, from the front (**left**) and the side (**right**) with a coordinate system. The angle α_B is the inclination angle of the blade, F is the total force acting on the ice, α_F is the angle of attack of the force F and the radius R_L is the longitudinal radius defining the blade geometry (note that the radius is sketched in an exaggerated manner). The blue line represents the ice surface.

Moreover, for basic speed skating geometries defined by only one longitudinal radius R_L, the effect of an inclined blade on the geometry in relation to the ice has to be considered. When a blade with one longitudinal radius is inclined in relation to the ice, two things will happen. Firstly, the projected curve of the blade in the x–z-plane will not be spherical anymore and will have a lower curvature than the original radius. Secondly, the projection of the inclined blade in the x–y-plane will have an effect on the contact which resembles a camber or curvature around the z-axis and has a steering effect. For visual representation of these effects see Figure 7.

For our case study, we looked at the crucial part of the stroke cycle on the straightaway, using the measured forces and angles from [9,10] as they are shown in Figure 8. By using the region of 30–100% of the stroke cycle, we considered 80% of the total force per stroke and neglected the time of the stroke when both blades were in contact with the ice.

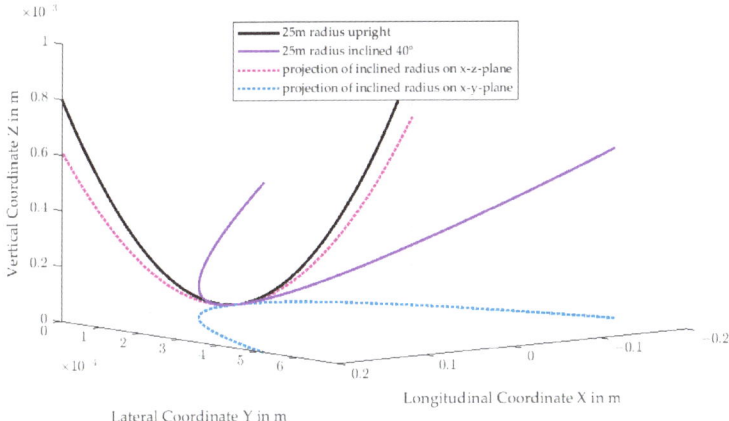

Figure 7. Projection of a spherical curve after inclination. This graph serves as an aid to explain the effect of an inclined speed skating blade on contact geometry.

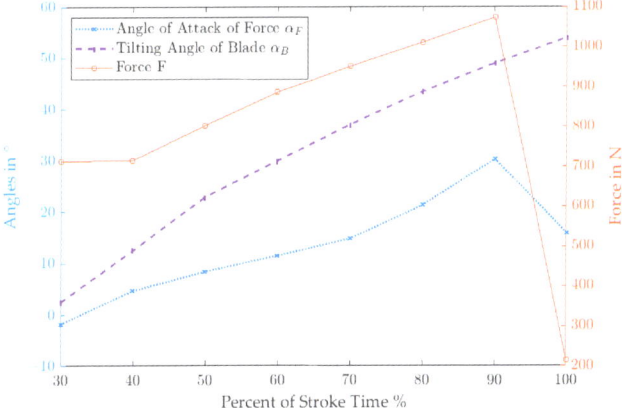

Figure 8. Angles of force and blade (left vertical axis) and force onto the blade (right vertical axis) during a speed skating stroke on a straightaway, values derived from [9,10]. The angles are measured from the vertical axis in space.

For material parameters, we used the properties of a commonly used type of blade material, namely, powder metallurgical high-speed steel. If not stated otherwise, we used a standard blade geometry of a 25 m longitudinal radius R_L, a skating speed of 8 m/s to comply with previous publications, and a standard ice temperature of $-5\ °C$, which is common in speed skating arenas. A basal ice temperature 4 K below surface temperature was assumed as realistic based on interviews with technical staff.

As a first analysis, we calculated the development of the contact area over one stroke cycle (see Figure 9). Several effects can be observed in this analysis. Through higher inclination angles, the contact area becomes narrower, which is due to the sharper angles of the blade's sides in contact. Simultaneously, the contact becomes longer, which is due to the fact that, by inclining a radius to the ice, the effective longitudinal radius increases (see Figure 7). This, again, leads to smaller indentation depths and, therefore, less plowing but also higher viscous drag due to the elongated contact. Depending on the relation between plowing and viscous contributions, the inclination can be beneficial to reduce overall friction.

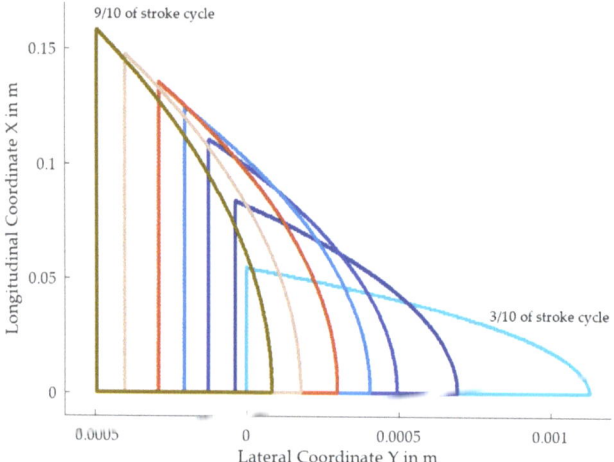

Figure 9. Edges of contact zones for a speed skating stroke cycle on the straightaway with varying forces and angles, moving from the 30% point of the cycle (first contact area to the right in light blue) up to the 90% point of the cycle (most left contact area in olive green). The blade is moving towards positive x values or upwards in this graph.

It is also important to note that the edge of the blade is not positioned at the left edge of the contact area but starts at (0,0) and curves up to the first contact point. From this, the steering effect of an inclined blade can be easily understood, as the above-shown right blade runs on its inner edge (skating upwards) and has a tendency to steer to the left.

We performed calculations using variations in temperature in reference to the literature. For this analysis, we always calculated a "full" stroke cycle (meaning 30–100%) and calculated a mean coefficient of friction over the eight states. The results can be seen in Figure 10.

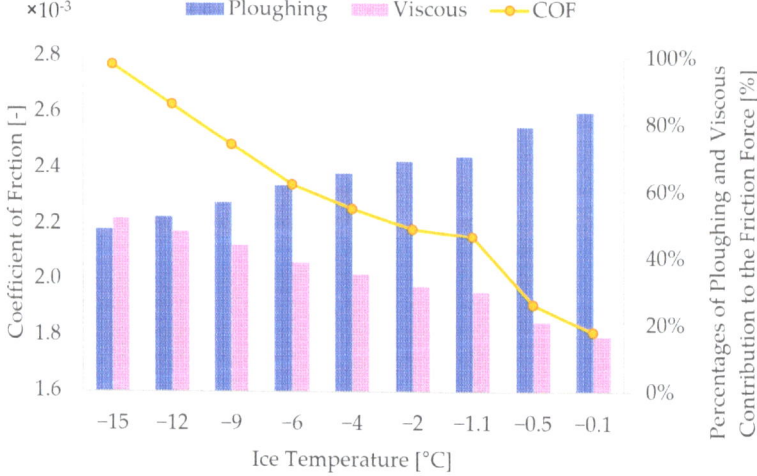

Figure 10. Coefficient of friction of a speed skating stroke cycle for varying ice surface temperatures and otherwise standard values and the corresponding percentages of contribution from plowing and viscous forces to the total friction force. For ice temperatures above −1.1 °C, which is the melting temperature under pressure, the blade temperature was set equal to the ice temperature.

The biggest difference to previous results by Lozowski et al. [4] is the lower COF value and much lower sensitivity to changes in temperature. This is due to the fact that our algorithm calculates similar values of plowing force but much lower values of viscous forces than previous studies. The percentage contribution of viscous forces to the friction force is around 30–40%, whereas, in previous studies, it was found to be around 60–80%.

In the further analysis of different effects, the 60% point of the stroke cycle was chosen as a point of reference. For this state, an additional variation in skating speed was performed to investigate the sensitivity. The results can be seen in Figure 11 and show a similar behavior to the previous studies: the changes in ice temperature have a greater effect at lower velocities and with decreasing ice temperatures. The COF changes from rising with the rising speed at higher ice temperatures to falling with the rising speed at low ice temperatures.

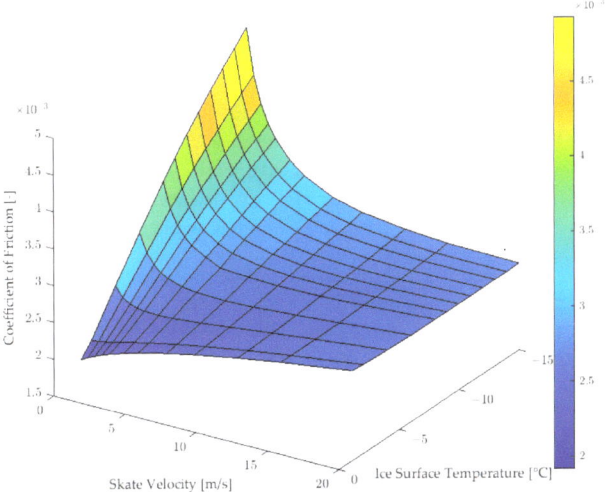

Figure 11. Coefficient of friction for the 60% point of the speed skating stroke cycle under variations of skate velocity and ice surface temperature.

When comparing our results to friction measurements in speed skating performed by de Koning et al. [11], two main things can be observed. Firstly, the coefficients of friction calculated using the algorithm are lower than the ones measured under realistic conditions. De Koning et al. measured COFs in the range of 4×10^{-3} to 6×10^{-3} under comparable conditions. This is partly explainable, as the algorithm neglects a few effects, e.g., imperfections and roughness of the ice surface and also the decelerating effects of active steering.

When looking at the more recent experimental results from Due et al. [7], which were obtained using a gliding vehicle, similarly higher COF values are measured. Secondly, the sensitivity of the COF to ice temperature and skating speed is much higher in de Koning's measurements.

Both issues suggest that the algorithm underestimates the viscous forces. One possible explanation for this can be found when considering the research by Canale et al. [12]. Through friction experiments using atomic force microscopy with an ice surface, they found that the encountered viscosity of the melt water layer is much higher than the viscosity of water at 0 °C from the literature. The F.A.S.T. code uses the standard value of 1.79×10^{-3} kg/m/s as dynamic viscosity η, whereas Canale et al. calculated a complex dynamic viscosity with real and imaginary parts, with values for the real part in the range of 2×10^{-3} kg/m/s to 80×10^{-3} kg/m/s depending on temperature and sliding speed [12]. If the above analysis from Figure 11 is redone with a viscosity of 40×10^{-3} kg/m/s, the results change considerably, not only in value but also in sensitivity to speed and temperature (see Figure 12).

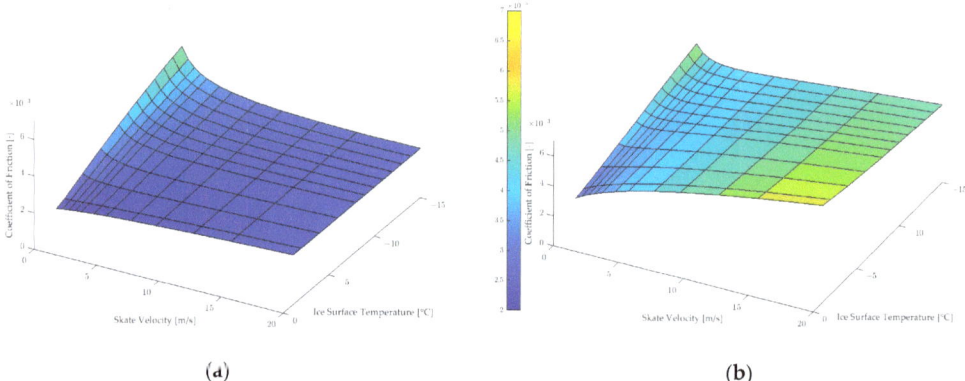

(a)   (b)

Figure 12. Coefficient of friction for the 60% point of the stroke cycle under variation of skate velocity and ice surface temperature with (**a**) dynamic viscosity of the water layer $\eta = 1.79 \times 10^{-3}$ kg/m/s as in Figure 11 and (**b**) dynamic viscosity of the water layer $\eta = 40 \times 10^{-3}$ kg/m/s in reference to [12]. Both (**a**) and (**b**) have the same axis limits and identical color ranges.

When we look at earlier publications for comparison, some research exists as a reference point, especially the publications [4,7]. However, due to a multitude of changes made to the code since then and further differences in the approach, we could not achieve comparability with these literature results.

3.2. Bobsleigh

As a second case study, the sport of bobsleigh was examined. Due to the above-mentioned additions to the code which allow for the definition of the blade or runner geometry through coordinates z(x) rather than a single radius, the calculation of real bobsleigh runner geometries becomes possible. Furthermore, the code now allows for curved ice surfaces, which allows the calculation of curves in the ice canal. In all ice canal sports, the loads are highest in curves and, therefore, the frictional losses are dominated by the curves.

The runner of bobsleighs can freely rotate around the lateral axis (see Figure 4). This is vital to maintain tangential contact with the ice while driving through curves. Our code is capable of delivering the angle of attack in equilibrium. As mentioned above, this addition comes with difficulties in convergence quality.

For a comparison of the overall behavior of the code in the sport of bobsleigh, we performed a variation over ice temperature and sliding speed using the following conditions.

For runner material parameters, we used Uddeholm Ramax HH, which is currently the only allowed material for bobsleigh runners. As longitudinal geometry, an older standard geometry of FES runners was used, which was developed in the early 2010s for the Altenberg track (see Figure 13). For lateral runner geometry, a radius of 7.5 mm is assumed, which is the allowed maximum and a common choice for two-man front runners. Concerning the normal force we assumed a mean normal acceleration on the sled of 1.4 g. With a total weight of 390 kg for a two-man bob and 44% of that load on the front runners (taken from [3]), we assume a load of 1220 N on a single front runner. For the track geometry, a mean curvature of a normal ice canal with a 90 m longitudinal radius and no lateral curvature is used. The sliding velocity was varied up to 35 m/s (126 km/h). These conditions are used as a standard for the bobsleigh analysis unless stated otherwise. For convergence reasons, we initially held the mounting rigid; therefore, we did not allow for a turning of the runner around the lateral axis, which is a valid assumption for open areas of the track (in contrast to narrow curves).

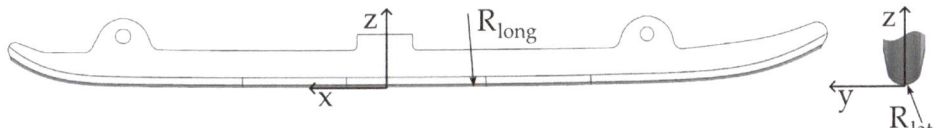

Figure 13. Sketch of a bobsleigh runner geometry in side view (**left**) and front view (**right**, enlarged) with coordinate systems. The longitudinal shape of the runner surface is defined by a multitude of radii $R_{long}(x)$ changing with coordinate x and a lateral radius R_{lat}, which may or may not change over x.

The field of the coefficient of friction, Figure 14, shows an overall similar picture to the speed skating results in Figure 11 but with lower values. The results are in general accordance with Poirier [3] but cannot be compared in detail, as Poirier usually calculates a front and a rear runner to obtain one value for a whole sled; however, he apparently adds both forces and applies them to one geometry. This is problematic due to two reasons: firstly, front and rear runners always have different longitudinal and lateral geometries and, secondly, as the relation between normal force and friction force is nonlinear, it is not the same to apply twice the load to one geometry as to apply once the load to two geometries.

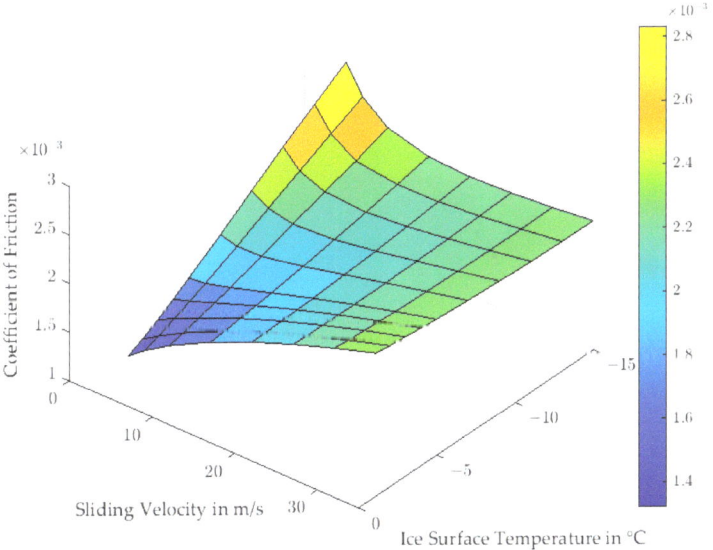

Figure 14. Coefficient of friction of a 2-man bobsleigh front runner with rigid mounting under variation of sliding speed and ice surface temperature.

Comparing these results to practical experience from competitive bobsleighing highlights a few points. Firstly, the overall COF values are much lower than in reality. From energetic considerations, we know that the frictional losses during a bob run correspond to the mean frictional coefficient in the region of ~0.015. The difference in COF values is more straightforward in bobsledding than in speed skating, since the ice quality and ice surface quality of the ice canal is much lower than in a speed skating arena. The ice in the canal is rough, wavy, and sometimes damaged, which will add to the coefficient of friction.

Secondly, we know from our experience in the sport that the COF variation with ice temperature shown by the model is faulty: "warm" ice, which is close to its melting point, makes horrible conditions for the sport, as warm ice makes a track slow. Additionally, warm conditions lead to high wear of the ice and as a result the high-speed loss of the track during one heat, which leads to a dependence of the starting order on finish time. If possible, such conditions should always be prevented using more freezing power. This means the insensitivity to temperature at higher sliding speeds cannot be found in real bobsleigh conditions.

As a second bobsleigh case study, we looked at entering narrow curves, as these are always critical situations in a bobrun. There is a fast change in contact geometry, a high risk of drift, and a fast-changing normal load. When passing through narrow curves with flat runners, two separate contact areas can develop. Poirier already expected this behavior and mentioned the need for a corresponding extension to the model. But this is only now relevant, as real runner geometries can be calculated, because it is in curves, where the complex geometry of real runners comes into play. For this exemplary study, we looked at the entry into the spiral of the Yanqing National Sliding Center in the Beijing region, where the 2022 Winter Olympics were held. Over a 10 m distance, the track bends from straight to a curve of a 27 m radius, the normal load more than doubles, and the sliding speed is ~32 m/s. For the five intermediate states of this entry, we calculated the contact situation using the above-mentioned conditions, except for speed and ice temperature, which was held at −4 °C. For this analysis, the rotation of the runner was left free.

The results of the contact area in Figure 15 show that the code indeed can find equilibrium conditions with two separate contact areas, which do not have to be equal in size but in rotational equilibrium. We also see that during the entry into the curve, the contact area becomes longer rather than wider. This aligns with the results for the coefficient of friction, which decreases during entry into the curve as seen in Figure 16.

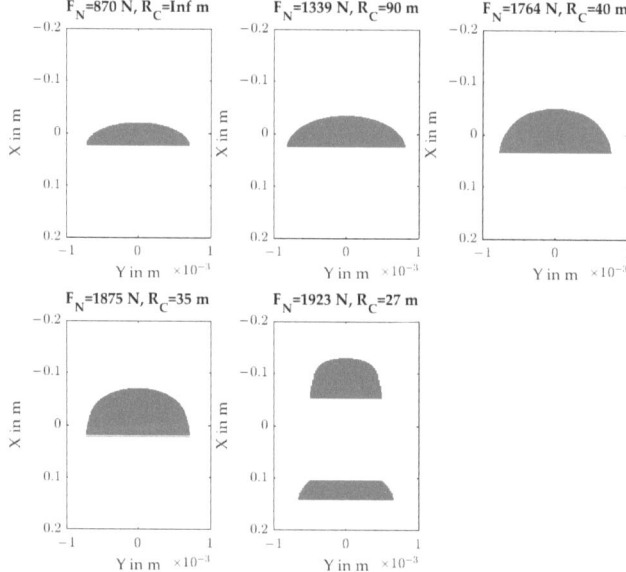

Figure 15. Change in contact area while entering a curve in the track, decreasing curve radius, and increasing normal forces from top left (straight) to bottom right (final curve radius). The runner moves in the direction of negative x values (upwards).

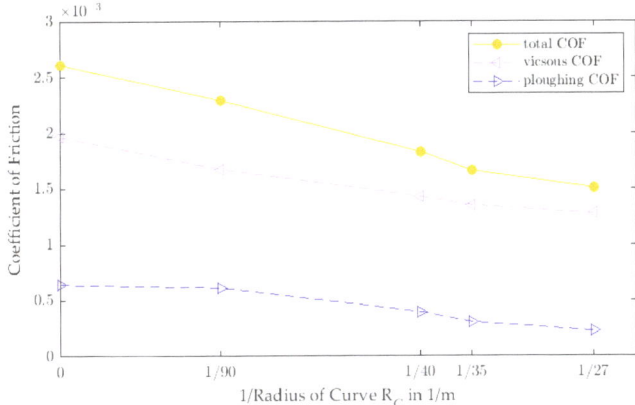

Figure 16. Development of the coefficient of friction for a 2-man front runner while entering into a curve of 27 m radius. The values corresponding to plowing and viscous force are also shown.

For this example, it is interesting to look at the development of the lubricating water film, which is depicted as a 3D plot in Figure 17 for the last calculated state of the curve entry. This last state is especially suitable to show the development of the melt water layer as it is a two-area contact. The figure shows how meltwater builds during the first contact. Between the two contacts, the melt water layer is preserved but slowly decreases in height as parts of it refreeze. Then, during the second contact, more ice is melted and the water layer increases again. This additional buildup leads to higher water layer thickness during the second contact compared to the first. Behind the second contact, the refreezing starts again and the water layer thickness decreases. In regions far behind the contact, the water layer will be refrozen completely (this region is not depicted in Figure 17).

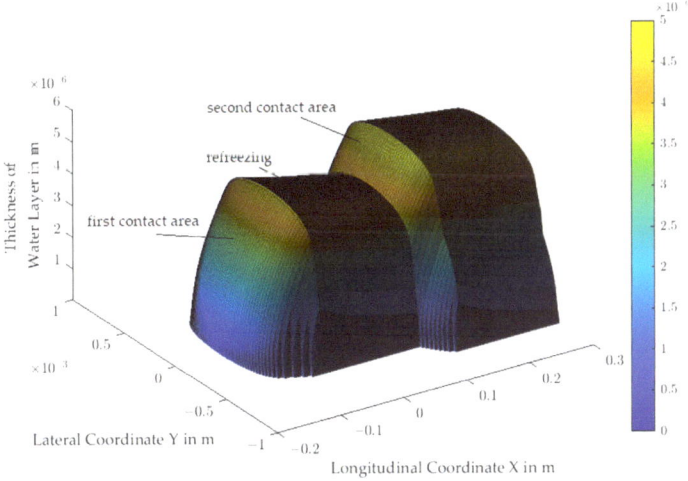

Figure 17. A 3D depiction of the water layer thickness for a 2-point contact as it occurs in narrow curves of the track (see Figure 15, bottom right). During passing, the water film increases and decreases again afterward. The runner moves in the direction of negative x values.

Furthermore, we looked at another specific issue derived from bobsleigh practice, which is a question of the effect of geometry. Poirier [3] determined that the F.A.S.T. code calculates much higher reductions in COF through flatter runners (i.e., a reduction in longitudinal radius) than from broader runners (i.e., an increase in lateral radius). Depending on the thermodynamic conditions, increasing the lateral radius can even increase friction in the model. This is because increasing the lateral radius reduces plowing but increases viscous forces and, depending on the other system conditions (temperature, velocity, and so on), one effect overweighs the other. This also highly depends on the normal load, as increasing normal loads increases the proportion of the plowing force. For high loads, e.g., 4 g of normal acceleration as in a curve, the gain through broader runners is higher, but still almost negligible compared to the gain through flatter runners.

This is another point where every person involved in the sport of bobsleigh would disagree because broader runners are found to be always faster given that there is no snow or hoarfrost on the track.

Whereas the use of flatter and broader runners is known to reduce frictional losses, in ice canal sports, it must always be balanced with control over the sled. Especially in curve entries and exits, pilots must have sufficient lateral grip to steer the sled in these strategically crucial situations. Losing control over the sled or encountering excessive drift can lead to side contact (i.e., time loss), hurt the ideal trajectory, and can set the sled up for a crash. Therefore, it is the responsibility of trainers and pilots to take into account their driving abilities, experience on a track, and weather conditions when choosing runners.

To get closer to a quantification of the control runners provide over a sled, we added the theoretically maximal lateral force the contact can hold as an output of the code. We did a variation of longitudinal and lateral radii over common ranges, now using a longitudinal geometry of one single radius. It can be seen in Figure 18 that increasing the lateral radius of the runner dramatically reduces the maximum lateral forces due to the decrease in indentation depth.

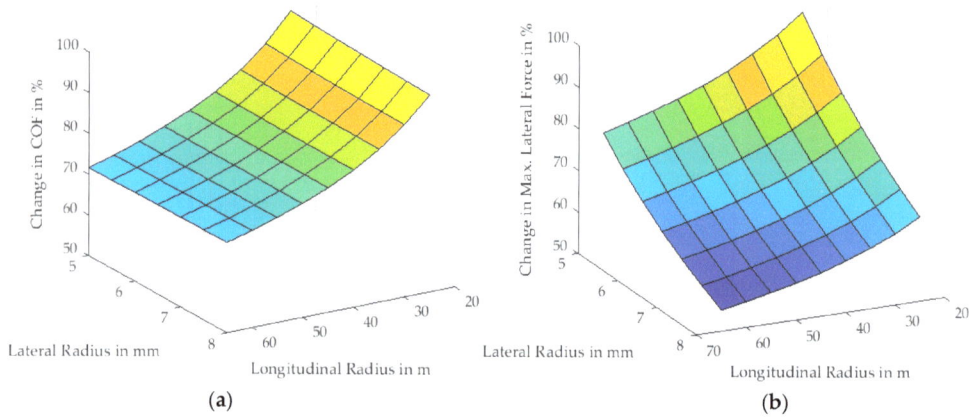

Figure 18. Influence of changing the longitudinal and lateral radii of a bobsleigh runner on (**a**) the coefficient of friction and (**b**) the maximum lateral force. Only percental changes to maximum values are shown. Both (**a**,**b**) have identical axes and color scales.

Still, deducing from the model, one would always suggest choosing the flattest runners (limitations to this apply due to performance in narrow curves) and reducing the lateral radius if needed for control.

4. Discussion

With the new extensions of the F.A.S.T. code, a wide range of winter sport-specific questions can be investigated in a more realistic manner. In speed skating, the representation of correctly inclined geometries and lateral force components add significant benefits to move towards more realistic modeling of the sport. For bobsleigh applications, the implementation of free runner geometries and ice curvature makes the code much more realistic and usable for practicable application in the development of sports equipment.

Adaptations for luge and skeleton are also possible and planned in the future. For luge, the new implementation of runner geometries is especially interesting, as the runners are mounted elastically with a camber; therefore, the contact geometry changes with the changing load as the runner turns around the x-axis. Moreover, for luge and skeleton, the implementation of free mounting is interesting, as the angle of the sled on the ice can be changed by the athlete through shifting weight.

Still, there are uncertainties in the implementation that should be further clarified through future research. This includes the ice hardness model, which was developed by Poirier [13] and recently reproduced by Du et al. [7] but is derived from measurements of drop tests with high scatter. The effect of deformation velocity on ice hardness, as well as the effect of pressure melting should be included in future ice hardness models. The publication by Liefferink et al. from 2021 [14] is an excellent starting point for this; however, it focuses on lower deformation velocities and is, therefore, not directly applicable. Furthermore, the temperature gradients in ice and blade are not well understood. The recent work by Du et al. [7] starts to tackle this issue, but due to the high sensitivity of the model to the blade temperature (as Poirier pointed out in [3]), a better representation could be beneficial, especially as we know in ice canal sports, the runner's temperature has a high effect on finish time, and all runner temperatures are checked by officials before the start of the race. At last, with the work of Canale et al. [12], the viscous properties of the meltwater are under question.

All F.A.S.T. models in general produce coefficients of friction that are surprisingly low. From comparison with empirical values from winter sports, we would have assumed values higher by factors 4 to 6. In the ice canal sports, part of this discrepancy may stem from the waviness of the ice canal from scraping, which could be added to the model in the future for analysis. Furthermore, the consideration of surface roughness in the system is neglected by the model and shall be briefly discussed here. The slider roughness is relevant, as we know it can be of the same order of magnitude as the lubrication layer thickness. Measurements of the ice surface roughness and waviness in a winter sports context have not yet been published to the best of our knowledge. The implementation of ice surface roughness would bring additional challenges with it, as it would need a time-resolved simulation as the slider moves through the ice. It can be expected that a rough contact will show higher COFs, as the indentation depth and, therefore, the plowing force, as well as the viscous forces, due to a patchy lubrication layer, will be increased.

There are also a few other limitations to the code for the application in winter sports. A smaller possible addition for speed skating would be the implementation of a curved trace which is calculated from the steering effect of the inclined blade. The next big step towards a more realistic representation would be the consideration of the elasticity of the runners. It is known that speed skating blades under high inclination bend and, therefore, change their shape. This is due to the fact that the contact with the ice is usually in the middle part of the blade, whereas the force by the athlete is applied via two distributed mounts at the shoe, one under the heel and one under the ball of the foot. In luge and skeleton, the runners themselves but also the whole sled can bend and twist, especially in curves and curve entries and exits. Of course, the integration of elastic behavior into the model can have many possible forms and is a complex task that would not be easily accomplished.

Author Contributions: Methodology, Software, Formal Analysis, R.P.; Investigation, Project Administration, Funding Acquisition, B.G.; Conceptualization, Validation, Resources, Data Curation, Writing—original draft preparation, Writing—review and editing, Visualization, R.P. and B.G. All authors have read and agreed to the published version of the manuscript.

Funding: This research was funded by the Federal Ministry of the Interior and Community of the Federal Republic of Germany. Budget Section 0601, Budget Item 684 22.

Data Availability Statement: The code we base our work on is freely available in [6] with our suggested corrections in Appendix A. Requests to access the extended code or datasets should be directed to the corresponding authors.

Acknowledgments: We thank the FES project leads in the sports of speed skating and bobsleigh, M. Büttner, and E. Zinn, for providing discipline-specific data and information.

Conflicts of Interest: The authors declare no conflicts of interest.

Appendix A. Corrected Code from Stell

The appendix of Stell [6] gives specific implementations of his previous F.A.S.T. code in the MATLAB programming language. This implementation contains errors and should not be employed.

Incorrect film thickness

When computing the film thickness, his code employs the following code segment a total of three times, either as dh or dhr:

```
dh = dz*(...
(u_w*v^2/h(i,j - 1)) ...
-(k_i*(T_i-T_b)/h_ice) ...
-(T_mf-T_i)*sqrt(v*p_i*c_i*k_i/(pi*abs(zo(1,1)-z(1,length(z))))) ...
+(T_sf-T_mf)*sqrt(v*p_s*c_s*k_s/(pi*abs(zo(i,1)-z(1,length(z))))) ...
)/(p_w*l_f*v);
```

The corresponding formulae are numbered (2.29) and (2.30) in Stell [6], and, in Poirier [3], they are numbered (4.25) and (4.26). The error is in the usage (twice) of "z(1,length(z))", which would give the coordinate at the end of the simulation domain. However, what is actually needed here is the running coordinate, as Poirier puts it, "The position along the length of the blade". In the code, we need to instead insert "z(1,j)" and obtain

```
dh = dz*(...
(u_w*v^2/h(i,j - 1)) ...
-(k_i*(T_i-T_b)/h_ice) ...
-(T_mf- T_i)*sqrt(v*p_i*c_i*k_i/(pi*abs(zo(i,1)-z(1,j)))) ...
+(T_sf-T_mf)*sqrt(v*p_s*c_s*k_s/(pi*abs(zo(i,1)-z(1,j)))) ...
)/(p_w*l_f*v);
```

This is consistent with the Python implementation of Poirier, which reads in the relevant section:

```
for (k = 1;k<zsteps;k++) {
z -= dz;
h_condf = -dz * delta_T_i / (rho_w*l_f) *...
...sqrt(k_i*rho_i*c_i / (v*pi*(z0[j]-z)));
h_conds = dz * delta_T_s / (rho_w*l_f) *...
...sqrt(rho_s*c_s*k_s / (v*pi*(z0[j]-z)));
}
```

the code in question being "z0[j]-z" with decreasing "z -= dz".

Missing case distinction for the contact area

There is an error in determining the contact area "A_r" of the rear runner when two runners are used. The Stell code reads on page 75 of [6]:

```
A_r = (y_max+y_maxr)*l_sr;
```

Which is only correct when both tracks overlap. Otherwise, the rear runner's contact area is determined in the same way as the front runner (being "A_r = pi*y_maxr*l_sr/2").

For the rear contact area, we thus need to discriminate between the two cases and write

```
if m == 1
A_r = pi*y_maxr*l_sr/2;
else % (being m == 0)
A_r = (y_max+y_maxr)*l_sr;
end
```

References

1. Penny, A.; Lozowski, E.; Forest, T.; Fong, C.; Maw, S.; Montgomery, P.; Sinha, N. Speedskate ice friction: Review and numerical model-FAST 1.0. In *Physics and Chemistry of Ice, Proceedings of the 11th International Conference in the Physics and Chemistry of Ice, Bremerhaven, Germany, 23–28 July 2006*; Kuhs, W.F., Ed.; RSC Publ.: London, UK, 2007; Volume 311, pp. 495–504.
2. Lozowski, E.P.; Szilder, K. FAST 2.0 Derivation and New Analysis of a Hydrodynamic Model of Speed Skate Ice Friction. In Proceedings of the Twenty-First International Offshore and Polar Engineering Conference, Maui, HI, USA, 19–24 June 2011.
3. Poirier, L. Ice Friction in the Sport of Bobsleigh. Ph.D. Thesis, University of Calgary, Calgary, AB, Canada, 2011.
4. Lozowski, E.; Szilder, K.; Maw, S. A model of ice friction for a speed skate blade. *Sports Eng.* **2013**, *16*, 239–253. [CrossRef]
5. Lozowski, E.P.; Szilder, K.; Maw, S.; Morris, A. A Model of Ice Friction for Skeleton Sled Runners. In Proceedings of the Twenty Fourth (2014) International Ocean and Polar Engineering Conference, Busan, Korea, 15–20 June 2014.
6. Stell, B. Thermal-Fluid Dynamic Model of Luge Steels. Master's Thesis, California Polytechnic State University, San Luis Obispo, CA, USA, 2017.
7. Du, F.; Ke, P.; Hong, P. How ploughing and frictional melting regulate ice-skating friction. *Friction* **2023**, *11*, 2036–2058. [CrossRef]
8. Döppenschmidt, A.; Butt, H.-J. Measuring the Thickness of the Liquid-like Layer on Ice Surfaces with Atomic Force Microscopy. *Langmuir* **2000**, *16*, 6709–6714. [CrossRef]
9. van der Kruk, E.; den Braver, O.; Schwab, A.L.; van der Helm, F.C.T.; Veeger, H.E.J. Wireless instrumented klapskates for long-track speed skating. *Sports Eng.* **2016**, *19*, 273–281. [CrossRef]
10. van der Kruk, E.; Schwab, A.L.; van der Helm, F.; Veeger, H. Getting the Angles Straight in Speed Skating: A Validation Study on an IMU Filter Design to Measure the Lean Angle of the Skate on the Straights. *Procedia Eng.* **2016**, *147*, 590–595. [CrossRef]
11. de Koning, J.J.; de Groot, G.; van Ingen Schenau, G.J. Ice friction during speed skating. *J. Biomech.* **1992**, *25*, 565–571. [CrossRef] [PubMed]
12. Canale, L.; Comtet, J.; Niguès, A.; Cohen, C.; Clanet, C.; Siria, A.; Bocquet, L. Nanorheology of Interfacial Water during Ice Gliding. *Phys. Rev. X* **2019**, *9*, 041025. [CrossRef]
13. Poirier, L.; Lozowski, E.P.; Thompson, R.I. Ice hardness in winter sports. *Cold Reg. Sci. Technol.* **2011**, *67*, 129–134. [CrossRef]
14. Lieferink, R.W.; Hsia, F.-C.; Weber, B.; Bonn, D. Friction on Ice: How Temperature, Pressure, and Speed Control the Slipperiness of Ice. *Phys. Rev. X* **2021**, *11*, 011025. [CrossRef]

Disclaimer/Publisher's Note: The statements, opinions and data contained in all publications are solely those of the individual author(s) and contributor(s) and not of MDPI and/or the editor(s). MDPI and/or the editor(s) disclaim responsibility for any injury to people or property resulting from any ideas, methods, instructions or products referred to in the content.

Article

Tribological Behavior of Hydrocarbons in Rolling Contact

Daniel Merk [1], Thomas Koenig [2], Janine Fritz [2] and Joerg W. H. Franke [2,*]

1 Schaeffler Technologies AG & Co. KG, 97421 Schweinfurt, Germany; daniel.merk@schaeffler.com
2 Schaeffler Technologies AG & Co. KG, 91074 Herzogenaurach, Germany; t.koenig@schaeffler.com (T.K.); janine.fritz@schaeffler.com (J.F.)
* Correspondence: joerg.franke@schaeffler.com; Tel.: +49-9132-82-3429

Abstract: In the analysis of tribological contacts, the focus is often on a singular question or result. However, this entails the potential risk that the overall picture and the relationships could be oversimplified or even that wrong conclusions could be drawn. In this article, a comprehensive consideration of test results including component and lubricant analyses is demonstrated by using the example of rolling contact. For this purpose, thrust cylindrical roller bearings of type 81212 with unadditized base oils were tested in the mixed-friction area. Our study shows that by using an adapted and innovative surface analysis, a deeper dive into the tribo-film is feasible even without highly sophisticated analytical equipment. The characterization of the layers was performed by the three less time-consuming spatially resolved analysis methods of μXRF, ATR FTIR microscopy and Raman spectroscopy adapted by Schaeffler. This represents a bridge between industry and research. The investigations show that especially undocumented and uncontrolled contamination of the test equipment could lead to surprising findings, which would result in the wrong conclusions. Simple substances, like hydrocarbons, are demanding test specimens.

Keywords: rolling bearing; tribology; polyalphaolefin; lubrication

Citation: Merk, D.; Koenig, T.; Fritz, J.; Franke, J.W.H. Tribological Behavior of Hydrocarbons in Rolling Contact. *Lubricants* **2024**, *12*, 201. https://doi.org/10.3390/lubricants12060201

Received: 30 April 2024
Revised: 21 May 2024
Accepted: 27 May 2024
Published: 3 June 2024

Copyright: © 2024 by the authors. Licensee MDPI, Basel, Switzerland. This article is an open access article distributed under the terms and conditions of the Creative Commons Attribution (CC BY) license (https://creativecommons.org/licenses/by/4.0/).

1. Introduction

Due to the complexity of tribological questions, it is often necessary to focus on single aspects. This could cause the risk of losing information and interdependencies and poses the risk of losing sight of the bigger picture. Therefore, we conducted a comprehensive study utilizing complementary surface analysis methods.

By using the example of rolling bearing tests with non-additized base oils, this publication aims to demonstrate the basic procedure and the possibilities of industrial practice. The focus is on the investigation of contacts that are susceptible to the wear of bearing components under mixed-friction conditions. The wear protection behavior of additive-based lubricants and their function has been extensively investigated by several researchers in the past.

The publications report on so-called tribo-films, which form in rolling bearing contacts, especially under mixed-friction conditions [1–3]. Sulfur and phosphorus carriers [4] are known as anti-wear or extreme pressure additives. Depending on the test conditions, these show an occupancy of the surface up to a sustainably wear-protecting tribo-film [5]. Burghardt [6] confirmed that lubricating oils which do not contain any additization could also build up a wear-protecting film. It was shown that under the selected test conditions there were significant differences in tribo-film formation between different base oils, which was ascribed to the chemical structure of the oils. In addition, it was demonstrated [7] that the tribo-film structure on a cylindrical thrust roller bearing is dependent on the slippage conditions of the contact zones. The DIN 51517-3 "Lubricating oils CLP, minimum requirements" [8] outlines the wear test according to DIN 51819-3 [9], the so-called FE8 low-speed wear testing, as an evaluation criterion for wear protection. The operating conditions of this standard test are very demanding; therefore, unadditized or insufficiently additized lubricating oils do not pass this test. This is certainly an extreme case for many applications; thus, considerable power reserves will be available in practice. It can be assumed that not all rolling bearing applications that are operated under mixing conditions have such high demands of the lubricant.

An approach to the bearing- and lubricant-independent evaluation of the rolling contact regarding the risk of surface-initiated damage, such as wear, was presented by Vierneusel [10]. Here, the specific friction energy flowing over the solid-state contact is evaluated by the characteristic value e_{SID}. This characteristic value is readily available in Schaeffler Technologies bearing design software Bearinx [11] for the evaluation of various rolling bearings under different operating and lubrication conditions.

The reported tests below and the analysis of the tested bearings show the correlation between the degree of surface stress and the formation of a stable tribo-film using pure base oils. The handling and recognition of unexpected influences are also considered here. The detailed investigation of the analytical results is discussed. Furthermore, the impact of the test setup on the results are explained in conclusion.

2. Materials and Methods

2.1. Bearings and Lubricants

The bearings used in the tests are cylindrical roller thrust bearings (CRTBs) of type similar to 81212, designed as a test bearing type. These test bearings have a polyamide cage with 15 rolling elements. The rolling elements and both washers are made of 100Cr6 steel, with each martensitic component hardened and stabilized, as shown in Figure 1.

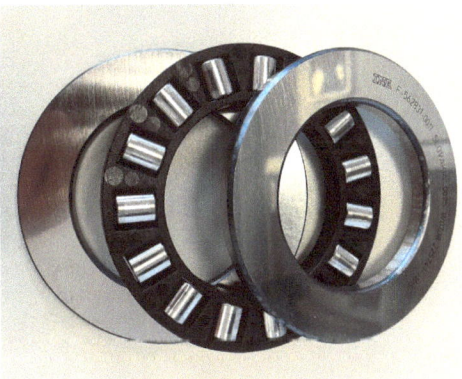

Figure 1. Cylindrical roller thrust bearing 81212.

The unadditized lubricants used were PAO 6 (OIL A) on the one hand and Dicarboxylic-acid ester (OIL B) on the other. The reference oil (OIL C) was a fully formulated gear oil with anti-wear additive. The data on the lubricating oils are listed in Table 1.

Table 1. Lubricant properties.

Lubricant	Oil Composition	Kinematic Viscosity [mm²/s]		Density [kg/m³]	Dynamic Viscosity [mPas] *
		40 °C	100 °C	15 °C	80 °C
OIL A	PAO 6 No additivation	30.9	5.82	827	7.1
OIL B	Bis(2-ethylhexyl) sebacate No additivation	11.3	3.3	917	4.1
OIL C	Synthetic (PAO) Fully formulated (P = 0.06%; S = 0.045%)	46.3	7.84	847	10.1

* Calculated value.

2.2. Testing Procedure

All tests were carried out by an FE8 test rig (Figure 2) according to DIN 51819, under various operating conditions, with different grades of solid body contact.

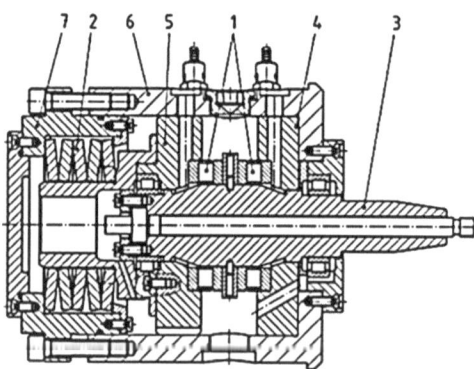

Figure 2. FE8 test rig setup [10]: 1—test bearing; 2—spring package; 3—shaft; 4—bearing housing drive side; 5—bearing housing spring side; 6—housing; 7—cup.

Each test bearing was weighed before and after the test run, to obtain information of weight loss and thus the resulting rolling bearing wear. For this purpose, the bearings were cleaned according to a standardized process and tested under test conditions, which are listed in Table 2 in detail.

Table 2. Rolling bearing test conditions.

Test Conditions	Test Name	FE8, 100 kN/7.5 rpm/80 °C	FE8, 50 kN/100 rpm/80 °C	FE8, 50 kN/800 rpm/80 °C
Test rig			FE8 (DIN 51819)	
Bearing			CRTB 81212 (F-562831.01)	
Cage		brass cage with 15 pockets	polyamide cage with 15 pockets	
Temperature			80 °C	
Load		100 kN	50 kN	
Resulting contact pressure *		≈2400 MPa	≈1750 MPa	
Speed		7.5 rpm	100 rpm	800 rpm
Oil flow			approx. 0.12 L/min	

* Calculated value [9].

In addition to the test conditions under the mixed-friction regime, the selected bearing type exhibits special kinematic conditions that lead to different values of frictional energy across the raceway [5–7,10]. The expected main failure mechanism of the chosen test setup is wear of the bearing components, which is measured in the form of a weight difference of the bearing components. The selected maximum runtime of 80 h is much shorter than the values of calculated fatigue life regarding ISO 281 [12]. The calculated contact conditions are listed in Section 3.1.

2.3. Surface Analysis

The selected three surface analysis techniques, X-ray fluorescence analysis (µXRF), attenuated total reflection fast Fourier transformation infrared (ATR FTIR) microscopy and Raman microscopy, as methods of so-called correlative spectroscopy, allow the areas of interest to be characterized via mapping in their entirety. All three methods are non-destructive, so it was possible to analyze the specimen generated in the FE8 tests without any mechanical sample preparation. Before the analysis, it was only necessary to clean the sample by rinsing off lubricant residues on the surface with a suitable organic solvent, e.g., n-heptane (CAS No. 142-82-5).

2.3.1. X-ray Fluorescence Analysis

In X-ray fluorescence analysis, high-energy radiation is used to excite the atoms of the sample to be examined. This method makes it possible to identify and determine the concentration of all elements from atomic number 11 (sodium). The physical principle of XRF analysis is based on the principle that the atoms in the sample are excited by high-energy X-rays. In these atoms, electrons from the inner shells are removed from the atom through interaction with the X-rays. In a very short time, these vacancies are filled with electrons from the outer shells. The free energy can be emitted as an AUGER electron or as an X-ray photon. The energy of the emitted X-ray photon depends on the difference of binding energies of both involved electron levels—the vacancy and the level from which the electrons jump into the vacancy. Because this difference is characteristic for every element, the excited specimen emits characteristic radiation (Figure 3). This fluorescence radiation can be used to analyze both the qualitative and quantitative elemental composition of the analyzed sample. µXRF is a special application of ED-XRF (energy-dispersive XRF), and it offers the possibility of a position-sensitive element analysis of non-homogeneous surfaces. The µ-XRF mappings were performed with a M4 Tornado (Bruker Nano GmbH, Berlin, Germany).

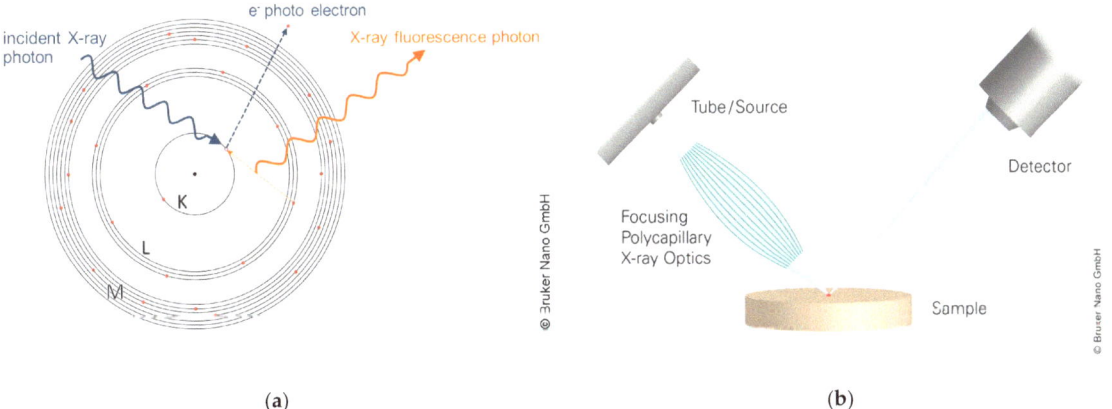

Figure 3. (**a**) Excitation of X-ray fluorescence radiation. (**b**) Schematic diagram of µXRF, M4 Tornado (images courtesy of Bruker Nano GmbH).

2.3.2. Infrared Spectroscopy

FTIR spectroscopy was used to determine the chemical structure of infrared active substances of the generated tribo-films on the washer surface. This method is a type of molecular spectroscopy (vibrational spectroscopy) and is mainly used in structural elucidation and for the identification of unknown, especially organic substances. FTIR spectroscopy is based on the physical effect that most molecules absorb light in the IR range of the electromagnetic spectrum and convert the absorbed energy into molecular vibrations. Vibrational spectroscopy is an energy-sensitive method. It is based on the periodic changes in the dipole moments of molecules or groups of atoms caused by molecular vibrations and the associated discrete energy transfers and frequency changes during the absorption of electromagnetic radiation. These molecular vibrations lead to an IR spectrum that serves as a characteristic "molecular fingerprint", from which statements about the chemical structure of the sample being examined can be made. In FTIR spectroscopy, the position and intensity of the absorption bands of a substance (structural groups or functional groups) are extremely substance-specific. The possibility of directly analyzing functional groups, which is often much more difficult or impossible with other analytical methods, is the essential feature of FTIR spectroscopy and explains its importance as one of the most important methods of instrumental analysis. In FTIR spectroscopy, there are various methods and recording techniques. Attenuated total reflection (ATR) is a non-destructive surface analysis technique and has become the most popular technique for measuring FTIR spectra. The ATR technique uses the effect that optical absorption spectra can be easily obtained by looking at the interaction of the totally reflected light emerging from the optically dense medium with the optically thin medium. As shown schematically in Figure 4, the sample surface is brought into contact with the ATR crystal, with IR radiation being absorbed by the sample at each reflection point. ATR crystals made of germanium (Ge) are best suited for the characterization of tribo-films, because of their low sample penetration depth of 0.66 µm due to their high refractive index.

ATR FTIR microscopy is a position-sensitive FTIR application which allows for the analysis of the surface of the raceway side of the washer by mapping a grid of 13 × 13 measurement points both circumferentially and transversally to the direction of the raceway of the thrust bearing surface. The 13 measurement points of each row (identical tribological conditions) were used to calculate a sum spectrum.

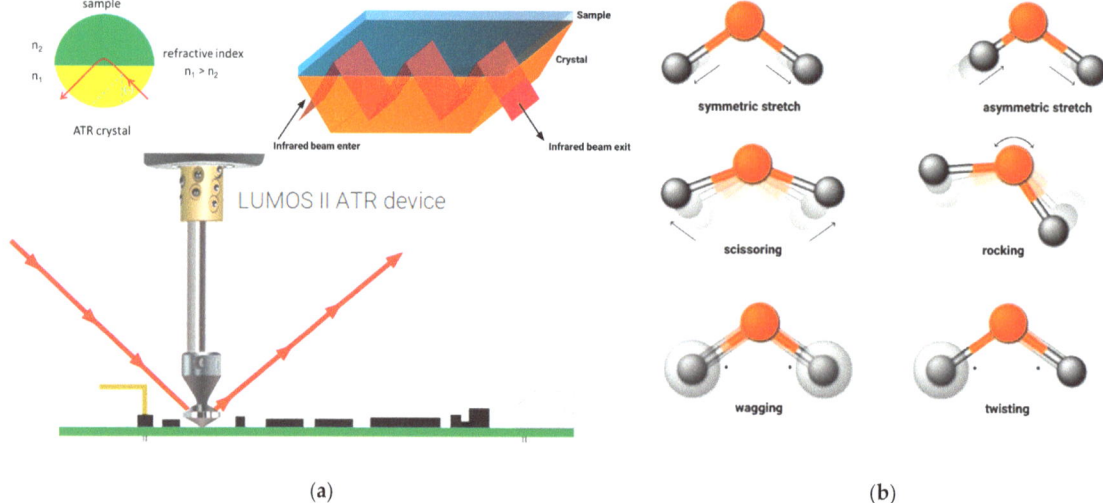

Figure 4. (**a**) Schematic diagram of ATR crystal. (**b**) Example of infrared vibration modes (images courtesy of Bruker Optics GmbH).

The ATR FTIR mappings were performed with a LUMOS II (Bruker Optics GmbH). The detailed equipment parameters of both the µXRF and ATR FTIR microscopy were identical to those described in [13].

2.3.3. Raman Spectroscopy

In this publication, we would like to use Raman microscopy as a third analytical method to analyze the IR-inactive compounds of tribo-films. The combination of these three spectroscopic analysis methods thus completes our correlative spectroscopy in industrial applications, which we use to characterize the tribo-films formed on the FE8 washers.

Raman spectroscopy is a molecular spectroscopy (vibrational spectroscopy) method and is mainly used in structure elucidation and for the identification of unknown, IR-inactive substances. Similar to FTIR spectroscopy, it provides information about the vibrational and rotational states of molecules. However, the physical principles and the excitation of the sample are different. The Raman effect is created by the interaction of electromagnetic radiation and the electron shell of the molecules and, in contrast to FTIR spectroscopy, is practically independent of the wavelength of the excitation radiation.

Excitation is achieved by monochromatic laser radiation. The main part of the laser light passes through the sample, while a small part is elastically scattered by the substance (the so-called Rayleigh scattering, same frequency as the laser). The part that is scattered inelastically (the so-called Raman scattering) contains the structural information. This is due to the deformability of the electron shell (polarizability) of the molecule during the oscillation process (periodic displacement of the bonding electrons in the molecule due to the oscillation of the core structure). (Figure 5).

Raman microscopy is a position-sensitive Raman application which allows, analogously to the applied ATR FTIR microscopy, for the analysis of the surface of the raceway side of the washer by mapping a grid of 13 × 13 measurement points both circumferentially and transversally to the direction of the raceway of the thrust bearing surface. The 13 measurement points of each row (identical tribological conditions) were used to calculate a sum spectrum. The Raman mappings were performed with a SENTERRA II (Bruker Optics GmbH). The technical specifications and the selected mapping parameters are shown in Table 3.

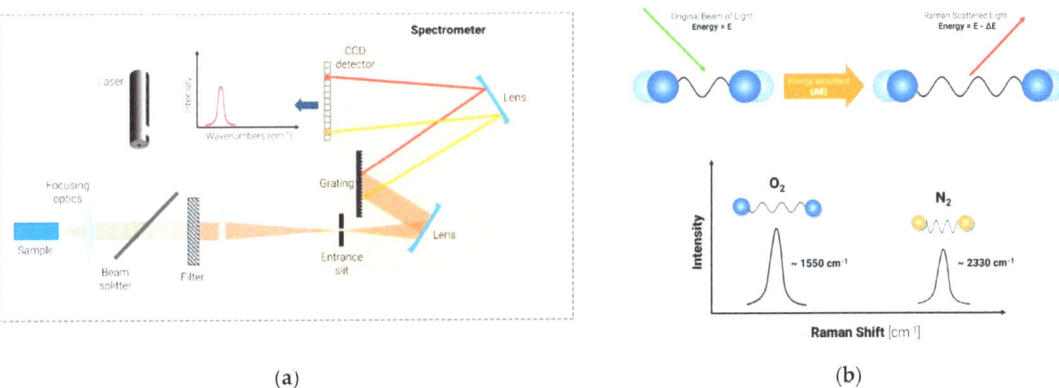

Figure 5. (a) Schematic diagram of Raman microscopy. (b) Example of Raman shift (images courtesy of Bruker Optics GmbH).

Table 3. Technical specifications and mapping parameters of Raman microscope, SENTERRA II (Bruker Optics GmbH).

Technical Specifications	Detailed Information
Laser	532 nm
Laser power	25 mW
Objective	20 × LD
Aperture	50 × 1000 µm
Integration time	20.000 ms/point
Co-addition	10
Spectral resolution	4 cm^{-1}
Mapping parameter	Grid of 13 × 13 measurement points

2.4. Lubricant Analysis

Analyzing the used lubricant in an adequate manner is as important as analyzing the tested parts. To interpret the used oil values, the analysis of the fresh oil is necessary.

The oil samples were analyzed by infrared spectroscopy (FTIR) in transmission mode—Bruker Tensor 27. The element content of the samples was measured by X-ray fluorescence spectroscopy (XRF)—Bruker AXS, S4 Explorer—and inductively coupled plasma optical emission spectrometry (ICP-OES)—Spectro Arcos eop. The viscosity and density were determined by a Stabinger Viscometer—Anton Paar Stabinger SVM 3000, Anton Paar, Graz, Austria. In addition, the Total Base Number (TBN) and Total Acid Number (TAN) were determined via titration—Metrohm TAN/TBN-Titrator. If necessary, the details of the methods of analysis can be requested from the author.

3. Results

3.1. Contact Conditions

The operating parameters were used to calculate the rolling contact conditions via bearing design software Bearinx [11], the rolling bearing calculation tool of Schaeffler Technologies. In Table 4, the resulting parameters are listed.

The rolling contact conditions at 7.5 rpm (FE8 low speed wear) are very harsh. The lubrication film parameter Kappa (κ) regarding ISO 281 is with ≤ 0.01, typical for boundary lubrication conditions. The friction energy is located at the solid-state contact, and no significant lubrication film is present. At 100 rpm, a thicker lubrication film forms, but the operation is in a strong mixed-friction regime. With the increase in bearing speed up to 800 rpm, the lubrication film grows. The surface asperities are still in contact, as shown by the maximal e_{SID} [10] value (e_{SIDmax}), which is slight decreased compared with the 100 rpm operation conditions. So, a mixed-friction regime is still present, but the overall surface strain is reduced, as shown by the e_{SID} value in Figure 6.

3.2. Test Results

The unadditized fluids were not testable under the harsh testing conditions of FE8, 100 kN/**7.5 rpm**/80 °C. The test rig was unable to examine the specimen because of torque cut-off. Reference OIL C could be tested three times under these conditions without any problems. The roller set wear of all six test bearings was below 10 mg. This was confirmed by the statistical average roller set wear of m_{50WK} = 3 mg. The milder test conditions were only conducted with unadditized OILS A and B. With a load of 50 kN and a speed of 100 rpm, the OIL A test runs produced a torque cut-off after short running times below 1 h. Despite the short service life, there was already considerable wear on the rolling bearing components. The test runs with OIL B fulfilled the test duration of 80 h. The increase in speed to 800 rpm resulted in wear-free test runs with both unadditized oils. To make the data comparable, Table 5 lists the roller set wear rates of the different tests in addition to the wear values.

Table 4. Calculated contact parameters for tested oils and conditions *.

Lubricant	FE8, 100 kN/7.5 rpm/80 °C	FE8, 50 kN/100 rpm/80 °C	FE8, 50 kN/800 rpm/80 °C
OIL A	e_{SIDmax} = 344.7 µJ/mm² κ = 0.01	e_{SIDmax} = 188.05 µJ/mm² κ = 0.08	e_{SIDmax} = 123.76 µJ/mm² κ = 0.43
OIL B	e_{SIDmax} = 346.9 µJ/mm² κ < 0.01	e_{SIDmax} = 188.07 µJ/mm² κ = 0.04	e_{SIDmax} = 156.34 µJ/mm² κ = 0.24
OIL C	e_{SIDmax} = 342.3 µJ/mm² κ = 0.01	e_{SIDmax} = 187.62 µJ/mm² κ = 0.11	e_{SIDmax} = 94.66 µJ/mm² κ = 0.60

* Values in gray just for comparison—no specimen created.

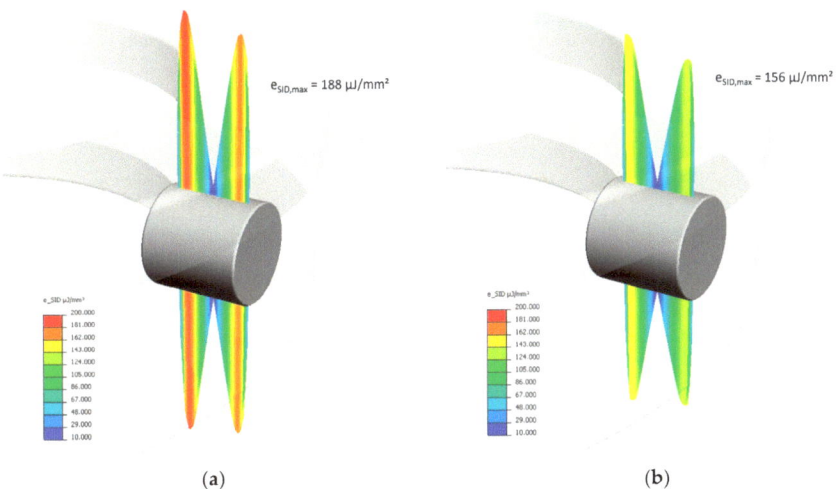

Figure 6. e_{SID} distribution over rolling contact for OIL B: (**a**) FE8, 50 kN/ **100 rpm**/80 °C; (**b**) FE8, 50 kN/ **800 rpm**/80 °C.

Table 5. Test duration, mean roller set wear and roller set wear rate.

Lubricant	FE8, 100 kN/7.5 rpm/80 °C	FE8, 50 kN/100 rpm/80 °C	FE8, 50 kN/800 rpm/80 °C
OIL A	not testable early torque cut-off	0.5 h ∣ 0.8 h (F) * 41 mg ∣ 43.5 mg 82 mg/h ∣ 54 mg/h	10 h ∣ 10 h (S) * <1 mg ∣ <1 mg <<1 mg/h
OIL B	not testable early torque cut-off	80 h/80 h (S) * 2 mg ∣ 2 mg <1 mg/h	10 h/10 h (S) * 1 mg ∣ <1 mg <<1 mg/h
OIL C	80 h ∣ 80 h ∣ 80 h (S) * 3 mg <<1 mg/h	-	-

* F = failure, test rig switched off automatically; S = suspended, manual switch off.

3.3. Oil Analysis

The analysis of the used oil samples after the test runs shows nothing really spectacular. Only the iron content of the test run "FE8, 50 kN/100 rpm/80 °C" with OIL A was slightly increased by a mean average of 8.5 ppm. No significant changes in element content were detectable in the other samples. The infrared spectra, TBN, TAN and viscosity data for all oil samples were identical compared to the respective unused oil. In none of the cases was there any recognizable oil degradation.

3.4. Surface Analysis

3.4.1. Optical

All specimens showed significant signs of use. However, wear tracks on the raceways were not optically visible compared with the unloaded areas, except for the **OIL A 100 rpm** specimen, where wear tracks could be seen. The discolorations on all other parts were typical of tribo-films. Table 6 displays the optical findings.

Table 6. Optical surface analysis.

Lubricant	FE8, 100 kN/7.5 rpm/80 °C	FE8, 50 kN/100 rpm/80 °C	FE8, 50 kN/800 rpm/80 °C
OIL A	Not testable Torque cut-off	Wear tracks (→) at slippage areas / Gray to brownish raceway	No wear visible / Gray-colored raceway, slight circumferential contrast differences "Saturn rings"
OIL B	Not testable Torque cut-off	No wear visible / Circumferential coloring, with brownish, grayish and black rings	No wear visible / Circumferential coloring with brownish, grayish and black dotted rings
OIL C	No wear visible / Blue-colored raceway, circumferential contrast differences "Saturn rings"	-	-

3.4.2. Infrared Spectroscopy (ATR-FTIR)

The infrared spectra of OIL C specimens are identical. Beside the raceways, there is nearly no IR activity. In the area of the raceways, strong absorption at the 1159 cm^{-1} wavenumber and weak absorption at 1435 cm^{-1} wavenumber are visible. This indicates carbonate (CO$_3$) and P-O structures on the surface [13].

The surfaces of the specimens tested with OIL A were less IR-active. The raceways of specimens out of test runs at **100 rpm** were virtually IR-inactive. Only partial O-H ($\approx$3350 cm^{-1}), X-O (1040 cm^{-1}) and Fe-O (630 cm^{-1}) typical bands are visible. At **800 rpm**, a stronger peak at 630 cm^{-1} dominates. This could be typical of iron oxide but is nearly at the limit of the spectrum scale and can be ambiguous.

This band is also detectable in all raceway spectra of the OIL B specimen. In addition, especially in the **800 rpm** specimen spectra, a sharp strong band at 1202 cm^{-1} is present. This could possibly be, i.e., a C-O or S=O bend; however, a more precise assignment is not possible. An overview is given in Table 7.

Table 7. ATR-FTIR surface analysis.

Lubricant	FE8, 100 kN/7.5 rpm/80 °C	FE8, 50 kN/100 rpm/80 °C	FE8, 50 kN/800 rpm/80 °C
OIL A	-	Weak * Fe-O typical bands	Weak * Fe-O typical bands
OIL B	-	Middle * C-O-C/S=O and Fe-O typical bands	Strong * C-O-C/S=O and Fe-O typical bands
OIL C	Strong * P-O-P typical bands	-	-

*—band intensity in infrared spectrum.

Figure 7 illustrates the infrared spectra from the inner raceway (slippage area) of the specimens under the three test conditions. There is a clear differentiation between the oil types used and smaller differences for the same oil under different test conditions.

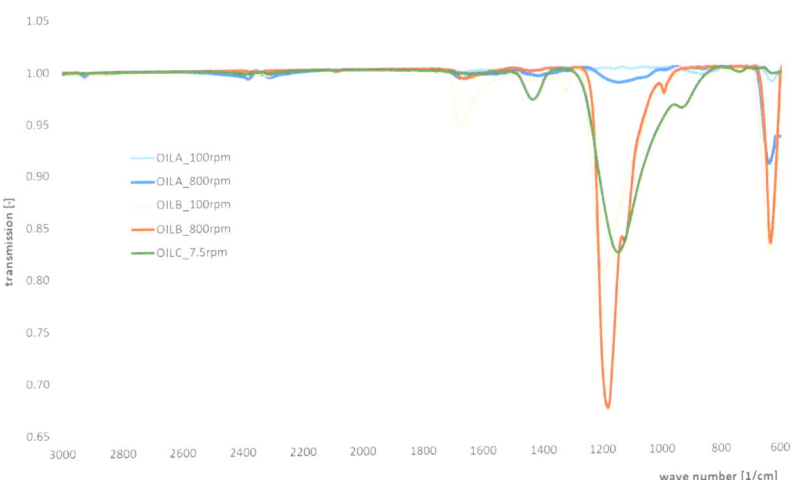

Figure 7. ATR-FTIR comparison of inner raceway area OIL A (blue), OIL B (orange) and OIL C (green).

3.4.3. Raman Spectroscopy

In contrast to the infrared spectra of OIL C specimens, the Raman spectra are unspecific. There are no differences between the raceways and the unstressed zones.

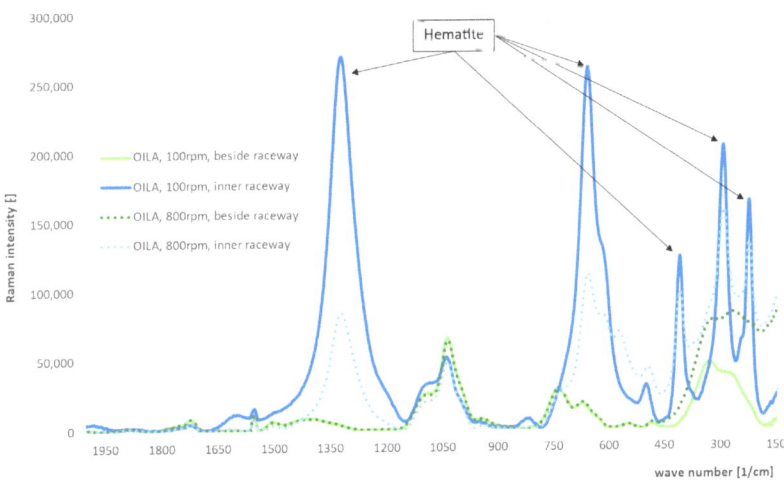

Figure 8. Raman analysis of OIL A samples: FE8, 50 kN/**100 rpm**/80 °C (solid line); FE8, 50 kN/**800 rpm**/80 °C (dotted line).

The OIL A specimens under both test conditions of 100 rpm and 800 rpm show similar spectra in the raceway area. All specimens were treated with an iron oxide layer (Fe_2O_3) [14] (Figure 8).

In addition, typical carbon bands are visible in some spectra taken from the slippage areas (Figure 9). This corresponds to optically visible dark-colored regions. Based on the positions of the D and G peaks, it is assumed that the structure of the carbon black is nano crystalline (NC-graphite) [15].

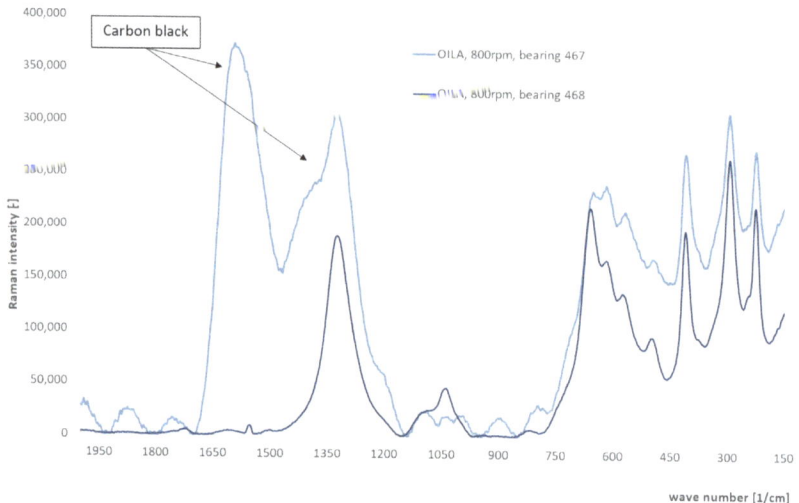

Figure 9. Raman analysis of OIL A sample. Area with additional carbon deposits.

Table 8 presents an overview of the Raman spectroscopy findings.

Table 8. Raman surface analysis.

Lubricant	FE8, 100 kN/7.5 rpm/80 °C	FE8, 50 kN/100 rpm/80 °C	FE8, 50 kN/800 rpm/80 °C
OIL A	-	Weak * Fe_3O_4 and Fe_2O_3 on raceway	Weak * Fe_3O_4 and Fe_2O_3, O-H (Carbon black) on raceway
OIL B	-	Middle * Fe_3O_4 and Fe_2O_3, O-H on raceway	Middle * Fe_3O_4 and Fe_2O_3, O-H on raceway
OIL C	None * No differences between stressed and unstressed areas	-	-

*—band intensity and respective wavenumber in Raman spectrum.

3.4.4. Element Analyses (μXRF and SEM-EDX)

Only in the OIL C specimens, increased phosphorous allocation in the raceway areas was detectable. In these parts, a significant increased phosphorous content, especially in the slippage areas, was visible. The OIL A specimens showed no specific surface deposits. In the "FE8, 50 kN/**800 rpm**/80 °C" specimens, slight traces of sulfur were detectable in the slippage zones of the raceway. The sulfur element was also detectable in all OIL B specimens at a low but significant concentration. This is surprising, because sulfur is not an intentional part of the OIL B chemistry. A possible origin of sulfur could be minimal oil residue in the test rig, equipment contamination or the bearing itself. Beside the raceways, neither sulfur nor phosphorus deposits were detectable. A short overview is given in Table 9.

Table 9. μXRF surface analysis; min (dotted) and max (solid) values of elements calcium, phosphor, sulfur and zinc are displayed.

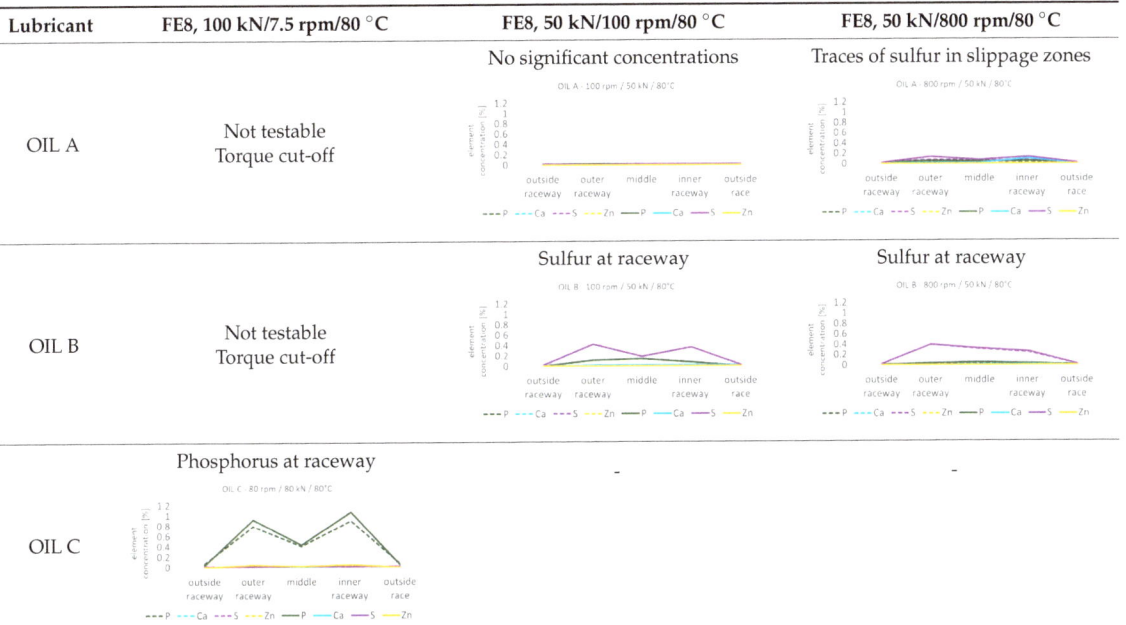

In the supporting SEM-EDX analysis, no spectacular result in element analysis was observed. On the surface layers of the washers, iron was the main element, as expected. Additionally, the OIL C specimen tribo-film contained the phosphorous and oxygen elements in the area of the raceways. On all other raceways, beside steel, only oxygen was present and, in some cases, also additional carbon. There were no additional elements detectable, beside the typical steel accompanying elements. The selected SEM-EDX method seems not to be sensitive enough to measure the sulfur concentration of the tribo-film of OIL B specimens as identified by μXRF analysis. The summary of SEM-EDX analysis is listed in Table 10.

Table 10. SEM-EDX surface analysis.

Lubricant	FE8, 100 kN/7.5 rpm/80 °C	FE8, 50 kN/100 rpm/80 °C	FE8, 50 kN/800 rpm/80 °C
OIL A	Not testable Torque cut-off	Carbon and oxygen	Traces of oxygen
OIL B	Not testable Torque cut-off	Oxygen	Oxygen
OIL C	Oxygen, phosphorous, (Na)	-	-

4. Discussion

By combining several analysis and measurement methods, it is possible to achieve an overall determination of the main influential factors for the wear test results. The surface analyses showed distinctive signals along with very good differentiability of the resulting tribo-films.

Despite the nominally poorer lubrication conditions, the OIL B specimens were less wear-affected than the OIL A specimens. Neither, however, was as sufficient as reference OIL C. Nevertheless, the respective oil analyses prove that unadditivated oils provide wear protection under medium mixed-friction conditions.

Contrary to the oil analysis, the surface analysis presented a more precise result. It was shown that different tribo-films were formed on the specimens dependent on the oil type used. With OIL A, as expected and like the brown-colored tribo-film regions reported by [7], an iron oxide-based tribo-film was detected. It seems that this kind of tribo-film is stable at ≈ 124 µJ/mm^2 specific friction energy values, as in test "FE8, 50 kN/800 rpm/80 °C", but obviously not wear-resistant enough for ≈ 188 µJ/mm^2 at 100 rpm.

In contrast, with OIL B, a mixture of an iron oxide-, carbon- and sulfur-containing layer was present. The sulfur in the tribo-films on OIL B specimens is in clear contradiction with the "detected as expected" tribo-films on the OIL A and OIL C specimens. Sulfur is a typical element in lubrication oils and known as a wear protection additive in several chemical structures. But OIL B is a pure dicarboxylic acid ester intentionally free of sulfur, as confirmed by the used oil analysis. It seems that a sulfur content of OIL B below the detection limit of 1 ppm promotes the formation of a sulfur-containing tribo-film. A possible source of sulfur could be minimal oil residue in the test rig, equipment contamination or the bearing itself. Further research is required here.

The identified phosphorous-based tribo-film of the OIL C specimens is in line with expectations.

An open issue is the determination of the high-pressure contact viscosity of the fluids, which was not measured here. The differences between polyalphaolefin and dicarboxylic acid ester published in [16] are not able to fully explain the observed differences in wear behavior. The sulfur exposure also appears to be relevant to the test results.

5. Conclusions

Firstly, the results clearly show that the wear behavior of a lubricated rolling bearing is not only dependent on the operation conditions. Neither the contact nor the nominal lubrication conditions can sufficiently explain the results. Beside the lubricant physics, the lubricant chemistry, including the lubricant additivation, plays a significant role. The analysis of the resulting tribo-film is vital to the investigation of tribological issues. Tribological contact appears to be more sensitive to surface-active substances than the state-of-the-art analysis methods for lubricants. Knowledge of the resulting tribo-films would significantly improve the interpretation of the results in the future.

Secondly, it was demonstrated that slippage including mixed-friction conditions generally does lead to a high wear level. The surfaces (steel substrate including tribo-film) of the rolling bearings are robust enough to tolerate a specific level of "insufficient lubrication". The specific friction energy flowing over the solid-state contact characterized by the value e_{SID} could be a first approach to quantify this specific level.

The following lists our most important findings:

- Different lubricant formulations lead to different tribo-films under the same test conditions.
- The same lubricant formulation leads to similar tribo-films under different mixed-friction conditions.
- Higher specific friction energy increases the wear risk by using the same oil formulation, as demonstrated with OIL A.
- The tribo-film is more contamination-sensitive than the oil analysis, as demonstrated with OIL B.

By adding a combination of innovative surface analysis techniques to the state-of-the-art analysis methods, the quality of tested and returned part forensic analyses will be significantly improved. This presents an opportunity for development and research in the field of rolling bearing technology.

Author Contributions: Conceptualization, J.W.H.F., J.F. and D.M.; methodology, J.F. and D.M.; validation, D.M., investigation, J.F.; resources, T.K.; data curation, J.W.H.F.; writing—original draft preparation, J.W.H.F.; writing—review and editing, J.F., D.M. and J.W.H.F.; visualization, J.F. and J.W.H.F.; supervision, T.K.; project administration, J.F. All authors have read and agreed to the published version of the manuscript.

Funding: This research received no external funding.

Data Availability Statement: Dataset available on request from the authors.

Acknowledgments: The authors would like to thank Bruker Nano GmbH and Bruker Optics GmbH for their technical support, including the provision of illustrations for this publication.

Conflicts of Interest: The test specimens examined in this study are the property of Schaeffler Technologies AG & Co. KG. The authors are employees of Schaeffler Technologies AG & Co. KG. The results were obtained in the course of their work.

References

1. Inacker, O.; Beckmann, P.; Oster, P. Abschlußbericht FVA Nr. 289 I und II Triboschutzschichten. In *FVA-Forschungsheft Nr. 595*; FVA-Forschungsreporte: Frankfurt, Germany, 1999.
2. Evans, R.D.; More, K.L.; Darragh, C.V.; Nixon, H.P. Transmission Electron Microscopy of Boundary-Lubricated Bearing Surfaces. Part I: Mineral Oil Lubricant. *Tribol. Trans.* **2004**, *47*, 430–439. [CrossRef]
3. Scherge, M.; Brink, A.; Linsler, D. Tribofilms Forming in Oil-Lubricated Contacts. *Lubricants* **2016**, *4*, 27. [CrossRef]
4. Forbes, E.S. Antiwear and extreme pressure additives for lubricants. *Tribology* **1970**, *3*, 145–152. [CrossRef]
5. Stratmann, A.; Burghardt, G.; Jacobs, G. Influence of Operating Conditions and Additive Concentration on the Formation of Anti-wear Layers in Roller Bearings. In Proceedings of the 20th International Colloquium Tribology, Stuttgart/Ostfildern, Germany, 12–14 January 2016; pp. 109–110. Available online: https://publications.rwth-aachen.de/record/672506 (accessed on 13 April 2024).
6. Burghardt, G.; Wächter, F.; Jacobs, G.; Hentschke, C. Influence of run-in procedures and thermal surface treatment on the anti-wear performance of additive-free lubricant oils in rolling bearings. *Wear* **2015**, *328–329*, 309–317. [CrossRef]
7. Gachot, C.; Hsu, C.; Suárez, S.; Grützmacher, P.; Rosenkranz, A.; Stratmann, A.; Jacobs, G. Microstructural and Chemical Characterization of the Tribolayer Formation in Highly Loaded Cylindrical Roller Thrust Bearings. *Lubricants* **2016**, *4*, 19. [CrossRef]
8. DIN 51517-3; Lubricants-Lubricating Oils-Part 3: Lubricating Oils CLP, Minimum Requirements 2018-09. Beuth Verlag: Berlin, Germany, 2018. [CrossRef]
9. DIN 51819-1; Testing of Lubricants-Mechanical-Dynamic Testing in the Roller Bearing Test Apparatus FE8-Part 1: General Working Principles 2016-12. Beuth Verlag: Berlin, Germany, 2016. [CrossRef]
10. Vierneusel, B.; Koch, O. Fast calculation method for predicting the risk of surface initiated damage in rolling bearings. In Proceedings of the World Tribology Congress, Beijing, China, 17–22 September 2017.
11. Bearinx High-Level Bearing Design. Available online: https://www.schaeffler-industrial-drives.com/en/news_media/media_library/downloadcenter-detail-page.jsp?id=68304192 (accessed on 16 April 2024).
12. ISO 281; Rolling Bearings-Dynamic Load Ratings and Rating Life 2007-02. Beuth Verlag: Berlin, Germany, 2007; ICS 21.100.20.
13. Franke, J.W.H.; Fritz, J.; Koenig, T.; Merk, D. Influence of Tribolayer on Rolling Bearing Fatigue Performed on an FE8 Test Rig—A Follow-Up. *Lubricants* **2023**, *11*, 123. [CrossRef]
14. Oh, S.J.; Cook, D.; Townsend, H. Characterization of Iron Oxides Commonly Formed as Corrosion Products on Steel. *Hyperfine Interact.* **1998**, *112*, 59–66. [CrossRef]
15. Ferrari, A.C.; Robertson, J. Resonant Raman spectroscopy of disordered, amorphous, and diamondlike carbon. *Phys. Rev. B* **2001**, *64*, 075414. [CrossRef]
16. Villamayor, A.; Guimarey, M.J.G.; Mariño, F.; Liñeira del Río, J.M.; Urquiola, F.; Urchegui, R.; Comuñas, M.J.P.; Fernández, J. High-Pressure Thermophysical Properties of Eight Paraffinic, Naphthenic, Polyalphaolefin and Ester Base Oils. *Lubricants* **2023**, *11*, 55. [CrossRef]

Disclaimer/Publisher's Note: The statements, opinions and data contained in all publications are solely those of the individual author(s) and contributor(s) and not of MDPI and/or the editor(s). MDPI and/or the editor(s) disclaim responsibility for any injury to people or property resulting from any ideas, methods, instructions or products referred to in the content.

 MDPI

Article

Practical Evaluation of Ionic Liquids for Application as Lubricants in Cleanrooms and under Vacuum Conditions

Andreas Keller [1], Knud-Ole Karlson [1], Markus Grebe [1,*], Fabian Schüler [2,*], Christian Goehringer [2] and Alexander Epp [2]

[1] Competence Center of Tribology, Mannheim University of Applied Sciences, 68163 Mannheim, Germany
[2] Materiales GmbH, Offakamp 9f, 22529 Hamburg, Germany; alexander.epp@materiales.de (A.E.)
* Correspondence: m.grebe@hs-mannheim.de (M.G.); fabian.schueler@materiales.de (F.S.);
Tel.: +49-621-292-6541 (M.G.); +49-40-5725-6735 (F.S.)

Abstract: As part of a publicly funded cooperation project, novel high-performance lubricants (oils, greases, assembly pastes) based on ionic liquids and with the addition of specific micro- or nanoparticles are to be developed, which are adapted in their formulation for use in applications where their negligible vapor pressure plays an important role. These lubricants are urgently needed for applications in cleanrooms and high vacuum (e.g., pharmaceuticals, aerospace, chip manufacturing), especially when the frequently used perfluoropolyethers (PFPE) are no longer available due to a potential restriction of per- and polyfluoroalkyl substances (PFAS) due to European chemical legislation. Until now, there has been a lack of suitable laboratory testing technology to develop such innovative lubricants for extreme niche applications economically. There is a large gap in the tribological test chain between model testing, for example in the so-called spiral orbit tribometer (SOT) or ball-on-disk test in a high-frequency, linear-oscillation test machine (SRV-Tribometer from German "Schwing-Reib-Verschleiß-Tribometer"), and overall component testing at major space agencies (ESA—European Space Agency, NASA—National Aeronautics and Space Administration) or their service providers like the European Space Tribology Laboratory (ESTL) in Manchester. A further aim of the project was therefore to develop an application-orientated and economical testing methodology and testing technology for the scientifically precise evaluation and verifiability of the effect of ionic liquids on tribological systems in cleanrooms and under high vacuum conditions. The newly developed test rig is the focus of this publication. It forms the basis for all further investigations.

Keywords: ionic liquids; cleanroom; vacuum application; component testing; model testing; screening tests

Citation: Keller, A.; Karlson, K.-O.; Grebe, M.; Schüler, F.; Goehringer, C.; Epp, A. Practical Evaluation of Ionic Liquids for Application as Lubricants in Cleanrooms and under Vacuum Conditions. *Lubricants* **2024**, *12*, 194. https://doi.org/10.3390/lubricants12060194

Received: 30 April 2024
Revised: 22 May 2024
Accepted: 24 May 2024
Published: 28 May 2024

Copyright: © 2024 by the authors. Licensee MDPI, Basel, Switzerland. This article is an open access article distributed under the terms and conditions of the Creative Commons Attribution (CC BY) license (https://creativecommons.org/licenses/by/4.0/).

1. Introduction

Cleanroom technology plays a crucial role in various industries, including semiconductor manufacturing, pharmaceutical production, biotechnology research, and aerospace engineering, where controlled environments are required for high-quality manufacturing. Stringent cleanliness requirements are necessary to prevent particles and other elemental or molecular contaminants from compromising sensitive processes or products. In specific cases, such as semiconductor or medical device manufacturing, vacuum is applied for a further improvement of process cleanliness and quality, or it is inevitably present, as in space applications [1,2]. A low volatility to prevent outgassing or evaporation is an initial requirement for all materials employed in a cleanroom environment.

Such conditions place special demands on tribology [3,4]. While the active surfaces of tribologically stressed components normally have the opportunity to form friction- and wear-reducing surface layers through chemical reactions with the gaseous ambient medium under atmospheric conditions, this is not possible in vacuum, where adhesion mechanisms dominate in the contact interfaces and can lead to malfunctions and the failure of tribological systems [5]. It is furthermore well known that in particular, solid lubricants, which are often parts of grease formulations, lose their lubricating functionality without the presence of oxygen or water. Finally, additives or low-boiling oil fractions can evaporate from lubricant oils, which can result in changes in properties such as chemical stability of viscosity. Therefore, general testing techniques and technologies have been developed to address the area of vacuum tribology [6–8].

The market for high-performance lubricants in the space, vacuum technology, cleanroom and semiconductor industries requires highly specified and well-understood materials, in particular due to the high cleanliness standards and the demand for long maintenance intervals, as the components are either difficult to access (space), or maintenance would lead to machine downtime and thus to high costs due to production losses (especially in the semiconductor industry). On the other hand, the demand for very small quantities in comparison with other industries is very characteristic. This niche is unattractive for major lubricant manufacturers that are used to supplying material multiple tons instead of grams to kilograms. As a result, the range of lubricant options is very limited, and users often have no choice but to use a commercially available standard product, to quantify the risk of failure at great expense and to plan and implement appropriate measures.

Two types of commercially available technical products dominate the field of clean and vacuum-compatible high-performance lubricants, perfluoropolyethers (PFPE) and multiple alkylated cyclopentanes (MAC). PFPE have a wide range of operating temperatures, but are virtually incompatible with additives due to the fluorinated base oils, i.e., only limited options for formulations with property-enhancing ingredients exist. In addition, they tend to degrade autocatalytically on ferrous surfaces, which drastically reduces the service life of tribological systems [9,10]. MAC lubricants, on the other hand, are known for long lifetimes, but are limited in the range of application temperatures due to their less inert chemical structure. Stability can be improved by the use of additives, which can, however, significantly enhance outgassing. In addition, this product group often exhibits high starting friction, which is problematic for the safe design of the drive technology [11]. There is also a risk that the MAC oils generate hydrogen through tribochemical reactions, which can lead to premature rolling bearing failures [12,13].

Ionic liquids (ILs) are organic salts; they consist of cations and anions that are (partially) built up from hydrocarbon structures and that are often liquid at room temperature and below. ILs display several advantages that make them suitable as promising compounds for clean and vacuum-compatible high-performance lubricants, in particular vacuum lubricants, such as a negligible vapor pressure, a high thermal and chemical stability, and a low flammability. Their advantageous tribological properties have been reported in various scientific studies [14–17] and in the context of vacuum lubrication for space application [11,18–20]. Commercial interest in ILs as lubricants has been mainly limited to their use as functional additives, for example, to improve conductivity in conventional lubricants to reduce the risk of electrocorrosion. This can mainly be related to the high cost in comparison to conventional mineral oil-based lubricants. ILs used as lubricant base oils display a promising alternative to the commercially available clean vacuum lubricants due to their potential to combine the thermal and chemical stability of PFPE-based lubricants with the lifetime of MAC fluids. While the potential is evident, and a high demand for such specialty lubricants exists, the development of IL-based lubricants represents a high financial risk. Due to the extreme consequences of failure of tribological systems, a very complex risk management is carried out that involves extensive and cost-intensive test campaigns [6]. Hence, and in particular for very new and innovative technologies, potential customers have to be convinced by good technical and scientifically sound arguments to consider the use of new products. In the tribological test chain, there is a large gap between initial model tests like the spiral orbit tribometer (SOT) [6–8] and dedicated component testing. The transferability of the results from the highly simplified tests to the behavior of original components such as roller bearings, plain bearings, linear guides, or spindle drives is not given [21]. This leads to major problems in the design of these construction elements and the selection of suitable material and lubricant combinations.

There is therefore an urgent need to develop an application-oriented test methodology for carrying out component or aggregate tests to evaluate the new high-performance lubricants under cleanroom and vacuum conditions. Such a multi-stage test methodology does not yet exist. Due to the storage and assembly of the components in an earth atmosphere, the tests must be carried out under both cleanroom and vacuum conditions. In addition, quick, simple tests, as well as more complex and application-related component and component tests, must be carried out as part of tribological testing.

This paper describes the systematic investigation of five ILs for use as base oils for cleanroom and vacuum lubricants, and in particular the development of a new type of component test rig and its integration into the material characterization process. In this rig, high-performance lubricants can be tested under cleanroom and vacuum conditions by replacing modular linear slideways, plain, and roller bearings as well as ball-screw drives in close proximity to the application. Initial measurement results show that the new test stand fulfills the expectations placed in it.

2. State of Research
2.1. Scientific Fundamentals of Ionic Liquids

Among the numerous publications about ionic liquids, many research papers and reviews discuss the applicability of ionic liquids as lubricants [16,22–25]. While the beneficial lubrication properties of pure ILs are commonly highlighted, their high cost and a potential corrosion risk are typical arguments for use as lubricant additives instead [16,17,26]. Indeed, most conventional applications could benefit from IL additives, but not fully exploit the potential that their specific properties offer. However, the following unique combination of features makes them very promising compounds for use in cleanroom (vacuum) high performance applications:

2.1.1. Vapor Pressure

Outgassing and evaporation are of particular concern for lubricants employed in cleanroom or vacuum applications. Firstly, volatile organic species mean a contamination risk for a clean environment, which can be particularly critical when sensitive (e.g., optical) equipment is present. In vacuum, evaporation of volatile compounds increases significantly, which can lead to vaporization either of low molecular mass fractions of an oil or functional additives—both mean a risk for the functionality of the lubricant. Ionic liquids are known for their very low vapor pressures, even at elevated temperatures [27–29]. It is challenging to determine reliable vapor pressures, since many ILs begin to decompose before their vapor pressure is measurable [27].

2.1.2. Thermal and Thermo-Oxidative Stability

The stability of ILs depends on the type of cations and anions, but also on their interaction. In particular, cations based on heterocyclic aromatic structures such as pyridinium, pyrrolidinium, or imidazolium are known for a high bond strength and overall stability. Less-stable cations are ammonium or phosphonium based aliphatic structures. For many ILs, the choice of anion is more critical than the choice of cation with regard to the thermal stability. In a recent study, the thermal stability of 66 ILs was compared by thermogravimetric measurements [30]; the most stable compounds from this study contained the fluorine-containing anions bis(trifluoromethyl)sulfonylimide, tetrafluoroborate, or hexafluorophosphate with onset temperatures higher than 400 °C. A relatively high decomposition temperature can, however, provide a wrong perception of the long-term stability. Fox et al. published a long-term study where the stability of ILs over 16 weeks of oven storage at elevated temperatures was examined [31]. During that period, the physical appearance and material properties of many tested substances changed drastically. It was shown that, in particular, the imidazolium ILs were most resistant to degradation, which is in line with the findings from other studies [32].

2.1.3. Tribological Properties

The potential of ILs for tribological use has been widely researched, with the ILs being used both as a base fluid and as additives [14–17]. It has been shown that ILs lubricating metallic contacts can tribologically outperform commercially available fully formulated oils. The ability to react with metal surfaces under formation of a tribofilm can significantly reduce friction and wear. However, the knowledge in this field is still quite undeveloped, and it has been noted as recently as 2013 [33] that there is still no clear understanding of the mechanism for lubrication, and selection of ionic liquids for tribological applications tends to be based on trial and error.

The use of ILs for tribological vacuum applications has been discussed in the context of the space industry [20,34–36]. While different cation types were discussed, it was an outcome of several studies that ILs that contained the bis(trifluoromethyl)sulfonylimide (BTA) anion showed the most advantageous properties.

2.2. Available Test Rigs

According to the current state of the art, only model test rigs are commercially available for testing tribological properties under vacuum conditions. This means that the tribological system consists of simply abstracted test specimens, such as balls, disks, or plates.

In the aerospace sector, tests on the so-called spiral orbit tribometer (SOT) are standard in the qualification phase of lubricants and coatings [37,38]. The spiral orbit tribometer (SOT) is a thrust bearing with only one ball and flat raceways [8]. As in an original rolling bearing, the ball is subject to rolling, spinning, and sliding motion. This is a much more realistic bearing simulation than other tribometers that only perform unidirectional sliding. Figure 1 shows the instrument's internal components in detail. A single ball is loaded between two flat plates. As the lower plate rotates, the ball is driven in a nearly circular orbit. The orbit is actually an opening spiral, with the spiral's pitch directly related to the friction coefficient. At the end of each orbit, the ball contacts the guide plate, which returns the ball to its original orbit radius. This orbit is repeatable during the entire test. A sensor measures the force exerted by the ball on the guide plate and this yields the friction coefficient. The instrument's kinematics are well understood [37].

For most of the orbit, the ball undergoes pure rolling with spin; however, when the ball comes into contact with the guide plate, there is a certain mechanical impact with increased sliding ratio. This region is colloquially termed by the inventors of the test bench the 'scrub', and the majority of lubricant degradation occurs here. An important assessment parameter in this test is lubricant degradation, which can be measured online using a gas element analyzer. Additionally, the SOT determines the friction coefficient of both the lubricated system and the non-lubricated specimens. The relative lifetimes of lubricant/material combinations correlate with full-scale ball bearing tests, and can be used to select the appropriate lubricant for an application.

In the SOT, Hertzian contact stresses between 0.50 and 5.00 GPa can be set using different normal forces and ball sizes, which corresponds to typical rolling bearing applications. Standard test plates are manufactured from 440C steel and polished to a surface roughness Ra < 0.05 microns. Typical balls used are either 12.7 mm (1/2") or 7.14 mm (9/32") diameter, manufactured from 400C and 52100 steel, respectively, depending upon the contact stress requirements of the particular test [37]. The instrument is designed to operate in ultrahigh vacuum or with a cover gas. Test acceleration is achieved by limiting the lubricant amount on the ball to tens of micrograms.

Figure 1. Setup spiral orbit tribometer (SOT) [8].

In addition to the SOT test, lubricants are often tested in a ball/plate contact arrangement under oscillating sliding friction. However, as described above, this rarely corresponds to the real loading in the application. The test is therefore even more abstract than the SOT. Typical tribometers for this model test arrangement are the high-frequency, linear-oscillation (SRV) Test Machine in accordance with DIN 51834 or ASTM D7421 from Optimol Instruments [19,39–41] or a classical pin- or ball-on-disk test rig [21,38].

3. Assessment of Imidazolium-Based ILs as High-Performance Lubricants

While the excellent lubrication properties of ILs have been demonstrated, it is also known that they can exert significant corrosion on metallic surfaces, in particular in the presence of water [42–44]. Making the molecular structure of an IL more hydrophobic could reduce water absorption and thus might reduce the corrosive impact on metals and alloys. This could, however, also impact other key properties for a lubricant, such as thermal properties, viscosity, or tribological functionality. In this investigation, five ILs that are based on an imidazolium cation and an fluoroalkyl sulfonylimide anion, but with differences in the length of the pendant alkyl groups of the imidazolium ion and the chain length of the fluoroalkyl group of the anion, were tested and compared (Figure 2):

1-Butyl-3-Methylimidazolium Bis(trifluoromethylsulfonylimide) (BMIM BTA),
1-Octyl-3-Methylimidazolium Bis(trifluoromethylsulfonylimide) (OMIM BTA),
1-Butyl-3-Methylimidazolium Bis(pentafluoroethylsulfonylimide) (BMIM BETA),
1-Butyl-3-Methylimidazolium Bis(nonafluorobutylsulfonylimide) (BMIM BBTA) and
1-Octyl-3-Methylimidazolium Bis(trifluoromethylsulfonylimide) (OMIM BBTA)

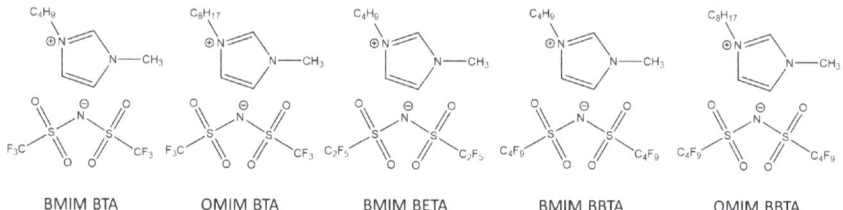

Figure 2. Chemical structure of the investigated ILs.

3.1. Water Uptake

The water absorption from air is an indicator for hydrophobicity and is thus critical for the corrosion of metallic materials. Also, highly hydrophilic ILs that absorb significant amounts of water show an outgassing risk and do not fulfill cleanliness requirements for vacuum applications. For space application, a water content <1 wt % is required [45]. It was expected that an increase in hydrocarbon group length at the imidazolium cation and an increase in fluorocarbon group length at the anion increase the hydrophobicity and reduce water absorption.

The candidate ILs were firstly dried by a bakeout in a glass flask in vacuum (100 °C, 10^{-1} mbar, 24 h). Then, 1 g of each fluid was allowed to stand open in laboratory air (22 °C, 50% relative humidity), and the weight was monitored over several days until a constant level was achieved. Results are depicted in Table 1.

Table 1. Water uptake of IL candidates.

Ionic Liquid	Water Uptake
BMIM BTA	0.20%
OMIM BTA	0.14%
BMIM BETA	0.12%
BMIM BBTA	0.15%
OMIM BBTA	0.05%

The water uptake was low for all tested ILs, and a tendency towards lower water absorption could be observed. With regard to the chemical structure, it was expected that BMIM BTA behaves least hydrophobic, and OMIM BBTA the most, which is reflected in the experimental findings.

3.2. Thermal Properties

The properties at high and low temperatures were analyzed with different methods. Melting points were characterized rheologically with an Anton Paar MCR 702 rheometer. The following test conditions were applied:

- Plate/plate measurement system.
- Test plate size: 25 mm diameter.
- Measurement gap/lubricant layer thickness: 1 mm.
- Test temperature range: −20–25 °C at 1 °C/min.
- Shear stress: 2 Pa.

The melting point was determined as the temperature at which a measurable rotation of the upper plate was detected. The thermal degradation was investigated with a NETZSCH Libra F1 thermo-microbalance (TGA). Samples were heated in air atmosphere from 20–700 °C at a rate of 10 °C/min; the thermal degradation was assessed as the temperature at which 2% of the original weight were lost. In addition, the long-term stability at constant temperature was analyzed. Therefore, 1 g of each fluid was stored in an oven for several weeks, and the mass loss was observed. Results are depicted in Table 2.

Table 2. Melting temperatures, degradation temperatures, and thermooxidative stability of IL candidates.

Ionic Liquid	T (Melt) [Rheology]	T (2% Mass Loss) [TGA, Air]	Mass Loss [Oven Storage]
BMIM BTA	<−20 °C	354 °C	0.8%/16 weeks/200 °C
OMIM BTA	<−20 °C	331 °C	2.0%/16 weeks/200 °C
BMIM BETA	<−20 °C	326 °C	1.4%/6 weeks/200 °C
BMIM BBTA	10 °C	319 °C	0.2%/16 weeks/150 °C
OMIM BBTA	20 °C	316 °C	2.5%/5 weeks/200 °C

Longer hydrocarbon groups in the cation and longer fluorocarbon groups in the anion increase the melting temperature due to their known tendency to crystallize. Also, and as expected, a higher susceptibility towards thermal degradation was observed with an increase in alkyl chain length. It can be seen that the range of operating temperatures would be smaller for these (more hydrophobic) ILs.

3.3. Viscosity

Viscosities and viscosity indices were determined rheologically with an Anton Paar MCR 702 rheometer. Tests at different shear rates and temperatures were performed in alignment with DIN 51810-1 [46]; the viscosity index was calculated according to ASTM D2270-10 [47]. The following test conditions were applied:

- Plate/plate measurement system.
- Test plate size: 25 mm diameter.
- Measurement gap/lubricant layer thickness: 1 mm.
- Test temperature range: 20–100 °C.
- 10 min holding time at target temperature before measurement.
- Shear rate: 0/s–100/s.

Results are shown in Table 3.

Table 3. Viscosities and viscosity indices of IL candidates.

Ionic Liquid	Viscosity (25 °C, 50/s) [Rheology]	Viscosity Index [Rheology]
BMIM BTA	52 mPa·s	205
OMIM BTA	96 mPa·s	152
BMIM BETA	111 mPa·s	178
BMIM BBTA	448 mPa·s	88
OMIM BBTA	484 mPa·s	97

Again, a behavior as expected from the chemical structures could be observed. The viscosity and the viscosity index correlate with the melting points and the hydrophobicity of the ILs.

3.4. Corrosion

Corrosion tests were carried out with SAE 52100 steel (1.3505, 100Cr6), as this is the standard material for rolling bearing parts; it is used in numerous tribological standard tests (such as SOT or SRV) and is also known to be very susceptible to corrosion. Therefore, round samples (height: 5 mm, diameter: 30 mm, surface roughness Rz: 6.3 µm) were manufactured. Samples were cleaned with acetone; afterwards, the surface was fully covered with IL. Samples were stored in a Memmert HCP 50 climate chamber for 2 weeks at 80 °C temperature and 60% relative humidity. The samples were compared visually. These tests only allowed a rough assessment (see Table 4), since visual characterization is very objective and test results showed variation.

Table 4. Qualitative assessment of steel corrosion by IL candidates.

Ionic Liquid	Corrosion on SAE 52100 Steel
BMIM BTA	Severe
OMIM BTA	Significant
BMIM BETA	Significant
BMIM BBTA	Significant
OMIM BBTA	Little

Test results indicate a correlation between corrosion and water absorption. While BMIM BTA had a severe corrosive impact on the steel, OMIM BBTA, in particular, showed only small signs of corrosive attack. An improved method would be required for a more detailed assessment.

3.5. Tribological Evaluation on Model Test Rigs

Parallel to the development of the new vacuum test bench, the first tests with the newly developed ionic liquids were already carried out under normal atmospheric conditions. Classic model tests on the SRV test bench [39] and the mini-traction-machine (MTM) [21] were used for this purpose.

First, the base fluid candidates without their thickener will be examined using an oscillating friction wear tribometer (SRV). Typically, lubricants are tested on the SRV under oscillating sliding friction, high pressure, high frequency, and small stroke in a ball/plate arrangement. The loads in the SRV represent a worst-case scenario, which probably only rarely occurs in real applications, but are regarded as a fast and consistent basis for comparing lubricants, which are also internationally recognized and standardized.

3.5.1. SRV-Test According to DIN 51834-2 [40] and ASTM D7421-1 [41]

In order to examine lubricants comparatively under model conditions in the laboratory, tests are often carried out today on the SRV oscillating friction wear test rig [21]. In the simplest test, a 10 mm steel ball (100Cr6, hardened and polished) oscillates on a steel disk (100Cr6, hardened, lapped). This test places high demands on the lubricant to be tested, as mixed friction conditions constantly prevail due to the load collective and the oscillation movement, and the pressures are high due to the point contact.

In the SRV, the ILs provided by the project partner are used to carry out endurance runs in accordance with DIN 51834-2 (300 N load, 50 Hz vibration frequency, 1 mm stroke, 50 °C sample temperature, 2 h running time) [40] to determine the friction and wear properties, as well as load increase runs in accordance with ASTM-D7421-1 [41] (80 °C sample temperature) to determine the maximum load-bearing capacity. A double determination is used as the test statistic. The SRV is particularly suitable for testing these base fluids, as only a very small amount of lubricant is required for the individual test.

In addition to the ILs, two vacuum lubricants (a perfluoropolyether (PFPE—Fomblin Z25), a multiply alkylated cyclopentane (MAC)—Nye2001A) and a commercially available engine oil (Castrol Edge 5W30) are tested as comparative references. Ideally, the ILs are able to beat all references in the standard test. The results of these tests are shown in the following diagrams (Figures 3 and 4).

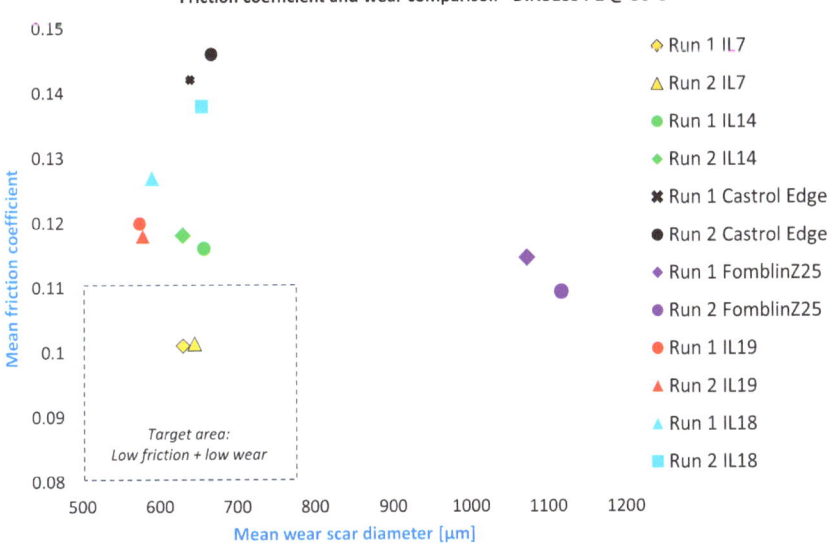

Figure 3. Coefficients of friction in the SRV test acc. DIN 51834-2.

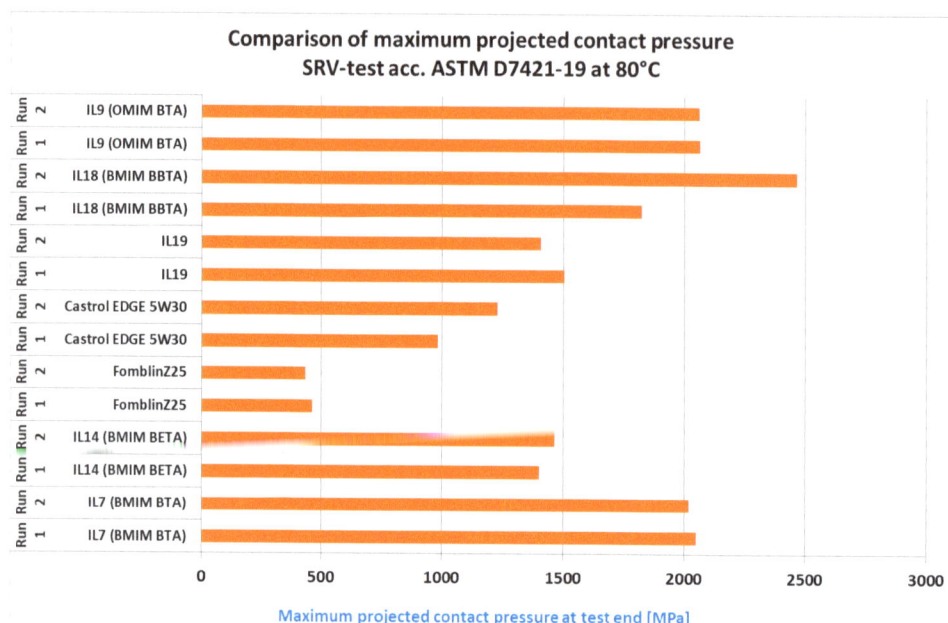

Figure 4. Maximum contact pressure reached in the SRV load progression test acc. ASTM D7421.

Figure 3 shows the results from the DIN standard runs. An ideal lubricant shows a low friction level in combination with low wear (small calotte diameter) and is therefore located in the lower left area of the graph. It can already be seen here that the IL candidates examined are close to each other and can undercut both the commercially available engine oil in terms of friction and the PFPE in terms of wear. BMIM BTA, in particular, shows great potential here. The second vacuum lubricant, Nye2001A, was not able to pass this test at all, which is why it was not included in the graph.

In the ASTM load step run (Figure 4), which particularly examines the high-pressure load-bearing capacity and protection against severe adhesive wear, similar results are shown. Here, even the weaker IL variants (e.g., OMIM BBTA, IL19) are on a par with the engine oil and shows a maximum surface pressure at least twice as high as the PFPE oil (the MAC also failed directly here again). The best IL variants exhibit maximum surface pressures of 2000 MPa (BMIM BTA, IL7)–2500 MPa (BMIM BBTA, IL18), which is equivalent to a high-quality gear oil in terms of load-bearing capacity. Due to its corrosion and temperature resistance, the project partner particularly favors variants of the BMIM BTA (IL7) sample for further development.

The good test values can also be seen in the microscopic images of the test specimens after the load increase run (see Figure 5). While the engine oil and PFPE show pronounced signs of adhesive wear on the test specimens, the wear pattern of BMIM BTA and OMIM BTA is primarily characterized by a rather soft tribofilm.

For the further course of the project, repetitions of the SRV tests of the most promising IL variants with stainless steel test specimens (material 1.4125) and under a nitrogen atmosphere are planned. Both represent an approximation of the conditions present in the real system. The nitrogen atmosphere prevents reactions with the ambient oxygen, and stainless steels are often used in vacuum systems for corrosion protection purposes.

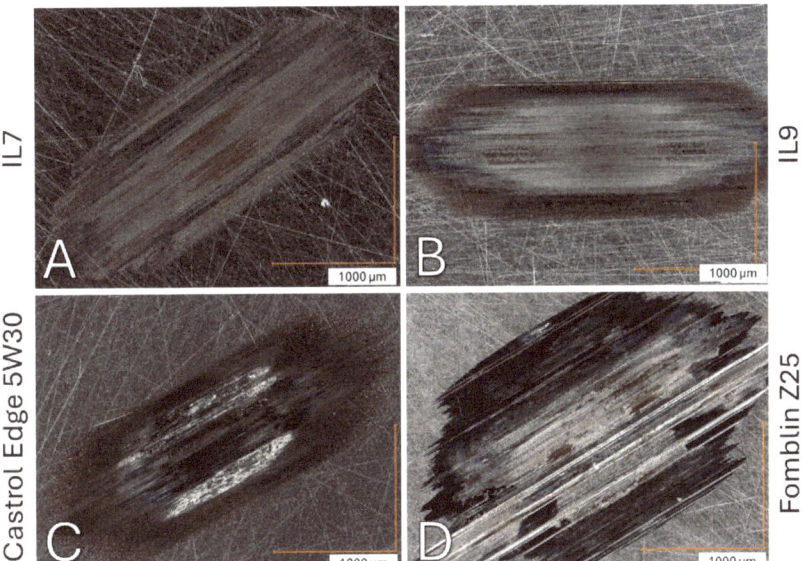

Figure 5. Wear patterns after the ASTM-D7421-1 test: (**A**) IL7, (**B**) IL9, (**C**) Castrol Edge 5W30, (**D**) Fomblin Z25.

3.5.2. MTM-Tests

As soon as grease variants of the most promising ILs are available, these will also be tested under rolling load using a mini-traction machine (MTM, see Figure 6). As described, many motion systems are designed as rolling systems under vacuum conditions, as the lowest possible friction and thus minimized heat input into the system can be achieved. In addition to the wear mechanisms of abrasion, adhesion, and tribo-oxidation present in the SRV tests, rolling friction also causes surface fatigue/breakdown. The plan here is to load the lubricating medium with repeated Stribeck and traction curves under 1 GPa Hertzian pressure until damage to the surfaces becomes noticeable due to increasing friction levels. The MTM tests are also to be seen as a preliminary stage or complementary to the rolling friction tests in the vacuum tribometer.

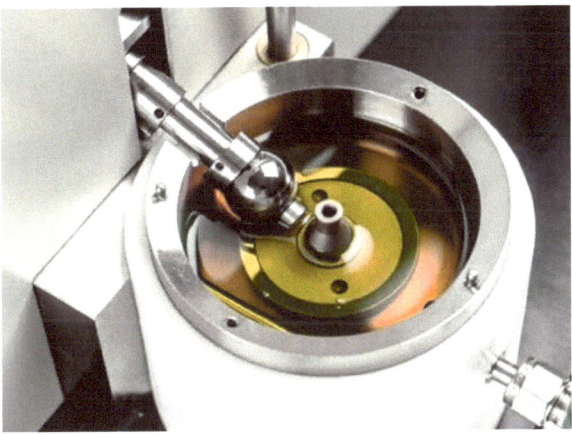

Figure 6. MTM setup.

Preliminary tests of the most promising base lubricants—as used in the SRV tests—showed good results. Tests were conducted with repeated speed variation runs (so-called Stribeck tests) with 500–20 mm/s rolling speed, 1 GPa Hertzian stress, 2% Slide-Roll-Ratio, and 50 °C pot/ambient temperature. The lubricants were applied as initial lubrication of 0.25 mL on top of the AISI SAE 52100 ball-and-disc specimen.

Figure 7 shows the results of all individual runs for the samples Fomblin Z25 and IL7.

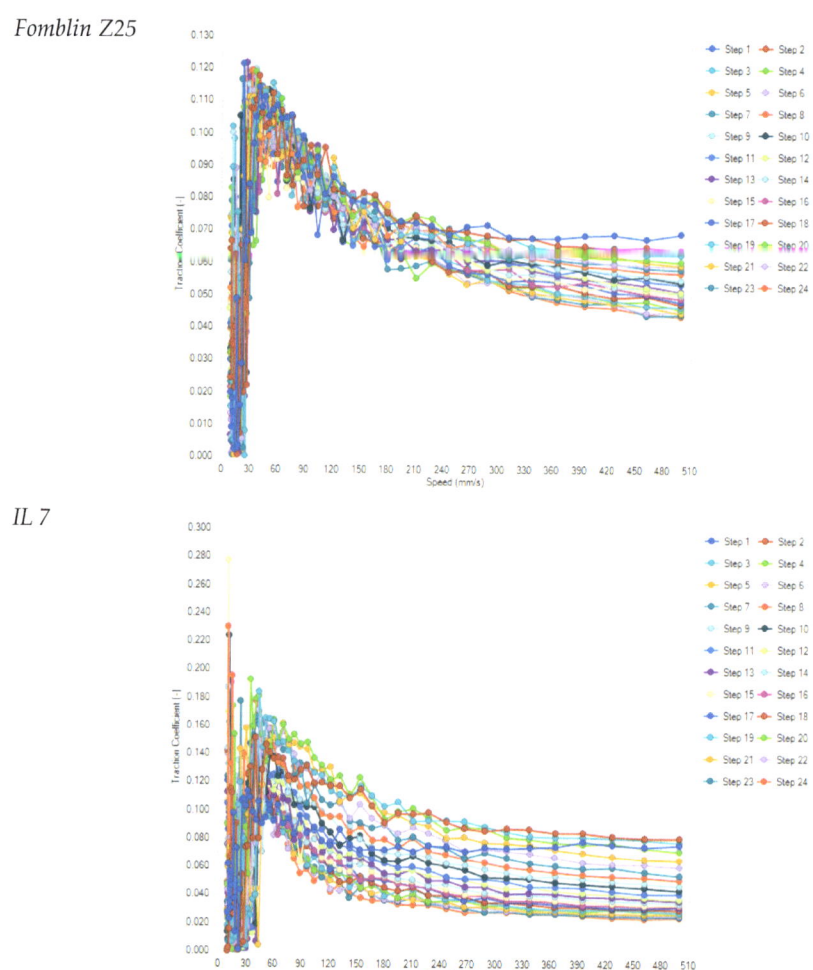

Figure 7. Comparison of Stribeck curve evolution depending on the lubricant.

All tested lubricants were able to survive the 24 repetitions of the Stribeck curves. Noticeable is the significant change in the Stribeck curve of IL7 with increasing number of repetitions, which decrease to levels below 0.040–0.015 for speeds higher than 150 mm/s (0.07 for Fomblin), which is probably due to changes in the surface structure of the ball/disc-pairing surfaces (see Figure 8). These show almost no smoothing wear for Fomblin but significant smoothing of the surface roughness for IL7, allowing a thinner specific lubrication film thickness and reduced friction.

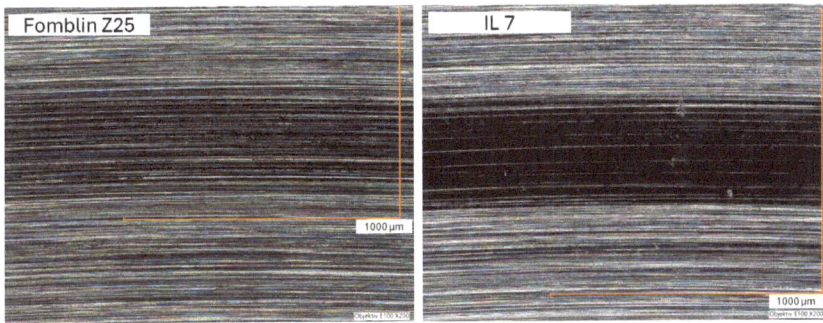

Figure 8. Comparison of wear tracks for (**left**) Fomblin Z25; (**right**) IL 7.

3.6. Summary of Ionic Liquid Assessment

The characterization of differently modified imidazolium fluoroalkylsulfonylimide ILs revealed the impact of chemical structure on different properties that are relevant for lubricant application. It could be shown that longer hydrocarbon groups at the imidazolium cation and longer fluorocarbon groups at the sulfonylimide anion led to less water absorption and thus less corrosion on 52100 steel. On the other hand, earlier degradation as well as higher melting temperatures and viscosities resulted. The least hydrophobic compound BMIM BTA showed the best behavior during tribological characterization under ambient conditions (Figure 9).

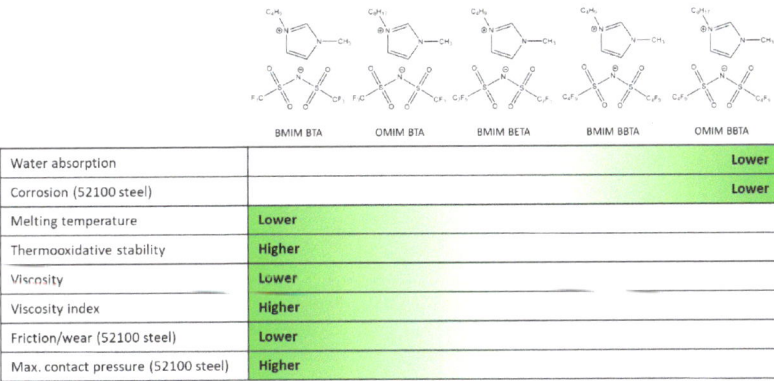

Figure 9. Observed structure–property relationships for investigated IL candidates.

4. Development of a Test Methodology for the Tribological Characterization of Ionic Liquids in Vacuum and Cleanroom Applications

The aim of the project was to develop a completely new type of component test rig that closes the gap between simple category VI model tests and component or aggregate tests (category III) [48]. For this purpose, the lubricants are tested in an interchangeable test cassette containing various design elements (Figure 10). Specifically, these are linear plain guides, linear roller guides, plain and roller bearings, and ball-screw drives. All components are placed in an easily replaceable cassette and subjected to frictional loads against each other. This test rig is an essential component of the application-oriented test strategy for new lubricants based on ionic liquids for cleanroom and vacuum applications, which is also to be developed.

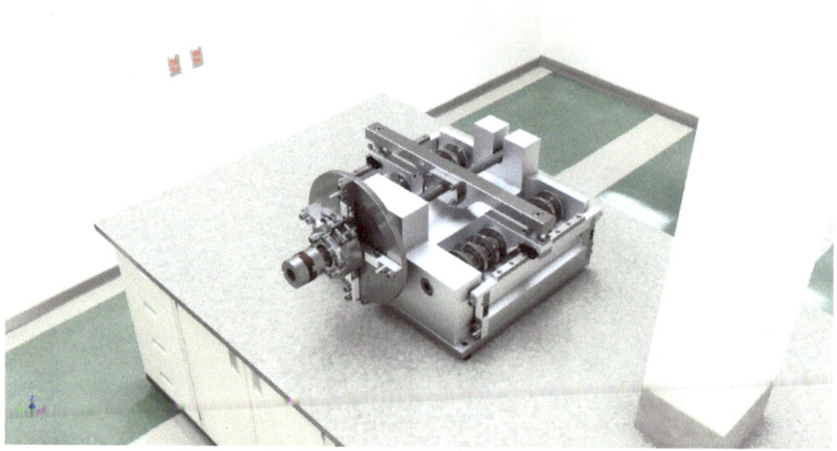

Figure 10. Test cartridge with the individually testable design elements.

The novelty of the planned test concept is that with just one powerful servo motor as the universal main drive (attached to the chamber by a magnetic coupling), the new ionic lubricants can be tested, analyzed, and evaluated simultaneously in a vacuum in the four tribological systems. The magnetic rotary coupling makes it possible to get rid of traditional mechanical rotary feedthroughs with higher leakage rates, outgassing sealing fluids, short service lives or low maximum speeds. If only individual systems are to be evaluated, the other systems required for the force flow must be replaced by sufficiently powerful reference systems (oversized, coated elements). The particularly compact design of the test bench (Figure 11) results from the conflicting technical requirements of testing as close as possible to the application (many different tribological systems) and, at the same time, efficiently in a high vacuum (10^{-3} to 10^{-6} bar), with the shortest possible evacuation times.

Figure 11. Test chamber of the new vacuum tribometer with pumps and motor (**left**: CAD model, **right**: real test rig).

The solution for determining the adhesion, sliding, and rolling friction properties (efficiency, aging condition, etc.) of different lubricant/friction bearing/material combinations in the new test bench setup (in a vacuum) consists of the defined application of force and the measurement of the reaction force and temperature development using temperature sensors (IR and/or thermocouple). The recording of the tribologically important parameters for force, speed, and temperature is carried out using the measurement techniques described below, on the basis of which the high-precision measurement and mathematical analyses of the tribological systems are made possible. In this way, the main measured variable—friction—can be analyzed and evaluated with high precision by measuring the frictional force and additionally via the temperature change. This enables a direct plausibility check of this important variable.

The force application and measurement of the reaction forces should be carried out as follows:

- Plain and roller bearings: Bracing of the bearing seats with a defined force to create a a a contact pressure of up to 1.5 GPa or by introducing a radial force through the ball screw. The reaction force on the bearing housing is measured using strain gauge sensors.
- Linear guide: Tensioning of the linear guide with a defined force to create a contact pressure of up to 1.5 GPa or by tilting the bushing. The reaction force on the guide shaft is measured using strain gauge sensors.
- Spindle drive (threaded nut/ball screw): Tensioning of the spindle nuts or ball screw with a defined force (FL). The reaction force is determined by the increase in torque on the drive motor. This is determined via a reaction force measurement on the motor.

Measuring the friction of the individual systems separately in a vacuum was a considerable design and metrological challenge. For this purpose, only the friction behavior of the spindle system is measured via the motor torque. For all other systems to be tested, reaction force measurements are used as the measuring principle to analyze the forces or torques acting directly on the component.

The numerous temperature measuring points, which record the integral friction work, help to validate the complex system. Thermocouples on each testing component record the temperature development near the friction point, and an optional infrared camera documents the heat distribution. In this way, interactions between the individual tribosystems can be visualized, and possible distortions in the measurement results can be reduced.

In a further expansion stage of the test rig, it is planned to install a camera system with which the lubricant distribution and any visual changes in the lubricant can be documented directly. In addition, the lubricant distribution and also the change in the color of the lubricant are monitored with a video inspection system. An IDS camera system with customized optics is used for this purpose, as these can be integrated directly into LabView. Interval recordings can be used to create a time-lapse video after the vacuum test bench test, which runs parallel to a progress display in the measurement diagram. In this way, changes to the tribological system, such as material transfer or wear particle generation and discharge, can be assigned to specific measured variables, and the interdependencies of the tribological system can be visualized.

Various heating and cooling scenarios can be simulated to test the operating temperature range of the lubricants and the cause-and-effect relationships resulting from constant and sudden temperature changes. The temperature is controlled (heating up to approx. 120 °C) using ceramic heater cartridges placed close to the friction contacts. The temperature distribution over the individual systems could be changed using shielding plates. This approach reflects the reality of satellite systems, for example, very well. To cool the entire system, the test cartridge contains a meander which can be supplied with liquid from an external cooling or heating unit via a fluid feed-through. Cooling is also required to dissipate the frictional heat generated during high-load tests.

Another feature of the test bench is an optional connection flange for a mass spectrometer for direct gas analysis. This allows potentially occurring outgassing to be fed directly to a mass analyzer in order to detect reaction processes of the lubricant with the construction materials or under energy supply. In addition, reactants and volatile additives in the lubricant formulation can be identified directly during use and reported to the project partner for redesign of the lubricant formulation.

A special measurement and control technology based on LabView was developed for automatic measurement and evaluation, customized for the test bench and the issues at hand (Figure 12). To ensure good signal/noise separation, high-performance DAQ modules from National Instruments, which have been tried and tested at the KTM, were used.

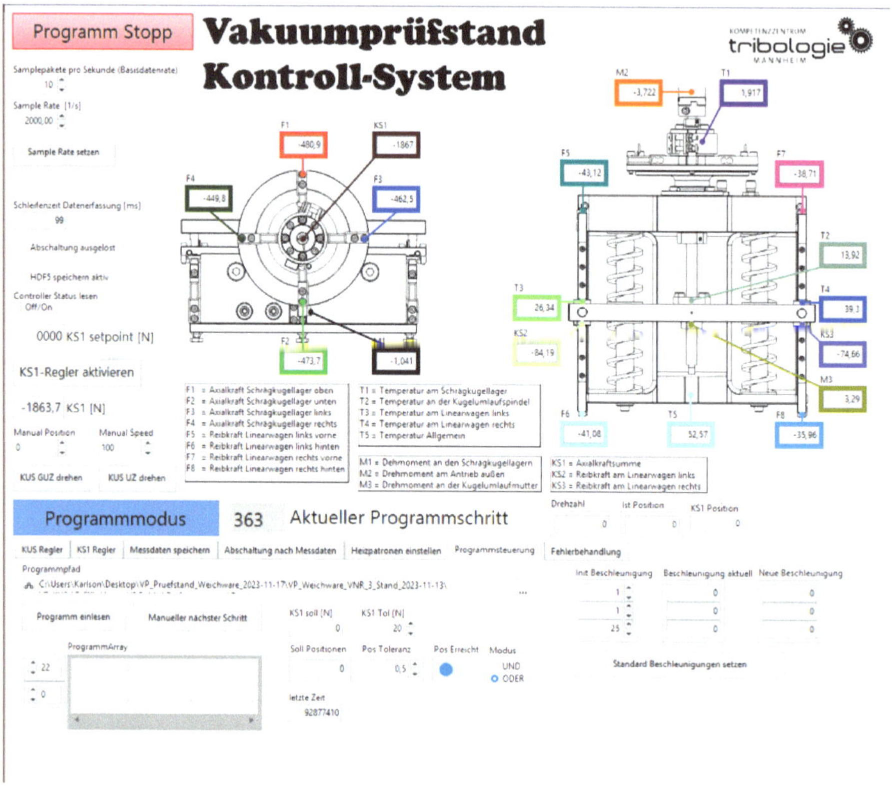

Figure 12. User-friendly (German-speaking) GUI of the new test rig (control part of the screen).

The test stand therefore meets our own high standard for modern tribometers, which we call "Tribometry 4.0" [49]. The aim of "Tribometry 4.0" is to understand tribological systems instead of just producing characteristic values. The necessary digitalization is achieved by recording numerous measured variables, using camera systems and, ultimately, AI-based analysis methods.

5. First Test Results on the New Vacuum Tribometer with Original Components

The first test results with the new vacuum component test rig are described below. In the test, a test cassette with the different design elements described above is subjected to dynamic stress in continuous operation. Specifically, two angular contact ball bearings of type B7200 (material EN 1.3505), two linear roller guides (type SSEBM10-155; stainless steel), and a ball screw (type Dold 1001-204-0300 (SFU1204), materials: Spindle EN 1.1219; nut 20CrMo) driven in a closed power flow by a servo-electric motor are located outside the vacuum chamber. The components work against coil springs during the stroke movement, which results in a dynamic load on the components. Different forces can be introduced through different strokes. The frictional forces or frictional torques and temperatures of the individual components can be measured separately. The test is completed when one component fails.

In the first test, a stroke of ± 35 mm was used, resulting in normal forces of max 2.1 kN. The test was started at room temperature and a vacuum of 3×10^{-5} mbar. The test duration was 20 h. Before the test, all components were dismantled, and the individual parts were cleaned in an ultrasonic bath using a multi-cycle cleaning procedure with boiling-point petrol and isopropanol. The grease quantities used were 200 µL per linear carriage, 200 µL for the spindle, and 100 µL per ball bearing. These very small quantities are necessary to keep contamination as low as possible.

Figure 13 shows the results of all measurement channels for the Braycote lubricant. In this run, the normal force on the ball screw was increased in four stages up to ±2.1 kN. This load increase cycle was repeated 21 times within 20 h (Figure 13, top left). The graphs at the bottom left show the temperatures of the individual components. Here, it can be seen that the temperature of the ball screw and the angular contact ball bearing are significantly higher than those of the two linear carriages. The temperature of the spindle rises again from the 16th hour, which indicates an imminent failure of this component. In contrast, the temperature of the angular contact ball bearing stabilizes, which is more indicative of normal operation. The load on the linear slide also results from the travel position. The force generates a load torque on the carriage. The friction force level on the left linear slide decreases slightly at the beginning (center, second diagram from the bottom). The friction force peaks on the right-hand linear carriage decrease slightly at the beginning, but also increase slightly again after approx. 17 h (center, bottom diagram).

Compared to the temperature measurements, the torques of the individual components and the overall drive are not as meaningful. Hardly any changes are recognizable here. The torque of the angular contact bearings (black curve, right, center) appears to decrease slightly over the test period, which correlates with the stabilization of the temperature. The torque on the spindle nut (green color) and also the total friction torque measured on the drive (orange curve) do not change.

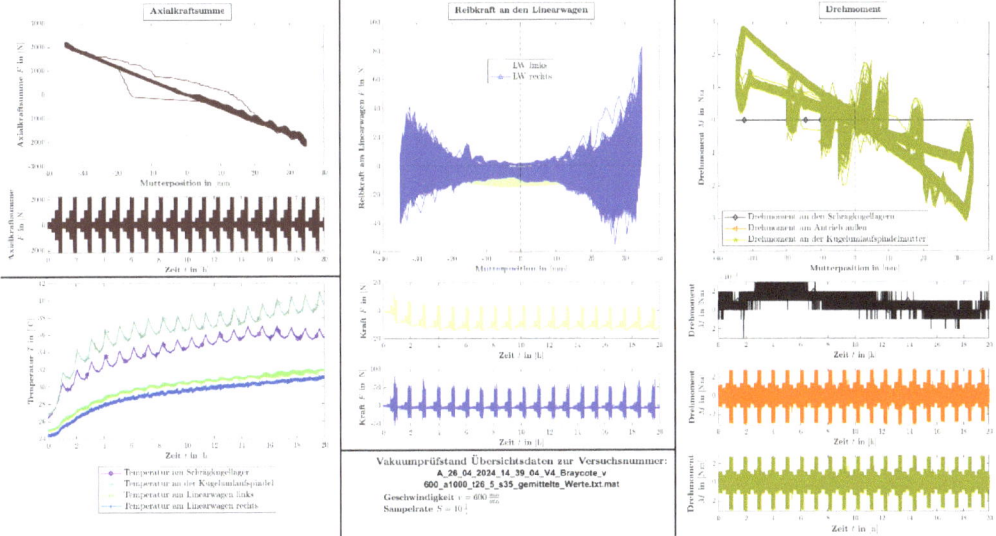

Figure 13. Overview of measurement signals of the first tests with Braycote 601 EF (Automated evaluation in German language).

The mass loss on the ball screw amounted to 55 mg at the end of the test. The mass of the two angular contact ball bearings decreased by 2.2 mg and 3.9 mg, respectively. Although the linear carriages were inconspicuous in terms of friction, the greatest mass losses were measured here at 106 mg (left) and 100 mg (right).

The test result shows that the spindle lubrication is the potential weak point. In a further test, the amount of grease on this component is now increased. If this is not effective, the load must be reduced, which can be achieved, for example, by increasing the size by one size.

6. Discussion, Conclusions, and Outlook

The testing of lubricants for applications in cleanrooms and under high vacuum (e.g., pharmaceuticals, aerospace, chip production) is a challenging task. The results of model tests (Cat. VI) are often not transferable to real applications, as the load collectives and boundary conditions in the components cannot be reproduced well enough. Complex aggregate tests (Cat. III) can only be carried out by a few specialized testing institutes and are extremely time-consuming and therefore expensive. As the volumes in this special niche segment are also very low, it is hardly possible to develop a cost-effective and yet application-oriented and individualized solution. The innovative vacuum test bench presented in this paper with a test cassette containing various mechanical design elements closes this gap.

For the first time, the tribological performance and characterization of high-performance lubricants based on ionic liquids can be documented, analyzed, and evaluated with high precision under cleanroom and vacuum conditions using a variety of measured variables on a universal component test bench (test class IV). The multifunctional design in combination with state-of-the-art measurement technology enables the desired transferability of the results to various practical applications.

Based on the new laboratory testing facilities now available, new types of ionic fluids are being investigated which represent a genuine alternative to the PFPE or MAC oils previously used in this area. In particular, the replacement of perfluoroorganic lubricants, materials, and coatings is urgently needed not only due to a potential restriction of PFAS-containing compounds in many of the above-mentioned areas, but also since end customers already expect PFAS-free solutions from suppliers. The physical–chemical and tribological tests presented here show that already the plain ionic liquids discussed here show the potential to outperform commercially available vacuum lubricants. The corrosive effects by ILs are a drawback that is investigated by employment of dedicated methods, and strategies for corrosion mitigation are currently elaborated. Overall, the blending of ILs offers an immense amount of fluid combinations for controlling the individual properties; the use of vacuum-compatible additives and the employment of rheological modifiers furthermore depict a huge toolbox for the development of tailor-made high-performance lubricants for all different kinds of lubricant applications that require outstanding material cleanliness.

Author Contributions: Conceptualization: A.K., F.S. and M.G.; methodology: A.K., M.G. and F.S.; investigation: K.-O.K., C.G. and A.E.; data curation: A.K., K.-O.K. and F.S.; writing—original draft preparation: M.G., A.K. and F.S.; writing—review and editing: M.G.; visualization: A.K. and K.-O.K.; supervision: M.G. and F.S.; project administration: M.G. and F.S.; funding acquisition: M.G. and F.S. All authors have read and agreed to the published version of the manuscript.

Funding: Funded by the German government's Central Innovation Program for SMEs (ZIM). Project number: KK5018704BR.

Data Availability Statement: The raw data supporting the conclusions of this article will be made available by the authors on request.

Conflicts of Interest: Authors Fabian Schüler, Christian Göhringer and Alexander Epp were employed by the company Materiales GmbH. The remaining authors declare that the research was conducted in the absence of any commercial or financial relationships that could be construed as a potential conflict of interest.

References

1. Fu, N.; Liu, Y.; Ma, X.; Chen, Z. EUV Lithography: State-of-the-art-review. *J. Microelectron. Manuf.* **2019**, *2*, 19020202. [CrossRef]
2. Kang, C.-W.; Fang, F.-Z. State of the art of bioimplants manufacturing: Part 1. *Adv. Manuf.* **2018**, *6*, 20–40. [CrossRef]
3. Roberts, E.W. *Space Tribology Handbook*, 5th ed.; ESR Technology Ltd.: Warrington, UK, 2013.

4. Kazuhisa, M. Considerations in vacuum tribology (adhesion, friction, wear, and solid lubrication in vacuum). *Tribol. Int.* **1999**, *32*, 605–616.
5. Fusaro, R.L.; Khonsari, M.M. *Liquid Lubrication for Space Application*; NASA Technical Memorandum 105198; NASA: Cleveland, OH, USA, 1992.
6. Lewis, S.; Buttery, M.; Poyntz-Wright, O.; Kent, A.; Vortsellas, A. Accelerated Testing of Tribological Components—Uncertainties and Solutions. In Proceedings of the 44th Aerospace Mechanisms Symposium, Cleveland, OH, USA, 16–18 May 2018; NASA: Cleveland, OH, USA, 2018.
7. Jansen, M.J.; Jones, W.R.; Preadmore, R.E.; Loewenthal, S.L. *Relative Lifetimes of Several Space Liquid Lubricants Using a Vacuum Spiral Orbit Tribometer (SOT)*; NASA Technical Memorandum 210966; NASA: Cleveland, OH, USA, 2001.
8. Pepper, S. *Spiral Orbit Tribometer—Friction and Lubricant Degradation Rate Can Be Quantified Rapidly*; NASA Tech Briefs; NASA: Cleveland, OH, USA, 2007.
9. Wolfberger, A.; Hausberger, A.; Schlögl, S.; Holynska, M. Assessment of the chemical degradation of PFPE lubricants and greases for space applications: Implications for long-term on-ground storage. *CEAS Space J.* **2021**, *13*, 377–388. [CrossRef]
10. Fukuchi, T. Degradation of perfluoropolyether fluids with metal ions. *ASME Trans.* **1999**, *121*, 348–351. [CrossRef]
11. Nyberg, E.; Schneidhofer, C.; Pisurova, L.; Dörr, N.; Minami, I. Ionic liquids as performance ingredients in space lubricants. *Molecules* **2021**, *26*, 1013. [CrossRef]
12. Kürten, D.; Khader, I.; Kailer, A. Tribochemical degradation of vacuum-stable lubricants: A comparative study between multialkylated cyclopentane and perfluoropolyether in a vacuum ball-on-disc and full-bearing tests. *Lubr. Sci.* **2020**, *32*, 183–191. [CrossRef]
13. Lu, R.; Minami, I.; Nanao, H.; Mori, S. Investigation of decomposition of hydrocarbon oil on the nascent surface of steel. *Tribol. Lett.* **2007**, *27*, 25–30. [CrossRef]
14. Minami, I. Ionic Liquids in tribology. *Molecules* **2009**, *14*, 2286–2305. [CrossRef]
15. Waheed, S.; Ahmed, A.; Abid, M.; Mufti, R.A.; Ferreira, F.; Bashir, M.N.; Shah, A.U.R.; Jafry, A.T.; Zulkifli, N.W.; Fattah, I.M. Ionic liquids as lubricants. An overview of recent developments. *J. Mol. Struct.* **2024**, *1301*, 137307. [CrossRef]
16. Liu, M.; Ni, J.; Zhang, C.; Wang, R.; Cheng, Q.; Liang, W.; Liu, Z. The application of ionic liquids in the lubrication field: Their design, mechanisms, and behaviors. *Lubricants* **2024**, *12*, 24. [CrossRef]
17. Zhou, Y.; Qu, J. Ionic liquids as lubricant additives: A review. *ACS Appl. Mater. Interfaces* **2017**, *9*, 3209–3222. [CrossRef] [PubMed]
18. Okaniwa, T.; Hayama, M. The application of ionic liquids into space lubricants. In Proceedings of the ESMATS, Noordwijk, The Netherlands, 25–27 September 2013.
19. Buttery, M.; Hampson, M.; Kent, A.; Allegranza, C. Development of advanced lubricants for space mechanisms based on ionic liquids. In Proceedings of the ESMATS, Hatfield, UK, 20–22 September 2017.
20. Dörr, N.; Merstallinger, A.; Holzbauer, R.; Pejakovic, V.; Brenner, J.; Pisarova, L.; Stelzl, J.; Frauscher, M. Five-stage selection procedure of ionic liquids for lubrication of steel-steel contacts in space mechanisms. *Tribol. Lett.* **2019**, *67*, 73. [CrossRef]
21. Grebe, M. *Tribometry—Application-Oriented Tribological Testing Technology as a Means to Successful Product Development*, 1st ed.; Expert Verlag: Tübingen, Germany, 2021; ISBN 978-3-8169-3521-6.
22. Werner, S.; Haumann, M.; Wasserscheid, P. Ionic liquids in chemical engineering. *Ann. Rev. Chem. Biomol. Eng.* **2010**, *1*, 203–220. [CrossRef] [PubMed]
23. Zhou, F.; Liang, Y.; Liu, W. Ionic liquid lubricants: Designed chemistry for engineering applications. *Chem. Soc. Rev.* **2009**, *38*, 2590–2599. [CrossRef] [PubMed]
24. Somers, A.E.; Howlett, P.C.; MacFarlane, D.R.; Forsyth, M. A review of ionic liquid lubricants. *Lubricants* **2013**, *1*, 3–21. [CrossRef]
25. Cai, M.; Lu, Q.; Liu, W.; Zhou, F. Ionic liquid lubricants: When chemistry meets tribology. *Chem. Soc. Rev.* **2020**, *49*, 7753–7818. [CrossRef] [PubMed]
26. Donato, M.T. A review on alternative lubricants: Ionic liquids as additives and deep eutectic solvents. *J. Mol. Liq.* **2021**, *333*, 116004. [CrossRef]
27. Ahrenberg, M.; Beck, M.; Neise, C.; Keßler, O.; Kragl, U.; Verevkin, S.P.; Schick, C. Vapor pressure of ionic liquids at low temperatures from AC-chip-calorimetry. *Phys. Chem. Chem. Phys.* **2016**, *18*, 21381–21390. [CrossRef]
28. Zaitsau, D.; Kabo, G.J.; Strechan, A.A.; Paulechka, Y.U.; Tschersich, A.; Verevkin, S.P.; Heintz, A. Experimental vapor pressures of 1-Alkyl-3-methylimidazolium bis(trifluoromethylsulfonyl)imides and a correlation scheme for estimation of vaporization enthalpies of Ionic Liquids. *J. Phys. Chem. A* **2006**, *110*, 7303–7306. [CrossRef]
29. Bier, M.; Dietrich, S. Vapor pressure of ionic liquids. *Mol. Phys.* **2010**, *108*, 211–214. [CrossRef]
30. Cao, Y.; Mu, T. Comprehensive investigation on the thermal stability of 66 ionic liquids by thermogravimetric analysis. *Ind. Eng. Chem. Res.* **2014**, *53*, 8651–8664. [CrossRef]
31. Fox, E.B.; Smith, L.T.; Williamson, T.K.; Kendrick, S.E. Aging effects on the properties of imidazolium-, quaternary ammonium-, pyridinium-, and pyrrolidinium-based ionic liquids used in fuel and energy production. *Energy Fuels* **2013**, *27*, 6355–6361. [CrossRef]
32. Salgado, J.; Villanueva, M.; Parajó, J.J.; Fernández, J. Long-term thermal stability of five imidazolium ionic liquids. *J. Chem. Thermodyn.* **2013**, *65*, 184–190. [CrossRef]
33. Kondo, Y. Chapter 5: Tribological Properties of Ionic Liquids. In *Ionic Liquids—New Aspects for the Future*; Intechopen: Rijeka, Croatia, 2013.

34. Street, K.W.; Morales, W.; Koch, V.R.; Valco, D.J.; Richard, R.M.; Hanks, N. Evaluation of vapor pressure and ultra-high vacuum tribological properties of ionic liquids. *Tribol. Transact.* **2011**, *54*, 911–919. [CrossRef]
35. Morales, W.; Street, K.W.; Richard, R.M.; Valco, D.J. Tribological testing and thermal analysis of an alkyl sulfate series of ionic liquids for use as aerospace lubricants. *Tribol. Transact.* **2012**, *55*, 815–821. [CrossRef]
36. Totolin, V.; Conte, M.; Berriozabal, E.; Pagano, F.; Minami, I.; Dörr, N.; Brenner, J.; Igartua, A. Tribological investigations of ionic liquids in ultra-high vacuum environment. *Lubr. Sci.* **2014**, *26*, 514–524. [CrossRef]
37. Buttery, M. Spiral Orbit Tribometer; Assessment of space lubricants. In Proceedings of the ESMATS, Vienna, Austria, 23–25 September 2009.
38. Buttery, M.; Cropper, M.; Wardzinski, B.; Lewis, S.; McLaren, S.; Kreuser, J. Recent observations on the performance of hybrid ceramic tribocontacts. In Proceedings of the ESMATS, Bilbao, Spain, 23–25 September 2015.
39. *DIN 51834-1*; Testing of Lubricants—General Principles for the Tribological Testing of Lubricants Using a Linear-Oscillation Test Machine. Beuth Verlag: Berlin, Germany, 2010.
40. *DIN 51834-2*; Testing of Lubricants—Tribological Test in the Translatory Oscillation Apparatus—Part 2: Determination of Friction and Wear Data for Lubricating Oils. Beuth Verlag: Berlin, Germany, 2017.
41. *ASTM D7421-19*; Standard Test Method for Determining Extreme Pressure Properties of Lubricating Oils Using High-Frequency, Linear-Oscillation (SRV) Test Machine. ASTM International: West Conshohocken, PA, USA, 2019.
42. Uerdingen, M.; Treber, C.; Balser, M.; Schmitt, G.; Werner, C. Corrosion behavior of ionic liquids. *Green Chem.* **2005**, *7*, 321–325. [CrossRef]
43. Arenas, M.F.; Reddy, M.G. Corrosion of steel in ionic liquids. *J. Min. Met.* **2003**, *39*, 81–91. [CrossRef]
44. Dilasari, B.; Jung, Y.; Sohn, J.; Kim, S.; Kwon, K. Review on corrosion behavior of metallic materials in room temperature ionic liquids. *Int. J. Electrochem. Sci.* **2016**, *11*, 1482–1495. [CrossRef]
45. ESA. *Space Engineering—Mechanisms*; ECSS-E-ST-33-01C Rev.2; ESA-ESTEC R&SD: Noordwijk, The Netherlands, 2019.
46. *DIN 51810-1*; Prüfung von Schmierstoffen—Prüfung der Rheologischen Eigenschaften von Schmierfetten—Teil 1: Bestimmung der Scherviskosität mit dem Rotationsviskosimeter und dem Messsystem Kegel/Platte. Beuth Verlag: Berlin, Germany, 2017.
47. *ASTM D2270-10*; Standard Practice for Calculating Viscosity Index from Kinematic Viscosity at 40 and 100 °C. ASTM International: West Conshohocken, PA, USA, 2010.
48. *GfT Worksheet No. 7: Tribology—Wear, Friction—Definitions, Terms, Testing*; Society for Tribology: Jülich, Germany, 2009.
49. Grebe, M. Tribology 4.0: From Da Vinci to digitalization. *Lube Mag.* **2023**, *178*, 15–17.

Disclaimer/Publisher's Note: The statements, opinions and data contained in all publications are solely those of the individual author(s) and contributor(s) and not of MDPI and/or the editor(s). MDPI and/or the editor(s) disclaim responsibility for any injury to people or property resulting from any ideas, methods, instructions or products referred to in the content.

Article

Improved Tribological Performance of a Polybutylene Terephthalate Hybrid Composite by Adding a Siloxane-Based Internal Lubricant

Shengqin Zhao [1], Rolf Merz [2], Stefan Emrich [2,*], Johannes L'huillier [2] and Leyu Lin [1,*]

[1] Chair of Composite Engineering, Rheinland-Pfälzische Technische Universität (RPTU) Kaiserslautern-Landau, 67663 Kaiserslautern, Germany; shengqin.zhao@mv.rptu.de

[2] Institute for Surface and Thin Film Technology (IFOS), 67663 Kaiserslautern, Germany; merz@ifos.uni-kl.de (R.M.); lhuillier@ifos.uni-kl.de (J.L.)

* Correspondence: emrich@ifos.uni-kl.de (S.E.); leyu.lin@mv.rptu.de (L.L.)

Abstract: To mitigate the environmental hazards aroused by fossil-based lubricants, the development of eco-friendly internal lubricants is imperative. Siloxane-based internal lubricants, widely applied as plasticizers in polymeric compounds, are a promising option. However, their impacts on the tribological properties of polymeric tribocomponents are still unclarified. Therefore, in the current study, a siloxane-based internal lubricant with the product name 'EverGlide MB 1550 (EG)' was dispersed into a polybutylene terephthalate (PBT)-based tribological composite to investigate whether the tribological properties of the composite can be optimized. A block-on-ring (BOR) test configuration was used for this purpose. It was found that the addition of EG to the composite significantly improved the tribological behavior; the improvement was particularly significant under lower load conditions (pv-product $\leq$ 2 MPa·m/s). Compared to the reference PBT composite, the addition of EG reduced the friction coefficient (COF) by about 30% and the specific wear rate by about 14%. An accompanying surface analytical investigation using photoelectron spectroscopy to elucidate the effective mechanisms at the molecular level showed the availability of tribologically effective and free EG after its addition to the composite in the relevant tribocontact.

Keywords: PBT; siloxane; lubricant; tribological properties; transfer film

Citation: Zhao, S.; Merz, R.; Emrich, S.; L'huillier, J.; Lin, L. Improved Tribological Performance of a Polybutylene Terephthalate Hybrid Composite by Adding a Siloxane-Based Internal Lubricant. *Lubricants* 2024, *12*, 189. https://doi.org/10.3390/lubricants12060189

Received: 16 April 2024
Revised: 21 May 2024
Accepted: 24 May 2024
Published: 28 May 2024

Copyright: © 2024 by the authors. Licensee MDPI, Basel, Switzerland. This article is an open access article distributed under the terms and conditions of the Creative Commons Attribution (CC BY) license (https://creativecommons.org/licenses/by/4.0/).

1. Introduction

With the thriving progress of aerospace, automotive, medical, and electronics technology, the demand for lightweight and durable wear-resistant materials is growing rapidly. Polymeric materials are lightweight, economical, and non-toxic, and they possess superior strength-to-weight ratios, corrosion resistance, and insulation, which are regarded as excellent alternatives to metallic materials [1–6]. In view of the lower processing temperature, as well as the flexible shape manufacturing of polymeric materials, polymeric materials had numerous advantages of production, such as lower energy consumption and costs than metallic materials, which has been widely favored in the industrial field [7,8]. Some polymeric materials demonstrate low friction noise, self-lubrication, and transfer film-forming mechanisms, which are considerable prospects to be adopted in the tribological industrial field, such as to gears, bearings, and seals [9,10]. However, the neat polymeric matrix generally presents inferior mechanical properties and load-bearing properties. The exhibition of their advantages on tribological performance relies heavily on the cooperation with reinforcing fillers [11]. For example, the fibers could be conferred superior tribological properties to make them suitable for more severe environments, which is regarded as one of the most important reinforcers for polymer materials [12,13].

As one of the most universally used engineering plastics, PBT offers excellent dimensional stability, desirable stiffness and strength, and superior heat aging behavior. It is widely utilized in industrial applications, such as in gears [14,15]. Nevertheless, PBT-based tribocomposites produced according to the traditional formulation still exhibit drawbacks of severe friction and intense wear, which cannot meet operation qualifications under some extreme/dangerous environments. Thus, to further improve the tribological performance and prolong the lifetime of the PBT-based tribological components, the proposal for an appropriate lubrication strategy is urged [16–18].

Traditional external lubricants utilized in metallic materials, i.e., mineral oil and mineral lubricating grease, can induce the swelling of plastic components, deteriorating their mechanical and tribological properties [19,20]. By means of this, only when choosing compatible lubricants can the performance and lifetime of plastic tribological components be effectively improved [21,22]. In actuality, internal lubricants, such as graphite-, Teflon-, and petroleum-based lubricants, have been extensively utilized in polymeric composites [23–25]. However, the production of these lubricants is heavily reliant on fossil fuels, and their application generates significant environmental pollution [26–28]. Therefore, it is imperative to explore eco-friendly alternatives to promote product innovation and industrial upgrading for traditional internal lubricants. Compared to petroleum products, siloxane internal lubricants do not contain harmful substances, such as heavy metals and polycyclic aromatic hydrocarbons. Except for elevating the tribological performance, the addition of siloxane has also proved to enhance the processability and dimensional stability of plastic tribological components, which possess intensive potential to be utilized as multifunctional additives in eco-friendly tribology components [29,30]. However, to date, there are few studies on the possibility of applying siloxane as a plasticizer/lubricant multifunctional additive, especially for PBT-based tribocomponents. The impacts of siloxane on the mechanical properties, tribological properties, and interface contribution are still unclarified. Additionally, the utilization of the combination between siloxane and the varying length scale fillers on the tribological performance of polymeric materials still exists in technical vacancy.

In this research, a siloxane-containing polymeric tribological composite was manufactured by applying a polybutylene terephthalate (PBT) matrix. The impacts of the siloxane internal lubricant on the mechanical and tribological properties of PBT composites were investigated. Accordingly, to demonstrate the potential tribological mechanisms within the introduction of siloxane, elemental compositions for the tribological transfer films of the composites were characterized by X-ray photoelectron spectroscopy (XPS). It is anticipated to provide insights into the plasticizing/lubricating effects induced by siloxane to polymeric materials, elucidate the potential mechanism of siloxane on the friction interface, and broaden the utilization of PBT-based tribocomponents in industry.

2. Experimental

2.1. Materials and Composites Manufacturing

PBT with the trade name Ultradur B 4520 was provided by BASF SE, Ludwigshafen, Germany. Short carbon fiber (C C6-4.0/240-T190) and graphite flake (RGC 39A) were kindly supported by SGL Carbon SE, Augsburg, Germany, and Superior Graphite Europe, Sundsvall, Sweden, respectively. Siloxane internal lubricant (50% siloxane dispersed in a proprietary polyethylene terephthalate resin, EverGlide (EG) MB1550, Polymer Dynamix) was kindly donated by Lehmann & Voss & Co., KG, Hamburg, Germany. In addition, different particles, i.e., submicron-sized zinc sulfide (Sachtolith HD-S, Venator Materials, Duisburg, Germany) and titanium dioxide (Kronos 2310, Kronos International, Inc., Leverkusen, Germany), as well as nano-sized silica (Aerosil R 9200, Evonik Industries, Hanau, Germany), were used as friction and wear reduction fillers within the PBT composites, which are all commercially available products. In this study, two kinds of PBT-based tribological composites were designed, namely PBT-TC and PBT-TC-EG. PBT-TC consists of short carbon fiber, graphite, zinc sulfide, titanium dioxide, and nanosilica. To study the effects of the siloxane on the friction and wear performance of this composite, EverGlide material was added to this composite, which is named PBT-TC-EG.

The designed composites were melting-compounded with the fillers by using a co-rotated twin-screw extruder (Leistritz ZSE 18 MAXX—40 D, Leistritz, Nürnberg, Germany) at a rotation speed of 200 rpm. The temperature from the feeder to the nozzle was selected between 120 and 260 °C. After extrusion, an ENGEL injection-molding machine (victory 200/80 spex, ENGEL, Schwertberg, Austria) was applied to injection-mold the composites into plates with a geometry of 50 mm × 50 mm × 4 mm, from which samples for mechanical and tribological investigations were prepared. At least five samples were tested for both characterizations in order to calculate the mean values and the standard deviations.

2.2. Mechanical Investigations

The tensile properties of PBT-TC and PBT-TC-EG were tested at room temperature (23 ± 1 °C) on a universal testing machine (RetroLine, Zwick GmbH & Co., KG, Ulm, Germany) according to DIN EN ISO 527-2:2012 within a dumbbell shape of type 1BB [31]. The measurements followed DIN EN ISO 527 using dumbbell-shaped specimens. The Young's modulus of samples was measured by the crosshead within a speed of 1 mm/min. The tensile strength and elongation at the break of samples were determined with a tensile speed of 50 mm/min.

2.3. Tribological Studies

A block-on-ring (BOR) tribometer was adopted to evaluate the tribological properties of the composites with a sample geometry of 4 mm × 4 mm × 10 mm (apparent contact area: 4 mm × 4 mm) under dry sliding conditions at room temperature according to standard ASTM G77-17 [32]. The sliding counterbody was the standard bearing inner ring (100Cr6, INA, Schweinfurt, Germany) with a diameter of 60 mm outside, 5 mm thickness, and a surface roughness R_a = 0.2 ± 0.04 µm. The 2D force transducer was utilized to define and record the normal and frictional forces during tests. Overall, 5 input load conditions, 1 MPa and 0.5 m/s, 1 MPa and 1 m/s, 2 MPa and 1 m/s, 3 MPa and 1.5 m/s, and 4 MPa and 2 m/s were selected to comprehensively analyze the tribological performance of the composites. Steady-state coefficients of friction (COF) and specific wear rates were calculated based on at least three measurements under each load condition. The specific wear rate (w_s, mm^3/Nm) was determined by the following equation:

$$w_s = \Delta h / (p \times s) \quad (1)$$

where Δh is the height loss measured by using a displacement transducer during the sliding test, p is the pressure, and s represents the sliding distance.

2.4. Chemical Characterization

Chemical characterization of the transfer film was carried out by X-ray photoelectron spectroscopy (XPS), using an Axis Nova small spot electron spectrometer (Kratos Analytical Ltd., Manchester, UK), equipped with 165 mm radius hemispherical analyzer, combined electrostatic and magnetic lenses, DLD detector and monochromatic Al K$^\alpha$ (1486.6 eV) X-ray source, working at an operating pressure lower 10^{-8} mbar. On each sample, three elongated measuring spots, with a size of 350 × 700 µm^2, were selected in the area of tribological loading and one outside. The concentrations given represent average values of the individual measurements. The detector axis was oriented parallel to the normal surface of the sample. Elemental concentrations were calculated from survey spectra (electron pass energy: 160 eV) using standard sensitivity factors given by Kratos Ltd. and Shirley shape fits for background subtraction. The binding states of carbon, oxygen, and silicon are characterized by C1s, O1s, and Si2p detailed core level spectra at an electron pass energy E_{pass} = 20 eV. Possible charging artifacts were compensated by shifting the binding energy scale with respect to a true C1s binding energy of 285 eV for aliphatic hydrocarbons. The information depth of this analysis is element- and material-dependent, smaller than 3–5 nm.

Chemical state analysis after subtraction of a Shirley-type background was performed by inscribing Gauss–Lorentz shaped partial peaks, localized at typical chemical shift, given by the NIST database [33].

3. Results and Discussion

3.1. Mechanical Properties

The impacts of the introduction of EG on the mechanical properties of the PBT-based material are illustrated in Figure 1. In comparison to the reference PBT-TC material, the incorporation of EG leads to a significant reduction in Young's modulus and ultimate tensile strength. However, a notable elevation of the elongation at break for the PBT-TC-EG is observed. This can be explained by the plasticizing effect of siloxane [34,35]. According to the research by Arzhakov et al., low-thermodynamic-affinity siloxane tends to localize at the local structural regions of the boundaries between supramolecular or suprasegmental structures for polymethyl methacrylate (PMMA), changing the mobility of the macromolecules or its segments in this region; then, the mechanical behavior of PMMA is affected [36]. In their work, the elastic modulus and the yield stress behaved sharply, cutting back with the high load of the siloxane. In addition, the flexible Si-O-Si bondage enriched in siloxane also proved to contribute to the ductility enhancement of the polymer matrix [37]. Hence, the high siloxane load in this research brings about a reduced modulus, raised tensile strain, and improved processability for the PBT matrix.

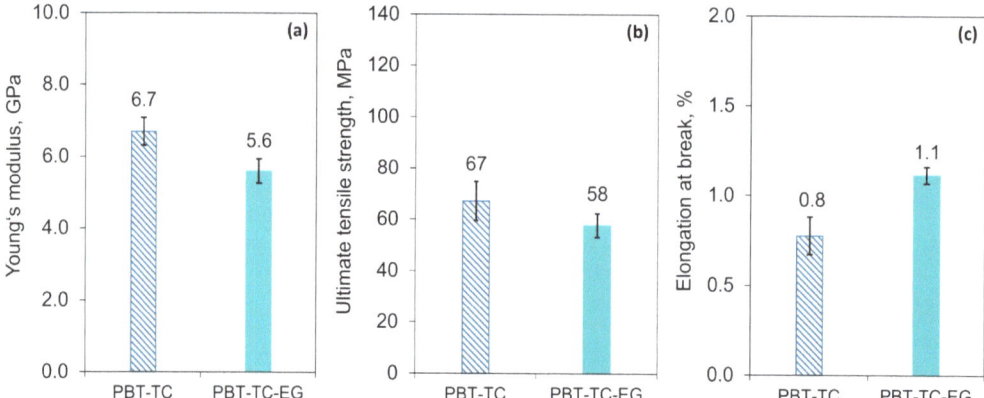

Figure 1. Young's modulus (**a**), ultimate tensile strength (**b**), and elongation at break (**c**) of the composites.

3.2. Tribological Properties

The COFs of PBT-TC and PBT-TC-EG under different load conditions are elucidated in Figure 2a. With the increment of input load conditions, the COFs of the materials undergo a maximum, which is independent of the composite's compositions. With respect to the influence of the siloxane addition on the tribological properties of PBT-TC, it is of great interest to observe that, except for extremely high load conditions (3 MPa and 1.5 m/s, 4 MPa and 2 m/s), the addition of EG could effectively reduce the COFs of the tribologically modified PBT-TC. Compared to PBT-TC, the COF reduction rates of PBT-TC-EG are 25% at 1 MPa and 0.5 m/s, 21.4% at 1 MPa and 1 m/s, and 30.6% at 2 MPa and 1 m/s, respectively. It is presumed that the introduction of EG could enhance the lubrication property of formed transfer films, which bring about lower friction forces during sliding under relatively low load conditions. However, once conducted under relatively high input load conditions (3 MPa and 1.5 m/s, 4 MPa and 2 m/s), the PBT-TC-EG presented higher real contact area between counterbody than PBT-TC as the softening due to EG addition, demonstrating higher COFs than the PBT-TC material.

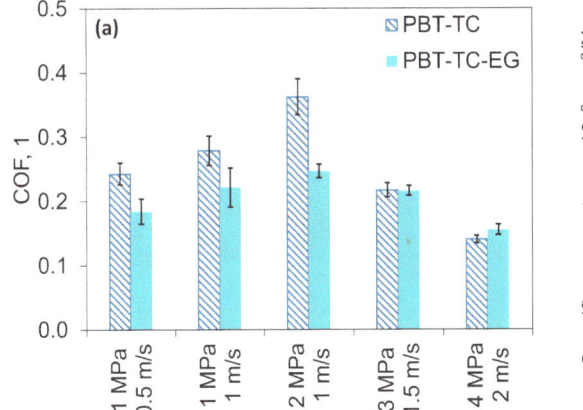
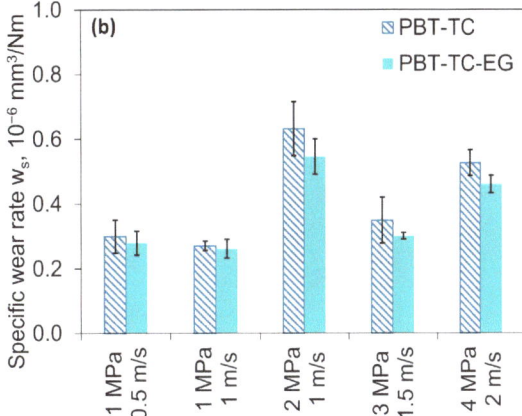

Figure 2. COFs (**a**) and specific wear rates (**b**) of the composites.

The lubrication property of the EG material can also be evidenced by the specific wear rates, as demonstrated in Figure 2b. Due to the introduction of EG, the specific wear rate of PBT-TC-EG slightly decreases under the load conditions studied. In comparison with PBT-TC, the mean specific wear rate of PBT-TC-EG declines to 6.6% at 1 MPa and 0.5 m/s, 3.7% at 1 MPa and 1 m/s, 12.7% at 2 MPa and 1 m/s, 14.3% at 3 MPa and 1.5 m/s, and 13.2% at 4 MPa and 2 m/s, respectively. Although PBT-TC-EG behaves in higher real contact areas under high load conditions, its EG-contained lubricative transfer film could effectively reduce the wear of the material, contributing to a relatively lower wear. Considering the tribological results, reasonable speculation for the lubricating mechanism of EG can be summarized as follows: During the sliding process of PBT-TC-EG, EG was released from the matrix, which was further mixed under mechanical force with the decomposed debris and compressed on the counterbody to form lubricative transfer films. The enriched flexible Si-O in EG is the smooth and soft segment that demonstrated a larger bond angle and bond length than C-O and then possessed ultra-low surface energy and was easy to slide. As a benefit of this, the lubricative transfer film containing EG can continuously mitigate the severe friction of PBT-TC-EG materials. Even under a similar load condition, the addition of siloxane can bestow comparable wear properties of PBT-TC-EG than the extraordinary wear-resistant polyether ether ketone or polyimide tribocomposites [38].

3.3. Chemical Characterizations

3.3.1. Elemental Concentrations of Composite Pins

To further elucidate the lubrication mechanisms of EG, the chemical characteristics of worn surfaces on the composite pins were analyzed by XPS survey spectra. The elemental concentrations of the samples are shown in Figure 3. Predominant elements on the surface are carbon and oxygen, as expected for polybutylene terephthalate (PBT), its carbon fiber, or graphite admixtures, respectively. However, it should be noted that the dominance of C and O is not a unique feature of the compound material. Practically all samples under atmospheric conditions are covered by a thin but omnipresent hydrocarbon adsorbate film ('adventitious hydrocarbon') [39].

Therefore, trace elements such as Si, Ti, Zn, Ca, and N found in much smaller proportions are more interesting for clarifying the relevant mechanisms. The occurrence of Si, Ti, and Zn is in accordance with known ingredients ZnS, TiO_2, and SiO_2 of the composites. Other detected elements, like Ca and N, are included in widespread fillers of plastic material. So, the remaining most significant difference between fresh PBT-TC and PBT-TC-EG pins is a higher silicon concentration of PBT-TC-EG.

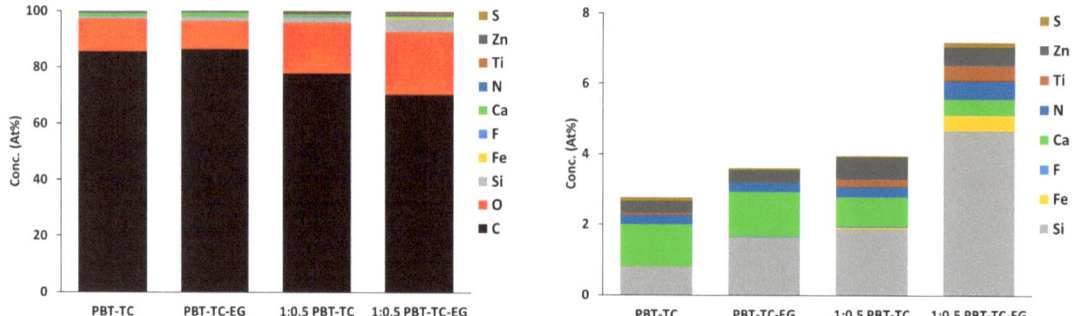

Figure 3. Elemental concentrations (at%) on unloaded and loaded composite pins after a tribological load of pv-collective 1 MPa and 0.5 m/s.

After tribological load, the detected carbon concentration decreases, and oxygen increases on the worn surface of the pin. This observation is consistent with various expected effects like enhanced oxidation processes, surface enlargement due to roughening and scoring, or removal of the adventitious hydrocarbon layer, covering the PBT bulk material with higher oxygen content. This also explains the increased concentrations of Si, Zn, and Ti at the worn surfaces of the pin and suggests their availability on the surface, with possible consequences for the tribological behavior. Additionally, transfer material from the ring counter body is detected on the worn composite pin, as evidenced by iron, and iron oxide, which is particularly significant with PBT-TC-EG. During the sliding friction process between the fiber-reinforced polymeric materials with metallic counterbody, direct contact between the end of the fiber and the metal ring engenders high flash temperature, which brings about the tribo-oxidation of the metal on the counterbody surface and leads to the formation of metal oxides on the counterface [40]. Via continuous mechanical mixing, the metal oxides can be peeled by the nanosilica and released in the interface [41]. These dissociated metal oxides could be pressed under force and embedded in the polymer matrix on the worn surfaces. As proved above, the PBT-TC-EG material possesses less modulus, indicating that it is easy to be embedded by the metal oxide, which leads to a higher iron concentration on the pin-worn surface.

But particularly striking is the significant increase in a detected silicon concentration on the friction surface for both cases (PBT-TC and PBT-TC-EG). At the same time, it is probably caused by SiO_2 in the case of PBT-TC. However, the silicon concentration on the worn surfaces of PBT-TC-EG is much more pronounced than that of PBT-TC materials. This implies the successful release of siloxane into the counterface, which may be a first hint for the availability of free, tribologically effective siloxane operated as a lubricant.

3.3.2. Elemental Concentrations of the Ring Samples

Figure 4 shows the elemental concentrations at the surface of the test rings, outside and inside the zone of tribological load. Dominant elements are carbon and oxygen, in accordance with the assumption of transfer material from the composite pins, although typical for the presence of an adventitious hydrocarbon adsorbate film, as discussed in the case of the pins.

Looking at the other elements, an almost identical element repertoire is observable like Si, F, Ca, N, Ti, Zn, and S, like in the case of the composite pins, which supports the assumption of pin transfer material. Remarkable is the detection of these elements even outside the friction track. It shows that transfer material and wear particles are not only deposited inside the zone of tribological load but also widespread in the direct surrounding.

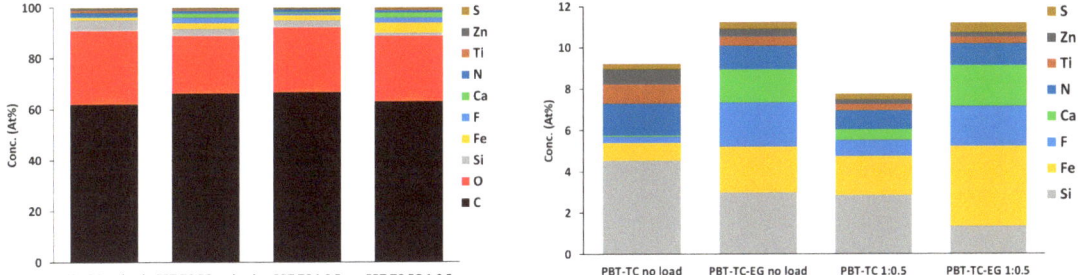

Figure 4. Elemental concentrations (at%) of rings outside (no load) and inside a zone of tribological load at a pv-collective of 1 MPa and 0.5 m/s.

A counterintuitive development is observed for silicon since the silicon concentration on the steel sample, in contrast to the composite sample, appears to decrease because of tribological load. Even significantly less silicon, as well as more iron, is detected on PBT-TC-EG wear tracks than on PBT-TC. A possible explanation for this is that siloxane is consumed during the process. At the composite pin, due to tribological load, continuously fresh siloxane is uncovered; at the ring, siloxane is consumed and has to be resupplied.

The assumption of siloxane consumption, even at this rather modest load (1 MPa and 0.5 m/s), may be supported by the observation that in the case of PBT-TC-EG, more iron is exposed, which is detectable by XPS. On the other hand, even in low load scenarios (1 MPa:0.5 m/s), it seems to be apparent that adding EG results in a softening and thus larger true contact area. This effect was already observed at high load conditions and resulted in a slightly higher friction coefficient.

When considering the total Si concentration of the corresponding composite pins, it can be noted that the available Si concentration PBT-TC-EG is much higher than that of PBT-TC (Cf. Figures 3 and 4), which correlates with the better tribological performance of PBT-TC-EG.

3.3.3. Chemical Binding Situation at Pins and Rings in Comparison

The C1s detail spectra in Figure 5 and resulting concentrations in Figure 6, resolving the chemical bond situation, show on pins and rings typical bonds for carbon, bound in aliphatic and aromatic hydrocarbons, without and with one, respectively, two oxygen atoms as the nearest neighbor of the carbon. These findings are in accordance with the presence of polybutylene terephthalate on both types of samples, but as discussed earlier, also for the presence of an adventitious hydrocarbon adsorption film. Interestingly, the proportion of hydrocarbon on the rings shows even higher oxygen participation than on the pin material. It identified the interpretation that the transferred polymer matrix passes through the decomposition/oxidation process via abrasion during sliding friction.

The O1s detail spectra (Figure 5) suffer from a superposition of electron binding energies. Unfortunately, oxygen* in the ester group of PBT (-C-(C=O*)-O-), as well as silicon dioxide (SiO_2) and siloxanes (-O-Si-O-), show partial peaks at nearly the same electron binding energy of about E_B = 532.3 eV and cannot be distinguished in O1s details.

The Si2p detail spectra (Figure 5) also provide little help in distinguishing between SiO_2 filling material and siloxane lubricant (-O-Si-O-) since both silicon atoms have two oxygen atoms as direct binding partners, resulting in a nearly identical electron binding energy E_B = 103.5 eV. A second contribution at E_B = 102 eV is in accordance with silicon bound as hexamethyl siloxane $Si(CH_3)_3$-O-$Si(CH_3)_3$ or silanol -Si(OH)-, compounds with only one oxygen atom as the binding partner. In view of the higher silanol -Si(OH)- concentration of the pins for PBT-TC-EG (1:0.5 PBT-TC-EG) than that of the pins for PBT-TC (1:0.5 PBT-TC), it can be interpreted the silanol -Si(OH)- is the main effective segments of siloxane. In this case, the higher silanol -Si(OH)- concentration on the ring of 1:0.5

PBT-TC-EG verified the effective release of siloxane from the matrix to the counterface during sliding friction.

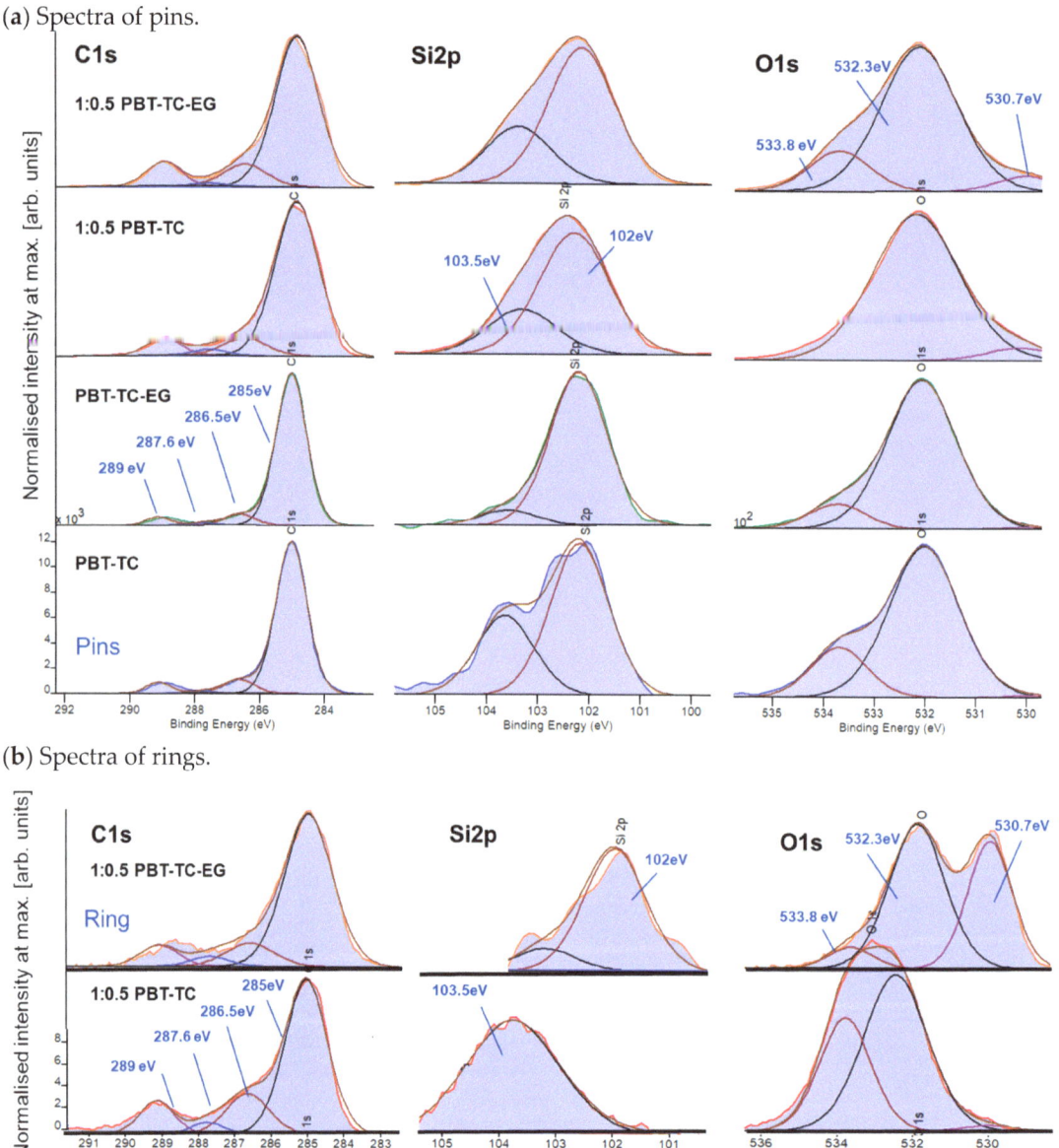

Figure 5. Detailed spectra (at%) of (**a**) pins at fresh and rubbed-off areas, respectively, and (**b**) rings inside a zone of tribological load (1:0.5) at a pv-collective of 1 MPa and 0.5 m/s.

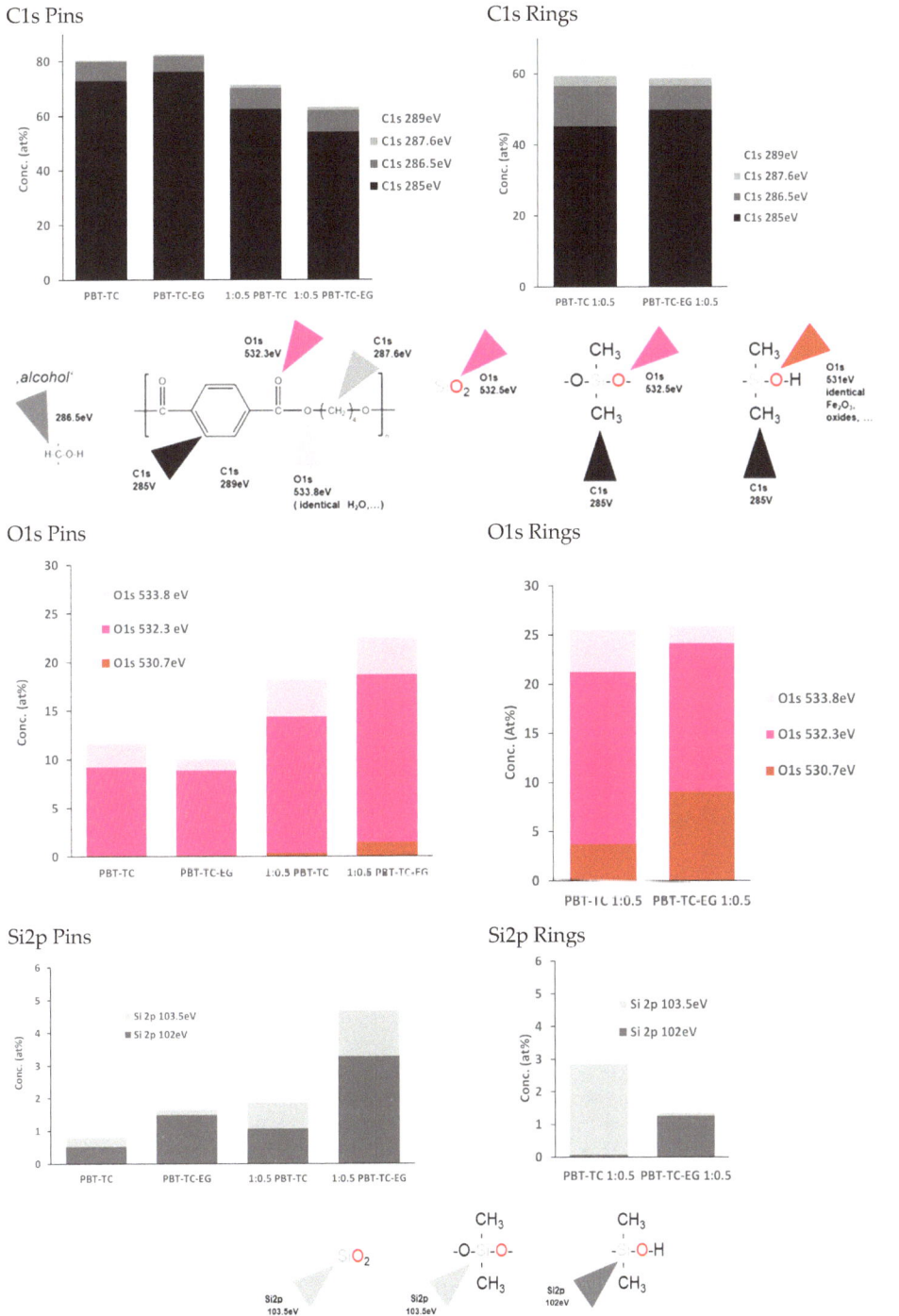

Figure 6. Carbon, oxygen, and silicon elemental concentrations (at%) of pins at fresh and rubbed-off areas and rings, broken down by their bonding situation.

4. Conclusions

The present study clarifies the impacts of EG on the mechanical and tribological properties of a PBT-based polymeric composite. The lubricating effect of EG, along with its potential mechanism, is further verified via XPS analysis. The introduction of EG dramatically elevates the elongation at break with the sacrifice of Young's modulus and ultimate tensile strength, thereby evidencing the plasticizing effect of EG on the PBT matrix. EG effectively improves the tribological properties of PBT-based materials, in particular under low load conditions (pv-product $\leq$ 2 MPa·m/s). The addition of EG can cut back 21.4–30.6% of COF and 3.7–14.3% specific wear rate of the PBT-based composite studied. The XPS findings indicate the dominant lubricating mechanism, which is the higher content of silicon-contained materials at the contact interface, which leads to a better lubrication of the mating pair. According to tribological/chemical investigations, the EG lubricating mechanism has been revealed. During the sliding friction process, the EG was released from the abrasion of the PBT-TC-EG matrix. With mechanical mixing and compressing, the EG-contained transfer films were formed. The flexible Si-O bond enriched in EG conferred the transfer film's impressive lubricating properties, which continuously serve to improve the tribological performance of PBT-TC-EG. The findings of this work can contribute to the development of siloxane-based high-performance tribological materials.

Author Contributions: Formal analysis, S.Z. and R.M.; Investigation, S.Z. and R.M.; Resources, J.L.; Writing—original draft, S.Z. and R.M.; Writing—review & editing, S.E. and L.L.; Visualization, S.Z. and R.M.; Supervision, J.L. and L.L.; Project administration, S.E. and L.L.; Funding acquisition, J.L. and L.L. All authors have read and agreed to the published version of the manuscript.

Funding: This research was funded by the Deutsche Forschungsgemeinschaft (DFG) for providing financial support to carry out the research work under the research grants project number: 508931230 and 499376717 and the State Research Center OPTIMAS at RPTU Kaiserslautern-Landau.

Data Availability Statement: The raw data supporting the conclusions of this article will be made available by the authors on request.

Acknowledgments: All the authors thank for BASF SE, Evonik Industries, Lehmann & Voss & Co., KG, Kronos International, Inc., SGL Carbon SE, Venator Materials, Germany, and Superior Graphite Europe, Sweden, for providing the experimental materials.

Conflicts of Interest: The authors declare no conflict of interest.

References

1. Dorri Moghadam, A.; Omrani, E.; Menezes, P.L.; Rohatgi, P.K. Mechanical and tribological properties of self-lubricating metal matrix nanocomposites reinforced by carbon nanotubes (CNTs) and graphene—A review. *Compos. Part B Eng.* **2015**, *77*, 402–420. [CrossRef]
2. Friedrich, K. Polymer composites for tribological applications. *Adv. Ind. Eng. Polym. Res.* **2018**, *1*, 3–39. [CrossRef]
3. Cooper, B.G.; Catalina Bordeianu Nazarian, A.; Snyder, B.D.; Grinstaff, M.W. Active agents, biomaterials, and technologies to improve biolubrication and strengthen soft tissues. *Biomaterials* **2018**, *181*, 210–226. [CrossRef]
4. Liu, H.; Su, X.; Tao, J.; Fu, R.; You, C.; Chen, X. Effect of SiO_2 nanoparticles-decorated SCF on mechanical and tribological properties of cenosphere/SCF/PEEK composites. *J. Appl. Polym. Sci.* **2020**, *137*. [CrossRef]
5. Yang, Z.; Yang, Y.; Wang, H.; Liu, F.; Lu, Y.; Ji, L.; Wang, Z.L.; Cheng, J. Charge Pumping for Sliding-mode Triboelectric Nanogenerator with Voltage Stabilization and Boosted Current. *Adv. Energy Mater.* **2021**, *11*, 2101147. [CrossRef]
6. Zhong, L.; Zhang, W.; Sun, S.; Zhao, L.; Jian, W.; He, X.; Xing, Z.; Shi, Z.; Chen, Y.; Alshareef, H.N.; et al. Engineering of the Crystalline Lattice of Hard Carbon Anodes Toward Practical Potassium-Ion Batteries. *Adv. Funct. Mater.* **2022**, *33*, 2211872. [CrossRef]
7. Privitera, V.; Scalese, S.; La Magna, A.; Pecora, A.; Cuscunà, M.; Maiolo, L.; Minotti, A.; Simeone, D.; Mariucci, L.; Fortunato, G.; et al. Low-Temperature Annealing Combined with Laser Crystallization for Polycrystalline Silicon TFTs on Polymeric Substrate. *J. Electrochem. Soc.* **2008**, *155*, H764. [CrossRef]
8. Bretos, I.; Jiménez, R.; Wu, A.; Kingon, A.I.; Vilarinho, P.M.; Calzada, M.L. Activated solutions enabling low-temperature processing of functional ferroelectric oxides for flexible electronics. *Adv. Mater.* **2014**, *26*, 1405–1409. [CrossRef] [PubMed]
9. Chernets, M.; Shil'ko, S.; Kornienko, A. Calculated assessment of contact strength, wear and resource of metal-polymer gears made of dispersion- reinforced composites. *Appl. Eng. Lett.* **2021**, *6*, 54–61. [CrossRef]

10. Marković, A.; Ivanović, L.; Stojanović, B. Characteristics, Manufacturing, and Testing Methods of Polymer Gears: Review. *Lect. Notes Netw. Syst.* **2024**, *866*, 269–282. [CrossRef]
11. Singh, P.K.; Siddhartha Singh, A.K. An investigation on the thermal and wear behavior of polymer based spur gears. *Tribol. Int.* **2018**, *118*, 264–272. [CrossRef]
12. Yang, Z.; Guo, Z.; Yuan, C. Tribological behavior of polymer composites functionalized with various microcapsule core materials. *Wear* **2019**, *426–427*, 853–861. [CrossRef]
13. Yang, Z.; Guo, Z.; Yang, Z.; Wang, C.; Yuan, C. Study on tribological properties of a novel composite by filling microcapsules into UHMWPE matrix for water lubrication. *Tribol. Int.* **2021**, *153*, 106629. [CrossRef]
14. Hale, W.R.; Pessan, L.A.; Keskkula, H.; Paul, D.R. Effect of compatibilization and ABS type on properties of PBT/ABS blends. *Polymer* **1999**, *40*, 4237–4250. [CrossRef]
15. Soudmand, B.H.; Shelesh-Nezhad, K. Study on the gear performance of polymer-clay nanocomposites by applying step and constant loading schemes and image analysis. *Wear* **2020**, *458–459*, 203412. [CrossRef]
16. Wang, Z.; Gao, D. Comparative investigation on the tribological behavior of reinforced plastic composite under natural seawater lubrication. *Mater. Des.* **2013**, *51*, 983–988. [CrossRef]
17. Stead, I.M.N.; Eckold, D.G.; Clarke, H.; Fennell, D.; Tsolakis, A.; Dearn, K.D. Towards a plastic engine: Low—Temperature tribology of polymers in reciprocating sliding. *Wear* **2019**, *430–431*, 25–36. [CrossRef]
18. Advincula, P.A.; Granja, V.; Wyss, K.M.; Algozeeb, W.A.; Chen, W.; Beckham, J.L.; Luong, D.X.; Higgs, C.F., III; Tour, J.M. Waste plastic- and coke-derived flash graphene as lubricant additives. *Carbon* **2023**, *203*, 876–885. [CrossRef]
19. Martín-Alfonso, J.E.; Valencia, C.; Sánchez, M.C.; Franco, J.M.; Gallegos, C. Evaluation of different polyolefins as rheology modifier additives in lubricating grease formulations. *Mater. Chem. Phys.* **2011**, *128*, 530–538. [CrossRef]
20. Bheemappa, S.; Gurumurthy, H.; Badami, V.V.; Hegde, P.R. *Tribological Behavior of Polymeric Systems in Lubricated Surfaces or Conditions*; Elsevier: Amsterdam, The Netherlands, 2022. [CrossRef]
21. Hlebanja, G.; Hriberšek, M.; Erjavec, M.; Kulovec, S. Durability Investigation of plastic gears. *MATEC Web Conf.* **2019**, *287*, 02003. [CrossRef]
22. Rodiouchkina, M.; Lind, J.; Pelcastre, L.; Berglund, K.; Rudolphi, Å.K.; Hardell, J. Tribological behaviour and transfer layer development of self-lubricating polymer composite bearing materials under long duration dry sliding against stainless steel. *Wear* **2021**, *484–485*, 204027. [CrossRef]
23. Samyn, P.; De Baets, P.; Schoukens, G. Influence of internal lubricants (ptfe and silicon oil) in short carbon fibre-reinforced polyimide composites on performance properties. *Tribol. Lett.* **2009**, *36*, 135–146. [CrossRef]
24. Ben Difallah, B.; Kharrat, M.; Dammak, M.; Monteil, G. Mechanical and tribological response of ABS polymer matrix filled with graphite powder. *Mater. Des.* **2012**, *34*, 782–787. [CrossRef]
25. Bashandeh, K.; Lan, P.; Meyer, J.L.; Polycarpou, A.A. Tribological Performance of Graphene and PTFE Solid Lubricants for Polymer Coatings at Elevated Temperatures. *Tribol. Lett.* **2019**, *67*, 99. [CrossRef]
26. Sajid, M.; Ilyas, M. PTFE-coated non-stick cookware and toxicity concerns: A perspective. *Environ. Sci. Pollut. Res.* **2017**, *24*, 23436–23440. [CrossRef] [PubMed]
27. Sony, M.; Naik, S. Green Lean Six Sigma implementation framework: A case of reducing graphite and dust pollution. *Int. J. Sustain. Eng.* **2020**, *13*, 184–193. [CrossRef]
28. Hazaimeh, M.D.; Ahmed, E.S. Bioremediation perspectives and progress in petroleum pollution in the marine environment: A review. *Environ. Sci. Pollut. Res.* **2021**, *28*, 54238–54259. [CrossRef] [PubMed]
29. Yu, C.; Liao, Y.; Zhang, P.; Li, B.; Li, S.; Wang, Q.; Li, W. Polyimide-based composite coatings for simultaneously enhanced mechanical and tribological properties by polyhedral oligomeric silsesquioxane. *Tribol. Int.* **2022**, *171*, 107521. [CrossRef]
30. Chen, Z.; Zhang, M.; Guo, Z.; Chen, H.; Yan, H.; Ren, F.; Jin, Y.; Sun, Z.; Ren, P. Synergistic effect of novel hyperbranched polysiloxane and $Ti_3C_2T_x$ MXene/MoS_2 hybrid filler towards desirable mechanical and tribological performance of bismaleimide composites. *Compos. Part. B Eng.* **2023**, *248*, 110374. [CrossRef]
31. DIN EN ISO 527-2:2012; Kunststoffe—Bestimmung der Zugeigenschaften—Teil 2: Prüfbedingungen für Form- und Extrusionsmassen. Deutsche Fassung EN ISO: Berlin, Germany, 2012.
32. ASTM G77-17; Standard Test Method for Ranking Resistance of Materials to Sliding Wear Using Block-on-Ring Wear Test. ASTM: West Conshohocken, PA, USA, 2022.
33. NIST X-ray Photoelectron Spectroscopy Database. NIST Standard Reference Database Number 20. *Natl. Inst. Stand. Technol.* **2023**. Available online: https://srdata.nist.gov/xps/ (accessed on 14 March 2024).
34. Piscitelli, F.; Lavorgna, M.; Buonocore, G.G.; Verdolotti, L.; Galy, J.; Mascia, L. Plasticizing and reinforcing features of siloxane domains in amine-cured epoxy/silica hybrids. *Macromol. Mater. Eng.* **2013**, *298*, 896–909. [CrossRef]
35. Mathew, A.; Kurmvanshi, S.; Mohanty, S.; Nayak, S.K. Preparation and characterization of siloxane modified: Epoxy terminated polyurethane-silver nanocomposites. *Polym. Compos.* **2018**, *39*, E2390–E2396. [CrossRef]
36. Arzhakov, M.S.; Arzhakov, S.A.; Gustov, V.V.; Kevdina, I.B.; Shantarovich, V.P. Structural and mechanical behavior of polymethylmethacrylate containing diethyl siloxane oligomer. *Int. J. Polym. Mater. Polym. Biomater.* **1998**, *39*, 319–333. [CrossRef]
37. Li, S.; Wang, H.; Liu, M.; Peng, C.; Wu, Z. Epoxy-functionalized polysiloxane reinforced epoxy resin for cryogenic application. *J. Appl. Polym. Sci.* **2019**, *136*, 46930. [CrossRef]

38. Qi, H.; Zhang, G.; Chang, L.; Zhao, F.; Wang, T.; Wang, Q. Ultralow Friction and Wear of Polymer Composites under Extreme Unlubricated Sliding Conditions. *Adv. Mater. Interfaces* **2017**, *4*, 1601171. [CrossRef]
39. Grey, L.H.; Nie, H.Y.; Biesinger, M.C. Defining the nature of adventitious carbon and improving its merit as a charge correction reference for XPS. *Appl. Surf. Sci.* **2024**, *653*, 159319. [CrossRef]
40. Hua, C.; Zhao, S.; Lin, L.; Schlarb, A.K. Tribological performance of a polyethersulfone (PESU)-based nanocomposite with potential surface changes of the metallic counterbody. *Appl. Surf. Sci.* **2023**, *636*, 157850. [CrossRef]
41. Zhao, S.; Hua, C.; Zhao, Y.; Sun, C.; Lin, L. Casting of regulable nanostructured transfer film by detaching nanofiller from the composite into the tribosystem—A strategy to customize tribological properties. *Tribol. Int.* **2024**, *195*, 109597. [CrossRef]

Disclaimer/Publisher's Note: The statements, opinions and data contained in all publications are solely those of the individual author(s) and contributor(s) and not of MDPI and/or the editor(s). MDPI and/or the editor(s) disclaim responsibility for any injury to people or property resulting from any ideas, methods, instructions or products referred to in the content.

Article

Transition between Friction Modes in Adhesive Contacts of a Hard Indenter and a Soft Elastomer: An Experiment

Iakov A. Lyashenko [1,*], Thao H. Pham [1] and Valentin L. Popov [1,2,*]

1 Department of System Dynamics and Friction Physics, Institute of Mechanics, Technische Universität Berlin, 10623 Berlin, Germany; pham.19@campus.tu-berlin.de
2 Samarkand State University, Samarkand 140104, Uzbekistan
* Correspondence: i.liashenko@tu-berlin.de (I.A.L.); v.popov@tu-berlin.de (V.L.P.); Tel.: +49-(0)30-314-75917 (I.A.L.)

Abstract: The tangential adhesive contact (friction) between a rigid steel indenter and a soft elastomer at shallow indentation depths, where the contact exists mainly due to adhesion, is investigated experimentally. The dependencies of friction force, contact area, average tangential stresses, and the coordinates of the front and back edges of the contact boundary on the indenter displacement are studied. It is found that first a stick–slip mode of friction is established, which is then replaced by another, more complex mode where the phase of a global slip of the elastomer on the indenter surface is absent. In both regimes, the evolutions of friction force and contact area are analyzed in detail.

Keywords: quasi-static tangential and normal contact; indentation; adhesion; elastomer; friction; contact area; experiment; tangential stresses

1. Introduction

Friction processes in the presence of adhesion play an important role in many technological processes. For example, due to the increasing use of electrical devices, MEMS (micro electric mechanical systems) have become a key technology in mechanical- and electrical-based domains. These systems are heavily influenced by adhesion, which causes "stiction", the unintentional adhesion of a compliant microstructure surface. In the transport sector, adhesion and friction are a key component in road safety, as with the knowledge of the contact characteristics between rubber and pavement interaction, braking processes can be improved [1,2]. In the last decades, robots have been developed to help humans in dangerous work, such as detection, monitoring, cleaning, searching, and rescuing. Additionally, they are used, among other things, in ships, pipelines, nuclear power plants, and wind power generation. To increase the functional work area of robots, climbing robots were developed for the first time in the 1960s by Nishi et al. [3,4]. For wall-climbing robots, bio-inspired adhesive microstructures are of great interest, because these adhesive adsorption methods do not require an extra energy supply [4]. Based on the smooth, adhesive pads of cockroaches or patches of microscopic hairs of beetles and Tokay geckos, new adhesives are being developed for wall-climbing robots. In a gecko's foot, there are up to five hundred thousand keratinous hairs or setae that each terminates in hundreds of spatula-shaped structures. Measurements of the adhesive force support the theory that the setae are operated by van der Waals forces [5]. For optimal functionality of the robots, not only the process of adhering to the surface is important; the detachment of the contacting parts is also as essential for locomotion of the robot as it is in biological organisms [6]. To break the strong adhesive forces of the entire adhesive pad of the connecting feet, much energy is needed. Therefore, a minimization of force expenditure during the detachment phase for reduced energy consumption is important. A shear component is present when the robot feet are moving along the wall. This provides a preload to the surface of the attaching feet. A similar shearing motion has been observed in the attachment mechanism

of a single gecko seta. During the attachment phase, toe uncurling, and during the detachment phase, toe peeling can be observed in geckos, which increase the effectiveness of setae. It has been observed that setal forces are dependent on three-dimensional orientation and preloading during the initial contact. Experiments have shown that perpendicular preloading, by pushing the seta towards the surface, alone is not sufficient to form an effective setal attachment when the seta is pulled away from the surface. The experiment demonstrated that in addition to the initial push into the surface, a parallel pull of the seta along the surface increases the pull-away forces over ten times compared to systems where the setal was only perpendicularly preloaded [5]. For energetic advantages, a peeling-like detachment mechanism has been applied in some robot designs with climbing abilities [7]. To improve those systems, a deeper understanding of the behavior of friction and adhesion under normal and tangential loads is needed.

Many studies have been conducted on the strength of the contact when only a normal force is applied to separate the contacting bodies [8–10]. This study concentrates on the contacts where a tangential force is applied. The contact strength and, associated with that, the breakaway forces are of interest. The interaction between friction and adhesion is to be investigated due to the fact that adhesion is a partial reason for friction, but, at the same time, it decreases under a friction force [11]. In the present work, it is being determined in relation to the friction mode in which the system is operating.

The extension of the JKR theory (Johnson–Kendall–Roberts) to tangential forces was already made by Savkoor and Briggs in 1977 [12]. It has been observed that the contact area shrinks when tangential forces are present in adhesive contacts. Like in the experiments conducted in this study, the normal load is held near zero, and due to adhesion, a finite contact area can be observed. When a tangential load is applied, the surfaces can peel apart so that elastic energy is released at the expense of creating surface energy. However, after the release of shear stress, the surfaces can re-adhere without altering the potential energy of the tangential load [12]. With increasing tangential forces, the contact area decreases through peeling and transitions from a circular shape into an elliptical shape due to decreasing adhesive interaction [13]. After reaching a critical value of tangential load, the indenter can no longer stick to the surface and the contact either breaks or sliding occurs, depending on the normal load. The transition can occur smoothly [14,15] or via mechanical instability [16]. The abrupt decrease in the contact area at the transition towards gross sliding can only be observed at low, normal loads. This results from the fact that at higher loads and therefore bigger contact areas the contact can sustain larger tangential stress before sliding starts. The larger tangential load results in a more significant deformation of the contact and therefore a reduction in adhesion [11]. From experiments with dry sliding, it has been discovered that the tangential pull-off force is proportional to the contact area, which leads to constant frictional stress and independence on normal forces. Therefore, for low, normal forces where adhesion dominates, the critical tangential force for detachment can be described by $T_c = \tau_0 A$, where τ_0 is the critical shear stress at pull-off and A is the contact area at the peak tangential force [17].

To understand the processes in adhesive contacts, a large number of both theoretical and experimental investigations are being carried out. However, many scientific groups who have studied the dependence of the contact area on the tangential force, where the indenter is displaced at constant velocity, have either only observed the dependence right after normal loading and not over a long period of time (or, to be more precise, not over many periods of increasing and decreasing tangential force) [18] or the dependence until gross slipping was observed [14,15]. Therefore, it is unknown if there will be a transition to a different friction mode when sliding for a long time.

The aim of this work is to study the transition of friction modes in the case of a rigid sphere being displaced with a constant velocity at shallow indentation depths over a soft elastomer layer with which the sphere is in adhesive contact.

2. Materials and Methods

The experimental setup described in detail in our previous paper [19] was used for the experiments presented in this work, so we will not go into detail on its design here. In each experiment, a spherical indenter made of steel was first indented to a depth of d_{max} = 0.2 mm into a soft gel sheet (thermoplastic polystyrene-type gel sheet, CRG N3005, TANAC Co. Ltd., Gifu, Japan [20]) with thickness h = 5 mm and linear dimensions 100 mm × 100 mm. Because the material being used possesses elastomeric properties (non-compressibility [21], low modulus of elasticity, and ability to withstand large elastic deformations), we will further refer to it as elastomer. After the end of the indentation phase, the indenter was pulled out of the elastomer to the zero-level d = 0 mm, and then tangential shear was performed at a fixed zero indentation depth d. However, it should be noted that we used precision linear stages PI M-403.2DG, which allow us to fix the indenter position with an accuracy of up to 1 μm. But, when changing the direction of motion, these devices have backlash, which, in our case, was about 6 μm (see [19] for a detailed explanation). Because in the experiment the indenter was first indented to a depth of d = 0.2 mm, which is multiple times greater than the backlash value, and then pulled out of the elastomer, there was only a single change in the direction of its motion. Therefore, the indentation depth d, at which the tangential shift of the indenter was realized, is not d = 0 mm (which corresponded to the instrument readings), but $d \approx 6$ μm, because the indenter in the pulling phase moves a distance shorter by the amount of backlash compared to the indentation phase. When using the more precise PI L-511.24AD00 linear stages, we discovered that in the case of exactly zero indentation depth d = 0 mm, the adhesive contact disappears completely during tangential shear. Therefore, the experimental results discussed below should be interpreted as corresponding to a shear indenter with a small, but non-zero indentation depth d. At this depth d, the contact exists mainly due to adhesion. It is worth noting that the experimental setup allows experiments to be conducted with a fixed indentation depth of indentation d, regardless of the value of the normal force, because a displacement-controlled mode (fixed grips) is used. Specifically, the indenter is displaced using actuators to which it is rigidly connected. Thus, in the experiment, force is not directly applied to the indenter; instead, its displacement is controlled, and all three components of the contact force are measured using a three-axis force sensor, as shown in Figure 1 by position (4).

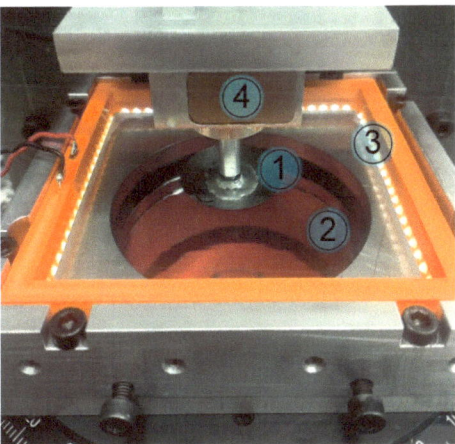

Figure 1. Real photo of the contact area between the spherical steel indenter (1) and the elastomer (2) illuminated by a surrounding LED (3). Contact forces are measured with the force sensor (4).

Figure 1 shows a real photograph of the contact area in ongoing experiments, where elements of the experimental setup are partially depicted.

Here, a spherical steel indenter (1) is pressed into a sheet of soft elastomer (2), which is placed on a silicate glass plate to which it is firmly attached by its own adhesion, without the use of glue or any additional fixing mechanisms. The contact area is uniformly illuminated using comprehensive LED lighting (3), and contact forces are measured using the ME K3D40 three-axis force sensor (4). The contact is observed from below the system using a digital camera through the aperture visible in the figure. Because the contact in the experiments has a relatively small size, a 5 MP physical resolution camera–microscope (model TOOLCRAFT TO-5139594, Conrad Electronic SE, Hirschau, Germany) was used to observe it.

To ensure quasi-static contact propagation conditions, the indenter velocity in the experiment, both when indenting in the normal direction and in tangential shear, was $v = 1$ µm/s. Because the optically transparent elastomer TANAC CRG N3005 [20] was used in the experiments, direct observation of the contact was possible. In order to clearly observe the contact area, the indenter surface was treated with P800 grit sandpaper prior to each experiment series to provide diffuse light scattering from the LED side-lighting system. After the treatment with sandpaper, the indenter surface was cleaned with alcohol and then quickly dried with a jet of compressed air before the experiment was carried out. In our previous work [21], the values of the elastic modulus of the elastomer used in this study $E \approx 0.324$ MPa and Poisson's ratio $\nu \approx 0.48$ were determined by generalizing a large amount of experimental data. Because the indenter was made of steel, with an elastic modulus of $E \approx 2 \times 10^5$ MPa, the indenter can be considered absolutely rigid, i.e., only the elastomer sheet could be deformed in the experiment.

In much earlier work by other authors [22], the experiments were conducted with a similar material, TANAC CRG N0505, but for a different contact geometry. Specifically, the study focused on sliding along the glass of an externally non-loaded elastomer sheet at various drawing velocities. However, the material CRG N0505 has a lower elasticity modulus, which means it can be deformed even more significantly under tangential shear, leading to more complex dynamic nonlinear effects; contact restructuring in the case of CRG N0505 is more complex than for stiffer CRG N3005 used in the present study. Our recent study [23] provides a comparison of the behavior of these two materials (the difference is particularly noticeable in supplementary videos attached to the article). Primarily, in our experiments, we prefer to use the stiffer material, CRG N3005. Stiffer gels contain less filler in the matrix (which can flow inside the material during deformation) and therefore exhibit properties closer to those of elastic solids, such as stable values of elasticity modulus and Poisson's ratio [21]. It should be noted that gels with similar properties are produced by other companies as well. For instance, in the experimental work [24], an optically transparent material "Super Gel" (Kihara Sangyo Co., Ltd., Osaka, Japan) is used.

3. Results

3.1. Transition between Friction Modes in Adhesive Contact as a Result of Degradation of Adhesive Properties of Contacting Surfaces

Figure 2 shows the experimentally measured dependencies of the friction force F_x on the value of tangential displacement x of the indenter, where the results of three consecutive experiments are depicted in different panels. In each experiment, the indenter was first immersed to a depth of $d_{max} = 0.2$ mm then pulled to a zero level of $d = 0.0$ mm, followed by tangential displacement. Moreover, the initial indentation to a depth of d_{max} in all experiments was carried out at the same location of the elastomer to ensure the same experimental conditions. Figure 2 shows the dependence plots corresponding to tangential shear without indentation and detachment phases in the normal direction, where the indenter was not horizontally displaced (i.e., $x = 0$ mm). In the first experiment (upper panel), after the motion starts, the tangential force first increases to some maximum and then decreases. This maximum is associated with additional strengthening of the contact during normal indentation when tangential shift did not occur; a similar situation was observed in [22] for another geometry of the experiment and much bigger sliding velocities.

At further shear, as it follows from the figure, a stationary mode of stick–slip friction mode is established, which we denote as "mode 1". In this stick–slip mode, once the friction force reaches its maximum value, it decreases sharply, which can only be due to gross slip, i.e., "mode 1" is a stick–slip mode in the classical sense. Such a mode can be theoretically described within the framework of models existing in the literature, such as, for example, [25]. However, in the present work, we limited ourselves to a detailed description of the experiment.

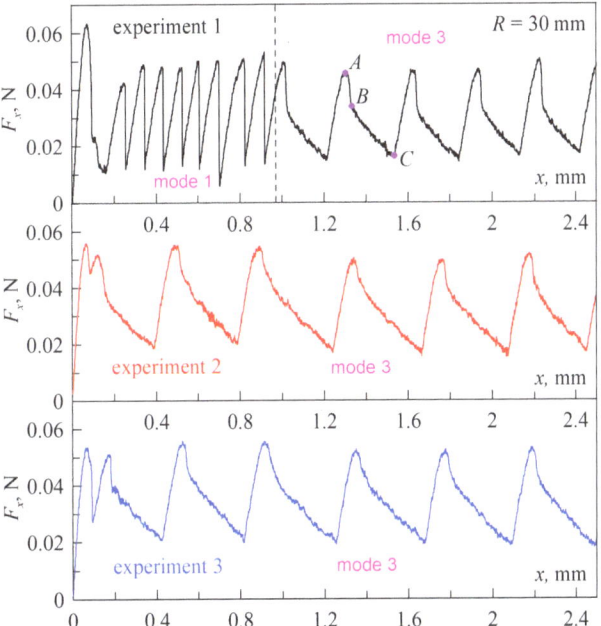

Figure 2. Dependencies of the friction force F_x on the tangential displacement x of the indenter in three consecutive experiments (panels from top to bottom) for an indenter with a radius $R = 30$ mm. A supplementary video (Video S1) is available for the figure.

In order to realize the stick–slip motion, it is necessary for the system to be able to store elastic energy that is released during the slip phase. The stick–slip mode is often studied using examples of systems in which one of the rubbing surfaces is displaced by a spring, the free end of which moves at a constant velocity [26,27]. In this case, the elastic energy is stored in the spring, and as soon as the contact stresses exceed a critical value, the rapid slip phase takes place. In the slip phase, the spring tension decreases, and in order for the tension to reach the critical value again, time is required during which the contacting surfaces come to rest. As a result, a stick–slip friction mode is established, in which periods surfaces stopping alternate with periods of rapid slipping. Any elastic body with a stiffness much smaller than the tangential contact stiffness, such as a cantilever in AFM (atomic force microscopy) experiments, in which the stick–slip mode is also observed, can act as a spring [28]. In the experiment, the results of which are shown in Figure 2, the indenter (upper rubbing surface) is rigidly connected to the driving actuator and is displaced at a constant velocity. In this case, the elastic energy is accumulated by deforming a part of the rubber layer in the contact area, because the rubber surface in the stick phase moves together with the indenter. When the shear stresses exceed the critical stresses, the rubber slips over the indenter with the release of elastic energy and the friction force decreases dramatically ("mode 1" in Figure 2). Therefore, in the described experiment, there is no rapid movement of the whole rubber sheet (lower rubbing surface) or the indenter (upper rubbing surface); the indenter always moves at a constant velocity, and the rubber rests on a rigidly fixed glass substrate.

It is worth noting that in the experiment, the indenter is shifted at a very low velocity $v = 1$ μm/s to ensure quasi-static contact conditions. During the "stick" phase, the contact can be considered quasi-static because the elastomer surface moves together with the indenter. However, during the "slip" phase, the elastomer slips over the indenter surface at a much higher velocity, exceeding the value at which the contact can be considered quasi-static. Therefore, to describe the observed contact processes, even with the indenter moving at very low velocities, it is necessary to take into account viscoelastic effects, which significantly complicates the theoretical description of the problem [29,30]. There are also studies that investigate the influence of velocity on contact properties, such as, for instance, [22,24].

The top panel in Figure 2 shows that over time, the stick–slip mode, denoted as "mode 1", is replaced by another mode, labeled "mode 3". In "mode 3", there is no phase of global slippage of the rubber on the indenter surface because of the absence of a sharp decrease in the friction force. Note that the form of friction force–displacement dependence in "mode 3" is typical for adhesive contacts at shallow indentation depths, and such dependencies were observed by us in previous works, such as, for example, [31,32]. Such a "mode 3" was thoroughly analyzed in our latest work [33] in which extensive analysis was conducted based on existing theories of tangential adhesive contact. In [33], it was shown that such a mode consists of alternating phases, such as attachment, stick, peeling, and sliding, which periodically repeat.

The different panels in Figure 2 show the results of three consecutive experiments, where only "mode 3" was observed in the second and third experiment. In order to eliminate the influence of individual elastomer surface characteristics, the indenter motion started at the same location on the rubber substrate in each experiment. From a comparison of all of the curves in Figure 2, an interesting feature emerges. On the dependencies $F_x(x)$, which correspond to "mode 3", after the friction force reaches its maximum value, its rapid decrease is observed for some time (section AB), although it is not instantaneous as in the stick–slip mode (also known as "mode 1"). After a rapid decrease in the friction force, it continues to decrease, although now much more slowly (section BC). When the friction force becomes minimal (point C), it starts to increase to its maximum value again. The above-mentioned peculiarity consists in the fact that as time passes, the section AB, where the friction force rapidly decreases after reaching the maximum value, becomes shorter, whereas the section BC, on the contrary, becomes longer. As a result, the period of transitions between maxima and minima of the friction force, observed in the $F_x(x)$ dependencies, increases.

It has already been mentioned above that in the first experiment (upper panel in Figure 2), after the start of indenter shear, the frictional force reaches some maximum value, which is significantly higher than the maximal force during further indenter shear. This maximum of F_x during the transition from static to kinetic friction is due to the fact that the initial contact is stronger than the contact after the first act of slip in the kinetic regime [34]. However, no such high peaks are observed in the next two experiments (middle and bottom panels in Figure 2), i.e., the contact before shear has the same strength as in the steady kinetic regime. This fact suggests that the adhesive strength of the contact decreases with time. The decrease in the shear strength of the contact was observed by us in previous works and was expressed as a decrease in the steady-state value of the friction force and shear stresses in the sliding mode [32]. The adhesive strength of the contact depends on the surface energy of the contacting bodies, which determines the specific work of adhesion $\Delta\gamma$. It has been repeatedly shown that freshly cleaned surfaces have the highest value of $\Delta\gamma$, which decreases with time due to the oxidation of friction surfaces, their contamination, etc. We attribute the transition between "mode 1" and "mode 3", observed in the upper panel of Figure 2, to the decrease in the specific work of adhesion $\Delta\gamma$ due to the contamination of the indenter surface during friction. Thus, large values of $\Delta\gamma$ should correspond to "mode 1". It should be noted that if the indenter is left exposed to air for some time before the experiment (for example, for one day), then "mode 3" will be observed immediately in the experiment. This is because the adhesive properties of the freshly treated surface of the indenter degrade rapidly, particularly due to surface oxidation. This is one of the reasons why we did not observe such a transition between modes earlier, as we did not conduct experiments immediately after cleaning the indenter. However, in any case, during interaction with the elastomer surface in the experiment, the adhesive properties of the indenter surface degrade much faster than when it is passively exposed to the air.

The assumption that the transition between "mode 1" and "mode 3" is caused by a decrease in the specific work of adhesion $\Delta\gamma$ is also supported by the results shown in Figure 3. This figure shows the dependencies of the normal force F_N on the indentation depth d in the indenter pulling phase after the initial indentation to a depth of $d_{max} = 0.2$ mm (in the figure, the dependencies do not reach the value of 0.2 mm because of the range of instrument backlash, which is described in Section 2 of the article). Figure 3a shows three curves corresponding to three consecutive experiments, the results of which are shown in Figure 2. All of the dependencies in Figure 3a correspond to the indenter pulling back to the zero level, and, after reaching that level, the indenter already starts to shift in the tangential direction.

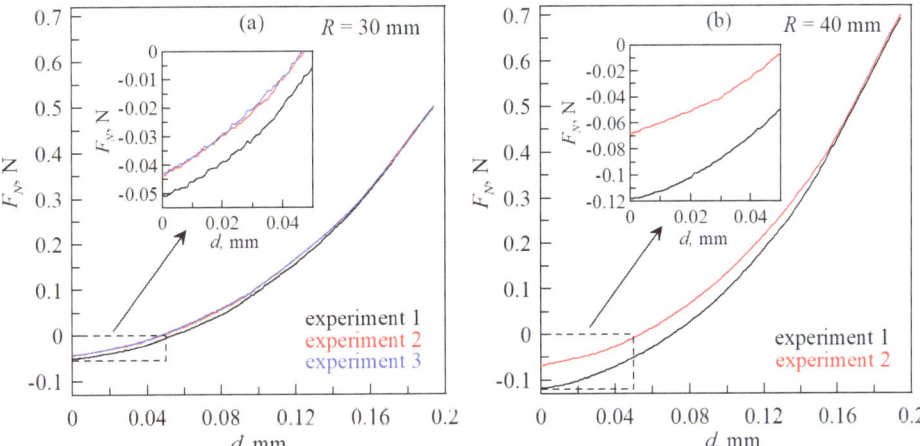

Figure 3. Dependencies of the normal force F_N on the indentation depth d corresponding to the pulling phase of the steel indenters from the CRG N3005 elastomer: (**a**) the results of three experiments with an indenter with a radius $R = 30$ mm are shown; (**b**) the results of two experiments with an indenter with a radius $R = 40$ mm are shown.

Figure 3 shows that the highest absolute value of the adhesion force $F_N < 0$ N is realized in the first experiment, while the following two experiments show similar adhesion forces. But, according to Figure 2, only in the first experiment the transition between friction modes takes place, while in the second and in the third experiments, only "mode 3" is realized. The data shown in Figure 3b show the dependencies of normal forces on indentation depth for two consecutive experiments, which are similar to those discussed above, only with an indenter with a radius $R = 40$ mm. Figure 3b also shows a decrease in the adhesive force in the second experiment compared to the first one. Thus, the data shown in Figures 2 and 3 complement each other and speak in favor of the fact that the change in friction modes is caused by contamination of the indenter surface, which occurs due to its interaction with the elastomer during the contact.

Figure 4 shows the dependence of the contact area A on the indenter shear x, which complements the experimental data shown in Figure 2. In order to show the initial decrease in area from a much larger value of $A \sim 7$ mm^2 along with the periodic stationary regime, in which $A \sim 1$ mm^2, the ordinate axis is presented on a logarithmic scale. Figure 4 shows that in stick–slip mode ("mode 1"), the contact area A varies within a small range, while in "mode 3" the area varies much more. Our previous work [31,32] studied in detail the nature of contact propagation in the stationary regime, which resembles "mode 3", but in the mentioned work, the case of an indentation depth d significantly different from zero was considered. Here (Figures 2 and 4), we study the situation when the indentation depth d is close to zero and the contact exists mainly due to adhesion. Therefore, this case requires additional analysis, which is carried out in the next section of the paper.

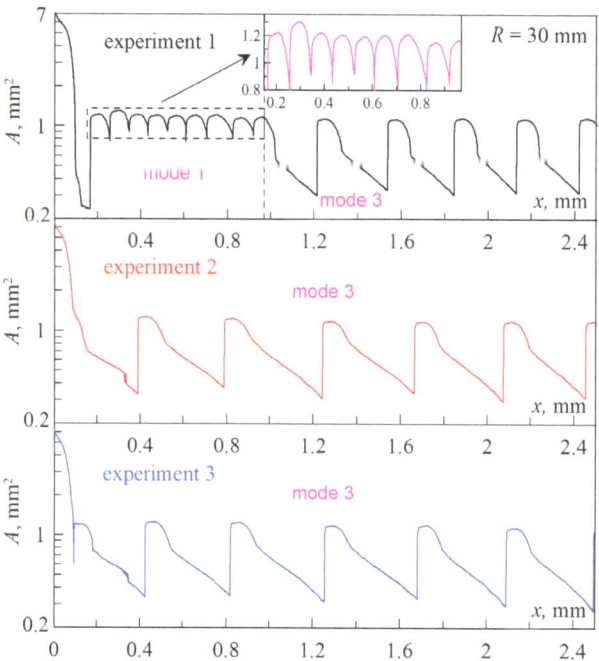

Figure 4. Dependencies of the contact area A on the indenter coordinate x corresponding to three consecutive experiments, the data of which are shown in Figure 2, for an indenter with a radius $R = 30$ mm. Supplementary videos (Videos S1–S3) are available for the figure.

3.2. Detailed Analysis of the Nature of Contact Propagation

Figure 5 shows the dependencies corresponding to the first experiment, the data of which are presented in Figures 2 and 4. In Figure 5, the region of the dependencies in which the transition between modes "mode 1" and "mode 3" takes place is depicted in detail. Additionally, Figure 5 shows the dependencies for the coordinates of the front and back contact edges, as well as the average value of the tangential stresses $\tau = F_x/A$. Note that the difference between the "front side" and "back side" coordinate values in Figure 5c at each fixed indenter coordinate x is the contact width measured in the direction of indenter motion. In the case of a circular contact, the width thus measured will coincide with its diameter.

Let us first consider the stick–slip mode ("mode 1"), which exists up to the vertical dashed line shown in Figure 5. In all dependencies in Figure 5, the numbers from one to seven show characteristic points corresponding to moments for which pictures of the contact area are shown in the top row in Figure 6. The bottom row in Figure 6 shows the same photos as in the top row, but the colors indicate differences between adjacent contact areas. Blue shows areas that have come out of contact and red shows areas that have come back into contact. This, however, does not apply to the first picture "1" in the bottom row, which is exactly the same as the corresponding image in the top panel.

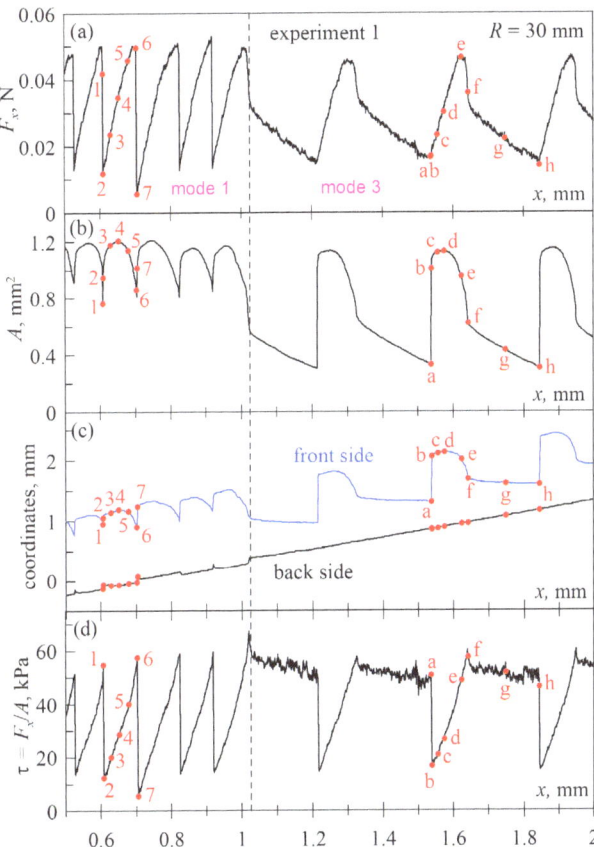

Figure 5. Dependencies on the tangential displacement of the indenter x: friction force F_x (**a**), contact area A (**b**), coordinates of the front and back contact boundaries (**c**), and average value of tangential stresses $\tau = F_x/A$ (**d**). All dependencies correspond to the first experiment, the results of which are shown in Figures 2 and 4. A supplementary video (Video S1) is available for the figure.

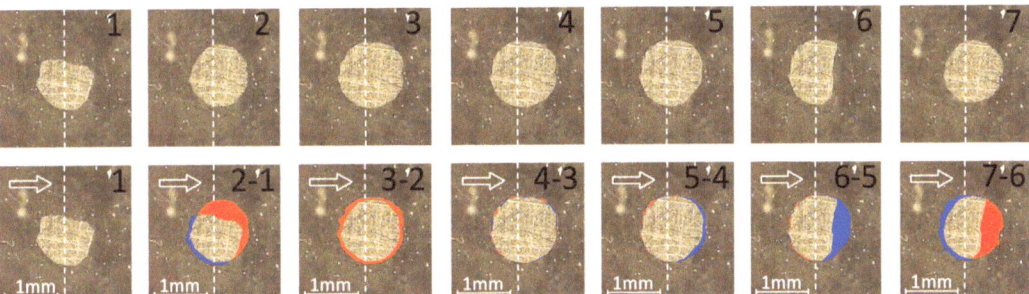

Figure 6. Top row of figures: photographs of the contact areas corresponding to points (1–7) shown in Figure 5. Bottom row: the same photographs as in the top row, but where colors show the differences between adjacent contact areas; blue shows the area that has come out of contact and red shows the area that has come into contact (except for the first figure "1", which is the same as in the top panel). For example, figure "2–1" shows the difference between the contact configurations "2" and "1" on the top panel, figure "3–2" shows the difference between the contact areas in figures "3" and "2", etc. A supplementary video (Video S1) is available for the figure.

Let us note one important point for understanding. In Figure 6, the contact configurations and their changes are shown in the coordinate system that is associated with the indenter, i.e., the indenter is assumed to be stationary. At the same time, the coordinate dependencies of the back and front of the contact boundary, which are shown in Figure 5c, are shown in the coordinate system associated with the rubber substrate, i.e., here, the elastomer into which the indentation is made is considered to be stationary. Therefore, for example, in the case where the rear edge of the contact is moving at the same speed as the indenter, Figure 6 will show no change in the rear edge, and Figure 5c will show a linear increase in the "back side" coordinate.

Let us now analyze the contact realignment process based on the figures presented above. Point 1 in Figure 5 is selected just before the next contact propagation. This point corresponds to the friction force F_x close to the maximum, as well as the maximum contact stresses τ. At the neighboring point 2, there is a jump in the contact area when the region at the front edge, shown in red in panel "2–1" in Figure 6, is joined. In addition to the discontinuous joining of a new contact area, there is a global slip of the rubber over the indenter surface in the entire contact area as the friction force decreases discontinuously, as demonstrated by trajectory 1–2 in the $F_x(x)$ dependence. As the contact slips and recovers, its size becomes slightly smaller on the back edge, as can be seen in panel "2–1" in Figure 6. The reason for this reduction is that part of the contact at the back edge, prior to slippage, existed due to normal adhesion, and it takes time to restore it. At global slip 1–2, the elastic energy stored in the deformed rubber layer sample is released, and for the next slip cycle (points 6–7), the rubber layer must deform again to the critical state to which the maximum values of friction force F_x and tangential stresses τ correspond. Note that the force F_x before global slippage always takes locally maximal values. The reason that the friction force at point 1 is not maximal is due to the fact that the contact characteristics (contact forces and photos of the contact area) are saved in the experiment with a small frequency of once per second, and the next preservation at point 1 occurred after the beginning of global slippage. After the act of slippage, the friction force and shear stresses increase monotonically (path 2–3–4–5–6). However, the contact area on this site behaves non-monotonically, which was observed earlier in [32]. The friction force in the stationary sliding mode is defined by the expression

$$F_x = \tau_0 A, \tag{1}$$

where τ_0 is the critical stress at which sliding is realized. According to (1), as the area A increases, the tangential force F_x also increases. However, this is only true for those systems in which the stresses τ_0 takes on constant values over the entire contact region. And, here we have a case with a complex inhomogeneous stress distribution. For example, in the photo "2–1" in Figure 6, the shear stresses in the "fresh" contact region at the front edge (shown in red) have close to zero values because it is a newly formed region after global slippage and has not yet been loaded by the tangential motion of the indenter. In the rest of the contact region, non-zero stresses τ are realized, and they are the ones that provide the non-zero force F_x. However, the stresses τ for this configuration (point 2) are significantly lower than the maximum value of τ_0, because global slip across the contact region (path 1–2) and the release of elastic energy have just occurred.

In addition to the above, it should also be taken into account that in a tangential contact, the maximum stresses are realized at its boundary [35]. Because contact failure begins when the stress reaches its maximum value, as the tangential force increases, contact failure will begin at the edge of the contact, as seen, for example, in panel "5–4" in Figure 6, which demonstrates the partial detachment of the rubber from the indenter at the front edge of the contact. The next panel of the figure "6–5" shows a significant reduction of the contact area due to its failure at the front edge, which corresponds to the moment of onset of global slippage at which the frictional force F_x and the average tangential stresses τ take their maximum values. During further slipping, the tangential stresses decrease across the entire contact area, with immediate contact regaining at its front edge when the tangentially unloaded area is joined (red area in panel "7–6" in Figure 6). The contact configuration in panel 7 in Figure 6 is similar to panel 2, as they correspond to the moments immediately after slip and the minimums of the force F_x. The described process is repeated periodically in time and corresponds to a stationary stick–slip motion in which, however, the minimum and maximum values of the tangential force F_x and the contact area A vary in some range. Such variations occur due to the fact that compared to the quasi-static shear of the indenter with a very small velocity, the process of global slippage of the rubber on the indenter surface is almost instantaneous. Therefore, an essential role here is played by the individual characteristics of the contact before the slip phase, which are always different in a real experiment (specific contact configuration, stress distribution, surface energy distribution of the contacting surfaces, presence of inhomogeneities, etc.).

Let us note one interesting detail that is visualized in panel "3–2" in Figure 6. Panel 2 of the figure is the contact configuration immediately after global slippage, and panel 3 is the next contact configuration that corresponds to the section of monotonic force buildup. What is common in panels 2 and 3 is that here the stresses τ are still far from the critical value at which slip is realized. The characteristic mentioned above is that the contact area on panel 3 is slightly larger than on panel 2. The newly acquired contact areas are shown in red in the "3–2" panel, indicating homogeneous contact propagation on all sides. It has been shown many times before that the contact area can only decrease with tangential shear (see, e.g., [36,37]). The reduction of the area in the tangential contact is caused by tangential stresses, which, however, are quite small immediately after slip. Therefore, the increase in contact area shown in panel "3–2" is due to contact propagation in the normal direction (normal adhesion). This spreading of the contact due to normal adhesion, however, leads to only a small increase in the contact area, so that already in panel 4 of Figure 6 the contact area is almost unchanged compared to panel 3 (see comparison panel "4–3"). The described contact rearrangement process is periodic in time and corresponds to a stable stick–slip mode, which we have labeled as "mode 1".

According to Figure 5, when the indenter is shifted further, the system switches from "mode 1" to "mode 3" and then continuously remains in "mode 3" (see also Figures 2 and 4). The characteristic points in "mode 3" are shown in the panels in Figure 5 with the letters a–h, and Figure 7 shows the corresponding contact configurations. The essential difference between "mode 3" and stick–slip "mode 1" is that in "mode 3" there is no global slipping of the rubber on the indenter surface and, therefore, the $F_x(x)$ dependence does not show areas with a sudden decrease in the friction force. Point (a) in Figure 5 corresponds to the minimal contact area A and the tangential force F_x. At this point, the maximum shear stresses τ are realized over the entire contact area, which correspond to the value τ_0 at which sliding is realized. Neighboring point (b) corresponds to the attachment of a large section of rubber at the front of the contact, which is shown in red in Figure 7, "b–a". This fresh contact section is not tangentially loaded immediately after attachment because the friction force F_x at points (a) and (b) takes on the same values (see Figure 5). Because the contact area A increases drastically at point (b), formally, the average tangential stresses τ decrease, as demonstrated by Figure 5d. In the section a–b–c–d–e, the friction force F_x increases monotonically, while the contact area A is non-monotonic. The reasons for this behavior have already been discussed above when describing "mode 1" and are due to the inhomogeneous distribution of tangential stresses in the contact zone.

The a–b–c–d–e section in "mode 3" has qualitatively similar features to the 2–3–4–5–6 section in "mode 1", which is also evidenced by the bottom rows of the photographs in Figures 6 and 7. Namely, in both cases, there is first a sharp increase in the contact area at the front edge, and then the contact area expands on all sides due to normal adhesion; after that, the contact area starts to decrease due to the rubber detaching from the indenter at the front edge. The corresponding dependencies of $F_x(x)$ and $A(x)$ in both modes are also visually similar. The main difference between the modes is observed at further indenter displacement. In "mode 1", after point 6 (maximum friction force), a global slip is observed with a sudden decrease in the friction force to a minimum value. However, in "mode 3", after the force F_x reaches a maximum at point (e), it begins to decrease smoothly, because in "mode 3" the decrease in F_x is not due to abrupt global slip but due to the continued detachment of the rubber at the front edge of the contact.

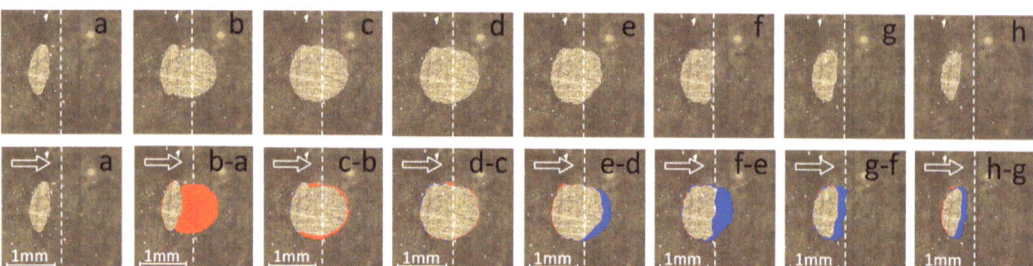

Figure 7. Top row of figures: photographs of the contact areas corresponding to points (a–h) shown in Figure 5. Bottom row: the same photographs as in the top row, but where colors show the differences between the contact areas, where blue shows the area that came out of contact and red shows the area that came into contact (except for the first panel "a", which coincides with the corresponding image in the top panel). For example, figure "b–a" shows the difference between figures "b" and "a" on the top panel, figure "c–b" shows the difference between the contact areas in figures "c" and "b", etc. A supplementary video (Video S1) is available for the figure.

The maximum stress $\tau = \tau_0$, at which the regime of homogeneous sliding of rubber on the indenter is established, is reached at the point (f), where the friction force is less than the maximum. Furthermore, in the sliding mode at section f–g–h, the stresses remain constant and the contact area A decreases, which leads to a decrease in the frictional force F_x (1). What is interesting here, however, is the nature of the contact area reduction. According to the dependencies shown in Figure 5c, in "mode 3", the back edge of the contact always moves linearly at the same speed as the indenter. This suggests that there is always a region of stationary slip in the contact region close to the back edge. Stationary sliding with constant velocity is observed in the indenter region, which is shown by the contact configurations with minimum areas A in panels (a) and (h) of Figure 7, and the maximum stress value $\tau = \tau_0$ is always realized in this contact region. The dependencies shown in Figure 5c indicate that in the f–g–h section, the elastomer-related coordinate of the front edge remains almost constant. Thus, here, the back-contact boundary shifts with the speed of the indenter while the front boundary stands still as the contact area decreases. In the coordinate system associated with the indenter, this situation corresponds to contact failure at its front edge, as can be seen in panels f–g–h in Figure 7 (see also the comparative photographs of the corresponding contact regions in the bottom row of Figure 7). After point (h), further contact propagation occurs, and the process described above is repeated again.

We note another difference between the friction regimes discussed in this paper. As mentioned above, the dependencies shown in Figure 5c demonstrate that in "mode 3" the back edge of the contact always moves at a constant velocity, which is the same as the shear rate of the indenter. This is due to the presence of a stationary slip zone at the back edge of the contact. However, in stick–slip "mode 1", the contact line configuration on the back edge undergoes a change, as can be seen in Figure 5c. The shape of the contact boundary changes during the next global slip of the rubber over the indenter surface, i.e., in the slip phase of the stick–slip mode. Next, in the stick phase, the back-contact boundary moves at the speed of the indenter but realigns again with another rubber slip. The process of rebuilding the contact can be clearly seen in the attached supplementary videos (Videos S1–S7).

3.3. Transition Mode between Regimes with "High" and "Low" Specific Work of Adhesion

Figure 5 shows the transition between "mode 1" and "mode 3" discussed in this paper. However, in the experiments, we also found a transient regime between these two stationary friction regimes. We labeled this mode as "mode 2" and observed it in the experiment with an indenter of a larger radius $R = 40$ mm. Figure 8 shows the successive transitions between all three modes. Here, the transient mode "mode 2" is something between "mode 1" and "mode 3", because in "mode 2" the system continuously transitions between the stick–slip ("mode 1") and "mode 3", in which there is no sudden reduction in tangential force due to global slippage of the rubber over the indenter surface. Note that here in "mode 1", the transition to "mode 3" also takes place once in the dependency region, the vicinity of which is shown by point A. However, in general, "mode 1" remains stable up to the first vertical line, where it is replaced by "mode 2", which in turn changes to the stable "mode 3" after the second vertical line.

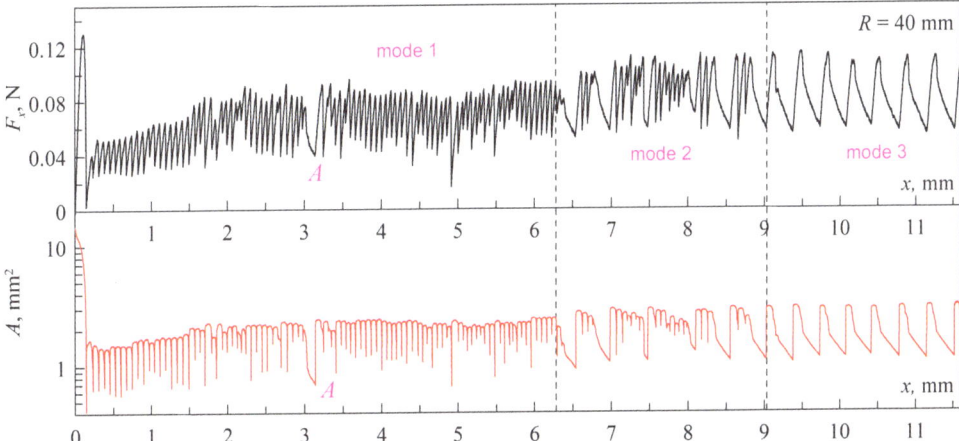

Figure 8. (**Top panel**) Dependence of friction force F_x on the indenter tangential displacement x, corresponding to the experiment with an indenter with a radius $R = 40$ mm; (**Bottom panel**) Corresponding dependence of the contact area A on the displacement x. A supplementary video (Video S4) is available for the figure.

The main difference observed between "mode 1" and "mode 3" is that in "mode 1" there are areas of abrupt reduction of friction force, while in "mode 3" there are no such areas. Therefore, the frequency of transitions between minimum and maximum values of tangential force in "mode 1" is much higher than the frequency of corresponding transitions in "mode 3". It follows that the value of the transition frequency can be used as a parameter that will determine in which mode the system is currently operating.

One particular observation worth noting is depicted in Figure 8. Here, we observe a tendency for an increase in friction force and contact area with the displacement of the indenter. This occurs because the elastomer surface in a real experiment is always inclined at a slight angle to the path of the indenter, as it is not possible to perfectly orient them in parallel. In this scenario, with the indenter shift, the indentation depth d increases due to this inclination. In Figure 8, the effect of the inclination becomes visually noticeable because the distance traveled by the indenter in this experiment is much greater than in the experiments described above (see, for example, Figure 2).

Above, Figure 3b shows the dependencies of the normal force F_N on the indentation depth d in the phase of pulling the indenter out of the elastomer volume, where the lower curve corresponds to the experiment, the data of which are shown in Figure 8. A second experiment was performed immediately after this experiment, and the upper curve in Figure 3b shows the normal force dependence in this experiment. From the comparison of the dependencies shown in Figure 3b, it follows that the adhesion force in the second experiment was significantly lower. Dependencies of friction force F_x and contact area A for the second experiment with indenter with a radius $R = 40$ mm are shown in Figure 9, from which it follows that here the stable "mode 3" was realized. This fact confirms the conclusions made above in the discussion of Figure 3a that the transition between "mode 1" and "mode 3" is caused by contamination of the indenter surface and a decrease in the specific work of adhesion.

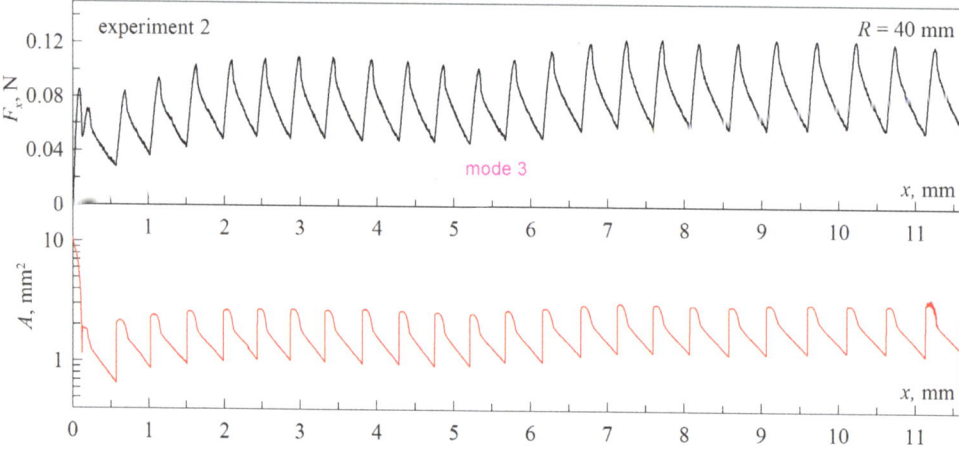

Figure 9. (**Top panel**) Dependencies of the friction force F_x on the indenter tangential displacement x, corresponding to the experiment with an indenter with a radius $R = 40$ mm; (**Bottom panel**) Corresponding dependence of the contact area A on the displacement x. Both dependencies correspond to the second experiment, which was performed immediately after the first experiment, the results of which are shown in Figure 8. Supplementary videos (Videos S4 and S5) are available for the figure.

For completeness of representation of the effect considered in this paper, we conducted an additional experiment with an indenter of an even larger radius $R = 50$ mm, the results of which confirm the conclusions above but do not introduce anything new; therefore, the data of this experiment are not included in the main text of the paper and are placed in the supplementary material (Figures S1–S3), which also contains videos of the experiment (Videos S6 and S7).

4. Conclusions

In this work, it has been shown experimentally that during friction of surfaces, between which there is a pronounced adhesive interaction, a transition between two different friction modes is observed. At the beginning of the motion, the system exhibits a stick–slip mode, in which global slippage of the rubber over the indenter surface occurs when the friction force and tangential stresses reach maximum values. As a result of this slippage, the friction force and stresses decrease dramatically and must again reach their maximum values for the next act of slippage. As the indenter moves further, another steady-state mode is established in which there is no sudden global slip of the rubber over the indenter. In this second mode, a stationary sliding region exists at the trailing edge of the contact where slip is realized at a velocity similar to the shear rate of the indenter. The tangential stresses in this region always take the maximum value corresponding to the rubber sliding. The presence of a region of stationary sliding leads to the fact that a sudden decrease in friction force becomes impossible. During friction in such a regime, there are acts of periodic discontinuous contact propagation at the front edge when new elastomer regions, which are initially unloaded in the tangential direction, come into contact. At further friction, there is a simultaneous loading of newly adhered rubber regions and their gradual withdrawal from the contact, which leads to a complex form of dependencies of friction force and contact area on the indenter shear. When the contact is narrowed down to the smallest size of the region in which steady-state sliding is always realized, contact spreads again. This mechanism leads to a significant reduction in the frequency of transitions between maximum and minimum values of friction force compared to the stick–slip mode. The first stick–slip mode is realized only briefly for a fresh, recently treated indenter surface. Experiments with indenters of different radii (30, 40, and 50 mm) indicate that the transition from stick–slip to another mode of friction occurs due to contamination of the indenter surface during its interaction with the elastomer. Despite the existence of a large number of papers on the study of adhesive tangential contact, we have not found such a transition effect between friction modes in the literature. Therefore, the proposed work is innovative and important from a fundamental point of view for a better understanding of the processes occurring in the contact between rubbing surfaces between which adhesive interaction is strongly pronounced.

Supplementary Materials: The following supporting information can be downloaded at: https://www.mdpi.com/article/10.3390/lubricants12040110/s1, Video S1: A spherical indenter with a radius $R = 30$ mm was immersed to a depth of $d_{max} = 0.2$ mm with a velocity of $v = 1$ μm/s into a layer of TANAC CRG N3005 elastomer with a thickness of $h = 5$ mm. After reaching the maximum depth, the indenter was pulled, with the same velocity, out of the rubber layer to a zero level where, in theory, the indentation should be $d_0 = 0$ mm. But, because of backlash of the motor, the indentation depth is about $d_0 = 6$ μm. After reaching the "zero" level, the indenter was moved in the tangential direction with a velocity of $v = 1$ μm/s. The video shows the first experiment conducted with the freshly machined indenter with a radius $R = 30$ mm. Separate panels in the video show the time dependencies of the normal (F_N) and tangential (F_x) forces, the contact area (A), and the average tangential stress ($<\tau> = F_x/A$). In addition, the lower left panel shows the evolution of the contact zone; it also shows the current values of the indentation depth (d), the tangential shift of the indenter (x), and the time (t) that has passed since the beginning of the indentation. The video relates to the first panels of Figure 2, Figure 4, and Figure 5 in the article. Video S2: It is similar to the Video S1 experiment, but with one difference, as, in this case, the indenter has already been contaminated after the first experiment. The video relates to the second panels of Figures 2 and 4 in the article. Video S3: It is similar to the Video S2 experiment. The video relates to the third panels of Figures 2 and 4 in the article. Video S4: It is similar to the Video S1 experiment, but for an indenter with a bigger radius of $R = 40$ mm; the tangential displacement of the indenter is also bigger. The video relates to Figure 8 in the article. Video S5: It is similar to the Video S4 experiment, but in this case, the indenter has already been contaminated after the first experiment. The video relates to Figure 9 in the article. Video S6: It is similar to the Video S4 experiment, but the indenter has a bigger radius of $R = 50$ mm. The video relates to Figure S1 in the supplementary material. Video S7: It is similar to the Video S6 experiment, with the difference that the indenter is now contaminated after the first experiment. The video relates to Figure S2 in the supplementary material. Figure S1: Dependencies of the tangential force (F_x) (top panel) and contact area (A) (bottom panel) on tangential shift of the indenter (x) in the first experiment with the indenter with a radius $R = 50$ mm. Supplementary Video S6 is also available (presented data are similar to dependencies obtained with indenter $R = 40$ mm that are shown in Figure 8 in the main article). Figure S2: Dependencies of the tangential force (F_x) (top panel) and contact area (A) (bottom panel) on tangential shift of the indenter (x) in the second experiment with the indenter with a radius $R = 50$ mm. Supplementary Video S7 is also available (presented data are similar to dependencies obtained with indenter $R = 40$ mm that are shown in Figure 9 in the main article). Figure S3: Dependencies of the normal force (F_N) on indentation depth (d) in the first and second experiments with an indenter with a radius $R = 50$ mm. Supplementary Videos S6 and S7 are also available (presented data are similar to dependencies obtained with indenter $R = 40$ mm that are shown in Figure 3b in the main article).

Author Contributions: Conceptualization, scientific supervision, project administration, writing—review and editing, V.L.P.; methodology, software, hardware, conducting of experiments, experimental data analysis and interpretation, validation, writing—original draft preparation, I.A.L.; software, experimental data analysis, visualization (supplementary videos), writing—original draft preparation, T.H.P. All authors have read and agreed to the published version of the manuscript.

Funding: This research was funded by Deutsche Forschungsgemeinschaft (Project DFG PO 810-55-3).

Institutional Review Board Statement: Not applicable.

Informed Consent Statement: Not applicable.

Data Availability Statement: The datasets generated for this study are available upon request from the corresponding authors.

Conflicts of Interest: The authors declare no conflicts of interest.

References

1. Zheng, B.; Tang, J.; Chen, J.; Zhao, R.; Huang, X. Investigation of Adhesion Properties of Tire—Asphalt Pavement Interface Considering Hydrodynamic Lubrication Action of Water Film on Road Surface. *Materials* **2022**, *15*, 4173. [CrossRef] [PubMed]
2. Al-Assi, M.; Kassem, E. Evaluation of Adhesion and Hysteresis Friction of Rubber–Pavement System. *Appl. Sci.* **2017**, *7*, 1029. [CrossRef]
3. Nishi, A.; Wakasugi, Y.; Watanabe, K. Design of a robot capable of moving on a vertical wall. *Adv. Robot.* **1986**, *1*, 33–45. [CrossRef]
4. Fang, G.; Cheng, J. Advances in Climbing Robots for Vertical Structures in the Past Decade: A Review. *Biomimetics* **2023**, *8*, 47. [CrossRef] [PubMed]
5. Autumn, K.; Liang, Y.A.; Hsieh, S.T.; Zesch, W.; Chan, W.P.; Kenny, T.W.; Fearing, R.; Full, R.J. Adhesive force of a single gecko foot-hair. *Nature* **2000**, *405*, 681–685. [CrossRef]
6. Federle, W.; Labonte, D. Dynamic biological adhesion: Mechanisms for controlling attachment during locomotion. *Philos. Trans. R. Soc.* **2019**, *374*, 20190199. [CrossRef] [PubMed]
7. Daltorio, K.A.; Gorb, S.; Peressadko, A.; Horchler, A.D.; Ritzmann, R.E.; Quinn, R.D. A Robot that Climbs Walls using Micro-structured Polymer Feet. In *Climbing and Walking Robots*; Tokhi, M.O., Virk, G.S., Hossain, M.A., Eds.; Springer: Berlin/Heidelberg, Germany, 2006; pp. 131–138. [CrossRef]
8. Greenwood, J.A.; Johnson, K.L. The mechanics of adhesion of viscoelastic solids. *Philos. Mag. A* **1981**, *43*, 697–711. [CrossRef]
9. Maugis, D.; Barquins, M. Fracture mechanics and the adherence of viscoelastic bodies. *J. Phys. D Appl. Phys.* **1978**, *11*, 1989–2023. [CrossRef]
10. Lin, Y.; Hui, C.Y. Mechanics of contact and adhesion between viscoelastic spheres: An analysis of hysteresis during loading and unloading. *J. Polym. Sci. Part B Polym. Phys.* **2002**, *40*, 772–793. [CrossRef]
11. Pérez-Ràfols, F.; Nicola, L. Incipient sliding of adhesive contacts. *Friction* **2021**, *10*, 963–976. [CrossRef]
12. Savkoor, A.R.; Briggs, G. The effect of tangential force on contact of elastic solids in adhesion. *Proc. R. Soc. Lond. A Math. Phys. Sci.* **1977**, *356*, 103–114. Available online: https://www.jstor.org/stable/79410 (accessed on 20 March 2024).
13. Sahli, R.; Pallares, G.; Papangelo, A.; Ciavarella, M.; Ducottet, C.; Ponthus, N.; Scheibert, J. Shear-Induced Anisotropy in Rough Elastomer Contact. *Phys. Rev. Lett.* **2019**, *122*, 214301. [CrossRef] [PubMed]
14. Mergel, J.C.; Sahli, R.; Scheibert, J.; Sauer, R.A. Continuum contact models for coupled adhesion and friction. *J. Adhes.* **2019**, *95*, 1101–1133. [CrossRef]
15. Sahli, R.; Pallares, G.; Ducottet, C.; Ben Ali, I.E.; Al Akhrass, S.; Guibert, M.; Scheibert, J. Evolution of real contact area under shear and the value of static friction of soft materials. *Proc. Natl. Acad. Sci. USA* **2018**, *115*, 471–476. [CrossRef] [PubMed]
16. Shen, L.; Glassmaker, N.J.; Jagota, A.; Hui, C.-Y. Strongly enhanced static friction using a film-terminated fibrillar interface. *Soft Matter* **2008**, *4*, 618–625. [CrossRef] [PubMed]
17. Das, D.; Chasiotis, I. Sliding of adhesive nanoscale polymer contacts. *J. Mech. Phys. Solids* **2020**, *140*, 103931. [CrossRef]
18. Waters, J.F.; Guduru, P.R. Mode-mixity-dependent adhesive contact of a sphere on a plane surface. *Proc. R. Soc. A* **2010**, *466*, 1303–1325. [CrossRef]
19. Lyashenko, I.A.; Popov, V.L.; Pohrt, R.; Borysiuk, V. High-Precision Tribometer for Studies of Adhesive Contacts. *Sensors* **2023**, *23*, 456. [CrossRef] [PubMed]
20. Electronic Resource: Innovation Company TANAC Co., Ltd. Available online: https://www.k-tanac.co.jp/crystalnone (accessed on 20 March 2024).
21. Lyashenko, I.A.; Popov, V.L.; Borysiuk, V. Experimental Verification of the Boundary Element Method for Adhesive Contacts of a Coated Elastic Half-Space. *Lubricants* **2023**, *11*, 84. [CrossRef]
22. Morishita, M.; Kobayashi, M.; Yamaguchi, T.; Doi, M. Observation of spatio-temporal structure in stick–slip motion of an adhesive gel sheet. *J. Physics Condens. Matter* **2010**, *22*, 365104. [CrossRef]
23. Lyashenko, I.A.; Pham, T.H.; Popov, V.L. Effect of Indentation Depth on Friction Coefficient in Adhesive Contacts: Experiment and Simulation. *Biomimetics* **2024**, *9*, 52. [CrossRef] [PubMed]
24. Yamaguchi, T.; Ohmata, S.; Doi, M. Regular to chaotic transition of stick–slip motion in sliding friction of an adhesive gel-sheet. *J. Phys. Condens. Matter* **2009**, *21*, 205105. [CrossRef] [PubMed]
25. Brochard-Wyart, F.; de Gennes, P.-G. Naive model for stick-slip processes. *Eur. Phys. J. E* **2007**, *23*, 439–444. [CrossRef] [PubMed]
26. Filippov, A.E.; Klafter, J.; Urbakh, M. Friction through Dynamical Formation and Rupture of Molecular Bonds. *Phys. Rev. Lett.* **2004**, *92*, 135503. [CrossRef] [PubMed]

27. Berman, A.D.; Ducker, W.A.; Israelachvili, J.N. Origin and Characterization of Different Stick—Slip Friction Mechanisms. *Langmuir* **1996**, *12*, 4559–4563. [CrossRef]
28. Yan, C.; Chen, H.-Y.; Lai, P.-Y.; Tong, P. Statistical laws of stick-slip friction at mesoscale. *Nat. Commun.* **2023**, *14*, 6221. [CrossRef] [PubMed]
29. Khudoynazarov, K. Longitudinal-Radial Vibrations of a Viscoelastic Cylindrical Three-Layer Structure. *Facta Univ. Ser. Mech. Eng.* **2024**, *online first*. [CrossRef]
30. Carbone, G.; Mandriota, C.; Menga, N. Theory of viscoelastic adhesion and friction. *Extreme Mech. Lett.* **2022**, *56*, 101877. [CrossRef]
31. Popov, V.L.; Li, Q.; Lyashenko, I.A.; Pohrt, R. Adhesion and friction in hard and soft contacts: Theory and experiment. *Friction* **2021**, *9*, 1688–1706. [CrossRef]
32. Lyashenko, I.A.; Filippov, A.E.; Popov, V.L. Friction in Adhesive Contacts: Experiment and Simulation. *Machines* **2023**, *11*, 583. [CrossRef]
33. Argatov, I.I.; Lyashenko, I.A.; Popov, V.L. Adhesive sliding with a nominal point contact: Postpredictive analysis. *Int. J. Eng. Sci.* **2024**, *accepted*.
34. Yoshizawa, H.; Israelachvili, J. Fundamental mechanisms of interfacial friction. 2. Stick-slip friction of spherical and chain molecules. *J. Phys. Chem.* **1993**, *97*, 11300–11313. [CrossRef]
35. Jäger, J. Axi-symmetric bodies of equal material in contact under torsion or shift. *Arch. Appl. Mech.* **1995**, *65*, 478–487. [CrossRef]
36. Papangelo, A. On the Effect of Shear Loading Rate on Contact Area Shrinking in Adhesive Soft Contacts. *Tribol. Lett.* **2021**, *69*, 48. [CrossRef]
37. Mergel, J.C.; Scheibert, J.; Sauer, R.A. Contact with coupled adhesion and friction: Computational framework, applications, and new insights. *J. Mech. Phys. Solids* **2021**, *146*, 104194. [CrossRef]

Disclaimer/Publisher's Note: The statements, opinions and data contained in all publications are solely those of the individual author(s) and contributor(s) and not of MDPI and/or the editor(s). MDPI and/or the editor(s) disclaim responsibility for any injury to people or property resulting from any ideas, methods, instructions or products referred to in the content.

Article

Calculation and Validation of Planet Gear Sliding Bearings for a Three-Stage Wind Turbine Gearbox

Huanhuan Ding [1,*], Ümit Mermertas [2], Thomas Hagemann [1] and Hubert Schwarze [1]

[1] Institute of Tribology and Energy Conversion Machinery, Clausthal University of Technology, 38678 Clausthal-Zellerfeld, Germany; hagemann@itr.tu-clausthal.de (T.H.); schwarze@itr.tu-clausthal.de (H.S.)

[2] Envision Energy CoE GmbH, 44629 Dortmund, Germany; uemit.mermertas@envision-energy.com

* Correspondence: ding@itr.tu-clausthal.de; Tel.: +49-5323-72-5007

Abstract: In recent years, the trend towards larger wind turbines and higher power densities has led to increasing demands on planet gear bearings. The use of sliding bearings instead of rolling bearings in planetary bearings makes it possible to increase the power density with lower component costs and higher reliability. Therefore, the use of planet gear sliding bearings in wind turbine gearboxes has become more common. However, the flexible structure and complex load conditions from the helical tooth meshes lead to highly complex elastic structure deformation that modifies the lubricant film thickness and pressure distribution and, thus, has to be considered in the calculation of the bearing's load-carrying capacity. This paper introduces a highly time-efficient calculation procedure that is validated with pressure measurement data from a three-stage planetary gearbox for a multi-megawatt wind energy plant. The investigations focus on three main objectives: (i) analyses of experimental and predicted results for different load cases, (ii) validation of the results of planet gear sliding bearing code, and (iii) discussion on mandatory modeling depths for the different planet stages. Results indicate the necessity of further research in this field of applications, particularly for the third-stage bearings.

Keywords: planet gear sliding bearing; validation; structural deformation; load carrying capacity; wind turbine gearbox

Citation: Ding, H.; Mermertas, Ü.; Hagemann, T.; Schwarze, H. Calculation and Validation of Planet Gear Sliding Bearings for a Three-Stage Wind Turbine Gearbox. *Lubricants* **2024**, *12*, 95. https://doi.org/10.3390/lubricants12030095

Received: 15 February 2024
Revised: 5 March 2024
Accepted: 9 March 2024
Published: 15 March 2024

Copyright: © 2024 by the authors. Licensee MDPI, Basel, Switzerland. This article is an open access article distributed under the terms and conditions of the Creative Commons Attribution (CC BY) license (https://creativecommons.org/licenses/by/4.0/).

1. Introduction

In recent decades, wind turbine applications for renewable power generation have rapidly grown. Current trends in wind turbine design tend to increase turbine capacity, particularly for offshore plants. While the first offshore farm was installed with 450 kW turbines in 1991, turbine capacities reached 8–10 MW in 2017 [1]. Recently, a 14 MW wind turbine offshore was developed, illustrating the rapid speed in research and market demand in this field [2]. Increasing turbine capacity incorporates new challenges for its components. This fact has led to the application of sliding bearings in planet gear stages, which offer new degrees of freedom for the gearbox design but incorporate load situations that do not exist in other sliding-bearing applications. The axial force components of the helical gear mesh lead to a moment load for the planet bearing that has to be restored by the lubricant film and causes a significant tilting movement between pin and planet that exceeds commonly known levels of misalignment in sliding bearing applications [3]. Moreover, wind turbine operating conditions are characterized by low rotational speeds and heavy loads, which lead to potential mixed friction domains in the entire load and speed range. These properties are accompanied by high demands on lubricant design [4] and require the consideration of wear [5,6]. Lucassen et al. [7] developed a methodology to identify critical operating conditions for planetary sliding bearings at different rotational speeds and torques based on elastohydrodynamic (EHD) predictions.

The extreme load conditions in wind turbine planet bearings are combined with significant structural deformation due to the compact, lightweight design of the gearbox. Prölß [8] investigated the impact of elastic structural deformation, which results in significantly decreased maximum pressure, higher film thickness, and, consequently, less contact intensity and wear. Hagemann et al. [3,9] present a comprehensive theoretical model for sliding planetary gear bearings based on the previous achievements of Prölß [8]. The results show that wear mainly occurs at the bearing edges, and the new resulting shape of the lubricant gap increases the bearing load carrying capacity due to wear. In addition, an optimized axial crowning of the contact partners reduces the local maximum contact intensity and maximum pressure. The elastic structural deformation provides film thicknesses with a low gradient in a circumferential direction of the load zone, which leads to a wider pressure distribution and a lower maximum pressure level. The deformation behavior depends on structural stiffness, which is predominantly influenced by rim thickness, in the case of the planet. This topic has been comprehensively investigated by researchers who study rolling element planet gear bearings. Dong et al. [10] investigate rim deformation for planet gears with rolling bearings and study its effect on the bearing load distribution. The authors obtained an optimal rim thickness for maximum bearing service life. Jones and Harris [11] and Fingerle et al. [12] pointed out that the ratio between mesh forces and rim thickness affects the magnitude of oval deformation of the planet and the service life of rolling element bearings. While general deformation phenomena are comparable for planets with sliding bearings, actual studies on the impact of rim thickness on their operating characteristics are missing.

In general, thermal effects have to be considered in the calculation of sliding bearings starting at a certain level of body temperature increase. Hagemann et al. [13] considered thermo-mechanical deformation in their analysis of a large high-speed turbine journal bearing. The results show a reduction in film thickness and an increase in maximum temperature and pressure for a large five-pad tilting-pad journal bearing that is validated with experimental data. Linjamaa et al. [14] pointed out that the effects of elastic and thermal deformation on journal bearing performance increase as the bearing is heavily loaded. Zhang et al. [15] investigated the relevance of thermal deformation on sliding planet bearings for wind turbine applications and proved their relevance for theoretical analysis. Gong et al. [16] analyzed the radial clearance for journal bearings supporting planet gears (JBSP), considering thermal deformation. The authors found that too tight clearances can lead to too thin films between pins and planets under large thermal deformation. They defined the optimization of the clearance as a task for the design procedure.

The confirmation of theoretical models for planet-bearing analysis requires validation with test results. For this purpose, tests have already been carried out on component test rigs [6,15,17,18]. However, the kinematics and the surrounding structure on these test rigs differ from practical gear units as they represent a replacement system of the actual arrangement and do not feature the carrier as an essential component of the planetary gear stage. In contrast to the test rigs, the planet also rotates around the sun gear, and the planet carrier has a different structural stiffness, which may influence the calculation. Currently, there is a lack of experimental data for sliding planet bearings in the literature. Therefore, this paper presents a planet-bearing analysis validated with its own measurements from a full-scale, three-stage multi-megawatt planetary gearbox. The bearing geometries were designed and optimized prior to testing based on the computational model from ref. [3,9]. All bearings feature an axial crowning to increase minimum film thickness and to ensure that the sliding bearings operate in the hydrodynamic regime in wide ranges of load-speed combinations. The validation procedure indicates that enhancements of the theoretical analyses are required for the third-stage bearings. In particular, local thermal deformation has to be considered as a novel aspect in planet-bearing design for wind turbine gearboxes.

2. Materials and Methods

2.1. Hydrodynamic Bearing Model

The lubricant film pressure in the bearing is calculated with a generalized average Reynolds Equation accounting for three-dimensional viscosity and cavitation, which is solved based on Elrod's algorithm [19]. For the calculation of the temperature distribution, a three-dimensional energy equation is solved in the gap. The energy equation is coupled with a heat conduction equation of the pin and the planet. A constant temperature distribution in a circumferential direction due to sufficiently high rotational speeds is assumed in the planet. A more detailed description of the bearing model is included in ref. [3].

2.2. Method for Calculation of Planet and Pin Deformation

A weak coupling between fluid and structure analysis is utilized for the simulation of bearing operating behavior under consideration of elastic structural deformation. Based on the assumption of a linear structure model, the constant structural elasticity information resulting from geometries, material properties, and boundary conditions are saved in an a priori-determined reduced stiffness matrix. According to Guyan's theory [20], the total stiffness matrix can be reduced at the master nodes, i.e., the nodes where the load is applied to the structure. This enables the calculation of deformations with a relaxation at many iteration steps in the planet gear-bearing code to achieve convergence. In this paper, the stiffness matrix is reduced to a sliding surface with 128 nodes in the circumferential direction and 32 nodes in the axial direction. The convergence criteria in the bearing code are set to a maximum local deformation change below 1.0 μm between two iterative steps, and the maximum local pressure modifies by less than 0.1 MPa. More details on the method for calculation of planet and pin deformation can be found in ref. [9].

2.3. Investigated Planet Gear Sliding Bearing

The basic parameters of the three-stage planet gear sliding bearings for a multi-megawatt wind turbine gearbox by the customer are listed in Table 1. The positions of the pressure sensors in the experiment are located in the zone of expected maximum load on the rotor side (RS), generator side (GS), and in the middle of the bearing (Mid). All pins of the three investigated stages feature an axial crowning to ensure a nearly homogeneous pressure distribution in an axial direction of the bearing and to reduce the high edge loading.

Table 1. Three-stage planet gear sliding bearing parameters.

Parameter	Stage 1	Stage 2	Stage 3
Bearing width/diameter, -	1.5	1.2	0.83
Nominal rotational speed, rpm	30	85	271
Nominal specific bearing load, MPa	13.5	12.7	10.4
Lubricant		ISO VG 320	
Lubricant density kg/m^3		853 @ 40 °C	
Lubricant specific heat capacity kJ/(kg·K)		2.0 @ 20 °C	
Lubricant thermal conductivity, W/(m·K)		0.13	

2.4. FEM Model for Structure Analysis: Material, Boundary Condition and Mesh Load

The planet gear bearings in the three stages have similar structural features. Figure 1 shows the CAD model of the investigated bearings, with five planets and pins as an example. This model is characterized by the periodic symmetry of the planet carrier. Neglecting any transmission errors, one-fifth of the model is utilized for the FE analyses to calculate the deformation. Additionally, the planet and pin are divided into three and two parts to enable different meshing of the structure, and the contact surfaces between the parts are defined as 'Bonded' to treat them as one physical body.

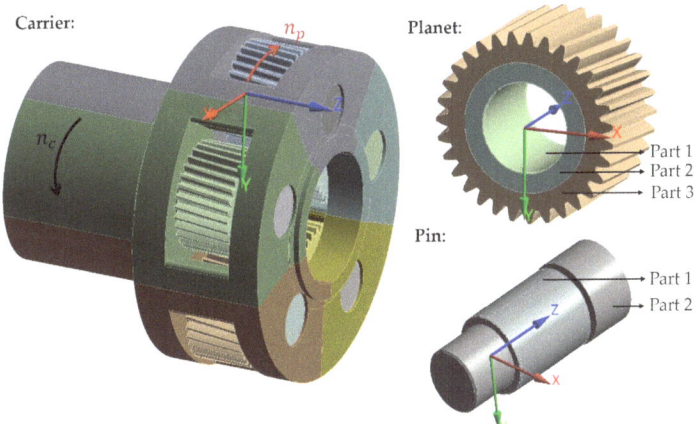

Figure 1. CAD model of planet gear sliding bearing.

Figure 2 shows the meshes of the planet and pin with carrier. The inner structure of the planet and the outer structure of the pin are both cylindrical geometries and, thus, are discretized hexahedral structural meshes with 128 in circumferential and 32 elements in axial directions. Tetrahedral meshes approximate the remaining structures. In this paper, the models of planet and pin with carrier are discretized to have 173,584 nodes and 391,536 nodes, respectively. Appendix A includes a grid convergence study. Its results indicate that the structural discretization used in this paper is appropriate and has a lower impact on the results than most other uncertainties in the entire simulation procedure.

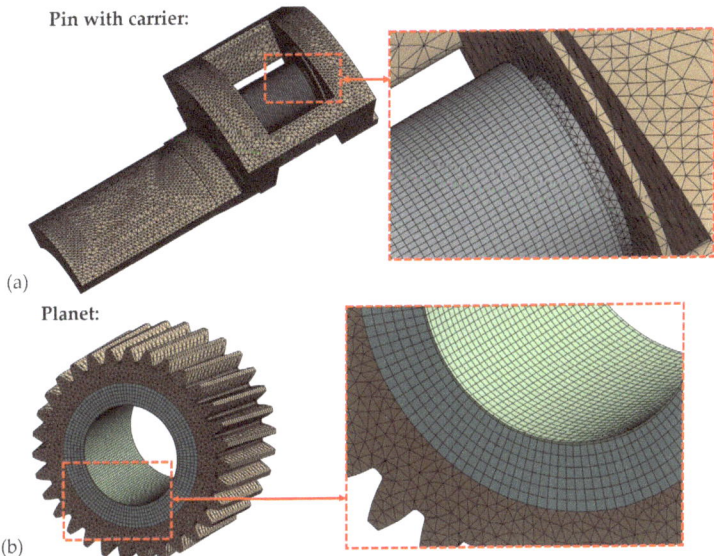

Figure 2. Mesh of pin with carrier (**a**) and planet (**b**).

Figure 3 explains the boundary conditions for the planet and pin with the carrier. For the planet, there are mesh forces via 'Remote Force' on the tooth flanks engaging with ring gear and sun gear which are in mechanical equilibrium with the gravity and oil film forces on the sliding surface. In order to avoid the rigid body movement of the planet due to numerical inaccuracies in the mechanical equilibrium, a 'Remote Displacement' with zero degree of freedom is defined on the inner surface of the outer structure of the planet (part 3) in Figure 3a while allowing a deformation on this surface. The connection between the pin and carrier in Figure 3b is defined as 'Bonded', simulating both components as one body. Furthermore, the tapered rolling element bearings supporting the carrier in the turbine housing structure restrict the radial movement of the carrier. The inner surface on the rotor side of the carrier is defined as 'fixed' to provoke the twist deformation caused by the transmitted torque between the rotor hub and generator. Periodic symmetry conditions are set on two cross sections of the one-fifth model of a carrier that interface to the neighboring part.

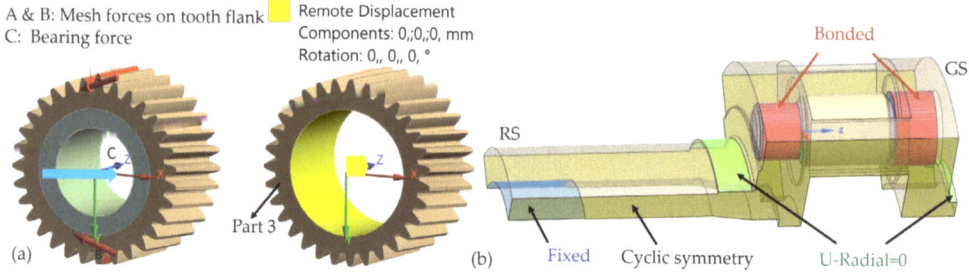

Figure 3. Boundary condition of planet (**a**), and pin with carrier (**b**).

Table 2 shows the material properties of the planet, pin, and carrier. Since the structural behavior is assumed linear, only Young's Modulus and Poisson's Ratio are required to determine the stiffness properties.

Table 2. Material properties.

Parameter	Planet	Pin	Carrier
Young's Modulus, MPa	210,000	210,000	176,000
Poisson's Ratio, -	0.3	0.3	0.275

Figure 4 shows the three components of the mesh forces on the tooth flanks on the sun gear and ring gear side, as well as their offset positions a and b from the planet center plane. The sum of the two tangential forces is equal to the oil film force, F_{sc}. The axial force caused by the helix angle of teeth generates an oil film moment of reaction $M_{sc,x}$ about the x-axis. As the load distribution on the tooth flanks of the investigated three-stage planetary bearings is not homogeneous in the lateral direction, additional oil film moments of reaction $M_{sc,y}$ are generated about the y-axis.

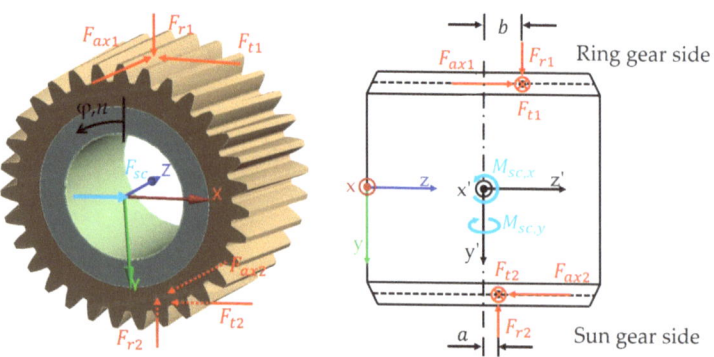

Figure 4. Mesh forces on the tooth flanks.

3. Results

In this section, T_r refers to the relative input torque and 100% T_r corresponds to the nominal load situation with a specific bearing load of 13.5 MPa, 12.7 MPa, and 10.4 MPa for the first to third stages, respectively. Based on a comparison between the predicted minimum film thickness and the roughness of the sliding surfaces, all bearings operate in the hydrodynamic regime for the loads investigated in this paper.

3.1. Predicted Film Thickness and Structure Deformation

The first-stage bearing is utilized as an example to compare the film thickness under part-load $T_r = 20\%$ and nominal load $T_r = 100\%$ conditions. The dimensionless axial coordinates $z = 0$ and $z = 1$ in Figure 5 represent the generator and rotor side of the bearing. As shown by the black point in Figure 5, the minimum film thickness is smaller for the nominal load $T_r = 100\%$, with a value of 8.9% of absolute radial clearance. In the area localized by the red line, the film thickness is below 21.1% of absolute radial clearance ΔR. A comparison of the results in Figure 5a,b indicates that this area is much larger for the nominal load $T_r = 100\%$, as expected. The film thickness in this load zone remains on a level slightly above the minimum film thickness, as shown by the blue line in an angular span ranging from $\phi = 214°$ to $338°$.

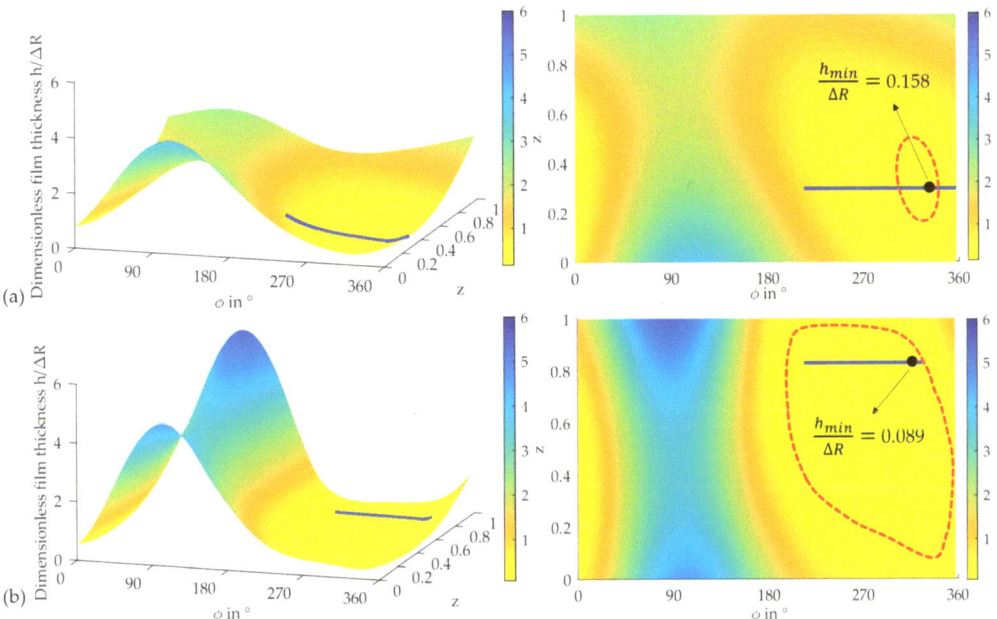

Figure 5. Film thickness for $T_r = 20\%$ (**a**) and 100% (**b**) loads of first-stage bearing.

The dimensionless total radial deformations on the sliding surface of the planet and pin for loads $T_r = 20\%$ and 100% are shown in Figure 6a,b, respectively. The region with the larger deformation is closer to the generator side (GS, $z = 0$) for the relative load $T_r = 20\%$ in Figure 6a due to the mesh force offset position, while an axially more homogenous deformation can be observed for the nominal load case in Figure 6b. Additionally, both deformation fields feature significant local maximum values in the load zone between $\phi \approx 240°$ and $310°$ as well as in the area 180 degrees offset from it, which exhibits a characteristic oval shape. This property can also be observed in Figure 7 and is caused by the highly flexible planet deformed by the combination of fluid film and mesh forces. From the left to the right, the axial position z in Figure 7 shifts from the generator side (GS) to the center (Mid) and the rotor side (RS) of the bearing individually. The radial bearing clearance is expressed in a dimensionless form relative to the absolute radial clearance ΔR, where ΔR_i is contour of pin, planet or resultant gap. Figure 7a displays the magnitude of the oval shape for load $T_r = 20\%$ is decreasing from GS to RS, since local pressure loads concentrate on the GS. In contrast, the resultant contour for load $T_r = 100\%$ shows the oval shape in Figure 7b over the entire bearing width due to the broader pressure distribution in the axial direction.

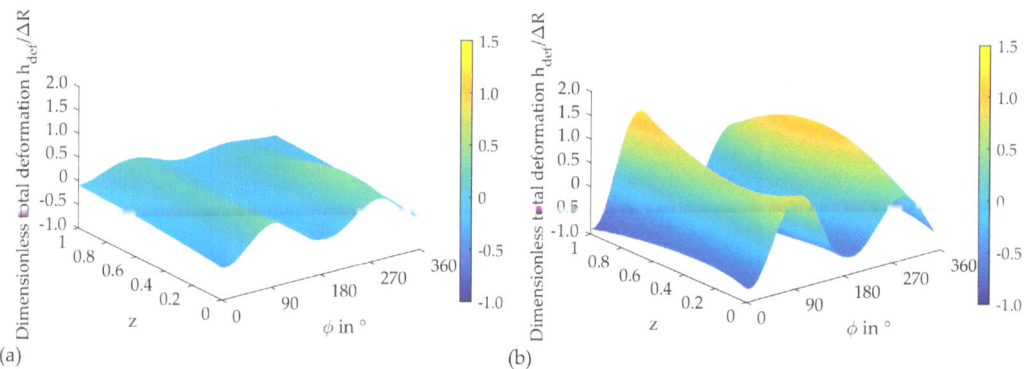

Figure 6. Total deformation field for $T_r = 20\%$ (**a**) and 100% (**b**) loads of first-stage bearing.

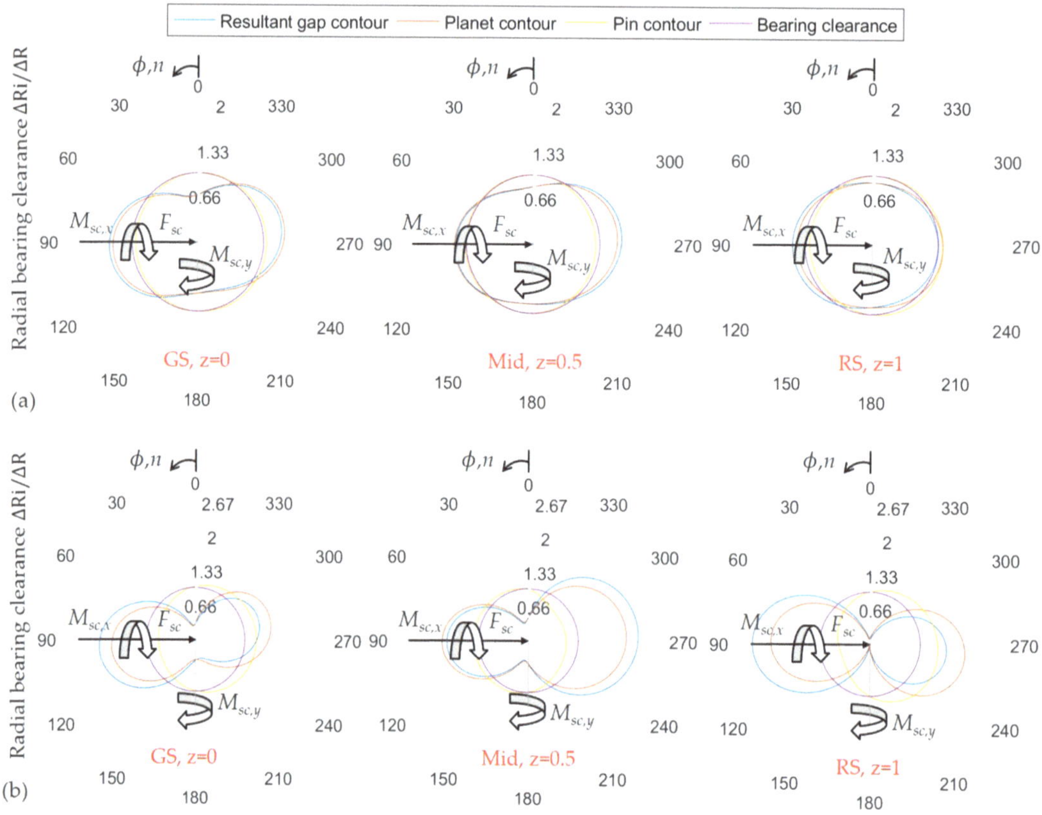

Figure 7. Gap contour for $T_r = 20\%$ (**a**) and 100% (**b**) loads of first-stage bearing.

3.2. Validation of Pressure Distribution

Figure 8 presents the predicted rising trend of the dimensionless maximum hydrodynamic pressure in the planet gear bearings of the three different gear stages for increasing relative loads, where $P_{max,123}$ means the maximum of all pressures. The maximum hydrodynamic pressure of third-stage bearing under relative load $T_r = 40–110\%$ is higher than that of the other two stages, although its nominal specific bearing load is lower.

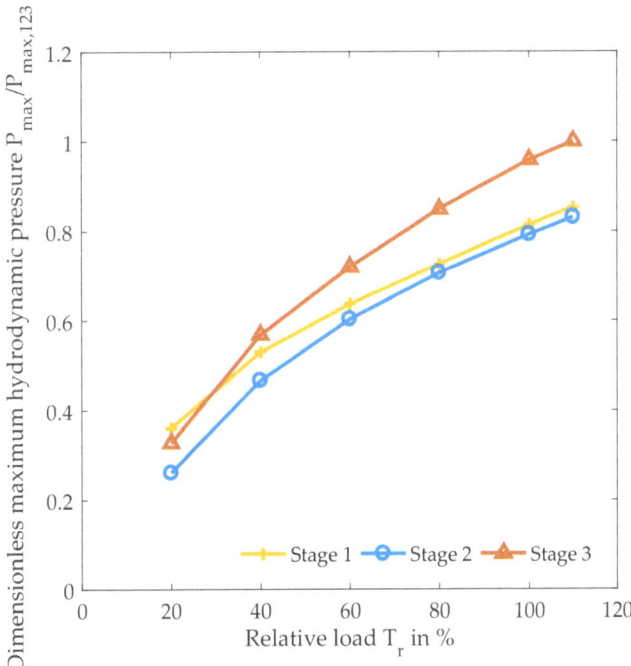

Figure 8. Maximum hydrodynamic pressure for variable relative loads $T_r = 20–110\%$ for the bearings in the three stages.

Figure 9 shows the predicted pressure distribution in the lubricant gap for relative loads of $T_r = 20\%, 60\%,$ and 100%, which is expressed in dimensionless form relative to the maximum pressure at $T_r = 100\%$ load in each stage. The locations of the three pressure sensors for each pin in the experiment are marked with red points to allow comparison between measured and simulated results. As already discussed for the predicted deformation and film thickness distribution, the pressure in the first-stage bearing for $T_r = 20\%$ load in Figure 9a is mainly concentrated on the GS due to the additional moment about the y-axis. With a rising load on the bearing, the range of high pressure becomes broader both in axial and circumferential directions. This tendency can be observed for all three stages. However, the circumferential growth of the load section is significantly lower for the third stage. In addition, the planet bearings in the first and second stages exhibit two peak pressure sections with local maxima in a circumferential direction at $T_r = 100\%$ load, while the third stage has a more homogeneous pressure distribution.

Table 3 includes the dimensionless axial offset positions of the mesh forces on tooth flanks relative to the bearing center for loads $T_r = 20\%, 60\%,$ and 100%, i.e., the lever arms for the moment are represented by a and b. Positive and negative values of a and b represent the offset position in the GS and RS direction, respectively. The resulting pressure distribution closer to the generator side for load $T_r = 20\%$ results from the fact that the values of a and b in Table 3 are both positive and comparably high, which leads to a moment on the planet about the y-axis. Therefore, it results in a larger elastic deformation of the pin and planet on the GS side, which explains the more significant oval shape on GS in Figure 7a. On the contrary, the absolute values of a and b for load $T_r = 100\%$ of the third-stage bearings are quite small, generating a very low additional moment about the y-axis, so the pressure distribution in Figure 9c is more homogeneous in the axial direction.

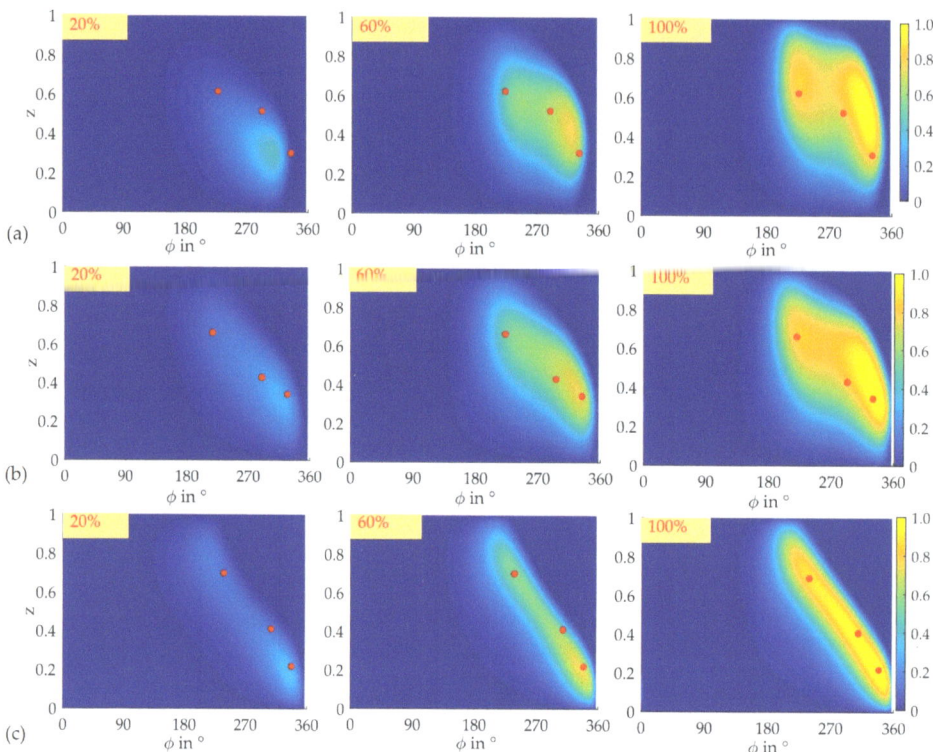

Figure 9. Pressure distributions for $T_r = 20\%, 60\%$, and 100% loads of first- (**a**), second- (**b**), and third-stage (**c**) bearing.

Table 3. Dimensionless axial offset position relative to the bearing center for three-stage planet gear sliding bearing.

	Stage 1		Stage 2		Stage 3	
	Offset a	Offset b	Offset a	Offset b	Offset a	Offset b
20%	0.38	0.19	0.13	0.14	0.021	0.35
60%	0.18	−0.09	0.09	−0.04	0.031	0.12
100%	0	−0.22	0.03	−0.11	0.021	0.021

Experiments were performed to investigate the load-carrying capacity of bearings for loads between $T_r = 20\%$ and 100% with increments of 10% load. The letters A to G are used to identify different pins of the same bearing stage. If pressure sensors fail during the experiments, their values are omitted. A comparison of the experimental and simulation results over the entire load range is summarized in Figure 10, where the pressures are expressed in dimensionless form relative to the maximum pressure of bearing with $P_{max,1}$, $P_{max,2}$, and $P_{max,3}$ for the respective stage. In addition to the measured and predicted sensor pressures, the maximum predicted pressure on the sliding surface is presented. In combination with Figure 9a,b, the characteristics in Figure 10 show that maximum predicted pressure is located close to a pressure sensor in the case of the second stage bearing while there is a bigger distance between the highest local pressure load and the sensor location in case of the first stage. The simulation only considers mechanically induced deformation and neglects thermal deformation based on the assumption of sufficiently low-temperature levels in this low-speed application. The predicted pressures at the sensor position for the first- and second-stage bearings in Figure 10a,b show very good agreement with the measurement data of all pins at the three axial positions over the entire load range. The deviations are in the range of 0.1 to 5.5% for first- and second-stage, except slightly higher values for the generator side at $T_r = 20\%$ to 60% load of the first-stage bearing. Combined with the calculation of the load carrying capacity of each bearing in less than ten minutes, this confirms the reliability as well as the efficiency of the planet gear bearing code.

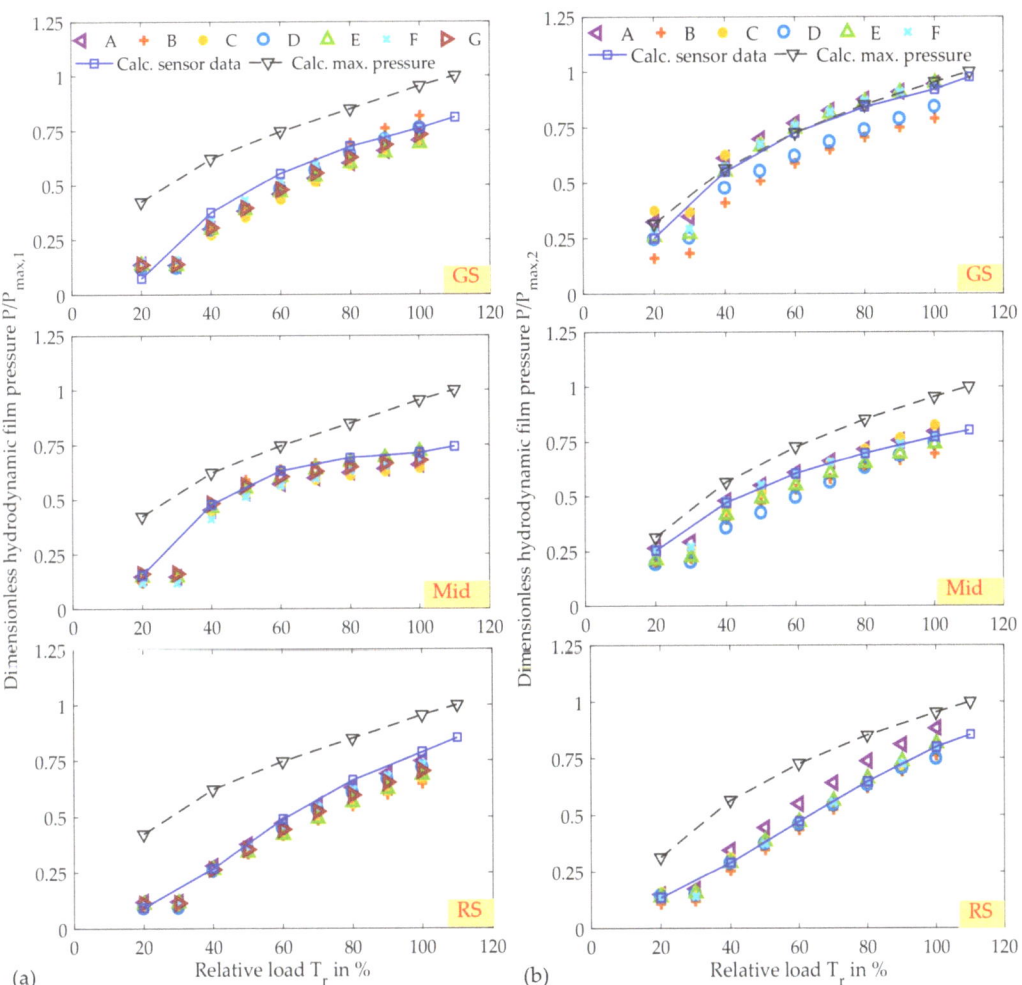

Figure 10. Comparison of pressure between measurement and prediction for the entire load range $T_r = 20$–110% of first- (**a**) and second-stage (**b**) bearing.

3.3. Extended Thermal Deformation Analysis for the Third Stage

More significant differences between measurement and prediction exist for the third-stage bearing. Figure 11a includes measured and predicted pressures for the third-stage bearing. Deviations become higher with increasing load, particularly for the mid-sensor position.

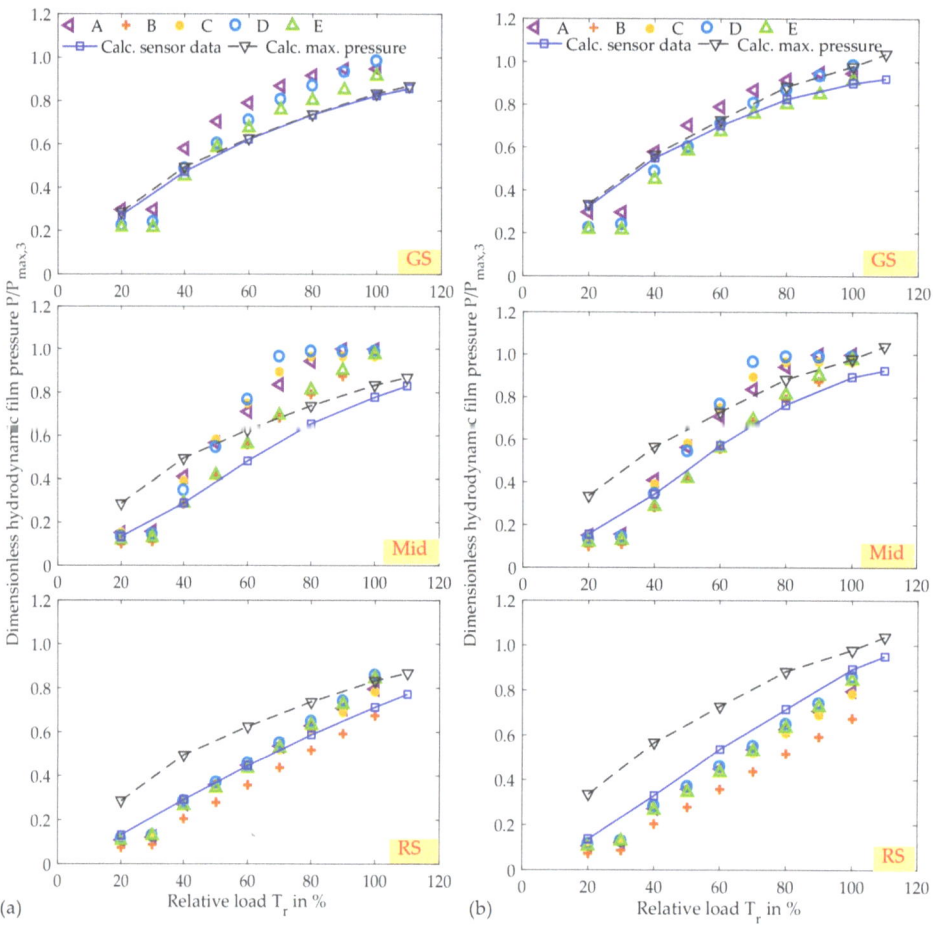

Figure 11. Comparison of pressure between measurement and prediction for the entire load range $T_r = 20$–110% of without (**a**) and with (**b**) thermal expansion of third-stage bearing.

Table 4 shows that the dimensionless experimental and predicted sliding surface temperatures relative to the oil supply temperature in °C at $T_r = 100\%$ load increase as the bearing rotational speed rises from a low speed in the first stage to a high speed in the third stage. The oil supply temperature of all bearings in the three stages is the same. The experimental and predicted temperatures of sliding surfaces in the first two stages bearing increase within a range of 18% and 42% of the oil supply temperature, respectively. The previous validation results show that the thermal expansion caused by the increasing temperature has no significant effect on the load-carrying capacity of the first- and second-stage bearings. However, the temperature of the third-stage bearing is much higher and reaches approximately 76% of the oil supply temperature. This temperature increase might incorporate non-negligible thermal deformation of the bearing components. To verify this conjecture, the temperature fields on sliding surfaces of the pin and planet simulated using the THD analysis are applied as external loads to calculate the thermal deformation of the third-stage bearing for loads $T_r = 20$–110%. Since the thermal deformation changes only slightly with modification of the pressure distribution, a one-time calculation of the thermal deformation at the beginning of the analysis is assumed to be sufficient. The shape of the radial thermal deformations is similar for all load cases. Figure 12 presents exemplary results for $T_r = 100\%$ in dimensionless form for two views. This three-dimensional thermal deformation is approximately parabolic in the axial direction, i.e., the bearing clearance at the bearing edge becomes larger through the positive value at $z = 0$ and 1, while it decreases in the bearing center through the negative value. The total thermal deformation of the pin and planet is regarded as an additional offset crowning, which is used together with the original crowning of the pin to recalculate the pressure for the third-stage bearing in the planet gear bearing code. As shown in Figure 11b, these pressures provide a significantly improved correspondence with the experimental pressure at the generator side (GS) and in the bearing center (Mid). Moreover, the results indicate that the position of maximum pressure at maximum torque load is slightly shifted away from the sensor location due to the consideration of thermal deformation as the deviation between the calculated sensor and calculated maximum pressure increases from Figure 11a to Figure 11b.

Table 4. Comparison of dimensionless temperature from sensor and prediction for $T_r = 100\%$ of three-stage planet gear bearings.

	Stage 1			Stage 2			Stage 3		
	GS	Mid	RS	GS	Mid	RS	GS	Mid	RS
Avg. exp. $T_{S,Exp}$, -	1.13	1.1	1.13	1.34	1.37	1.42	1.67	1.74	1.76
Calc. $T_{S,Exp}$, -	1.18	1.17	1.15	1.4	1.38	1.36	1.64	1.69	1.65
Deviation Δ, %	4.4	6.2	1.8	4.6	0.6	3.7	1.4	3.0	6.3

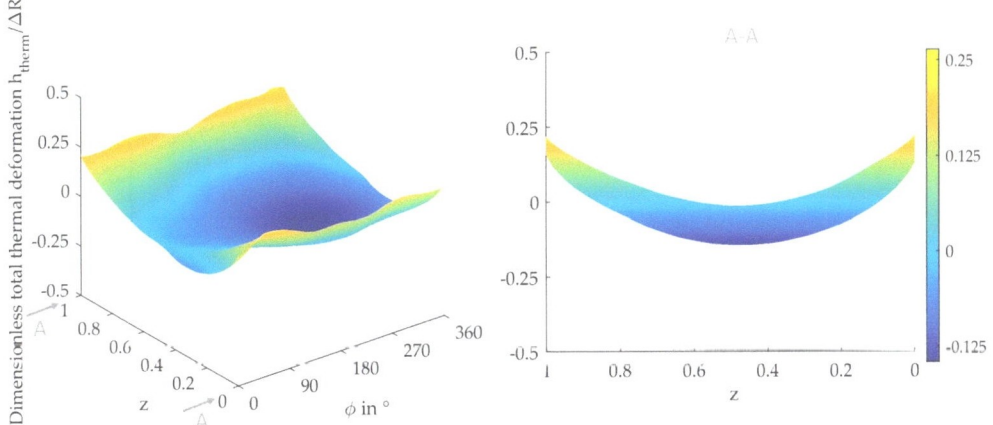

Figure 12. The sum of dimensionless radial thermal deformation on the sliding surface of pin and planet for nominal load $T_r = 100\%$ for third-stage bearing.

Results indicate that an enhancement of the modeling depth by thermal deformation is required for the third-stage bearings. Although the validation results in Figure 11b still show some deviation on the rotor side (RS), there is a significant improvement in the agreement of measured and predicted results.

4. Discussion and Conclusions

This paper investigated the load-carrying capacity of the planet gear sliding bearings for three gear stages of a multi-megawatt wind turbine gearbox. For this purpose, the bearings are analyzed both experimentally and theoretically within a detailed thermo-elasto-hydrodynamic simulation considering the structural deformation. The deformation, oil film thickness, and pressure distribution are analyzed for the entire load range. Comparisons of measured and predicted hydrodynamic pressures provide very good agreement for the first- and second-stage bearings if only mechanical deformation induced by the film pressure is modeled. Results indicate that this approach is not sufficient for the third stage bearings, as deviations between simulation and experiment occur. Here, the additional consideration of approximative thermal deformation that predominantly shows a shape similar to an axial crowning of the pin leads to a significant improvement in simulation results. The simplified approach for the consideration of thermal deformation assumes that a one-time calculation of thermal deformation at the beginning of the analyses is sufficient as thermal deformation only slightly changes by the modification of the pressure distribution. However, the deviations between measurement and prediction remain on an unacceptable level, indicating that a more detailed thermal deformation analysis might be necessary to close the gap between experimental and theoretical results for the third stage. An alignment of the consideration of thermal deformation with the procedure for mechanical one seems to be promising, but the entire calculation procedure should be kept on a complexity level that fulfills the real-time expectations of industrial practice.

Author Contributions: Conceptualization, methodology, T.H. and H.D.; software, H.D.; validation, H.D. and Ü.M.; investigation, writing, and visualization, H.D.; supervision and funding acquisition, H.G. and Ü.M. All authors have read and agreed to the published version of the manuscript.

Funding: This research received no external funding.

Data Availability Statement: Data are contained within the article.

Conflicts of Interest: Author Ümit Mermertas was employed by the company Envision Energy CoE GmbH. The remaining authors declare that the research was conducted in the absence of any commercial or financial relationships that could be construed as a potential conflict of interest.

Appendix A. Grid Convergence Study

The grid convergence study is carried out by calculating the radial deformation of a sliding surface at $\phi = 270°$ across the bearing width for different grid densities of planet and pin with carrier in third-stage bearing as in Figures A1 and A2. The radial deformations are shown in a dimensionless form relative to their maximum value of planet and pin. The results in Tables A1 and A2 demonstrate that the deviation of the selected mesh density from the maximum mesh density is 0.4% and 0.01% for planet and pin with carrier, respectively.

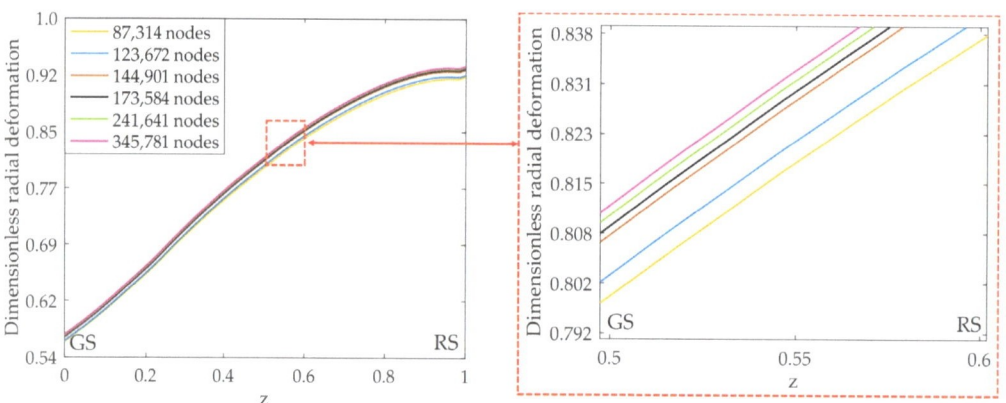

Figure A1. Dimensionless radial deformation of sliding surface at $\phi = 270°$ across the bearing width for different grid densities of a planet in third-stage bearing.

Table A1. Deviation of dimensionless radial deformation of the sliding surface at $\phi = 270°$ and $z = 0.55$ for different grid densities from the maximum mesh density of the planet in third-stage bearing.

Nodes Number	Radial Deformation, -	Deviation, %
345,781	0.8329	-
241,641	0.8314	0.18
173,584	0.8297	0.39
144,901	0.8283	0.55
123,672	0.8220	1.30
87,314	0.8187	1.70

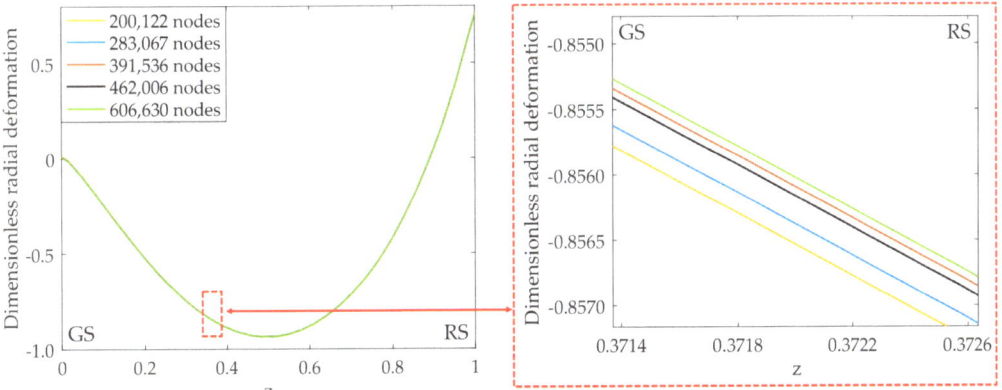

Figure A2. Dimensionless radial deformation of sliding surface at $\phi = 270°$ across the bearing width for different grid densities of pins with a carrier in third-stage bearing.

Table A2. Deviation of dimensionless radial deformation of the sliding surface at $\phi = 270°$ and $z = 0.3718$ for different grid densities from the maximum mesh density of pin with a carrier in third-stage bearing.

Nodes Number	Radial Deformation, -	Deviation, %
606,630	−0.8558	-
462,006	−0.8559	0.02
391,536	−0.8559	0.01
283,067	−0.8561	0.04
200,122	−0.8563	0.06

References

1. Díaz, H.; Guedes Soares, C. Review of the Current Status, Technology and Future Trends of Offshore Wind Farms. *Ocean Eng.* **2020**, *209*, 107381. [CrossRef]
2. Offshore Wind Turbine SG 14-222 DD | Siemens Gamesa. Available online: https://www.siemensgamesa.com/products-and-services/offshore/wind-turbine-sg-14-222-dd (accessed on 13 February 2024).
3. Hagemann, T.; Ding, H.; Radtke, E.; Schwarze, H. Operating Behavior of Sliding Planet Gear Bearings for Wind Turbine Gearbox Applications—Part I: Basic Relations. *Lubricants* **2021**, *9*, 97. [CrossRef]
4. Muzakkir, S.M.; Hirani, H.; Thakre, G.D. Lubricant for Heavily Loaded Slow-Speed Journal Bearing. *Tribol. Trans.* **2013**, *56*, 1060–1068. [CrossRef]
5. Xiang, G.; Han, Y.; Wang, J.; Wang, J.; Ni, X. Coupling Transient Mixed Lubrication and Wear for Journal Bearing Modeling. *Tribol. Int.* **2019**, *138*, 1–15. [CrossRef]
6. Lehmann, B.; Trompetter, P.; Guzmán, F.G.; Jacobs, G. Evaluation of Wear Models for the Wear Calculation of Journal Bearings for Planetary Gears in Wind Turbines. *Lubricants* **2023**, *11*, 364. [CrossRef]
7. Lucassen, M.; Decker, T.; Gutierrez Guzman, F.; Lehmann, B.; Bosse, D.; Jacobs, G. Simulation Methodology for the Identification of Critical Operating Conditions of Planetary Journal Bearings in Wind Turbines. *Forsch. Im Ingenieurwesen* **2023**, *87*, 147–157. [CrossRef]

8. Prölß, M. Berechnung langsam laufender und hoch belasteter Gleitlager in Planetengetrieben unter Mischreibung. *Verschleiß Und Deform.* **2020**. [CrossRef]
9. Hagemann, T.; Ding, H.; Radtke, E.; Schwarze, H. Operating Behavior of Sliding Planet Gear Bearings for Wind Turbine Gearbox Applications—Part II: Impact of Structure Deformation. *Lubricants* **2021**, *9*, 98. [CrossRef]
10. Dong, P.; Lai, J.; Guo, W.; Tenberge, P.; Xu, X.; Liu, Y.; Wang, S. An Analytical Approach for Calculating Thin-Walled Planet Bearing Load Distribution. *Int. J. Mech. Sci.* **2023**, *242*, 108019. [CrossRef]
11. Jones, A.B.; Harris, T.A. Analysis of a Rolling-Element Idler Gear Bearing Having a Deformable Outer-Race Structure. *J. Basic Eng.* **1963**, *85*, 273–278. [CrossRef]
12. Fingerle, A.; Hochrein, J.; Otto, M.; Stahl, K. Theoretical Study on the Influence of Planet Gear Rim Thickness and Bearing Clearance on Calculated Bearing Life. *J. Mech. Des.* **2019**, *142*. [CrossRef]
13. Hagemann, T.; Kukla, S.; Schwarze, H. Measurement and Prediction of the Static Operating Conditions of a Large Turbine Tilting-Pad Bearing Under High Circumferential Speeds and Heavy Loads. In Proceedings of the ASME Turbo Expo 2013: Turbine Technical Conference and Exposition, San Antonio, TX, USA, 3–7 June 2013. [CrossRef]
14. Linjamaa, A.; Lehtovaara, A.; Larsson, R.; Kallio, M.; Söchting, S. Modelling and Analysis of Elastic and Thermal Deformations of a Hybrid Journal Bearing. *Tribol. Int.* **2018**, *118*, 451–457. [CrossRef]
15. Zhang, K.; Chen, Q.; Zhang, Y.; Zhu, J.; Wang, M.; Feng, K. Numerical and Experimental Investigations on Thermoelastic Hydrodynamic Performance of Planetary Gear Sliding Bearings in Wind Turbine Gearboxes. *Tribol. Int.* **2024**, *191*, 109081. [CrossRef]
16. Gong, J.; Liu, K.; Zheng, Y.; Meng, F. Thermal-Elastohydrodynamic Lubrication Study of Misaligned Journal Bearing in Wind Turbine Gearbox. *Tribol. Int.* **2023**, *188*, 108887. [CrossRef]
17. Goris, S.; Ooms, M.; Goovaerts, M.; Krieckemans, K.; Bogaert, R. Plain Bearings for Wind Turbine Gearboxes-Trajectory towards Technology Readiness. In Proceedings of the Conference for Wind Power Drives 2017: Tagungsband zur Konferenz, Aachen, Germany, 7–8 March 2017; p. 327.
18. Meyer, T. Validation of Journal Bearings for Use in Wind Turbine Gearboxes l Wind Systems Magazine. Available online: https://www.windsystemsmag.com/validation-of-journal-bearings-for-use-in-wind-turbine-gearboxes/ (accessed on 31 January 2024).
19. Elrod, H.G. A Cavitation Algorithm. *J. Lubr. Technol.* **1981**, *103*, 350–354. [CrossRef]
20. Guyan, R.J. Reduction of Stiffness and Mass Matrices. *AIAA J.* **1965**, *3*, 380. [CrossRef]

Disclaimer/Publisher's Note: The statements, opinions and data contained in all publications are solely those of the individual author(s) and contributor(s) and not of MDPI and/or the editor(s). MDPI and/or the editor(s) disclaim responsibility for any injury to people or property resulting from any ideas, methods, instructions or products referred to in the content.

Review

Changes in Surface Topography and Light Load Hardness in Thrust Bearings as a Reason of Tribo-Electric Loads

Simon Graf * and Oliver Koch

Chair of Machine Elements, Gears and Tribology (MEGT) RPTU Kaiserslautern-Landau, Gottlieb Daimler Straße 42, 67663 Kaiserslautern, Germany
* Correspondence: simon.graf@rptu.de; Tel.: +49-0631-205-3726

Abstract: The article focuses on the findings of endurance tests on thrust bearings. In addition to the mechanical load (axial load: $10 \leq C0/P \leq 19$, lubrication gap: $0.33~\mu m \leq h0 \leq 1.23~\mu m$), these bearings are also exposed to electrical loads (voltage: $20~Vpp \leq U0 \leq 60~Vpp$, frequency 5 kHz and 20 kHz), such as those generated by modern frequency converters. In a previous study, the focus was on the chemical change in the lubricant and the resulting wear particles. In contrast, this article focuses on the changes occurring in the metallic contact partners. Therefore, the changes in the surface topography are analysed using Abbott–Firestone curves. These findings show that tests with an additional electrical load lead to a significant reduction in roughness peaks. A correlation to acceleration measurements is performed. Moreover, it is shown that the electrical load possibly has an effect on the light load hardness. An increase in the occurring wear could not be detected during the test series. Also, a comparison with mechanical reference tests is made. The article finally provides an overview of different measurement values and their sensitivity to additional electrical loads in roller bearings.

Keywords: tribo-electric contact; bearing currents; surface topography; mechanical and electrical loads; light load hardness

1. Introduction

Rolling bearings are critical components in mechanical engineering, facilitating independent rotational movements between two machine parts. They are exposed to a variety of loads (e.g., mechanical, thermal, chemical) under different operating conditions. These influence the service life of the rolling bearings and can be largely classified and considered by means of service life calculations.

As a result of the increased use of electric motors in combination with fast-switching frequency inverters with IGB (insulated-gate bipolar) or SiC (carbon silicide) transistors [1–3], the motor bearings, for example, may be exposed to electrical loads in addition to mechanical loads. This phenomenon is known as parasitic electrical current passage. This current passage causes accelerated damage to the rolling bearings [4,5] and lubricants [6,7], which can lead to component failure within a short time. Different damage mechanisms can be identified. In the case of lubricants, for example, accelerated oxidation [8,9] or molecular changes in the lubricant chemistry [6,10] can occur. Grey running marks [11–13], accumulations of individual discharge craters [11–13] or electromechanically caused lenticular protrusions known as fluting [14–16] appear on the metallic raceways of the rolling bearings. To date, these electromechanical damage events can only be considered to a limited extent in the design of rolling bearings or other machine elements. The dimensioning parameters currently used in design do not yet provide sufficient reliability in practical applications [13,17–19]. Furthermore, no distinction is made between the types of damage (lubricant oxidation, grey track or fluting) to be dimensioned against. This leads to overdimensioning and unexpected early failures of the components [17–21]. Alternatively, expensive protective measures are taken to limit the parasitic current flow [17,21–23].

Citation: Graf, S.; Koch, O. Changes in Surface Topography and Light Load Hardness in Thrust Bearings as a Reason of Tribo-Electric Loads. *Lubricants* **2024**, *12*, 303. https://doi.org/10.3390/lubricants12090303

Received: 21 June 2024
Revised: 22 August 2024
Accepted: 23 August 2024
Published: 28 August 2024

Copyright: © 2024 by the authors. Licensee MDPI, Basel, Switzerland. This article is an open access article distributed under the terms and conditions of the Creative Commons Attribution (CC BY) license (https://creativecommons.org/licenses/by/4.0/).

By focussing on the damage to the metallic components of a rolling bearing, a correlation between component vibrations and the onset of damage phenomena [23–26] could be demonstrated. The vibration excitation caused by fluting in particular is used to reliably identify these at an early stage and initiate maintenance [27,28]. Furthermore, in [29,30], the component vibration was correlated to the occurring discharge energies of the individual breakdowns. This behaviour was also observed in [18].

Further studies on components damaged by the passage of current focus on the changes that occur in the material as described by microscopy [30–34], metallurgy [30,33,34] or spectroscopy [35]. These publications describe the damage that occurs in more general terms.

In addition, there are also publications that prevent or delay the occurrence of electrical damage through the use of targeted, expensive measures (such as the use of insulation [20] and adaptation of the electrical control system [35]) so that the calculated L10 bearing service life is achieved.

The following investigation continues the results of article [36]. While the changes in the lubricant were considered in detail in [36], the focus here is on the influence on the metallic surface. Different measurement methods are used, which present the test series regarding the changes in the surface topography, the influence of the electrical load on the light load hardness and the results of a vibration analysis.

2. Test Setup

The Chair of Machine Elements, Gears and Tribology (MEGT) has specialised test benches to investigate the electrical loads in rolling bearings of inverter-fed electric motors. These test rigs make it possible to vary electrical and mechanical operating conditions independently of each other, which enables a detailed investigation of the influence of electrical currents on rolling bearings [11,22,36].

The GESA (ger. Gerät zur erweiterten Schmierstoffanalyse—device for advanced lubricant analysis/developed by MEGT) test rig used in the investigations presented here allows the specific investigation and measurement of bearing currents in thrust bearings, among others. The purely vertical load (see Figure 1) ensures uniform tribological conditions at the rolling contacts. This makes it possible to precisely analyse the electrical properties of the rolling bearings and lubricants.

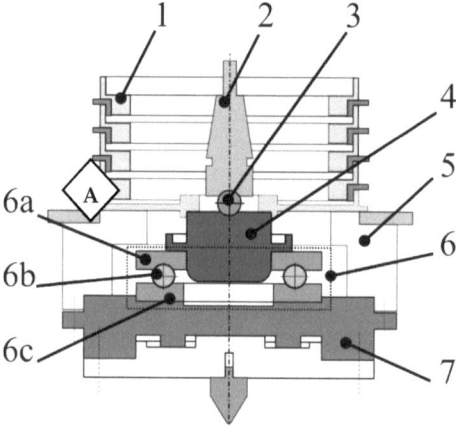

Figure 1. Sectional view of GESA (1 distributing ring/2 driving shaft/3 centring ball/4 shaft/ 5 housing/6 tested bearing/6a rotating ring/6b rolling element/6c stationary ring/7 bearing ring holder/A acceleration sensor) [36].

In the tests, axial deep groove ball bearings (type 51208) are subjected to a combined mechanical and electrical load. These mechanical and electrical boundary conditions can be

found in Tables 1 and 2. To make it easier to differentiate between the test regimes, they are labelled as a function of the axial force applied. This results in the following relationship:
- Test series A—axial load 4000 N ($C0/P$ 19);
- Test series B—axial load 6000 N ($C0/P$ 13);
- Test series C—axial load 8000 N ($C0/P$ 10).

Table 1. Boundary conditions specified at the test bench.

Designation	Force/N	Rotation Speed/rpm	Temperature/°C	Common-Mode-Voltage/Vpp	Switching Frequency/kHz
A-m1	4000	1000	40	-	-
A-m2			80		
A-e1	4000	1000	40	60	20
A-e2			40	40	
A-e3 *			40	20	
A-e4			80	60	
B-m1	6000	1000	40	-	-
B-e1	6000	1000	40	60	20
B-e2			40		5
B-e3			80		20
C-m1	8000	1000	40	-	-
C-e1	8000	1000	40	60	20

* including retry with -a and -b denoted.

Table 2. Mechanical parameters at the test bench.

Parameter	Unit	Test Series A	Test Series B	Test Series C
Contact force	N	4000	6000	8000
C0/P	-	19	13	10
Hertzian pressure	MPa	1494	1710	1883
Single contact area	mm^2	0.27	0.35	0.43
Lubrication gap height *	µm	0.79 @40 °C	0.76 @40 °C	0.72 @40 °C
		0.22 @80 °C	0.21 @80 °C	-
Specific lubrication gap	-	1.23 @40 °C	1.19 @40 °C	1.13 @40 °C
		0.34 @80 °C	0.33 @80 °C	-

* according to [37].

Using the lubricant used here (non-additive mineral oil identical to [36]), the following mechanical loads and tribological lubricant film heights are present in the tests.

3. Test Procedure and Measured Variables

The test bearings are first subjected to a 16 h mechanical run-in. This results in individual roughness peaks being smoothed out and constant tribological and electrical behaviour being achieved. This running-in is carried out at 2400 N, 1000 rpm and a lubricant temperature of 40 °C. The bearing is then dismantled, and the surface topography of the bearing raceways is measured. This is performed using the method described in Section 3.2, which has already been successfully tested in [11,21–23,38]. The test specimen is then remounted and loaded for 168 h. The load collectives correspond to the boundary conditions listed in Table 1. After completion of the test run, the surface topography of the bearing raceways is measured again. The measured variables listed below are analysed.

3.1. Acceleration Measurement

To gain insights into the relationship between running noise, vibrations and surface topography of the bearing raceway, continuous vibration measurements were performed during the test runs. For this purpose, an acceleration sensor is located on the housing of the test cell (Figure 1 A), and the rotation speed of the bearing is also recorded via a Hall sensor. A piezoelectric sensor (manufacturer: Dytran, Los Angeles, CA, USA/designation: 3056B5) is used; it has a resolution of 50 mV/g and covers a measuring range between 0 and 100 g amplitude at a frequency of 1 Hz to 10 kHz. The subsequent evaluation is carried out by visualising the order spectra over the test time.

3.2. Confocal and Light Microscopy of the Rolling Elements and Raceway Surfaces

Two different methods are used to measure the surfaces of the rolling elements at selected test points. A selection of different microscopes and macroscopes is used to visually compare the surfaces. The 3D surface measurement is carried out using a confocal microscope (vertical resolution of the lens used is 0.006 µm). A developed test fixture ensures that the sample (such as the bearing rings) is always positioned and aligned identically to the microscope. This allows the almost identical surface section to be monitored at different times. In order to enable comparability of the measured surface sections at the different test times, a special measurement recording was developed. This is shown schematically in Figure 2. To ensure optimum and reproducible positioning, the test bearing is provided with a chamfer before the start of the actual test series. After inserting the bearing ring on the base surface, the position is determined via the contact surface and the positioning bolt. In addition, the chamfer is pressed against the contact surface via the clamping screw and thus fixed in position.

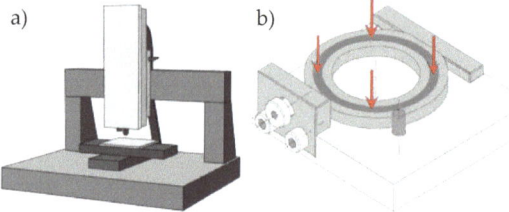

Figure 2. Surface measuring equipment: (**a**) confocal microscope with positioning stage on which the sample is placed; (**b**) sample holder with visualisation of possible measurement points [11].

The reproducibility of the surface measurement using this image was confirmed in [11,18]. The objective used in the measurements on the confocal microscope (manufacturer: NanoFocus, Oberhausen, Germany) achieves a resolution of 1.55 µm in the plane and 0.006 µm in the height.

The post-processing and preparation of the 3D surfaces is conducted using the commercial programme MountainsMap (manufacturer: Digital Surf, Besancon, FRA/version: premium 6.0). In this step, so-called artefacts, such as reflections or strong measurement deviations of the surface measurement, are first searched for and eliminated by means of interpolation. Based on this, the shape is then separated from the roughness for further evaluation, resulting in a plane with superimposed roughness and corrugation. The evaluation is carried out using Abbott–Firestone curves [39], histograms of the surface heights and a selection of surface characteristics in accordance with [40].

3.3. Light Load Hardness

As a result of the electromechanical load, the surface is massively reshaped in parts. This can be observed and documented in the form of a change in the topography or a change in individual surface parameters). In addition to this influence on the surface topography, it can be assumed that the mechanical properties of the rolling bearing raceways close to the surface are also influenced. In line with this hypothesis, the light load hardness of the bearing raceways was measured as part of a separate series of tests. A Micro-Vickers hardness tester (manufacturer: Mitutoyo Cooperation, Kawasaki, Japan/designation: HM-112) with a test weight of 1 kg ($\mapsto$ light load hardness) was used for this purpose. Microindentation was carried out on each bearing ring in the centre of the raceway. In order to include any scatter in the hardness measurement in the assessment of the results, five measurements were taken along the raceway for each bearing ring. Furthermore, sufficient spacing between the individual indentation points ensured that there was no mutual interference between the individual measurements.

3.4. Determination of Wear Weights

A high-precision balance (manufacturer: Ohaus, Parsippany, NJ, USA/designation: Explorer EX225D) is employed to measure the electroerosive wear of the analysed rolling bearings. This has a readability of 0.01 mg with a repeatability of 0.02 mg. It is used to weigh the individual rolling bearing rings as well as the rolling element sets (including the cage) in selected tests.

4. Results

4.1. Electrical Load over Time

The results of test series A (see Figure 3a) show the expected influence of the amplitude of the source voltage on the resulting load characteristics. Furthermore, tests A-e1 and A-e2 show an almost constant behaviour of apparent bearing current density and bearing apparent power over the test period. Irrespective of this, A-e1 continues to show a slight increase in the load level and an increase in scatter from a running time of around 72 h. This effect is more pronounced for the bearing apparent power than for the average bearing current density, for example. The tests with identical loads (A-e3-a and A-e3-b) show the repeatability of the energisation over the test time at a comparable load level. In addition, both tests have a short isolating phase (A-e3-a $\rightarrow$ 30 h/A-e3-b $\rightarrow$ 48 h), after which the load values are almost identical until the end of the test. Test A-e4 was carried out within an increased lubricant temperature of 80 °C but with a source voltage of 60 V (pk to pk) comparable to A-e1. This is also confirmed by the trend of the load variables, as shown in Figure 3a. However, after around 120 h, there is a sudden drop in the average load value in both characteristic values of this test. The measurement data do not provide conclusive information regarding the cause of this drop. Test series B is shown together with the electrical test of test series C in Figure 3b. For test series B, despite the identical amplitude of the source voltage, there are clearer differences in the resulting apparent bearing current density and bearing apparent power than in test series A, for example. However, this difference in the electrical load can still be categorised as small, at least in this type of evaluation. Both test B-e1 and test C-e1 scatter more strongly at the beginning of the tests. This scatter decreases after a test duration of about 72 h. As in test series A, this change is also more clearly recognisable in the diagrams of the average storage appearance performance due to the more pronounced gradients. It should also be noted that in test B-e1, there was no further electrical load from 158 h onwards. This occurred due to a single failure of the voltage source. The other tests in test series B, which were carried out with a reduced switching frequency (B-e2) and increased lubricant temperature (B-e3), show an almost constant electrical load over the test period and are unremarkable. When comparing the test series with each other, it should be noted that the load levels of the electrical rating parameters differ. Thus, due to the lack of reference to a mechanical load variable, the average bearing apparent power at identical source voltage is in a similar range (at 60 V (pk to pk) approx. 4 VA). Using the bearing current density, the electrical load levels are dependent on the Hertzian contact area, which is why test series A has the highest load and test series C the lowest load at identical source voltage.

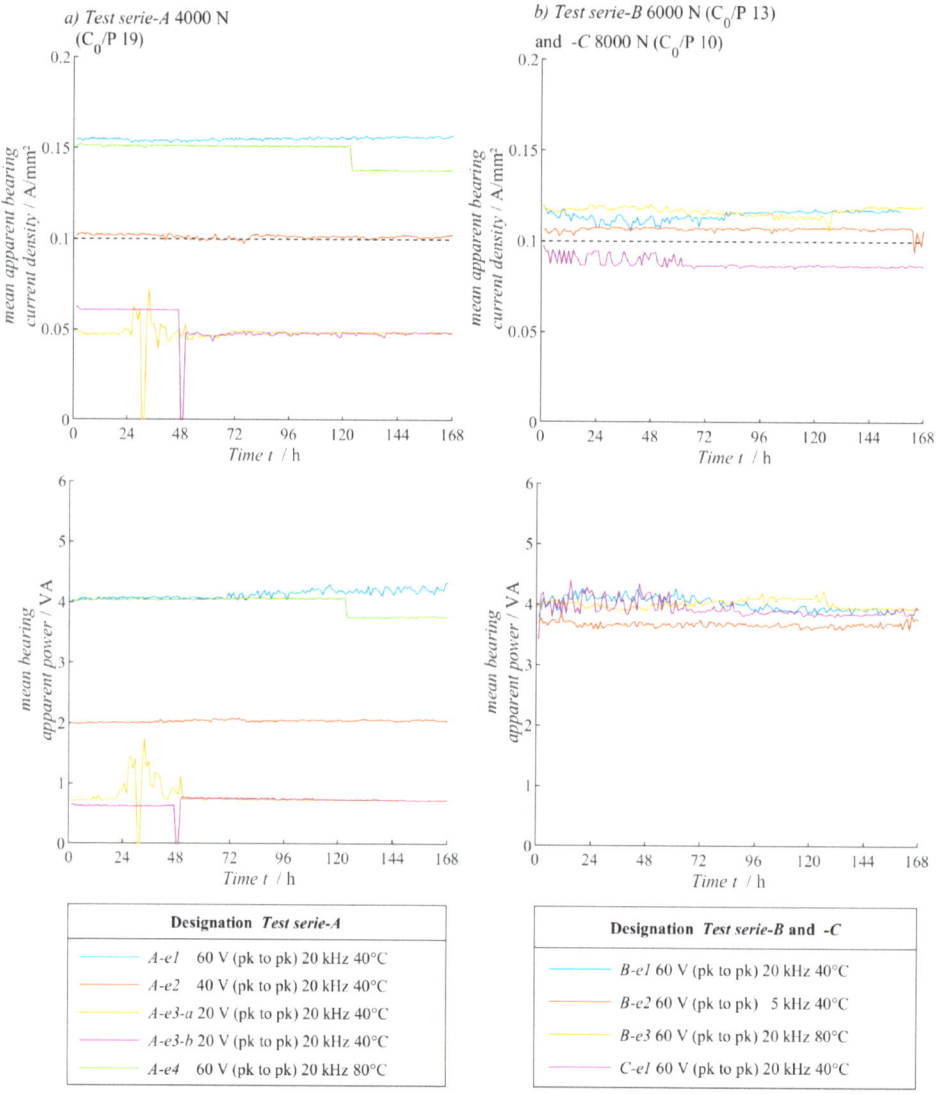

Figure 3. Mean apparent bearing current density and mean bearing apparent power for the test series with visualisation of the critical load values in the measuring range (0.1 A/mm^2) for test series -A (**a**) and -B (**b**).

4.2. Surface Topography

To supplement the visual assessment of the surface changes [36], a quantifying measurement of the roughness of the bearing raceways was carried out using a confocal microscope. The method presented in [11] was used here, in which the bearing rings are provided with a mirror coating, which then allows them to be positioned and aligned exactly to the microscope used in a specially developed test fixture. This allows the almost identical surface section to be measured at different times. Using this image, the bearing raceways were scanned after running-in for 16 h and after the load phase of 168 h. Figure 4 shows the Abbott–Firestone curves resulting from these measurements after the running-in and loading phases for all tests.

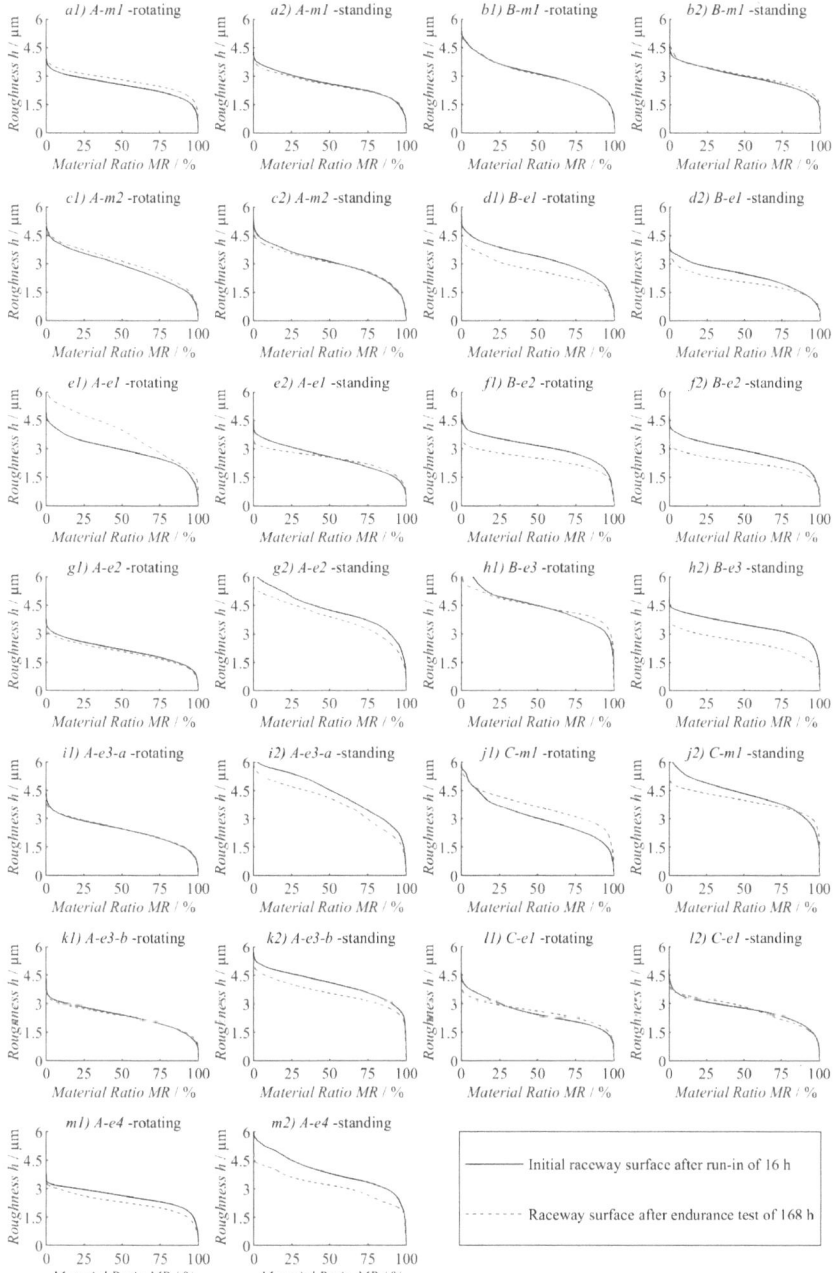

Figure 4. Abbott–Firestone curves of the bearing raceways at the beginning and end of the endurance tests for different operating conditions.

Figure 4 shows that there is no significant surface change in the purely mechanical reference tests of test series A and B. The material area curves on the rotating and stationary ring after the running-in and loading phase of the respective tests are almost identical (see Figure 4(a1–c2)). This behaviour does not correspond to the results of the reference test

C-m1. This is due to the severe surface damage that occurred in this test and must be considered when interpreting the results.

Furthermore, this type of visualisation clearly shows the influence of the electrical load. Based on the microscopic assessment of the surface [36], a smoothing effect was observed. This is shown by a clear drop in the profile heights after the electromechanical load compared to the initial level after the bearing run-in. The results from test A-e1 deviate from this (see Figure 4e1 A-e1-rotating). Here, there was an increase in the profile height of the rotating ring. This is a direct consequence of the fluting created here (see Figure 4c), which results in a noticeable deepening of the surface. In addition, this type of surface evaluation shows that the corrugation shading or bearing current marks are not noticeable.

Looking at the other results of test series A with lower source voltage and increased temperature, it is noticeable that there was no or only a slight drop in the profile heights on the rotating ring. In contrast, the influence of the electrical load and the smoothing that occurs is clear on the stationary ring. An effect of the different heights of the source voltage between 20 V (pk to pk) and 40 V (pk to pk) (see figure parts g2, i2 and k2) cannot be determined in this form of visualisation. While only minor changes occurred on the rotating ring in test series A, reductions in the profile height can be observed on both bearing rings in test series B.

Furthermore, no difference in the surface smoothing that occurs as a result of the varied switching frequency from test B-e1 (see Figure 4(d1,d2)) to B-e2 (see Figure 4(f1,f2)) can be determined. The evaluation of the material contact ratio curves of test series C cannot be clearly assigned to a purely mechanical or electrical load as a result of the severe surface damage occurring here in the form of pronounced material breakouts (pitting).

The tests with an increased lubricant temperature of 80 °C (A-m2, A-e4 and B-e3) should be emphasised separately. While the mechanical reference test showed no changes in the distribution of the profile height even at this temperature, the tests with additional electrical load resulted in a significant reduction in roughness. This is increased in comparison to the other electrically and mechanically loaded tests carried out in the respective test series. Accordingly, the tests with electrical load and increased temperature resulted in the greatest smoothing of the bearing raceway on the rotating ring. The stationary ring is also smoothed at elevated temperatures, but these changes are rather small compared to the rotating ring. This is also evident in a direct comparison with tests performed at an identical source voltage with a lubricant temperature of 40 °C. Here, in test A-e1, there is a pronounced ripple formation, which did not occur when the lubricant film height was reduced by tempering the lubricant to 80 °C (test A-e4).

4.3. Acceleration Detection

Parallel vibration measurement proved to be a useful tool during the tests for recording the interaction between the electrical and mechanical load. The continuous recording of measured values allows strong changes in the contact partners to be limited in time, which is not possible with the surface examination based on measurements at the start and end of the test.

As part of the vibration analysis that follows the vibration measurement or is carried out in parallel, the order spectrum is formed from the measured vibrations. The magnitude of the amplitude is represented as a multiple of the excitation frequency (speed) of the so-called order. The representation above the order allows the identification of characteristic multiples of the excitation frequency and an analysis of the measured vibrations. Figure 5 shows the temporal development of the acceleration amplitude over the order to investigate the operating behaviour over the test time. A proven tool for converting the measurement data recorded at the acceleration sensor into an order spectrum is the Fourier transformation, which is also used here. The order spectrum is formed and stored over a vibration measurement interval of 20 min. This means that the temporal change in the order spectrum is also available via the measurement of the endurance test, which is also mapped. To analyse the order spectrum, the characteristic frequencies of the test specimen must be known. In particular, the kinematic conditions of the thrust bearing must be considered. Due to the identical raceway diameter of the stationary and rotating ring, it can be assumed that the cage rotates at approximately half the angular velocity of the driving ring. In addition, the rollover speed of the rolling elements is the multiple of the number of rolling elements multiplied by the speed of the cage. The frequency of the test fixture must also be taken into account. During the evaluation, it was shown that the design of the contact surface of the stationary ring (see Figure 1 pos. 7) has an influence on the vibrations of the test cell. The contact surface of the stationary ring is provided with eight recesses to facilitate the removal of the ring and increase the volume of the oil sump. These interruptions in the contact surface led to jumps in rigidity when the rolling elements rolled over the stationary ring. This results in periodic excitation, which is recognisable in the order spectrum. The position of these individual frequencies and details of their respective order are marked separately in Figure 5. In the evaluation, the continuous individual measurements are combined to form a three-dimensional temporal curve and displayed for each test. The respective test series were arranged one below the other to facilitate direct comparability in the respective load situation. Furthermore, recurring characteristic frequencies, such as the triple rollover frequency of the rolling element set $f_{\text{Rolling Element}}$ or the multiples of the passing frequency of the rolling element set over the support surface $f_{\text{Support Surface}}$, were identified. The reason for the clear indication of the triple rollover frequency of the rolling element set is the way in which the test cell is connected to the four-ball apparatus. As a result of the manufacturing tolerances, the coaxiality of the drive unit to the test cell cannot be maintained exactly, resulting in a slight radial and angular offset of the two axes. These offsets are absorbed and compensated for by the centre point and the centring ball. As a result, however, the multiples of the first characteristic frequency of the bearing are excited more strongly and, therefore, appear more clearly in the order spectrum. In test A-e1, in contrast to all other endurance tests, not the triple but the double rolling element set frequency occurs. One possible explanation for this is a stochastic better coaxiality of the two axes of rotation, which is why there was a double excitation.

The occurrence of rollover frequencies with several multiples shows that the test cell has natural oscillations in the frequency range under consideration. In addition to the passing frequency of the rolling elements, including their multiples, frequency bands could also be determined that are caused by jumps in stiffness when rolling over the bearing surface of the stationary ring, which is provided with pockets. Irrespective of this, the occurrence of pronounced surface changes, such as the fluting in test A-e1 (see Figure 5e) or the bearing failure in test series C, could be localised in time by means of the changes in the order spectra. Minor surface changes such as fluting shading were also characterised by an increase in individual spectra but are more difficult to identify. Overall, vibration monitoring of mechanically and electrically loaded tests is a good complementary measurement method for localising individual phenomena over time and quantifying their effects.

4.4. Light Load Hardness

The results of the five individual measurements and the resulting mean value are visualised in Figure 6 and listed in Table 3. It should be noted that in test A-e1, in which corrugations occurred on the rotating ring, the small load hardness measurement was carried out differently. Five individual measurements were taken between the corrugations, and five further individual measurements were taken in the corrugation valley. These results are also listed in Table 3. For better comparability, Figure 6 shows only the small load hardnesses between the corrugations for this test.

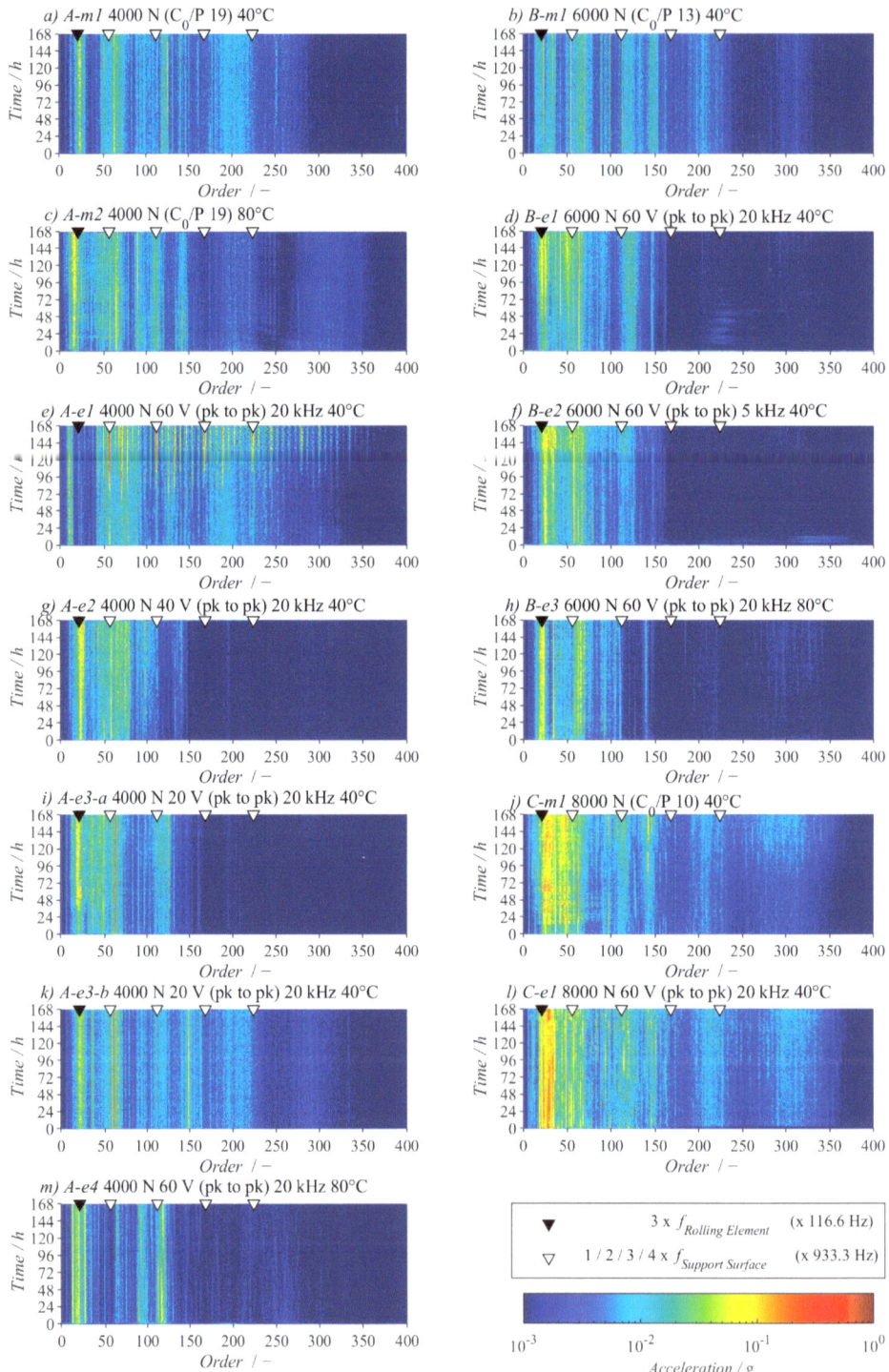

Figure 5. Development of the order spectra (related to a rotation speed of 1000 rpm) over the test period.

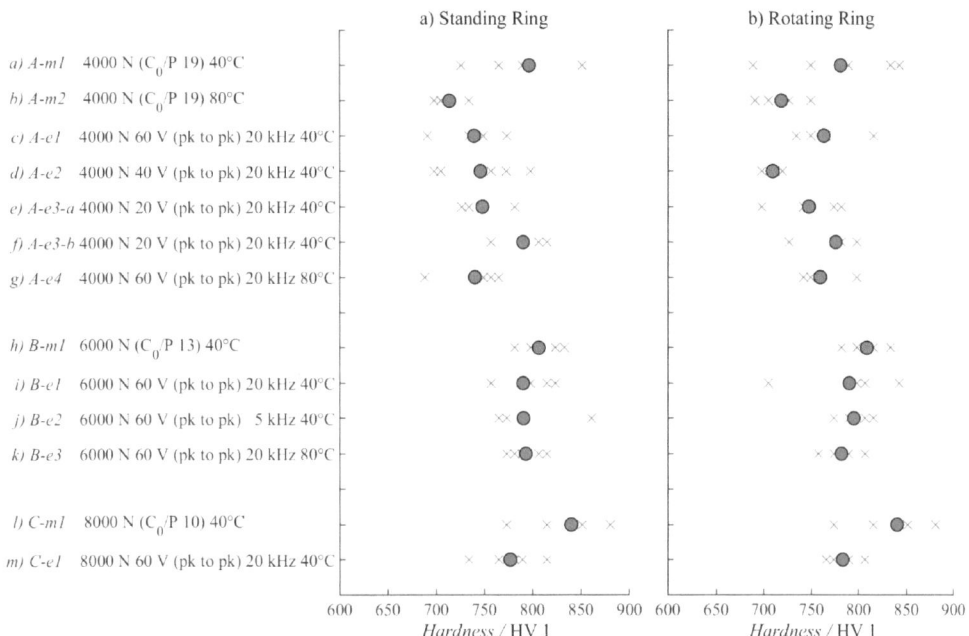

Figure 6. Light load hardness measured on the (**a**) stationary and (**b**) rotating bearing ring after a loading phase of 168 h. Representation of the five individual measurements (×) and the respective mean value (·) in accordance with Table 3.

Table 3. Results of the hardness measurements using Vickers microindentation for the stationary and rotating bearing rings of the bearing rings of the test bearings after the loading phase of 168 h.

Designation	Light Load Hardness @Standing Ring (in HV 1) Mean (Single Measurement)	Light Load Hardness @Rotating Ring (in HV 1) Mean (Single Measurement)
A-m1	796.6 (765/851/851/726/790)	780.4 (749/689/842/833/789)
A-m2	713.8 (734/698/698/734/705)	718.4 (705/691/720/749/727)
A-e1	739.2 (749/749/734/691/773)	762.4 (749/734/815/765/749)
A-e1 in fluting area		681.2 (665/678/652/720/691)
A-e2	746.2 (773/698/798/705/757)	709.6 (698/698/720/712/720)
A-e3-a	748.0 (749/749/727/781/734)	747.2 (773/781/742/742/698)
A-e3-b	790.2 (757/815/757/815/807)	775.4 (798/727/781/773/798)
A-e4	740.2 (688/765/757/749/742)	759.0 (742/749/798/749/757)
B-m1	806.8 (798/798/781/824/833)	808.4 (815/781/815/798/833)
B-e1	790.2 (824/757/757/815/798)	790.0 (842/798/705/798/807)
B-e2	790.8 (765/861/773/790/765)	795.0 (815/790/807/773/790)
B-e3	793.2 (781/773/790/807/815)	781.6 (781/773/790/757/807)
C-m1	-	840.0 (880/773/852/815/880)
C-e1	777.0 (734/781/815/790/765)	783.2 (765/790/781/773/807)

Overall, the individual measurements in the respective tests show a scatter of around 100 degrees of hardness (<15% of the average hardness). Due to the nature of the test using a minimum load, scattering of this magnitude is to be expected and is, therefore, not particularly conspicuous. Furthermore, it can be seen that as the contact force increases in the individual test series, the hardness increases. This behaviour is independent of whether an additional electrical load is present or a purely mechanical reference test was carried out. This effect can be explained by the greater plasticisation of the surface roughness because of the increase in contact pressure (see Section 4.2). This means that a larger proportion of roughness peaks are plasticised and compacted, which is accompanied by a corresponding increase in hardness. Figure 6 shows that the average light load hardness of the bearing raceways is reduced compared to the mechanical reference when an additional electrical load is applied. This behaviour can be observed in all three test series. One exception is the mechanical reference test A-m2 (Figure 6b), which was carried out with an increased lubricant temperature of 80 °C. Here, the hardness is lower than in the other tests. Overall, however, these deviations or reductions in hardness under additional electrical stress are within the scatter of the mechanical tests and, therefore, not so clear that a compelling correlation can be recognised. Further tests need to be carried out in order to obtain the necessary statistical validation.

However, there is a clear difference between the hardness in the corrugated valley and between the corrugations. This is reduced by an average of 80 degrees of hardness in the fluting valley compared to the raceway hardness. This may be due to the changed material structure or the changed topography of the surface (see also Figure 4(e1)). A pronounced sponge structure can be seen in the fluting valley, while the raceway was electrically smoothed in the areas in between.

Overall, the influence of the electrical load on the near-surface hardness cannot yet be conclusively determined from this series of tests. However, it seems appropriate to investigate this parameter in the future.

4.5. Wear Weight

Table 4 summarises the weights of the two bearing rings and the rolling element, including the cage, before and after the tests. Only minimal changes in the single-digit milligram range were observed between the measurements; no correlation with the current flow can be recognised. It should also be noted that even in test A-e1, despite the pronounced fluting, no detectable difference in weight could be recorded.

Table 4. Change in weight (in grams) after the loading phase of 168 h with comparison to the weights of the components after the run-in phase.

Designation	Mass after Run-In/g (Figure 1—6a/6b/6c)	Mass after Load Phase/g (Figure 1—6a/6b/6c)	Delta/g (Figure 1—6a/6b/6c)
A-m1	89.413/76.958/83.486	89.411/76.937/83.486	$-0.002/-0.021/0.001$
A-m2	-	-	-
A-e1	88.846/76.935/83.817	88.847/76.905/83.818	$0.001/-0.030/0.001$
A-e2	89,221/76,828/83,453	89,219/76,760/83,455	$-0.002/-0.068/0.002$
A-e3-a	-	-	-
A-e3-b	89,317/76,525/83,718	89,317/76,477/83,719	$0.001/-0.047/0.001$
A-e4	-	-	-
B-m1	88,751/76,331/83,350	88,750/76,328/83,351	$-0.001/-0.003/0.001$
B-e1	-	-	-
B-e2	89,512/76,924/83,814	89,510/76,872/83,810	$-0.002/-0.052/-0.004$
B-e3	89,938/76,804/83,502	89,934/76,810/83,495	$-0.004/0.007/-0.006$
C-m1	89,380/76,408/83,614	89,380/76,380/83,619	$-0.001/-0.028/0.005$
C-e1	88,916/77,009/83,429	88,916/76,971/83,431	$0.000/-0.038/0.002$

5. Discussion of the Findings

Based on the investigations conducted on the electrical load and its interaction with the changes occurring in the components due to current flow, the following partial conclusions and evaluations can be drawn.

5.1. Surface Topography—Significant

The use of a topographical measurement (in this case by means of a confocal microscope) of the surface shows a decisive added value in the classification of the change in surface roughness as a result of an electrical load. The comparison of the Abbott–Firestone curves of the running-in phase with those of the loading phase clearly shows the influence of the electrical–mechanical load (see Figure 4). The comparison with the mechanical reference tests, in particular, once again shows the significant influence of the additional electrical load. Based on the evaluation of the surface characteristics, it appears advisable in the context of the present tests to use the surface parameters of the reduced peak height (Spk) and the reduced valley height (Svk) to classify the effect of the electromechanical load on the surface topography.

5.2. Acceleration Analysis—Assisting Interpretation

With the help of the vibration analysis, the first occurrence of pronounced surface changes, such as the fluting in test A-e1 or the bearing failure in test series C, could be localised in time based on changes in the order spectra. It also confirms the smoothing observed in the surface examination, which is characterised by a decreasing amplitude of the vibrations. At the same time, localised track changes led to the formation of new bands, and the ripple shading observed in some tests manifested itself in a comparable form. By analysing the rollover frequencies, it was possible to determine that the test setup has vibrations in the evaluated frequency range during operation. In addition to the passing frequency of the bearing rings, including several multiples, frequency bands could also be determined that are caused by jumps in stiffness when rolling over the bearing surface of the stationary ring, which is fitted with pockets.

Furthermore, a possible cross-correlation between the occurring electrical load, the occurring surface topography and the temporal occurrence of individual vibration amplitudes in the order spectrum could be demonstrated.

5.3. Light Load Hardness—Further Research Necessary

As expected, the surface hardness of the bearing raceways also increases with an increase in contact pressure. This effect is due to the increasing compaction and plasticisation of the roughness peaks because of the increased contact force. It was also observed that the minimum hardness in the tests with additional electrical load was reduced by up to 5% compared to the mechanical reference tests. Due to the high scatter of the individual measurements, however, it is not possible to clearly determine whether this effect is a consequence of the electrical load or whether it is a statistical uncertainty. However, the significantly lower hardness in the centre of the corrugation (11% lower) than in the areas between the individual corrugations, where the initial raceway is still present, seems clear.

The extent to which the change in the topography of the surface or the increased thermal load on the material is the cause of this could not be conclusively clarified. However, the series of measurements suggests that the material change caused by the electrical load should be investigated in more detail.

5.4. Wear Mass—Possibly Too Short Test Duration

The results of the weight determination allow several possible interpretations; for example, the selected test time may not have been sufficient for detectable wear, or the temperature associated with the electrical load may have led to a forming process of the surface, which is why the roughnesses are only plastically deformed.

6. Conclusions

This study examined the effects of electrical loads on the surface topography and material properties of thrust ball bearings under various operating conditions. The results demonstrate that electrical loads can cause significant changes in surface roughness and hardness of the bearing raceways, leading to accelerated component damage. These findings highlight the need to consider electrical loads in the design of rolling bearings to avoid unexpected failures and over-dimensioning.

It is important to note that these results are based on a series of measurements that show the effects of current passage on metallic components under real load conditions. However, since only a limited number of tests were conducted, including a single repeat test, no statistically robust conclusions can be drawn regarding the correlation between surface changes and electrical load. Instead, the results illustrate more general effects induced by electrical influences.

In conclusion, it can be stated that, in addition to the precise analysis of the electrical loads and the vibrations developing over the test period, the changes in the surfaces of the contact partners are one of the most meaningful characteristics of the electromechanical loads. The influence of the electrical load on the light load hardness and wear could not be fully clarified within the scope of this article. The associated impact of these effects on the service life of the components will be the subject of future research.

Author Contributions: Conceptualization, S.G. and O.K.; methodology, S.G.; investigation, S.G.; writing—original draft preparation, S.G.; writing—review and editing, O.K. All authors have read and agreed to the published version of the manuscript.

Funding: This work was carried out within the framework of the projects "Model for determining the thermal stress of lubricants as a result of mechanical and electrical loads in rolling contact" (Project No. SA898/25-1/407468812) and "Determination of ball bearing impedances under steady-state operating conditions by means of a further developed rolling contact model at full film lubrication" (Project No. SA898/32-1/470273159). Both are financially supported by the Deutsche Forschungsgemeinschaft (DFG) e.V. (German Research Foundation).

Data Availability Statement: No new data were created or analyzed in this study.

Conflicts of Interest: The authors declare no conflict of interest.

References

1. Ammann, C.; Reichert, K.; Joho, R.; Posedel, Z. Shaft voltages in generators with static excitation systems-problems and solution, Energy Conversion. *IEEE Trans.* **1988**, *3*, 409–419.
2. Kerszenbaum, I. Shaft currents in electric machines fed by solid-state drives. In Proceedings of the 1992 IEEE Conference Record of the Industrial and Commercial Power Systems Technical Conference, Pittsburgh, PA, USA, 4–7 May 1992; pp. 71–79.
3. Preisinger, G. Cause and Effect of Bearing Currents in Frequency Converter Driven Electrical Motor, Investigations of Electrical Properties of Rolling Bearings. Ph.D. Thesis, Technische Universität Wien, Vienna, Austria, 2002.
4. Recker, C.; Weicker, M. Einfluss der elektrischen Schmierfettleitfähigkeit auf die Ausbildung von Lagerströmen bei umrichterbetriebenen 1.5 kW-Asynchronmotoren. In Proceedings of the 5th VDI-Fachkonferenz—Schadensmechanismen an Lagern, Aachen, Germany, 28–29 June 2022.
5. Zika, T. Electric Discharge Damaging in Lubricated Rolling Contacts. Ph.D. Thesis, Technische Universität Wien, Vienna, Austria, 2010.
6. Spikes, H.A. Triboelectrochemistry: Influence of Applied Electrical Potentials on Friction and Wear of Lubricated Contacts. *Tribol. Lett.* **2020**, *68*, 1–27. [CrossRef]
7. García Tuero, A.; Rivera, N.; Rodríguez, E.; Fernández-González, A.; Viesca, J.L.; Hernandez Battez, A. Influence of Additives Concentration on the Electrical Properties and the Tribological Behaviour of Three Automatic Transmission Fluids. *Lubricants* **2022**, *10*, 276. [CrossRef]
8. Durkin, W.; Fish, G.; Dura, R. Compositional Effects on the Electrical Properties of Greases. In Proceedings of the ELGI Eurogrease Q2, Hamburg, Germany, 22 March 2022.
9. Erdemir, A.; Farfan-Cabrera, L.; Anderson, W.B. Comparative Tri Comparative Tribological Properties of Commercial Drivetrain Lubricants under Electrified Sliding Contact Conditions. In Proceedings of the 2nd STLE Tribology & Lubrication for E-Mobility Conference, San Antonio, TX, USA, 30 November–2 December 2022.
10. Holweger, W.; Bobbio, L.; Mo, Z.; Fliege, J.; Goerlach, B.; Simon, B. A Validated Computational Study of Lubricants under White Etching Crack Conditions Exposed to Electrical Fields. *Lubricants* **2023**, *11*, 45. [CrossRef]
11. Graf, S.; Sauer, B. Surface mutation of the bearing raceway during electrical current passage in mixed friction operation. *Bear. World J.* **2021**, *5*, 137–147.
12. Schneider, V.; Stockbrügger, J.O.; Poll, G.; Ponick, B. *Abschlussbericht FVA 863 I—Stromdurchgang am Wälzlager–Verhalten Stromführender Wälzlager*; Forschungsvereinigung Antriebstechnik e.V.: Frankfurt, Germany, 2022.

13. Tischmacher, H. Systemanalysen zur Elektrischen Belastung von Wälzlagern bei Umrichtergespeisten Elektromotoren. Ph.D. Thesis, Gottfried Wilhelm Leibniz University, Hannover, Germany, 2017.
14. Kohaut, A. Riffelbildung in Wälzlagern infolge elektrischer Korrosion. *Z. Für Angew. Phys.* **1948**, *1*, 197–211.
15. Capan, R.; Graf, S.; Koch, O.; Sauer, B.; Safdarzadeh, O.; Weicker, M.; Binder, A. *Abschlussbericht FVA 650 III—Stromdurchgang in Wälzlagern—Untersuchung von Oberflächenveränderungen und Folgeschäden an Wälzoberflächen durch Stromdurchgang*; Forschungsvereinigung Antriebstechnik e.V.: Frankfurt, Germany, 2024.
16. Muetze, A. *Bearing Currents in Inverter-Fed AC Motors*; Shaker Verlag: Darmstadt, Germany, 2004.
17. Gemeinder, Y. Lagerimpedanz und Lagerschädigung bei Umrichtergespeisten Antrieben. Ph.D. Thesis, Technische Universität Darmstadt, Darmstadt, Germany, 2016.
18. Graf, S. Charakterisierung und Auswirkungen von Parasitären Lagerströmen in Mischreibung. Ph.D. Thesis, University of Kaiserslautern-Landau, Landau in der Pfalz, Germany, 2023.
19. Kriese, M.; Wittek, E.; Gattermann, S.; Tischmacher, H.; Poll, G.; Ponick, B. Influence of bearing currents on the bearing lifetime for converter driven machines. In Proceedings of the 2012 XXth International Conference on Electrical Machines, Marseille, France, 2–5 September 2012; pp. 1735–1739. [CrossRef]
20. Boyanton, H.E.; Hodges, G. Bearing fluting [motors]. *IEEE Ind. Appl. Mag.* **2002**, *8*, 53–57. [CrossRef]
21. Bechev, D. Prüfmethodik zur Charakterisierung der Elektrischen Eigenschaften von Wälzlagerschmierstoffen. Ph.D. Thesis, Technische Universität Kaiserslautern, Kaiserslautern, Germany, 2020.
22. Gonda, A.; Capan, R.; Bechev, D.; Sauer, B. The Influence of Lubricant Conductivity on Bearing Currents in the Case of Rolling Bearing Greases. *Lubricants* **2019**, *7*, 108. [CrossRef]
23. Graf, S.; Capan, R.; Koch, O. Wechselwirkung von Tribologie und Elektrisch Induzierter Oberflächenmutation in Wälzlagern. In Proceedings of the 5th VDI-Fachkonferenz—Schadensmechanismen an Lagern, Aachen, Germany, 28–29 June 2022.
24. Biswas, S.; Bose, S.C.; Bhave, S.K.; Pramila Bai, B.N.; Biswas, S.K. Study of corrugations in motor bearings. *Tribol. Int.* **1992**, *25*, 27–36. [CrossRef]
25. Tsypkin, M. Induction motor condition monitoring: Vibration analysis technique—A practical implementation. In Proceedings of the 2011 IEEE International Electric Machines & Drives Conference (IEMDC), Niagara Falls, ON, Canada, 15–18 May 2011; pp. 406–411. [CrossRef]
26. Zuercher, M.; Heinzler, V.; Schlücker, E.; Esmaeili, K.; Harvey, T.J.; Holweger, W.; Wang, L. Early failure detection for bearings in electrical environments. *Int. J. Cond. Monit.* **2018**, *8*, 24–29. [CrossRef]
27. Hemati, A. A Case Study: Fluting Failure Analysis by Using Vibrations Analysis. *J Fail. Anal. Preven.* **2018**, *19*, 917–921. [CrossRef]
28. Kudelina, K.; Asad, B.; Vaimann, T.; Belahcen, A.; Rassõlkin, A.; Kallaste, A.; Lukichev, D.V. Bearing Fault Analysis of BLDC Motor for Electric Scooter Application. *Designs* **2020**, *4*, 42. [CrossRef]
29. Romanenko, A.; Mütze, A.; Ahola, J. Incipient Bearing Damage Monitoring of 940-h Variable Speed Drive System Operation. *IEEE Trans. Energy Convers.* **2017**, *32*, 99–110. [CrossRef]
30. Ma, J.; Xue, Y.; Han, Q.; Li, X.; Yu, C. Motor Bearing Damage Induced by Bearing Current: A Review. *Machines* **2022**, *10*, 1167. [CrossRef]
31. Didenko, T.; Pridemore, W.D. Electrical Fluting Failure of a Tri-Lobe Roller Bearing. *J Fail. Anal. Preven.* **2012**, *12*, 575–580. [CrossRef]
32. He, F.; Xie, G.; Luo, J. Electrical bearing failures in electric vehicles. *Friction* **2020**, *8*, 4–28. [CrossRef]
33. Becker, A.; Abanteriba, S. Electric discharge damage in aircraft propulsion bearings. *Proc. Inst. Mech. Eng. Part J J. Eng. Tribol.* **2014**, *228*, 104–113. [CrossRef]
34. Ost, W.; De Baets, P. Failure analysis of the deep groove ball bearings of an electric motor. *Eng. Fail. Anal.* **2005**, *12*, 772–783. [CrossRef]
35. Ghosh, B.; Cao, J.; Zhu, J. A Study on Heatsink Cooling Fan Lifetime Evaluation. In Proceedings of the 2023 Annual Reliability and Maintainability Symposium (RAMS), Orlando, FL, USA, 23–26 January 2023; pp. 1–5. [CrossRef]
36. Graf, S.; Koch, O.; Sauer, B. Influence of Parasitic Electric Currents on an Exemplary Mineral-Oil-Based Lubricant and the Raceway Surfaces of Thrust Bearings. *Lubricants* **2023**, *11*, 313. [CrossRef]
37. Hamrock, J.B.; Dowson, D. Isothermal elastohydrodynamic lubrication of point contacts III: Fully Flooded Results. *J. Lubr. Technol.* **1977**, *99*, 264–275. [CrossRef]
38. Gonda, A.; Paulus, S.; Graf, S.; Koch, O.; Götz, S.; Sauer, B. Basic experimental and numerical investigations to improve the modeling of the electrical capacitance of rolling bearings. *Tribol. Int.* **2024**, *193*, 109354. [CrossRef]
39. Abbott, E.; Firestone, F. Specifying Surface Quality—A Method Based on Accurate Measurement and Comparison. *Mech. Eng.* **1933**, *55*, 569–572.
40. DIN-EN-ISO-25178-2; Geometrische Produktspezifikation (GPS)—Oberflächenbeschaffenheit: Flächenhaft—Teil 2: Begriffe und Oberflächen-Kenngrößen. Deutsches Institut für Normung (DIN): Berlin, Germany, 2012.

Disclaimer/Publisher's Note: The statements, opinions and data contained in all publications are solely those of the individual author(s) and contributor(s) and not of MDPI and/or the editor(s). MDPI and/or the editor(s) disclaim responsibility for any injury to people or property resulting from any ideas, methods, instructions or products referred to in the content.

MDPI AG
Grosspeteranlage 5
4052 Basel
Switzerland
Tel.: +41 61 683 77 34

Lubricants Editorial Office
E-mail: lubricants@mdpi.com
www.mdpi.com/journal/lubricants

Disclaimer/Publisher's Note: The title and front matter of this reprint are at the discretion of the Guest Editor. The publisher is not responsible for their content or any associated concerns. The statements, opinions and data contained in all individual articles are solely those of the individual Editor and contributors and not of MDPI. MDPI disclaims responsibility for any injury to people or property resulting from any ideas, methods, instructions or products referred to in the content.

www.ingramcontent.com/pod-product-compliance
Lightning Source LLC
LaVergne TN
LVHW072313090526
838202LV00019B/2277